Springer Proceedings in Materials

Volume 34

Springer Proceedings in Materials publishes the latest research in Materials Science and Engineering presented at high standard academic conferences and scientific meetings. It provides a platform for researchers, professionals and students to present their scientific findings and stay up-to-date with the development in Materials Science and Engineering. The scope is multidisciplinary and ranges from fundamental to applied research, including, but not limited to:

- Structural Materials
- Metallic Materials
- Magnetic, Optical and Electronic Materials
- Ceramics, Glass, Composites, Natural Materials
- Biomaterials
- Nanotechnology
- Characterization and Evaluation of Materials
- Energy Materials
- Materials Processing

To submit a proposal or request further information, please contact one of our Springer Publishing Editors according to your affiliation:

European countries: **Mayra Castro** (mayra.castro@springer.com)
India, South Asia and Middle East: **Priya Vyas** (priya.vyas@springer.com)
South Korea: **Smith Chae** (smith.chae@springer.com)
Southeast Asia, Australia and New Zealand: **Ramesh Nath Premnath** (ramesh.premnath@springer.com)
The Americas: **Michael Luby** (michael.luby@springer.com)
China and all the other countries or regions: **Mengchu Huang** (mengchu.huang@springer.com)

This book series is indexed in **SCOPUS** database.

R. Velmurugan · G. Balaganesan · Naresh Kakur ·
Krishnan Kanny
Editors

Dynamic Behavior of Soft and Hard Materials Volume 1

Proceedings of 13th International Symposium on Plasticity and Impact Mechanics 2022

Springer

Editors
R. Velmurugan
Department of Aerospace Engineering
Indian Institute of Technology Madras
Chennai, India

G. Balaganesan
Department of Mechanical Engineering
Indian Institute of Technology Madras
Chennai, India

Naresh Kakur
Department of Materials Science
and Engineering
University of California
Irvine, CA, USA

Krishnan Kanny
Department of Mechanical Engineering
Durban University of Technology
Durban, South Africa

ISSN 2662-3161 ISSN 2662-317X (electronic)
Springer Proceedings in Materials
ISBN 978-981-99-6029-3 ISBN 978-981-99-6030-9 (eBook)
https://doi.org/10.1007/978-981-99-6030-9

This Springer imprint is published by the registered company Springer Nature Singapore Pte Ltd.
The registered company address is: 152 Beach Road, #21-01/04 Gateway East, Singapore 189721, Singapore

Paper in this product is recyclable.

Contents

About the Editors

Dr. R. Velmurugan is Senior Professor of Aerospace Engineering at Indian Institute of Technology (IIT) Madras. His areas of research include composite materials, nanomaterials, finite element analysis, structural crashworthiness, and impact mechanics. He has completed many consultancy and sponsored projects from DRDO Labs, ISRO centres, Government agencies, and private industries. He has made numerous research articles in international journals and conferences and guided many students for Ph.D., M.S., and M.Tech. degrees. He has delivered invited lectures in many international and national conferences. He is Fellow of Aeronautical Society of India and Institution of Engineers and served as Member of the Editorial Board, for several journals of national and international repute.

Dr. G. Balaganesan pursued his bachelor's and master's in mechanical engineering and Ph.D. in aerospace engineering from the Indian Institute of Technology (IIT) Madras. He has more than 30 years of experience in the field of manufacturing and teaching. His active research is in the field of impact loading and repair of oil and gas pipelines using advanced composite materials. He has published 40 international peer-reviewed journal papers. He has presented his research work in more than 30 international conferences. He has filed six patents, and some of them are products of start-up companies incubated at IIT Madras. He is instrumental in setting up laboratory facilities of new IITs such as IIT Hyderabad in 2008, IIT Palakkad and Tirupati in 2015, and IIT Jammu in 2018. He has served as Faculty Member on deputation at the Department of Mechanical Engineering, IIT Jammu, from 2018 to 2020.

Dr. Naresh Kakur is currently Postdoctoral Scholar at the Department of Materials Science and Engineering, University of California, USA. He completed his dual degree (M.S. + Ph.D.) course at the Department of Mechanical Engineering, Indian Institute of Technology (IIT) Madras. His area of interest is on impact properties of bio-inspired glass/epoxy composites for wind turbine blade applications in collaboration with Helicoid Industries Inc. He has more than 40 research and review articles published in journals and conferences of national and international repute.

Prof. Krishnan Kanny holds a Ph.D. in Engineering and Science from TU, Alabama, USA, and an M.Sc. from UKZN, South Africa. He is professionally registered with the Engineering Council of South Africa (ECSA) and is Member of SAIMechE, ASME, and AIAA. His areas of research are lightweight materials and nanotechnology. His interests include designing, processing, and testing of composite materials systems, reinforced thermosets, thermoplastics, and nano-infused structures for aerospace, naval, and automotive applications. Additionally, he is involved in computational and analytical modelling and characterization and morphological analysis including scanning electron microscopy, transmission electron microscopy, and X-ray photoelectron spectroscopy. He is the recipient of many awards, among them is Chair Professorship from MG University in India.

Impact Studies

Finite Element Analysis of Ballistic Impact on Dissimilar Multi-Layered Metal Targets

Palak Bhagoria and Manoj Kumar

Abstract Impact problems have always been a field of interest for researchers. Practical problems involve large amount of heat generation due to plastic deformation. In this paper, we have studied monolithic and multi-layered target being impacted by rigid projectile using finite element method. Two different materials—1100-H14 aluminium and Weldox 700 E steel—are layered up as target. The influence of thickness of layer, number of layers, and temperature softening on the ballistic phenomena are studied using Finite Element Package ABAQUS/ Explicit. To include the thermal softening effect, the dynamic temperature displacement explicit model is used for the finite element analysis. The targets are modelled using Johnson–Cook material model and Johnson–Cook fracture model. Projectile shapes such as ogival and conical are used for comparison of ballistic resistance of each configuration. The analysis with monolithic target showed that the failure mechanism in both the cases—conical and ogival-shaped projectile—involved hole enlargement and petal formation. Multi-layered configuration of aluminium layer in front and steel layer at back showed more obstruction. For ogival projectile, the target failure involves ductile hole enlargement and petalling, whereas conical projectile caused target to fail by ductile hole enlargement and plugging. As the frontal portion of projectile changes from conical to blunter shape, failure mechanism transforms from petalling to plugging. It was observed that with decrease in impact velocity and decrease in number of layers, velocity drop increased. © 2016 The Authors. Published by Elsevier Ltd. Peer-review under responsibility of the organizing committee of Implast 2016. © 2016 The Authors. Published by Elsevier Ltd. Peer-review under responsibility of the organizing committee of Implast 2016.

Keywords Ballistic impact · Finite element methods · Multi-layered · Thermal softening

P. Bhagoria · M. Kumar (✉)
Dr. B. R. Ambedkar National Institute of Technology, Jalandhar 144011, India
e-mail: kumarm@nitj.ac.in

R. Velmurugan et al. (eds.), *Dynamic Behavior of Soft and Hard Materials Volume 1*,
Springer Proceedings in Materials 34, https://doi.org/10.1007/978-981-99-6030-9_1

1 Introduction

Penetration mechanics, a field of applied mechanics involves the study of interaction of a projectile with a target. The experiments done in order to study this are known as ballistic experiments that deal with the effect of projectile when launched, its flight behaviour and impact when striking a target. Ballistic impact involves impact of a small mass projectile on a target at high velocity within fractions of milliseconds. The failure of target is influenced by various parameters including nose shape of projectile, size of projectile, type of target-monolithic or polylithic, target thickness, apex angle of projectile etc. The change in failure mode results in difference in the ballistic resistance of the target. Number of studies have been conducted in this area to study the effect of these parameters on ballistic resistance capability of material.

Gogolowski et al. [1] did ballistic experiments on 2024 aluminium targets by fragment simulant projectile, circular cylinder projectile with different fineness ratios. They found that the failure mode changes from petalling to plugging as plate thickness increases. Børvik et al. [2] did simulations for the evaluation of the constitutive relation and fracture criterion for Weldox 460 E steel where strain rate, temperature and stress triaxiality changes during plastic deformation. It was observed that the force increases with decreasing notch radius. Teng et al. [3] numerically studied three fracture modes—confined fracture inside the cylinder, the shear cracking on the lateral surface, and petalling. Iqbal et al. [4] conducted numerical simulations to study the influence of internal nose angle and calibre radius head of projectiles when impacted on ductile targets—12 mm thick Weldox 460 E steel target and 1 mm thick 1100-H12 aluminium target respectively. It was observed that the ballistic limit increases linearly with decrease in internal nose angle of projectile. Kpenyigba et al. [5] studied failure modes in case of different shaped projectiles wherein the failure modes correlated to the projectile shapes. Also to study the effect of thickness they made use of sandwich configuration and concluded that the energy absorbed and the nonlinear increment of ballistic resistance by multi-layered plate is high compared to monolithic plate. Palta et al. [6] through their studies analysed that layered configurations including hybrid plates highly influence the ballistic performance. Ballistic resistance is enhanced in case of layered and hybrid plates when compared with ballistic resistance of monolithic plates when impacted by different shaped projectiles.

Practical ballistic impact problems involve large amount of heat generation due to plastic deformation. The amount of heat generated is very high in comparison with the amount of heat dissipated by conduction and convection. Thus, it becomes a necessity to consider parameters which include effect of temperature rise in the regions of deformation near the perforation of projectile as this temperature rise reduces the target material flow stress which results in more deformation to take place. This is known as thermal softening effect. Including this parameter in the numerical simulation will help in getting results which will be in close agreement with the experimental results.

The present work involves study of multi-layered or sandwiched configuration modelled using dynamic temperature displacement explicit model so as to include the thermal softening effect. Two different nose-shaped projectile—ogival and conical—are considered. Investigation of temperature variation that occurs near the region of fracture is done. Numerical simulations are performed using ABAQUS/Explicit finite element code, and ballistic curves are drawn for the different cases.

2 Computational Modelling

For the present finite element analysis, target and projectile are modelled by dynamic temperature displacement explicit model using ABAQUS. For conducting the study, three target configurations are considered—6 mm monolithic 1100-H14 aluminium target, 6 mm monolithic Weldox 700 E steel target, sandwiched target with 1100-H14 aluminium layer in front, and Weldox 700 E steel layer at the back, each layer of 6 mm thickness. The target plate has been modelled as 3D deformable body. A circular partition is made at the centre of the plate. The partition is been made so as to focus the study in the region where stresses and strains localize the most. Two types of projectile—ogival and conical nose-shaped are considered and modelled as 3D analytic rigid body. Mass of the projectile is taken as 195 g. The target plate thickness is taken 6 and 12 mm with length 100 × 100 mm, and projectile has diameter 20 mm and length 62 mm as shown in Fig. 1.

Here surface-to-surface contact between projectile and target is modelled. In the present analysis, projectile is taken as analytical rigid body, hence it is chosen to be the master body, whereas the target plate is meshed finely and is also considered as deformable body, and therefore, target plate is chosen as slave body. The target plate was fixed at the periphery using "encastre" boundary condition. The projectile was given an initial velocity, and the observations.

Meshing of the target was done to obtain accurate results within optimum time as shown in Fig. 2. After creating parts, assigning properties, assembly, and load allotment, various mesh attributes are applied on the model. These attributes include seeds, mesh techniques, and element types. With small mesh sizes, more accurate solutions can be obtained, but in order to attain higher accuracy, larger simulations will have to be performed which increases the total computational time. Hence,

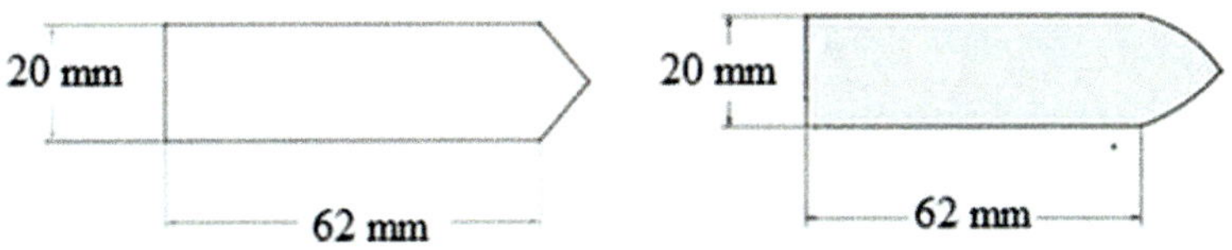

Fig. 1 Geometric dimensions of (**a**) conical (**b**) ogival projectile for residual velocity were taken by varying initial velocity as 250, 300, 400, 450, and 500 m/s

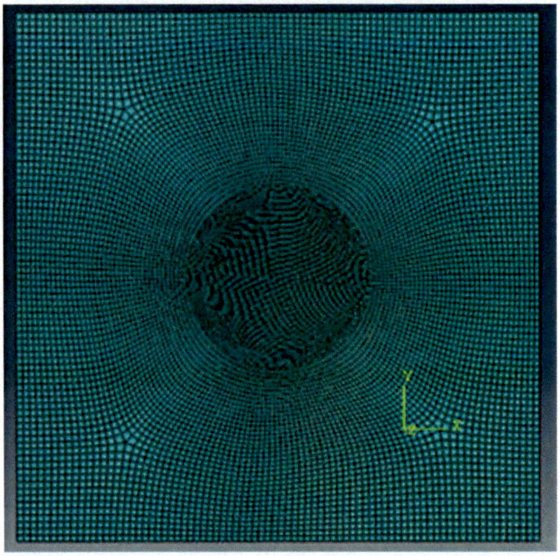

Fig. 2 Discretization of the target plate

convergence study is performed to obtain optimal balance between accuracy and simulation time.

The target plate was meshed with C3D8R, hexahedral elements. Hexahedral elements are usually preferred for conducting 3D nonlinear analysis. Reduced integration was used in order to reduce the total computational time without compromising with the accuracy of results. Reduced integration makes use of one integration point less than that used in full integration in each direction. This is used in order to reduce total computational time. 80 elements were taken along the outer edge and 200 elements along the partition edge on each plate. Four elements were taken along the target thickness. Adaptive meshing was adopted in order to reduce number of elements distorted. Fine meshing was particularly done in regions of high stress gradients which is created through partitioning. Table 1 gives the details of number of nodes and elements for different target configurations used in the present analysis.

Table 1 Mesh details of different target configurations

S. No.	Target configuration	Nodes	Elements
1	6 mm monolithic aluminum plate	81,790	64,784
2	6 mm monolithic steel plate	81,555	64,600
3	Multi-layered plate	123,480	97,492

3 Constitutive Material Modelling

To study the fracture behaviour of target, Johnson–Cook elasto-viscoplastic material model was used which takes into account the effect of linear thermo-elasticity, yielding, plastic flow, isotropic strain hardening, strain rate hardening, and softening due to adiabatic heating and damage [7, 8]. The equivalent von Mises stress $\overline{\sigma}$ of the Johnson–Cook model is given by

$$\tilde{\sigma}\left(\overline{\varepsilon}^{pl}, \dot{\overline{\varepsilon}}^{pl}, \widehat{T}\right) = \left[A + B\left(\overline{\varepsilon}^{pl}\right)^{n}\right]\left[1 + Cln\left(\frac{\overline{\varepsilon}^{pl}}{\dot{\overline{\varepsilon}}^{pl}}\right)\right]\left[1 - \widehat{T}^{m}\right] \quad (1)$$

where A, B, n, and C are materials parameters as

A = yield stress.

B = strain hardening parameter.

C = strain rate sensitivity parameter.

n = strain hardening exponent.

$\overline{\varepsilon}^{pl}$ is equivalent plastic strain, $\dot{\overline{\varepsilon}}^{pl}$ is equivalent plastic strain rate, $\dot{\varepsilon}_0$ is reference strain rate, and $\widehat{T}$ is non-dimensional temperature defined as:

$$\hat{T} = (T - T_0)/(T_{\text{melt}} - T_0), \; T_0 \leq T \leq T_{\text{melt}} \quad (2)$$

where T is the current temperature, T_{melt} is the melting point temperature, and T_0 is the room temperature. The fracture model proposed by Johnson–Cook takes into account the effect of stress triaxiality, strain rate, and temperature on the equivalent fracture strain. The equivalent fracture strain $\overline{\varepsilon}_f^{pl}$ is expressed as

$$\overline{\varepsilon}_f^{pl}\left(\frac{\sigma_m}{\tilde{\sigma}}, \dot{\overline{\varepsilon}}^{pl}, \widehat{T}\right) = \left[D_1 + D_2\exp\left(-D_3\frac{\sigma_m}{\tilde{\sigma}}\right)\right]\left[1 + D_4ln\left(\frac{\dot{\overline{\varepsilon}}^{pl}}{\dot{\varepsilon}_0}\right)\right]\left[1 + D_5\widehat{T}\right] \quad (3)$$

where D_1–D_5 are material damage constants. These constants can be obtained through tensile tests performed at quasi-static and high strain rates. $\frac{\sigma_m}{\overline{\sigma}}$ is the stress triaxiality ratio, and σ_m is the mean stress.

Parameter D can be stated as

$$D = \frac{\sum \Delta\epsilon^{p}}{\epsilon_f^{p}} \quad (4)$$

where $\Delta\epsilon^{p}$ tells about the increment of equivalent plastic strain that occurs in an integration cycle, and ϵ_f^{p} is the failure strain of the target material. Fracture will take place when the value of D within an element reaches unity.

The rise in temperature due to adiabatic heating conditions is calculated as follows

$$\Delta T = T - T_r = \int_0^{\epsilon_{eq}} \varkappa \frac{\sigma_{eq}}{\rho C_p} d\epsilon_{eq} \quad (5)$$

Here,

C_p-Specific heat, ρ-density of material, ϰ-Taylor-Quinney coefficient which is the percentage of plastic work converted to heat due to thermal softening effect.

Johnson Cook Damage initiation criterion is a special case of ductile damage criterion model for predicting the onset of damage due to nucleation, growth and coalescence of voids in ductile metals.

Hence all material properties of 1100-H14 aluminium and Weldox 700 E steel are modelled in simulations through Johnson–Cook material model. The target plate is double layered wherein first layer is made up of 1100-H14 aluminium and second layer is of Weldox 700 E steel. The properties of both the materials are enlisted in the Table 2. Thermal softening effect and high strain rate condition is taken into consideration.

Table 2 Material parameters [9, 10]

Material parameters	1100-H14 aluminium	Weldox 700 e steel
Modulus of elasticity, E (N/mm^2)	68,948	210,000
Poiss on's ratio, ν	0.33	0.33
Density ρ (kg/m^3)	2712.6	7850
Expansion (μ m/m–°**C**)	23.6	1.2
Conductivity, K (W/m–K)	222	44.5
Yield stress, A (N/mm^2)	102.82	859
B (N/mm^2)	49.79	329
n	0.197	0.57
Reference strain rate, $\dot{\varepsilon}_0$ (s^{-1})	1.0	0.005
C	0.001	0.011
m	0.859	1.071
T_{melt} (K)	893	1800
T_0 (K)	293	293
Specific heat, C_p (J/kg K)	920	452
Inelastic heat fraction, α	0.9	0.9
D_1	0.071	0.361
D_2	1.248	4.768
D_3	−1.142	−5.107
D_4	0.147	−0.008
D_5	0.0	1.33

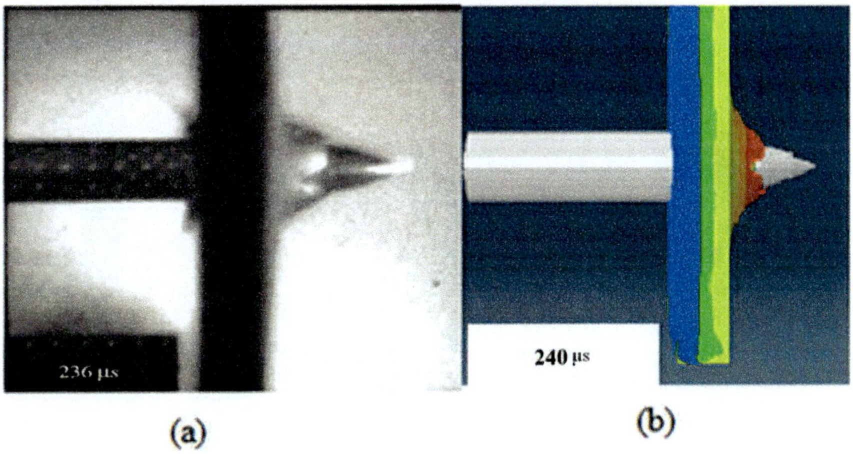

Fig. 3 Validation of failure of 12 mm monolithic target plate impacted by ogival nose-shaped projectile (**a**) experimental [10] (**b**) numerical simulation

4 Model Validation

For conducting validation, model considered by Dey et al. [10] is considered, and the impact is made using ogival-shaped projectile on 12 mm steel target plate. The impact velocities used in their experimental study were used for conducting simulations. After performing certain simulations by varying friction coefficient, 0.1 was chosen as coefficient of friction. It was found that the present finite element model was able to predict well the residual velocities, and the mechanisms observed were in close agreement with the experimental results. Figure 3 shows the experimental and simulation results.

5 Results and Discussion

The failure mechanism of target plate when impacted by ogival and conical-shaped projectile involves ductile hole enlargement and petal formation in all the three cases as presented here. It was observed that in the case of monolithic aluminium plate more stretching of the material occurs because aluminium is more ductile compared to steel as shown in Fig. 4.

Figure 5 shows that when conical projectile makes an impact on target, failure occurred involved ductile hole enlargement and plug formation. It was observed that the failure mechanism changed to plugging as the frontal portion of the projectile changed from conical to blunt. Also, plug is formed as a result of shear localization where constricted zones of intense shear are simulated.

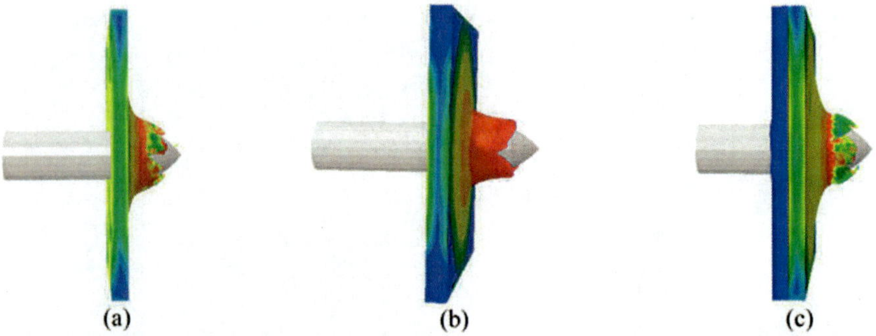

Fig. 4 Deformation and fracture pattern during penetration of ogival projectile at different time on (**a**) monolithic steel target plate, (**b**) monolithic aluminium target plate, (**c**) multi-layered target plate

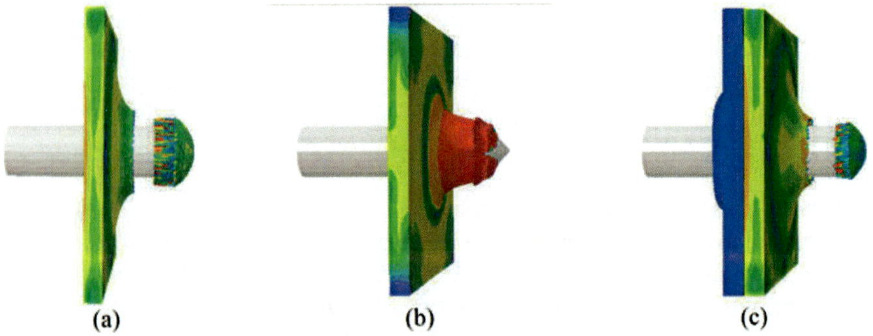

Fig. 5 Deformation and fracture during penetration of conical projectile at different time on (**a**) monolithic steel target plate, (**b**) monolithic aluminium target plate, (**c**) multi-layered target plate

5.1 Temperature Variation in Various Target Configurations Due to Thermal Softening

In reality, ballistic impact problems involve large amount of heat generation when projectile makes an impact on target. The rate of heat generation is much greater than the amount of heat dissipation by conduction and convection. Hence, temperature will be enhanced in the local regions of deformation which is basically the impact zone of projectile which is shown in Table 3. This reduces the target material flow stress resulting in more deformation to take place. This phenomenon is called thermal softening effect.

Table 3 Maximum temperature rise in various target configurations

S. No.	Target configuration	Maximum temperature rise for different type of projectile (in K)	
		Ogival	Conical
1	Monolithic steel target plate	933.3	825.3
2	Monolithic aluminium target plate	571.2	266.2
3	Multi-layered target plate	1017	546

5.2 *Velocity Drop for Various Targets*

When projectile is impacted on the target plate, large amount of kinetic energy is released which gets converted into different deformation processes. The mechanisms include membrane stretching, global deformation, and local plastic deformation at the impact zone. These mechanisms result in petal or plug formation. Velocity drop for various target configurations impacted by different shaped projectiles is shown in Fig. 6. It can be seen that the velocity drop is more in monolithic steel target plate compared to monolithic aluminium plate in both the cases, i.e., impact by ogival and conical projectile, which shows that steel has better ballistic resistance properties compared to aluminium.

For the case of multi-layered target, it is seen that when impact is made through ogival projectile, velocity drop increases in comparison with monolithic aluminium

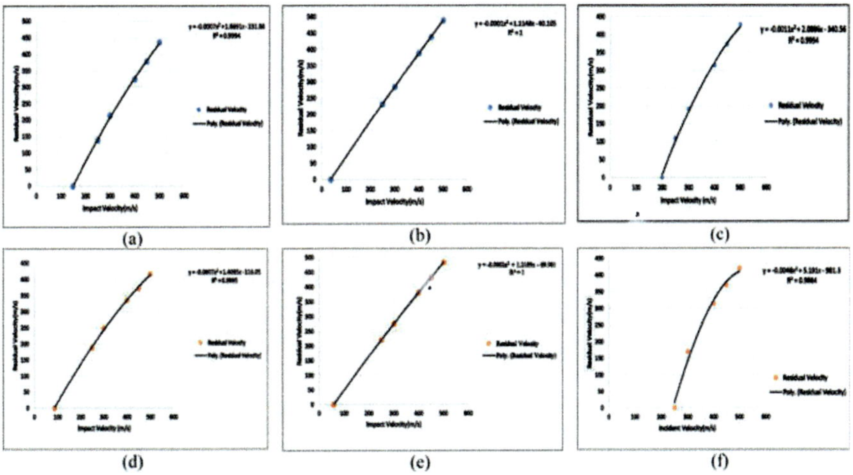

Fig. 6 Ballistic resistance curves for ogival projectile (**a**) monolithic steel (**b**) monolithic aluminium (**c**) multi-layered and for conical projectile (**d**) monolithic steel (**e**) monolithic aluminium (**f**) multi-layered target

and steel plate but when impacted by conical projectile, it remains quite close to that observed in monolithic steel target.

5.3 *Ballistic Resistance Curves*

It can be seen that the current ballistic impact problem and the finite element model predicts well the residual velocities for the given impact velocities. Figure 6 represents the ballistic resistance curves for ogival and conical nose-shaped projectile, respectively. These curves have impact velocity on the x-axis and residual velocity on the y-axis. Ballistic resistance curves are drawn using least square fitting method. From the curves for both conical and ogival-shaped projectile, it can be concluded that the multi-layered target is superior in terms of ballistic resistance as compared to monolithic targets because the slope of the curve is steeper for multi-layered which shows that velocity drop is more drastic in this case.

6 Conclusions

The paper focuses on studying the ballistic resistance of monolithic and sandwiched target impacted by two different shaped projectiles—ogival and conical nose-shaped. Thermal softening effect is taken into consideration to study the temperature variation near the fracture zone. Dynamic temperature displacement explicit model is used in the ABAQUS/CAE platform to perform the numerical simulations. For conducting the study, three types of targets were considered—6 mm monolithic aluminium target, 6 mm monolithic steel target, and 12 mm multi-layered target. The multi-layered target was a sandwiched target, combination of aluminium layer in the front and steel layer at the back with 6 mm thickness of each layer. Conclusions from the study are as follows.

1. When ogival projectile makes an impact on either monolithic or sandwiched target, the failure mechanism involves ductile hole enlargement and petal formation.
2. When conical projectile makes an impact on either monolithic or multi-layered target, the failure mechanism involved ductile hole enlargement and plug formation. As the frontal portion changes from conical to blunt, the failure mechanism changes from petalling to plugging.
3. It was observed that with decrease in impact velocity the drop in velocity increases.
4. When comparing monolithic aluminium target and steel target, it was noted that velocity drop is greater in case of steel target which concludes that steel has better ballistic resistance capability as it reduces the impact velocity to a greater extent.
5. It was observed that with increase in number of layers velocity drop decreased.

6. In case of both monolithic and multi-layered target, it was seen that the maximum temperature rise occurs when ogival projectile makes an impact on the target. For monolithic steel target, monolithic aluminium target, and multi-layered target, temperature rise on impact by ogival projectiles were 933.3, 571.2, and 1017 K, respectively.
7. Maximum temperature rise was observed in case of monolithic steel target plate (933.3 K) followed by multi-layered target and monolithic aluminium target.

For both ogival and conical projectile, multi-layered target is superior in terms of ballistic resistance as compared to monolithic targets.

References

1. Gogolowski RP, Morgan BR (2002) Ballistic experiments with titanium and aluminium targets, FAA Rep. No. DOT/FAA/AR-01/21, Federal Aviation Administration, Washington, DC
2. Borvik T, Hopperstad OS, Berstad T (2003) On the influence of stress triaxiality and strain rate on the behaviour of a structural steel. Part II. Numerical study. Euro J Mech 22:15–32
3. Teng X, Wierzbicki T, Hiermaier S, Rohr I (2005) Numerical prediction of fracture in the Taylor test. Int J Solids Struct 42:2929–2948
4. Iqbal MA, Gupta G, Diwakar A, Gupta NK (2010) Effect of projectile nose shape on the ballistic resistance of ductile targets. Euro J Mechan 29:683–694
5. Kpenyigba KM, Jankowiak T, Rusinek S (2013) Influence of projectile shape on dynamic behavior of steel sheet subjected to impact and perforation. Thin Walled Struct 65:93–104
6. Palta E, Gutowski P, Fang H (2018) A numerical study of steel and hybrid armor plates under ballistic impacts. Int J Solids Struct 137:279–294
7. Johnson GR, Cook WH (1983) A constitutive model and data for metals subjected to large strains, high strain rates and high temperatures. In: Proceedings of the seventh International symposium on Ballistics. American Defense Preparedness Association, Washington DC, USA. pp 541–547
8. Johnson GR, Cook WH (1985) Fracture characteristics of three metals subjected to various strains, strain rates, temperatures and pressures Eng Fract Mech 21:31–48
9. Kumar R, Kumar M, Kumar P (2019) Finite Element Analysis of Ballistic Impact on monolithic and multi layered target plate with and without air gap. Adv Mech Eng Springer publication. https://doi.org/10.1007/978-981-15-3639-7_71
10. Dey S, Borvik T, Teng X, Wierzbicki T, Hopperstad OS (2017) On the ballistic resistance of double-layered steel plates: an experimental and numerical investigation. Int J Solids Struct 44:6701–6723

Drop Analysis of Plate-Type Fuel Assembly in a Research Reactor

G. Verma, Samiran Sengupta, and S. Bhattacharya

Abstract In nuclear research reactors, plate-type fuel assemblies are often employed to achieve high neutron flux. During loading/unloading of these assemblies, a tight control over fuel assembly movement is maintained with established procedure. However, any mishandling during this procedure could result in an accidental drop, leading to mechanical damage to the fuel assembly. It may further result in fission product release to the environment in case of mishandling of a spent fuel assembly. The present work investigates the extent of the damage to the fuel assembly during a drop accident and evaluates the fuel plate integrity. In this paper, drop accidents are classified into two categories based on the location where the fuel assembly drop takes place. If an accident occurs within the reactor core region, impact on other fuel assemblies placed on the grid plate is considered. In case it occurs in other region of the reactor pool, a direct impact of the fuel assembly on the reactor pool bottom is envisaged. Numerical simulations are performed to simulate drop accidents using finite element analysis, and quantitative assessment of the damage is presented in terms stress, deformation, and energy history in the present work.

Keywords Plate-type fuel · Postulated initiating event · Drop accident

Nomenclature

m	Mass of the fuel assembly
m_{add}	Added mass of the fuel assembly
t	Time
g	Acceleration due to gravity
V	Volume of the fuel assembly

G. Verma (✉) · S. Sengupta
Research Reactor Design and Projects Division, Bhabha Atomic Research Centre, Mumbai 400085, India
e-mail: gaurav91verma@gmail.com

S. Bhattacharya
Homi Bhabha National Institute, Mumbai 400085, India

R. Velmurugan et al. (eds.), *Dynamic Behavior of Soft and Hard Materials Volume 1*, Springer Proceedings in Materials 34, https://doi.org/10.1007/978-981-99-6030-9_2

ρ	Fluid density
F_D	Drag force
v	Fluid assembly velocity
C_D	Drag coefficient
A_p	Orthographic projection area
v_t	Terminal velocity
Δ_t	Time step increment
w_{max}	Element maximum eigenvalue
P_m	primary membrane stress intensity
P_b	Primary bending stress intensity
S_m	Design stress intensity
S_u	Ultimate tensile strength

1 Introduction

Plate-type fuel assemblies are often used in research reactors for a compact core to obtain very high neutron flux. During loading, unloading, and shuffling operations, any mishandling of the fuel assemblies could result in fuel assembly drop accident, which is one of the important postulated initiating events (PIEs) in reactor design. A drop accident can result in mechanical damage to the fuel assemblies and may result in fission product release (in case of mishandling of a spent fuel assembly). Although, the geometry of the plate-type fuel assembly is such that the fuel plates are mechanically well protected against such impacts and damage causing radiation product release is normally not expected. Nonetheless, a fuel assembly drop accident needs to be considered in the reactor design to prove this design aspect.

Many researchers around the world have performed quantitative assessment of accidental fuel assembly drop for different types of fuel geometry. Wu et al. [1] performed drop analysis for a BWR fuel assembly dropped from the crane hook hitting the rack bottom plate of the spent fuel pool. The damage to the fuel assembly was assessed using finite element simulation. Petkevich et al. [2] carried out experiments involving a series of drop tests of the fuel assembly of different complexity to validate the numerical analysis performed. Kim et al. [3] carried out drop impact analysis of plate-type fuel assembly considering different orientations during the drop accident. Kim et al. [4] also investigated the drop behavior of a plate-type fuel assembly through a dropping test in water of a full-scale dummy fuel assembly. Kim et al. [5] proposed a methodology to predict the axial impact behavior on a fuel structure by using the commercial finite element codes ANSYS and DYNA3D. Kim and Yoon [6] predicted the dynamic failure behavior of a Spacer Grid using dynamic impact analysis method with ABAQUS.

1.1 Geometry of the Fuel

A typical plate-type fuel assembly consists of box-type geometry (Verma et al. [7]). It comprises a fuel box consisting of fuel plates (swaged to the fuel box) and a lower end-fitting. The lower part of the fuel box is joined to the upper part of the end-fitting either through welding or screwing. In the core, the fuel assembly is arranged on the grid plate such that the fuel assembly toe is inserted into it.

2 Impact Characteristics of the Fuel Assembly

In order to characterize the impact characteristics of the fuel assembly, a drop accident can be basically categorized into two situations based on drop accident location (i) outside the reactor core (on the reactor pool) and (ii) within the reactor core (onto the other fuel assemblies). For the first case, the fuel assembly will impact the pool bottom or any other mechanical components which are within the reactor pool. In the other case, the falling fuel assembly will impact the other fuel assemblies loaded in the reactor core.

2.1 Classification of Drop Accidents and Their Maximum Drop Height

For drop occurring outside the core, it is assumed that the pool bottom is rigid, and at the moment of impact, maximum velocity of the fuel assembly is achieved. For drop on to the loaded fuel assemblies (within the core), it is assumed that the end-fitting of the falling fuel assembly could impact either on the side plate or on the fuel handling pin of the standing fuel assembly and subsequently damaging the fuel plates. Any fracture of the fuel handling pin may also result in damage to the side plates at the swaged location. The maximum drop height considered for drop outside the core is 7.65 m and within the core as 5.2 m. In addition, most severe falling orientation is assumed when the fuel assembly drops.

2.2 Design Criteria

The design criteria of ASME Sec-III, Subsection-NB Level-D service limit are used because fuel assembly drop accident is an unlikely event to occur categorizing it as a postulated accident event [8]. Level-D service limits allow gross general deformation with some consequent loss of dimensional stability and damage requiring repair. Table 1 summarizes the stress limits of the drop impact analysis.

Table 1 Stress limits of the drop impact analysis

Design limits	Stress limits
P_m < Min (2.4 S_m or 0.7 S_u)	P_m < 106.18 MPa
$P_m + P_b$ < Min (3.6 S_m or 1.05 S_u)	$P_m + P_b$ < 159.26 MPa

2.3 Impact Velocity Calculation

The forces acting on the fuel assembly when it falls freely in the reactor pool are inertial force, buoyancy force, and drag force, as per Kim et al. [3]. Then, the force balance is given as

$$(m - \rho V)g - F_D = (m + m_{add})\frac{dv}{dt} \tag{1}$$

The drag force is given as

$$F_D = \frac{1}{2}\rho A_p C_D v^2 \tag{2}$$

With increase in velocity, the resisting drag force increases. Then, the governing differential equation is given as

$$\int_0^v \frac{dv}{(m - \rho V)g - \frac{1}{2}\rho A_p C_D v^2} = \frac{1}{(m + m_{\text{add}})}\int_0^t \mathrm{d}t \tag{3}$$

Solving the differential equation, the velocity is evaluated as

$$v = \sqrt{\frac{(m - \rho V)g}{\frac{1}{2}\rho A_p C_D}} \tanh\left(t\frac{\sqrt{(m - \rho V)g.\frac{1}{2}\rho A_p C_D}}{m + m_{\text{add}}}\right) \tag{4}$$

The fuel assembly accelerates until the force due to gravity is balanced by the resistance forces of the surrounding fluid. When the fuel assembly reaches its terminal velocity, kinetic energy is maximum, and at that moment, acceleration is zero. The terminal velocity is then defined as

$$v_t = \sqrt{\frac{(m - \rho V)g}{\frac{1}{2}\rho A_p C_D}} \tag{5}$$

2.4 Numerical Analysis Methodology

For the present fuel drop numerical analysis, an explicit method is used to solve the finite element equation using an explicit time integration scheme. In this scheme, the nodal accelerations are directly solved as the inverse of the diagonal mass matrix times the net nodal force vector. The force vector takes contributions from applied pressure; body forces, etc., and other parameters like damping, viscosity, etc. In this scheme, accelerations are known at time t, then, the velocities are calculated at time $t + \Delta t/2$, and displacements at time $t + \Delta t$. Subsequently, strain and stress are calculated from displacements. The issue with the explicit scheme is that it is conditionally stable and the size of the time step is limited. Hence, for stability, the maximum time step must be less than the critical value (also known as the Courant time step) which depends on the element size or the element order. This critical value is the time taken by the sound wave to travel across an element.

2.5 Finite Element Modeling

The fuel assembly at the moment of impact is modeled using finite element mesh, as shown in Fig. 1. For out-of-core drop, the pool bottom is modeled with coarse meshes, having high stiffness, perfectly rigid frictionless surface. Bilinear elastic–plastic material properties are assumed for the fuel assembly structure at room temperature. An initial impact velocity of 4.728 m/s is the initial condition for vertical orientation drop and 0.72 m/s for horizontal orientation drop. The total numbers of elements are 7,91,295 eight-noded brick elements. For drop within core, the impact velocity of 4.529 m/s is imposed on the falling fuel assembly as the initial condition. The total number of elements used in the analysis are 12,50,541 eight-noded brick elements.

3 Fuel Assembly Drop Accident onto the Reactor Pool

For drop accidents occurring outside the reactor core, the falling fuel assembly will directly impact the pool bottom. Since most of the impact energy is absorbed by the lower end-fitting of the fuel assembly, the stresses induced in the fuel plates are considerably reduced. A quantitative assessment of the fuel plate damage is carried out, and the stresses induced due to impact condition are compared with the design criteria in Sect. 2.2. It is important to note that in all the analyses only the first impact is simulated as it is significant from structural integrity point of view; hence, subsequent impact on rebounds is not simulated (Yim et al. [9]).

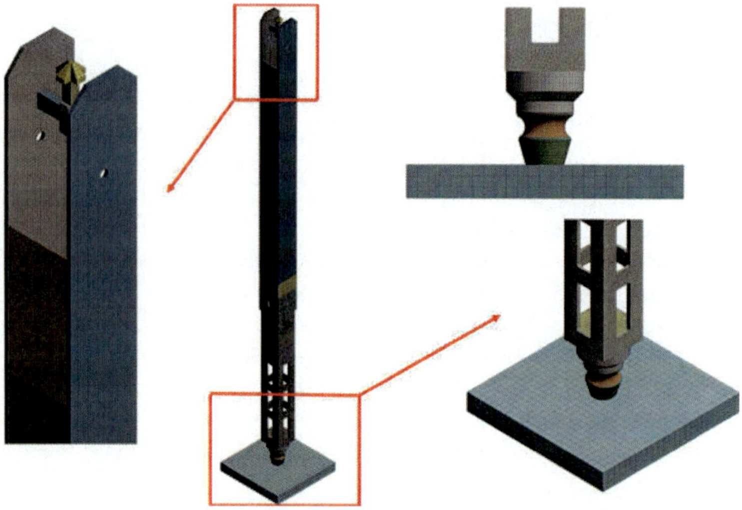

Fig. 1 Finite element model of fuel assembly at the moment of impact

3.1 *Vertical Orientation*

In the vertical orientation drop, the stress is primarily concentrated at the fuel toe region, at the beginning of the impact, which later propagates throughout the structure and diffuses with time. Therefore, the stress oscillations observed are attributes of the wave propagation and attenuation, not of subsequent impacts. Figure 2 shows the maximum stress history at the fuel plate clad. The maximum membrane stress (P_m) at the fuel plate clad is 101.8 MPa (<106.18 MPa), whereas the maximum membrane plus bending stress ($P_m + P_b$) is 104.55 MPa, which is lower than the design limit of 159.26 MPa for the accident case. Thus, the structural integrity of the fuel plate is maintained.

Figure 3 shows the internal, kinetic, and total energy history of the system. Figure 4 shows the maximum stress intensity (membrane and membrane plus bending) at the fuel handling pin location. From Fig. 4, it is observed that P_m is 25.14 MPa, whereas ($P_m + P_b$) is found to be 107.89 MPa. It can be seen that the both the P_m and ($P_m + P_b$) are within allowable limits. Figure 5 shows the maximum stress intensity at the end plate and transition box junction. It is observed that P_m is 98.23 MPa whereas ($P_m + P_b$) is 100.63 MPa which is also within allowable limits. The maximum allowable strain for Al 6061(O) (fuel assembly structural material) varies from 16 to 18% with increasing thickness as per ASME Section-II, part-B [10]. For conservativeness, the maximum allowable strain is taken to be 16%. Figures 6 and 7 display the maximum effective plastic strain. From Figs. 6 and 7, it can be observed that the maximum effective plastic strain over time at the fuel plate clad location is 2.04% whereas the maximum effective plastic strain over time observed for the complete structure is 6.47% in the localized region of the transition box.

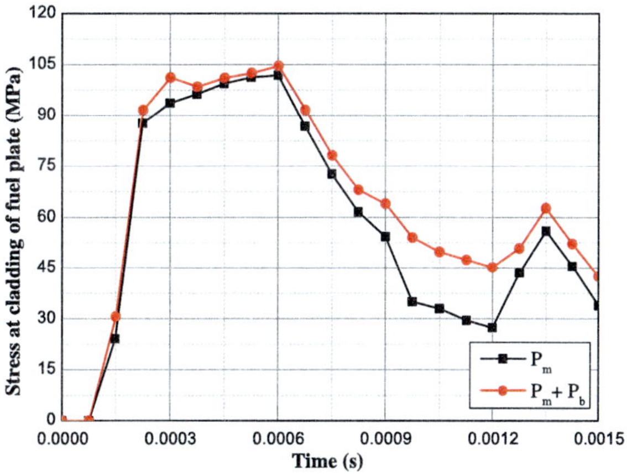

Fig. 2 Maximum stress history at fuel plate clad for vertical orientation drop on the reactor pool bottom

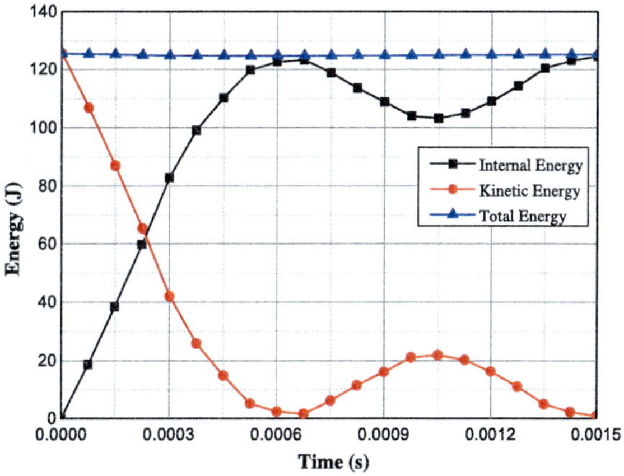

Fig. 3 Internal, kinetic, and total energy history for vertical orientation drop on the reactor pool bottom

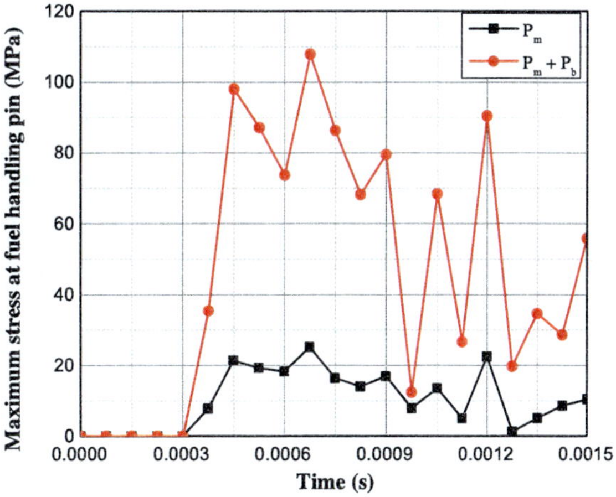

Fig. 4 Maximum stress history at fuel handling pin for vertical orientation drop on the reactor pool bottom

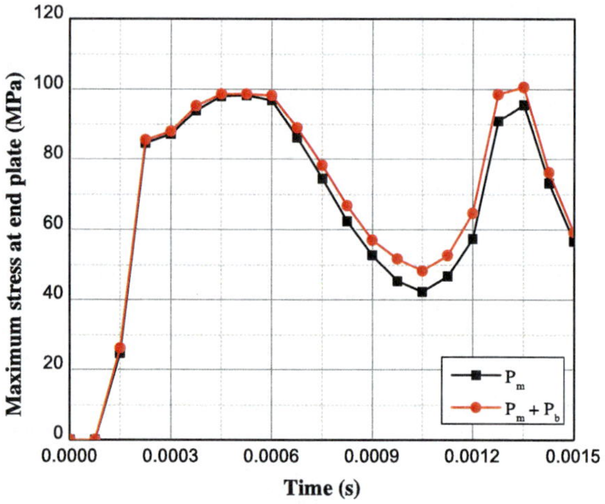

Fig. 5 Maximum stress history at the end plate transition box junction for vertical orientation drop on the reactor pool bottom

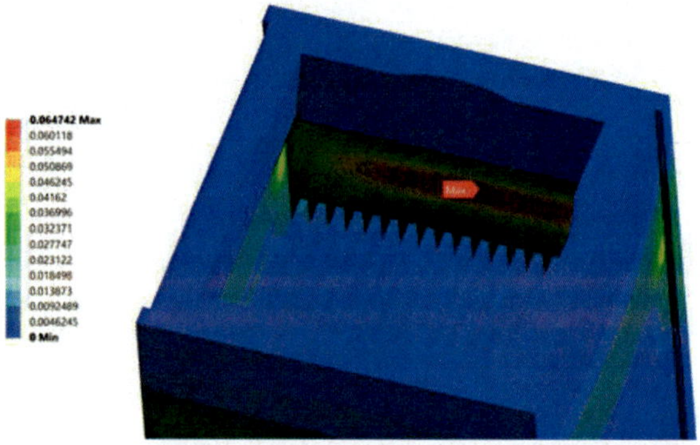

Fig. 6 Effective plastic strain contour in the transition box region for vertical orientation drop on the reactor pool bottom

Fig. 7 Effective plastic strain in the fuel plates for vertical orientation drop on the reactor pool bottom

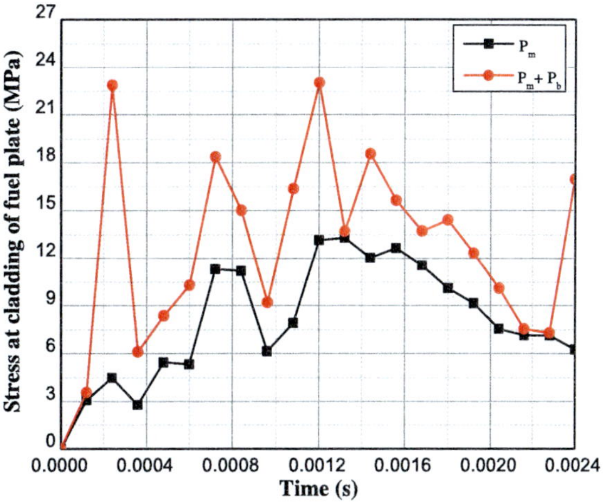

Fig. 8 Maximum stress history at the fuel plate for horizontal orientation drop on the reactor pool bottom

3.2 Horizontal Orientation

In this case, the fuel is assumed to fall in horizontal orientation and impact the reactor pool bottom. In this orientation, maximum velocity (least drag) is achieved when the fuel is in 45° rotation about the longitudinal axis. In such situation, the initial impact is expected to occur diagonally in horizontal orientation. Since the impact velocity is relatively small for this orientation, it is expected that the stresses and strains are relatively small compared to the vertical orientation case. Figure 8 shows the stress intensity history at the fuel plate clad. It is observed that P_m at the fuel plate location is 13.29 MPa whereas $(P_m + P_b)$ is 23.06 MPa. No effective plastic strain is observed in the fuel plates, whereas maximum effective plastic strain is 0.63%. Figure 9 displays the internal, kinetic, and total energy history for horizontal orientation drop.

4 Fuel Assembly Drop Accident Within the Reactor Core

In this scenario, the fuel is expected to drop inadvertently onto the other fuel assemblies arranged in the grid plate. In such cases, the falling fuel assembly could directly hit the fuel handling pin or the side plates during initial impact. From numerical simulations, it is observed that direct impact of the falling fuel on the fuel handling pin is more severe, hence, considered for analysis. The fuel handling pin, which is located above the fuel plates, is used for handling purposes. Also, this fuel handling pin protects the fuel plates from external impacts. Hence, most of the kinetic energy is expected to be absorbed by the fuel handling bar during impact. However, it is necessary to determine if fracture of the fuel handling bar may occur during such an impact as additional simulations for the impact on the fuel plate needs to be performed if

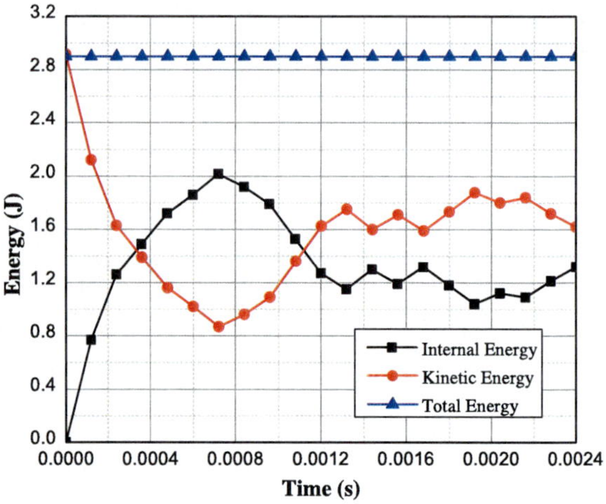

Fig. 9 Internal, kinetic, and total energy history for horizontal orientation drop on the reactor pool bottom

some kinetic energy still exists with the falling fuel post impact. If such is not the case, the fuel plate structural integrity is guaranteed.

Figure 10 shows the energy history. The total, kinetic, and internal energy of system are shown. After impact, the fuel handling bar internal energy increases because the bar absorbs the kinetic energy of the falling fuel assembly. Figure 11 shows the stress distribution history in which a fracture of the fuel handling bar has occurred. From Fig. 11, it can be observed that the P_m is within allowable limits, however, $(P_m + P_b)$ for the handling bar exceeds the allowable limits. Hence, fracture due to bending is expected. Also, from Fig. 12, it is observed that the maximum effective plastic strain is 19.41% which is beyond the allowable range. This indicates that the fuel handling bar undergoes plastic deformation and fracture at the moment of impact as it crosses the allowable limit. It is important to note that during this process; almost 99% of the kinetic energy of the falling fuel is dissipated due to the fracture of the handling bar. With the remaining ~1% of the kinetic energy, the velocity of further impact to the fuel plates is very small. Thus, it can be assumed that even after fracture of the handling bar, the damage to the fuel plates is highly unlikely. Figure 13 shows the maximum effective plastic strain near the handling bar region.

5 Conclusions

In the present work, an analysis methodology for an accidental drop of fuel assembly for a research reactor was established. The drop accidents were primarily classified into two categories based on the location of the fall and numerical simulation for each accident condition was carried out using finite element procedures. An explicit analysis scheme was used for the analysis. For an accident occurring in the reactor

Fig. 10 Internal, kinetic, and total energy history for the vertical orientation drop of fuel-on-fuel

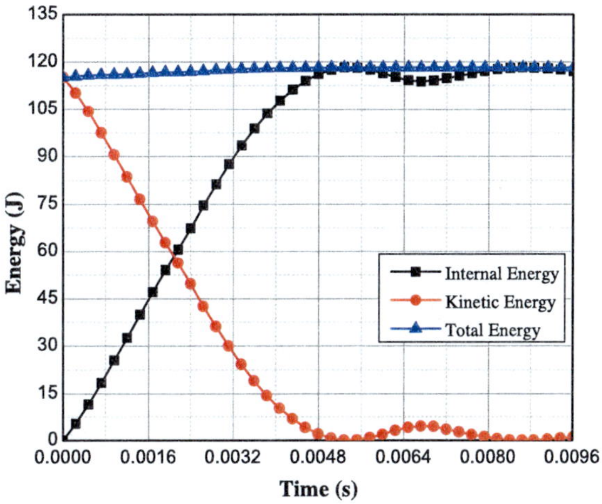

Fig. 11 Maximum stress history at handling pin bar for vertical drop on top of fuel for drop within reactor core

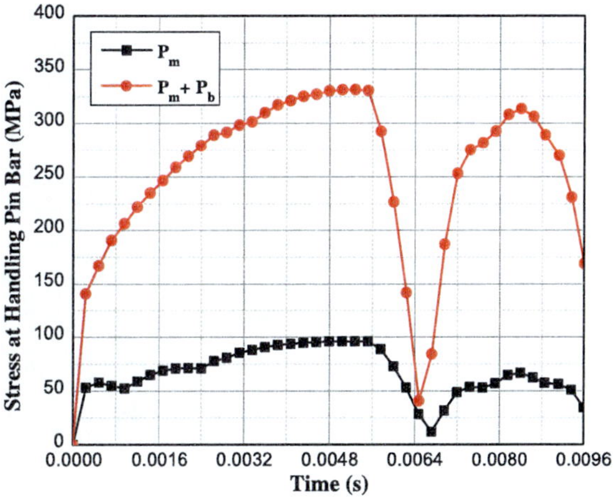

pool, a direct fall of the fuel assembly on the pool was considered to be the most conservative. Two different orientations of drops were envisaged and analyzed. The maximum stress at the fuel plate clad was found to be lower than the design limits, concluding that the fuel structural integrity was maintained. For drop accident occurring within the reactor, the fracture of the fuel handling bar was investigated. In this simulation, fracture of the fuel handling bar was observed due to the impact of the falling fuel assembly, which additionally advocates the protection of the fuel plates from direct impacts when fuel assemblies placed on the core. It was also observed that during this process; almost all of the kinetic energy of the falling fuel was dissipated during the fracture of the handling bar, and the damage to the fuel plates was highly unlikely.

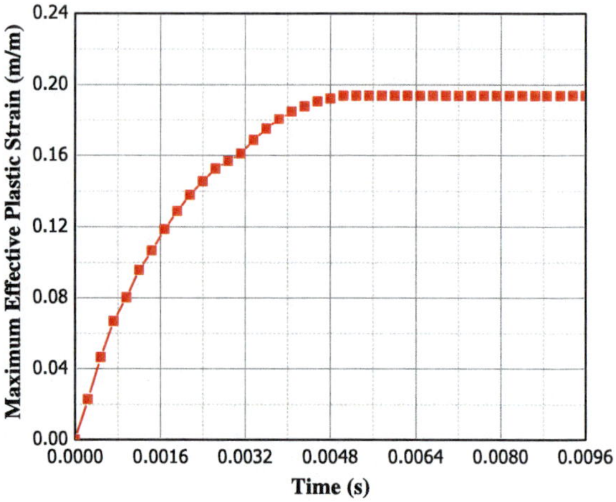

Fig. 12 Maximum effective plastic strain history at handling pin bar for vertical drop on top of fuel for drop within reactor core

Fig. 13 Maximum effective plastic strain contour in the fuel handling bar for vertical orientation drop in the reactor core on top of fuel

References

1. Wu HJ, Tseng CC, Cheng SC (2006) A numerical analysis for a BWR fuel assembly drop event. J Nuc Sci Tech 43:1068–1073. https://doi.org/10.1080/18811248.2006.9711196
2. Petkevich P, Abramov V, Yuremenko V, Piminov V, Makarov V, Afanasiev A (2011) Simulation of the nuclear fuel assembly drop test with LS-Dyna. SMiRT-21 Trans 2011, New Delhi, India
3. Kim HJ, Yim JS, Lee BH, Oh JY, Tahk YW (2014) Drop impact analysis of plate-type fuel assembly in research reactor. Nucl Eng Tech 46(4):529–540. https://doi.org/10.5516/NET.09.2013.103
4. Kim HJ, Yim JS, Tahk YW, Oh JY, Kong EH (2019) Drop behaviors of a plate-type fuel assembly used in research reactor for a drop accident. Prog Nucl Energy 113:255–262. https://doi.org/10.1016/j.pnucene.2019.01.020
5. Yoon KH, Kim JY, Kim HK (2007) Axial Impact Analysis of a 16 by 16 fuel assembly by the FE method. SMiRT-19 Trans, Toronto, Canada
6. Kim JY, Yoon KH (2007) A repeated dynamic impact analysis for 7x7 spacer grids by using ABAQUS/standard and explicit. SMiRT-19 Trans, Toronto, Canada

7. Verma G, Eswaran M, Sengupta S, Reddy GR, Mammen S (2017) Dynamic characteristics of immersed plate type fuel assemblies under seismic excitation. Nuc Eng Des 314:11–28. https://doi.org/10.1016/j.nucengdes.2017.01.005
8. ASME BPVC Section III (2011)
9. Yim JS, Kim HJ, Tahk YW, Oh JY, Lee BH (2012) Evaluation of fuel plate integrity during a fuel assembly drop accident using an energy method. Proc Int RERTR, Poland, pp 14–17
10. ASME BPVC Section II (2011)

Effect of CRH Value on Damage Resistance of Aluminum Plate Under Bullet Impact

Rajeev Kumar and Vimal Kumar

Abstract Ballistic performance of ductile targets is the most important factor for the safety of defense/protective structures such as shelters for armed conflict, bunkers for military liveware etc. In the present study, a three-dimensional numerical simulation is carried out using ABAQUS to study the effects of caliber radius head (CRH) value on the ballistic resistance of aluminum plates under normal bullet impact. The Johnson-Cock (JC) constitutive model was used to carry out the numerical simulations. The monolithic AA-7075 target of uniform thickness having size 100 × 100 mm were impacted by two different bullets with CRH value 1.0 and 2.5. The weight and radius of the bullet body were kept identical in all the simulations. The target plate was restrained from all four sides. The impact velocity of the bullet varied from 0.65 to 1.5 km/sec and the response of the target i.e., energy dissipation, damage, reaction force, etc. were obtained and compared. For a given impact velocity, the residual velocity of the projectile with CRH 1.0 was noticed higher than that for the projectile with CRH 2.5. In general, the reaction and energy absorption of the plate was noticed smaller against the projectile with CRH 2.5 when the impact velocity was closer to the ballistic limit. Projectile with CRH 1.0 and CRH 2.5 failed the ductile target through enlargement of the hole. The ballistic impact was critical in terms of hole enlargement and failure mode for a projectile with CRH 1.0 because of the sharper nose and smaller contact area and also longer ballistic length.

Keywords Numerical simulation · Normal impact · CRH · Ballistic limit · Aluminum

R. Kumar (✉) · V. Kumar
Department of Civil Engineering, National Institute of Technology Hamirpur, Hamirpur 177005, India
e-mail: rajeev.kr1893@gmail.com

V. Kumar
e-mail: panchariya.vimal@gmail.com

R. Velmurugan et al. (eds.), *Dynamic Behavior of Soft and Hard Materials Volume 1*, Springer Proceedings in Materials 34, https://doi.org/10.1007/978-981-99-6030-9_3

1 Introduction

From the civilian and military point of view, there is a great need for the safety of defense/protective structure such as shelters for armed conflict and bunkers for military liveware; most of the studies on ballistic impact was focused on the normal impact conditions because the normal incidence condition is one of the worst critical situations for predicting the target performance. Several researchers [1–4] discussed the effects of different parameters like projectile nose shape, diameter, target thickness, and impact velocities on the older alloys of aluminum which are either not used or rarely used nowadays for designing defense/protective structures. The aluminum 7075-T6 is most widely used because of better important properties like greater strength, light in weight, etc. It has very good mechanical properties and exhibits good ductility, toughness, and good fatigue resistance. Other advantages of this material are that it is free from low-temperature embrittlement and provides greater rigidity, hence this material is most commonly used for highly stressed structural applications. In the present study, the effects of CRH value 1.0 and 2.5 were studied on the response of the target, i.e., energy dissipation, ballistic limit, residual velocity, reaction force, and damage profile against the impact velocity varied from 0.65 to 1.5 km/sec.

2 Constitutive Modeling

The finite element analysis was carried out for studying the material behavior of 7075-T6 Aluminum alloy [5] using the Johnson–Cook elasto-viscoplastic model [6, 7]. The model considers high strain rates sensitivity, larger deformation, yielding, plastic flow, and thermal softening. The equivalent von-Mises stress ($\tilde{\sigma}$) in the Johnson–Cook model is defined in Eq. (1)

$$\overline{\sigma}\left(\overline{\varepsilon}^{Pl}, \dot{\overline{\varepsilon}}^{Pl}, \overline{T}\right) = \left[A + B\left(\overline{\varepsilon}^{Pl}\right)^n\right]\left[1 + C\ln\left(\frac{\dot{\varepsilon}^{Pl}}{\dot{\varepsilon}_0}\right)\right]\left[1 - \left(\overline{T}\right)^m\right] \tag{1}$$

where A, B, C, n, and m represent the yield stress, hardening constant, strain rate sensitivity hardening exponent, and thermal softening exponent, respectively. $\overline{\varepsilon}^{Pl}$ is equivalent plastic strain, $\dot{\overline{\varepsilon}}^{Pl}$ is equivalent plastic strain rate, $\dot{\varepsilon}_0$ is a reference strain rate, and $\overline{T}$ is non-dimensional temperature can be written in Eq. (2)

$$\overline{T} = \frac{(T - T_0)}{(T_m - T_0)}, \ T_0 \le T \le T_m \tag{2}$$

where T is the current temperature, T_m is the melting point temperature, and T_0 is the room temperature. The fracture model proposed by the Johnson–Cook model includes the effect of stress triaxiality, strain rate, and temperature on the equivalent

Table 1 Johnson–Cook parameter used in the present study [5]

JC parameters	AA7075-T6	AISI4340
A (MPa)	546	792
B (MPa)	678	510
C	0.024	0.014
n	0.71	0.26
M (k)	1.56	1.02
Reference strain rate	1	1
d_1	−0.068	0.05
d_2	0.451	3.44
d_3	−0.952	−2.12
d_4	0.036	0.002
d_5	0.697	0.61
Melting temp (k)	925	1520
Transition temp (k)	293.2	293.2
Density (tonne/mm^3)	2.81e-09	7.85e-09
Young's modulus (MPa)	717,000	205,000
Poisson's ratio	0.33	0.33
Yield strength (MPa)	503	710
Ultimate strength (MPa)	560	985
% elongation	11	10–12

failure strain. The equivalent fracture strain $\overline{\varepsilon}_f{}^{Pl}$ is expressed in Eq. (3)

$$\overline{\varepsilon}_f^{Pl}\left(\frac{\sigma_m}{\overline{\sigma}}, \dot{\overline{\varepsilon}}^{Pl}, \overline{T}\right) = \left[d_1 + d_2\exp\left(-d_3\frac{\sigma_m}{\overline{\sigma}}\right)\right]\left[1 + d_4 \ln\left(\frac{\dot{\overline{\varepsilon}}^{Pl}}{\dot{\varepsilon}_0}\right)\right]\left[1 + d_5\overline{T}\right] \quad (3)$$

where $d_1 tod_5$ are material parameters, $\frac{\sigma_m}{\overline{\sigma}}$ is the stress triaxiality ratio, and σ_m is the mean stress. The Johnson–Cook parameter and material properties used in this study are given in Table 1.

3 Finite Element Method

The finite element model of the target and the projectile was developed using three-dimensional modeling using ABAQUS/EXPLICIT. The projectile and target plate were modeled as the solid deformable body. The caliber radius heads (CRH) of ogive nosed projectiles was varied keeping the weight and radius constant for all the simulations. The surface-to-surface contact between the projectile and the target was modeled considering kinematic contact and employing friction at the interface.

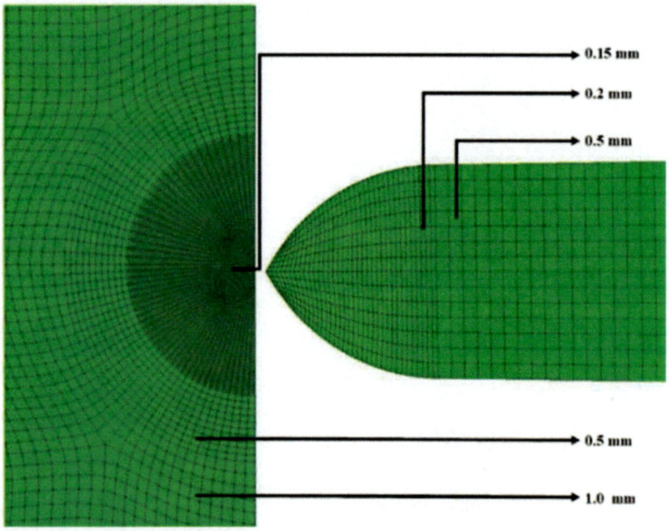

Fig. 1 Detail of meshing in finite element model

The projectile was considered as the master surface and the contact region of the plate as the slave surface. The target plate is restrained from all four sides while the projectile was given an impact velocity. The dimensions of the target plate are 100 × 100 mm with a thickness of 6 mm. The meshing in the finite element model is shown in Fig. 1. The total number of elements in the model was approximately 396 k. The ballistic limit was studied for different impact velocities which were ranging from 650 to1500 m/sec.

4 Results and Discussion

4.1 Residual Velocity and Energy Dissipation During Impact

The relation between the residual and impact velocity is shown for both the CRH values in Fig. 2a. The residual velocities were noticed 1314 and 1280 m/sec for CRH 1.0 and CRH 2.5, respectively, at the impact velocity 1500 m/sec. The residual velocity decreased with a decrease in the impact velocity exhibiting almost linear proportionality. The ballistic limit for CRH 1.0 was noticed to be relatively higher compared to CRH 2.5 of the target noticed. The ballistic limits were 750 and 700 m/sec for CRH 1.0 and 2.5, respectively (see Fig. 3). In general, the energy dissipation was nominally reduced with a decrease in the impact velocity. The energy dissipated by the plates were noticed at 2.07 and 1.87 kJ for CRH value 1.0 and 2.5, respectively, at the ballistic limits, see Fig. 2b.

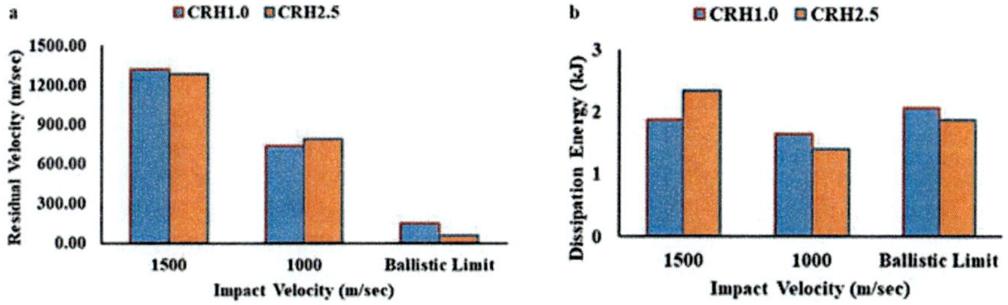

Fig. 2 **a** Residual velocity vs impact velocity curve **b** energy dissipation

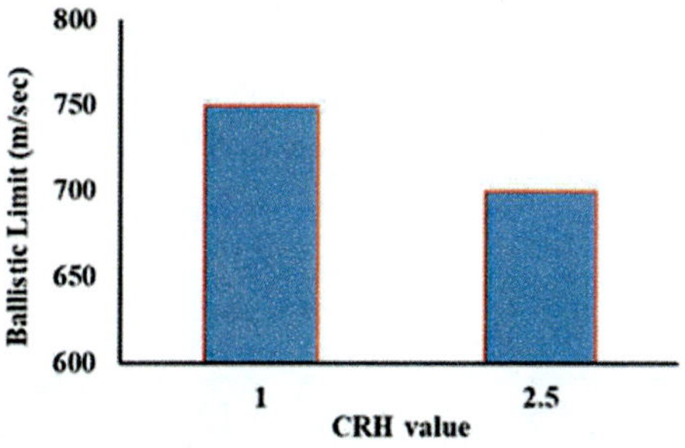

Fig. 3 Ballistic limit of different projectile shapes

4.2 *Computation of Reaction Force During Impact*

The plots of reaction force vs time are shown in Figs. 4a and b for CRH values 1.0 and 2.5, respectively. In general, the reaction force decreased with a decrease in the impact velocity. For CRH value 1.0, maximum values of reaction force were noticed 253.61, 198.39, and 149.67 kN for the impact velocity of 1500, 1000 m/sec and at ballistic limit, respectively. Similarly, for CRH value 2.5, the magnitude of maximum reaction force was 258.50, 190.16, and 132.16 kN for the impact velocity of 1500, 1000 m/sec, and ballistic limit, respectively.

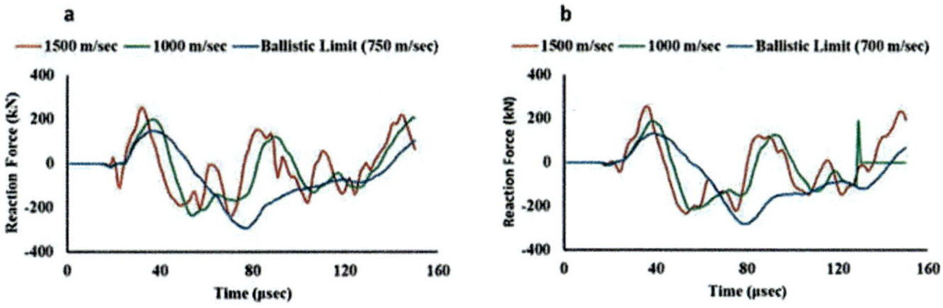

Fig. 4 Reaction and time response for **a** CRH 1.0 and **b** CRH 2.5

4.3 Failure Modes

Figures 5a and b shows the failure modes of the monolithic aluminum target impacted by the projectile with different impact velocity for CRH 1.0 and 2.5, respectively. The mode of failure is influenced by the caliber radius head of the projectile. CRH 1.0 projectile failed the ductile target through enlargement of the hole which were 7.69, 8.19, and 8.98 mm in size for impact velocity of 1500, 1000, and 750 m/sec, respectively. Similarly, CRH 2.5 projectile failed the ductile target through enlargement of the hole which were 7.70, 7.52, and 8.50 mm in size for impact velocity of 1500, 1000, and 700 m/sec, respectively. Also, it is found that the projectile with CRH 1.0 creates a critical role in terms of hole enlargement as compared to CRH 2.5 at impact velocity of 1000 m/sec and irrespective of their ballistic limit. This is because the projectile nose became sharper with decrease in CRH value from 2.5 to 1.0; therefore, it is more quickly to hit the object with a longer ballistic length. However, there is no significant change (almost constant) in terms of hole size at impact velocity of 1500 m/sec due to the high impact velocity.

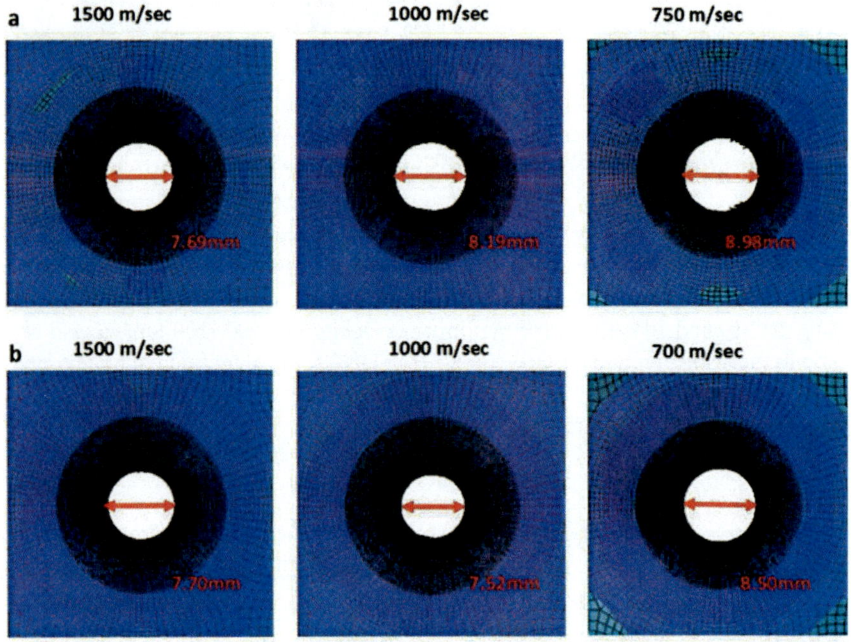

Fig. 5 Failure modes of the target at different impact velocities for **a** CRH 1.0 **b** and CRH 2.5

5 Conclusions

The ballistic performance of 7075-T6 aluminum alloy targets is studied under varying impact velocity under the normal impact of projectile (dia. 7.62 mm). Numerical simulations are carried out to determine the ballistic performance using Johnson–Cook constitutive model. The study showed that the ballistic limit of the aluminum plate decreased with an increase in CRH value from 1.0 to 2.5. The ballistic limit was 700 and 750 m/sec for the projectile with CRH 2.5 and 1, respectively. For a given impact velocity, the residual velocity of the projectile with CRH 1.0 was noticed higher than that for the projectile with CRH 2.5. In general, the reaction and energy absorption of the plate was noticed smaller against the projectile with CRH 2.5 when the impact velocity was closer to the ballistic limit. Projectile with CRH 1.0 and CRH 2.5 failed the ductile target through enlargement of the hole. The ballistic impact was critical in terms of hole enlargement and failure mode for a projectile with CRH 1.0 because of the sharper nose and smaller contact area and also longer ballistic length.

Acknowledgements The authors sincerely thanks Dr. M. A. Iqbal, IIT Roorkee, India, and Dr. Venkatesan J, SERC Chennai, India, for helping in carrying out numerical simulations.

References

1. Khan SH, Azeem M, Ansari R (2010) Numerical simulation of ogive nose projectile impact on aluminum plates of different thicknesses, pp 27–29. https://doi.org/10.1243/09544062jmes1523
2. Iqbal MA, Gupta G, Gupta NK (2010) 3D numerical simulations of ductile targets subjected to oblique impact by sharp nosed projectiles. Int J Solids Struct 47:224–237. https://doi.org/10.1016/j.ijsolstr.2009.09.032
3. Gupta NK, Iqbal MA, Sekhon GS (2008) Effect of projectile nose shape, impact velocity and target thickness on the deformation behavior of layered plates. Int J Impact Eng 35:37–60. https://doi.org/10.1016/j.ijimpeng.2006.11.004
4. Iqbal MA, Gupta G, Diwakar A, Gupta NK (2010) Effect of projectile nose shape on the ballistic resistance of ductile targets. Euro J Mech A/Solids 29:683–694. https://doi.org/10.1016/j.euromechsol.2010.02.002
5. Sundaram SK, Bharath AG, Aravind B (2022) Influence of target dynamics and number of impacts on ballistic performance of 6061–T6 and 7075–T6 aluminum alloy targets. Mech Based Des Struct Mach 50:993–1011. https://doi.org/10.1080/15397734.2020.1738245
6. Johnson GR (1983) A constitutive model and data for materials subjected to large strains, high strain rates, and high temperatures. In: Proceeding 7th information symposium ballistics, pp 541–547
7. Johnson GR (1985) Fracture characteristics of three metals subjected to various strains, strain rates, temperatures and pressures, pp 31–48

Impact Response of RC Beams Through Contacting and Non-contacting Sensors

S. Kumar, M. M. Dalbehera, V. Kumar, and A. K. Mittal

Abstract This paper presents the performance of reinforced concrete beams which were tested under an instrumented drop-weight impact testing facility. The reinforced concrete beams of size 2.5 × 0.25 × 0.15 m were made with M40 grade concrete. The beams were having two different quantities of reinforcement, i.e., 2.74 and 4.12%. A rigid steel mass of 200 kg was dropped on the beams from 1 to 1.5 m heights. The deflection, strain in steel reinforcement, reaction force, etc., were measured using various contacting and non-contacting sensors during the impact event. For the measurement of deflection, the linear variable differential transformers (LVDTs) and laser displacement sensors (LDS) were used as contacting and non-contacting sensors, respectively. The response obtained using these two sensors was also compared and discussed. From the experimental investigation, it was observed that the behaviour of these two sensors was completely different under impact loading though both were linear positioning sensors. It was observed that the accurate measurement of deflection was impossible in the case of LVDTs and the results were also less promising compared to LDS. The deflection measured by LVDT was about 5–7 times the deflection measured by LDS. Thus, the non-contacting sensor was noticed to have performed better under impact loading compared to contacting sensor LVDT. The peak impact force increased by 45.5% with the increase in the drop height. The cracks in the RC beams initiated from their soffit under impact loading; however, the failure mode was noticed differently. Beam A1 witnessed flexural deformation while beam A3 had shown splitting damage.

Keywords Contacting sensor · Non-contacting sensor · Impact · Beam · Reinforced concrete · Damage · Deflection

S. Kumar (✉) · M. M. Dalbehera · A. K. Mittal
Structural Engineering Group, Central Building Research Institute, Roorkee 247667, India
e-mail: sksachinraj60@gmail.com

V. Kumar
Department of Civil Engineering, National Institute of Technology Hamirpur, Hamirpur 177005, India
e-mail: vimalkumar@nith.ac.in

R. Velmurugan et al. (eds.), *Dynamic Behavior of Soft and Hard Materials Volume 1*, Springer Proceedings in Materials 34, https://doi.org/10.1007/978-981-99-6030-9_4

1 Introduction

The important reinforced concrete structures need to be designed for impact loads, which are exposed to impact load by moving objects, such as falling rocks in mountain areas and falling heavyweight dealt with in factories and warehouses due to accidents, aircraft impact on nuclear containments, and vehicular impact on bridges. At present, most of these structures have been designed by statically surcharging the maximum impact forces following the allowable stress design concept or the maximum impact forces with a construction coefficient following the ultimate limit state design concept. Therefore, the safety level of such structures should be determined using a more accurate evaluation of the characteristic impact-resistant capacity. This requires the study of the impact behaviour of structural components with a focus on structural response parameters such as deflection, residual deflection, and strains in steel reinforcement with an acceptable degree of accuracy in a real impact event. May et al. [1] studied the crack pattern of RC beams under impact using high-speed photography to measure strains in reinforcement. Fujikake et al. [2] observed that the amount of longitudinal reinforcement and the drop height significantly influence the local and global failure mode of the structure. Chen and May [3] found that the stiffness of the impact zone had more influence on the response of the beam compared to the support conditions. Tachibana et al. [4] found that the impulse resulting from the impact is directly proportional to the momentum of the impactor and impulse. Kishi et al. [5] studied the impact response of RC beams and found that maximum and residual deflections were more or less proportional to the input impact energy. As multiple parameters are required to analyse an RC element under impact loading that requires proper instrumentation too. Earlier the instrumentation and data acquisition were used to be done manually and were labour-intensive. But nowadays, modern transducers and data acquisition help in a much simpler and more precise way. As compared to static loading, it is difficult to measure precise response under impact loading due to the strain rate dependence of the material and vibrations. Therefore, high-quality sensors with good linearity, high-frequency response, quick response time and good reproducibility are needed in the instrumentation process for the measurement of dynamic response [6]. The displacement under impact depends on velocity, the mass of the impactor and the materials of the specimen [7]. So an appropriate synchronization is necessary between the impact event and instrumentation. In this investigation, the RC beams were tested under drop-weight impact from 1.0 and 1.5 m height. The impact force, reaction, deflection and damage were studied and compared. The reaction was measured using load cells and the displacement response was studied with the help of contacting (inductive) and non-contacting (electronic) transducers.

2 Experimental Setup and Instrumentation

The experimental programme consists of the casting of RC beams of size 2.5 × 0.25 × 0.15 m using M40 grade concrete. The maximum size of the coarse aggregate used in the mix proportion was 20 mm. The average compressive and splitting tensile strength of designed M40 grade concrete tested in the laboratory were 38.43 and 4.13 MPa, respectively. The experimental results for typical RC beams are presented and discussed in this paper. An instrumented impact loading test setup for dropping the weight at a variable height of up to 2.5 m and recording the impact response was designed. A specially designed support system Fig. 1 for loading the beam and installing the load cell was fabricated. A 200 kg drop weight was raised at the desired height and dropped to impact the beam. The impact loading system along with the data acquisition system and instrumented beam tested in the drop-weight experiment is shown in Fig. 2. To compare the behaviour of inductive and electronic transducers under impact event, laser displacement sensors and LVDT were attached to the beam. The instrumented beam specimen with the sensors is shown in Fig. 3. Four beams (beam A1, A2, A3, and A4) having two different reinforcement arrangements were subjected to impact by 200 kg steel weight, respectively. Beams A1 and A2 were having 2.74% reinforcement, while beams A3 and A4 were having 4.12% reinforcement. Beams A1 and A3 were impacted from 1.5 m height, while beams A2 and A4 were impacted from 1.0 m height. During the impact event, the time history of reaction forces measured with the load cell, deflections measured with contacting/non-contacting sensors, and strains in reinforcement have been analysed. All the response data were recorded in a high-speed data acquisition system with its various modules.

Fig. 1 Experimental setup

Fig. 2 DAQ system with the experimental setup

Fig. 3 Different positions of LDS

3 Experimental Results and Discussion

3.1 Impact Force and Reaction Response

The behaviour of an RC beam under impact loading consists of two response phases. First, the local response due to the stress wave occurs at the loading point during a very short period after impact. Second, the overall response including the free vibration effect due to the elastic–plastic deformation occurs over a long period in the whole structural member after impact. In this study, to gain confidence in the instrumentation scheme and the measured data, two samples of the beam was tested and the repeatability of the recorded data was also studied. While analysing the data, the value of the reaction forces and impulse were studied. The impact and reaction

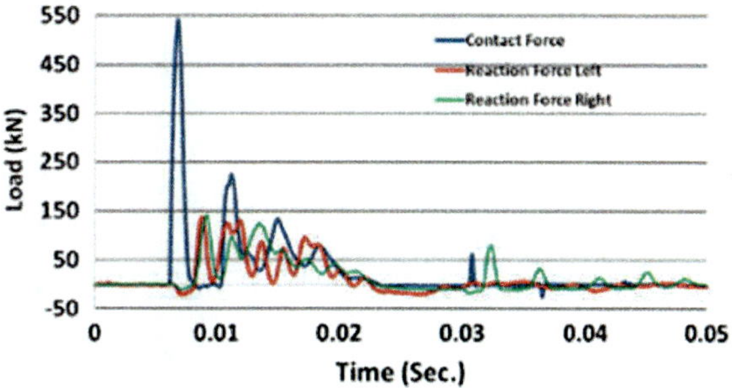

Fig. 4 Impact and reaction forces for Beam A1

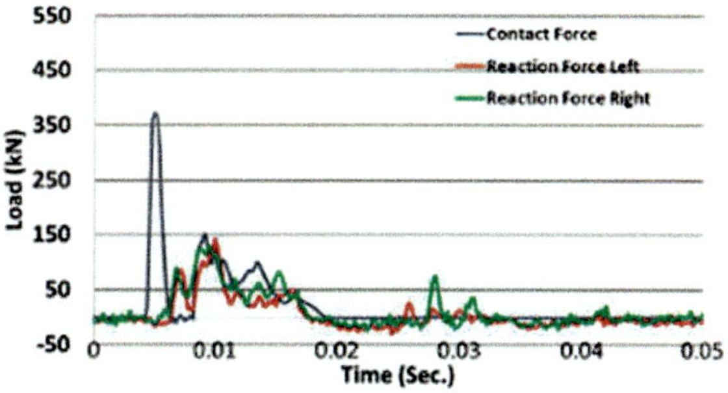

Fig. 5 Impact and reaction forces for Beam A2

forces for beams A1 and A2 were measured with the installed impactor and reaction-type load cells and analysed. The generated data by the load cells during the impact event is plotted in Figs. 4 and 5 for impact force and reaction, respectively. The magnitude of impact force is compared (in Fig. 6) for beams A1, A2, A3, and A4. The peak magnitude of impact force was measured 541.1 and 371.8 kN, respectively, for beams A1 and A2. The magnitude was reduced for beams A3 and A4. Similarly, the average reaction from the left and right support was measured 139.8 and 134.2 kN, respectively. The increase in the drop height increased the impact force and reaction by approximately 45.5 and 4.1%, respectively. Thus, it could be concluded that the reaction is less sensitive to the impact event compared to the impact force.

3.2 *Deflection Response*

The deflection at the middle span of the beam was obtained during the impact event using laser displacement sensors (LDS) and LVDT. The deflection time history obtained using LDS is shown in Figs. 7 and 8, respectively, for beams A1 and A2. The response suddenly increased after impact and then reduced to a significant extent. The response then witnessed the second peak of smaller magnitude due to the bouncing of the hammer. As seen in Figs. 7 and 8, during the impact event the

Fig. 6 Comparison between impact force for beams A1, A2, A3, and A4

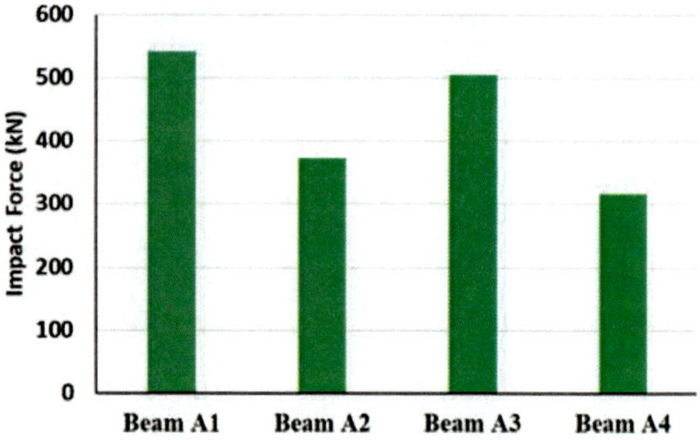

Fig. 7 Displacement at mid-span in Beam A1

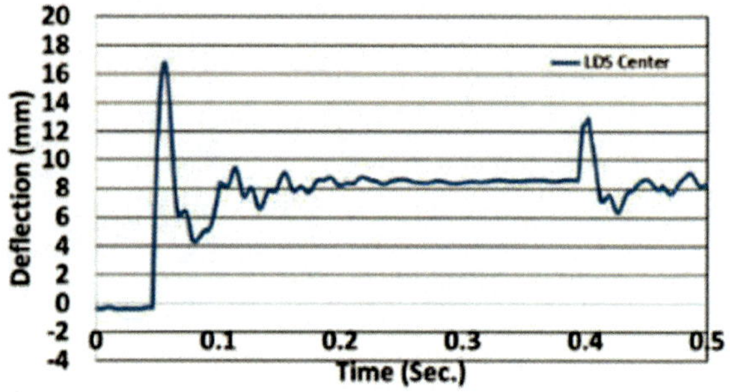

Fig. 8 Displacement at mid-span in Beam A2

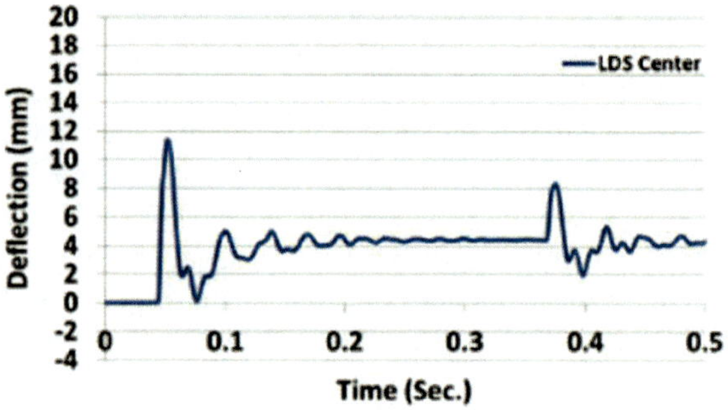

peak value of mid-span deflection for beams A1 and A2 is 16.77 and 11.37 mm, respectively.

It should be noted that the deflection of the specimen when measured with the help of a non-contacting sensor (LDS), it is necessary to synchronize the frequency of impact with the sampling rate of LDS. Thus, an appropriate sampling rate has to be filled in the configuration table of the operating software. Because the frequency of impact events was very high, a high sampling rate DAQ hardware was used to record this deflection. The laser displacement sensor generated the voltage signal according to the deflection experienced in the specimen under impact event. After

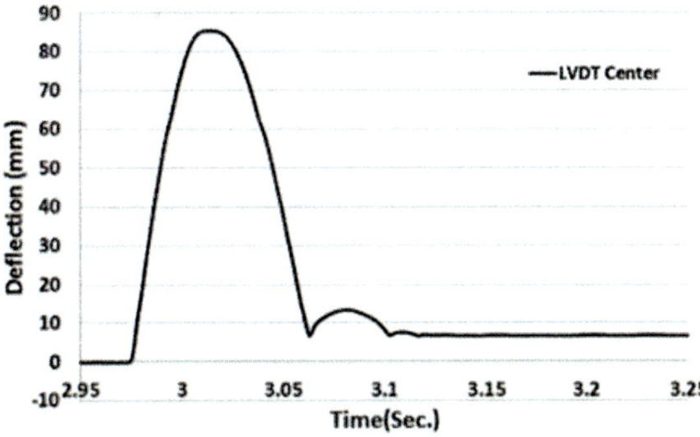

Fig. 9 Deflection in beam A1 measured by LVDT

the impact event, the recorded voltage correlated with the deflection to multiply using a calibration factor to obtain the actual deflection. Simultaneously, when the deflection was measured using a contacting sensor (LVDT) in beams, it was found that the magnitude of deflection recorded by the LVDT was multiple times (5–7 times) than the deflection measured by the LDS (non-contacting sensor). The deflection was 86.1 and 77.7 mm in beams A1 and A2, respectively. The graph between deflection vs. time measured using LVDT is shown in Figs. 9, 10, 11, and 12, respectively, for beams A1, A2, A3, and A4. The higher magnitude of deflection was primarily due to the inertia effects and induced momentum in the moving component of LVDT. The deviation in the response of LVDT and LDS was because the LVDT consists of a primary winding and two secondary windings in the cylindrical structure [8]. A soft iron core that has been attached to the specimen is free to move in the cylindrical structure developing the voltage difference between secondary windings which gives the measurement of deflection [9–11]. The LVDTs are sensitive to variations in the amplitude of the operating voltage and the internal frequency generators for the LVDT require stability [12]. On the other hand, the laser displacement sensors are based on position sensitive detector (PSD) and complementary MOSFET (CMOS)/ charged coupled devices (CCD) detectors. PSD detector is exposed to a spot of light, and this exposure causes a change in local resistance and then electron flow in the electrodes. The currents are used to determine the location of the light spot [13]. The charge-coupled devices are silicon-based array detectors of light that are working on the Raman spectroscopy principle which shows the variation in storing the charge due to the displacement of the light spot. Similarly, the deflection of the light spot can be measured with the help of a complementary MOSFET (CMOS) detector [14–16].

3.3 Damage Pattern

The damage/cracks in beams A1 and A3 are presented in Figs. 13 and 14, respectively. The cracks in beam A1 were noticed wider, longer, and more in numbers compared to beam A3. Also, beam A1 was noticed to have flexural deformation, whereas beam

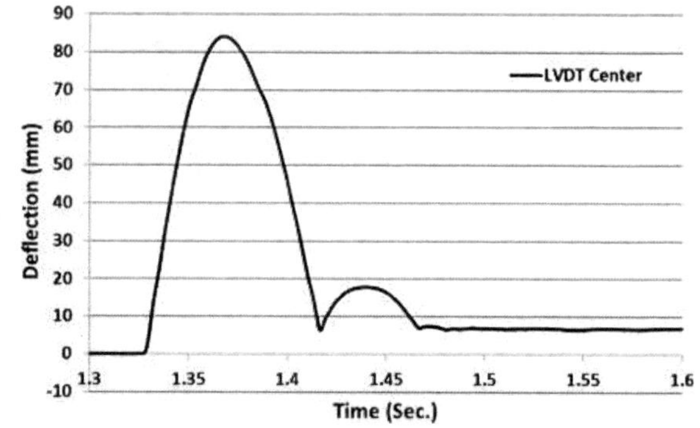

Fig. 10 Deflection in beam A2 measured by LVDT

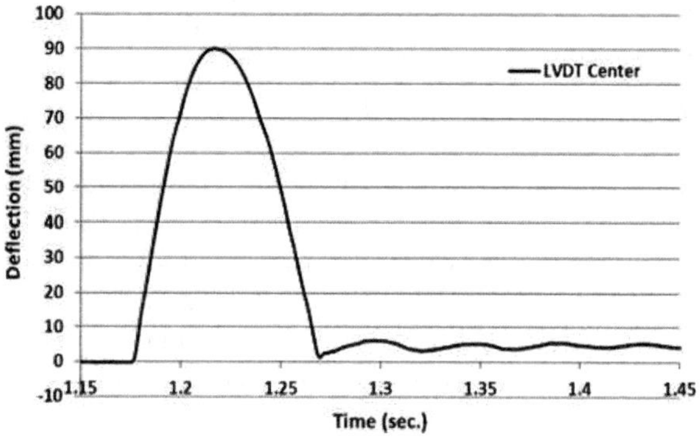

Fig. 11 Deflection in beam A3 measured by LVDT

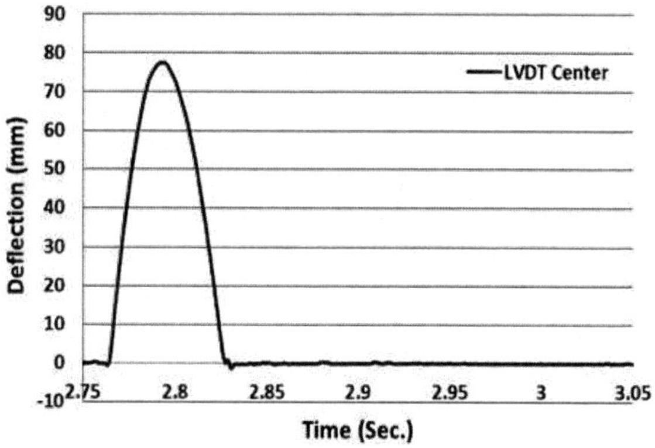

Fig. 12 Deflection in beam A4 measured by LVDT

A3 showed splitting kind of damage initiated from the bottom. The length of the crack in beams A1 and A3 was 133 and 115 mm, respectively. The maximum width of the developed crack was noticed up to 6 and 3 mm in beams A1 and A3, respectively.

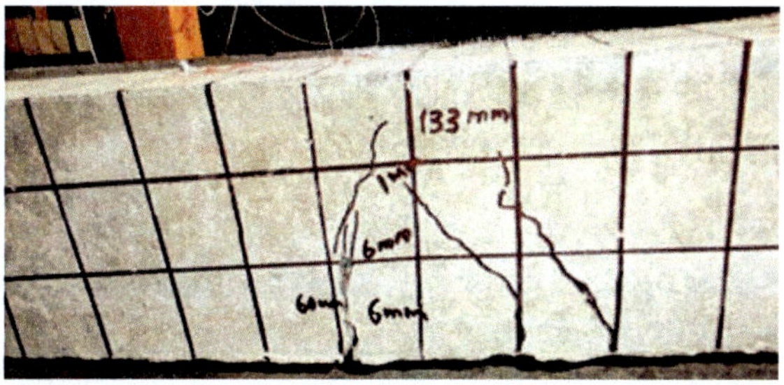

Fig. 13 Cracking in beam A1

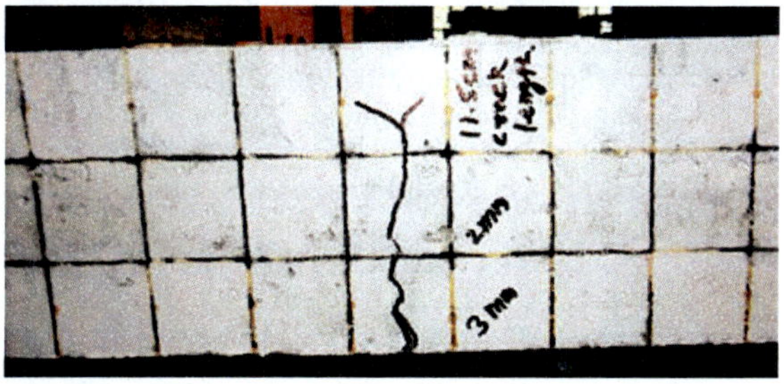

Fig. 14 Cracking in beam A3

4 Conclusions

The experimental program was made to study the impact response of RC beams using a drop-weight impact testing facility. Sufficient instrumentation was done to record various responses during the impact event. Impact tests performed on RC beams indicate that reaction forces increase as the flexural stiffness is increased by increasing the drop height though it was nominal (4%). The deflection–time response during the impact event indicated the laser displacement sensors (non-contacting) performed well with respect to LVDT (contacting) due to the absence of movable components. The inductive transducer (LVDT) was not found suitable to measure the response under dynamic events like impact wherein exist large vibrations or inertia play a significant role. The frequency of impact events is very high, so it is difficult to measure the deflection of any target with the help of inductive transducers because they have mechanical parts in their constructions which influence by inertia. Laser displacement sensors are non-contact sensors based on optical displacement measurement which is ideal for measuring the displacement under impact loading. The magnitude of deflection measured by LVDT was 5–7 times the deflection measured by LDS. The cracks in the RC beams initiated from their soffit under impact loading; however, the failure mode was noticed differently. Beam A1 witnessed flexural deformation, while beam A3 had shown splitting damage.

Acknowledgements The authors gratefully acknowledge the financial support provided by CSIR-Central Building Research Institute, India. The technical assistance rendered by the Project Assistants and other staff of the Structural Engineering group in casting specimens and performing the experiment is acknowledged.

References

1. May IM, Chen Y, Owen DRJ, Feng YT, Thiele PJ (2006) Reinforced concrete beams under drop weight impact loads. Comput Concr 3:79–90
2. Fujikake K, Li B, Soeun S (2009) Impact response of reinforced concrete beam and its analytical evaluation. J Struct Eng ASCE 135(8):938–950
3. Chen Y, May IM (2009) Reinforced concrete member under drop weight impact. Inst Civil Eng Struct Build 162(2):45–56
4. Tachibana S, Masuya H, Nakamura S (2010) Performance based design of reinforced concrete beams under impact. Nat Hazard Earth Syst Sci 10:1069–1078
5. Kishi N, Mikami H (2012) Empirical formulas for designing reinforced concrete beams under impact loading. ACI Struct J 109(4):509–519
6. Chen Y, May IM (2009) Reinforced concrete members under drop-weight impacts. Struct Build 162:45–56
7. Dinic G, Perry SH (1990) Shear plug formation in concrete slabs subjected to hard impact. Eng Fract Mech 35:343–350
8. Masi A, Danisi A, Losito R, Martino M, Spiezia G (2011) Study of magnetic interference on an LVDT: FEM modeling and experimental measurements. J Sens
9. Martino M, Danisi A, Losito R, Masi A, Spiezia G (2010) Design of a linear variable differential transformer with high rejection to external interfering magnetic field. IEEE Trans Magn 46:674–677
10. Spiezia G, Losito R, Martino M, Masi A, Pierno A (2011) Automatic test bench for measurement of magnetic interference on LVDTs. IEEE Transact Instrument Measur 60:1802–1810
11. Masi A, Danisi A, Losito R, Martino M, Spiezia G (2010) Study of Magnetic Interference on a LVDT Prototype. In: paper presented at IEEE instrumentation & measurement technology conference
12. Tariq H, Takamori A, Vetrano F, Wang C, Bertolinia A, Calamai G, DeSalvo R, Gennai A, Holloway L, Losurdo G, Marka S, Mazzoni M, Paoletti F, Passuello D, Sannibale V, Stanga R (2002) The linear variable differential transformer (LVDT) position sensor for gravitational wave interferometer low-frequency controls. Nucl Instrum Methods Phys Res Sect A 489:570–576
13. Cui S, Soh YC (2010) Linearity indices and linearity improvement of 2-d tetralateral position-sensitive detector. IEEE Transact Elect Dev 57:2310–2316
14. Janesick JR (2001) Scientific charge-coupled devices, SPIE Press. The International Society for optical Engineering Bellingham, Washington USA
15. Hainaut O (1996) Basic CCD image processing. European Southern Observatory
16. Baker RJ (2011) CMOS: circuit design, layout, and simulation In: Tewksbury SK, Brewer JE (eds) IEEE press series on microelectronic system

Numerical Investigation of Debris Impact on Multi-layer Kevlar Shield

Parameswaran Sivakumar, Rajeev Chaturvedi, and Jay Prakash Tripathi

Abstract The numerical simulations of hypervelocity impact on multi-layer dry Kevlar fabrics are presented in this paper. A macroscopic finite element model of Kevlar fabric is implemented with 3D solid elements and defined using homogenized properties. Simulations were performed with erosion model in ABAQUS software. The erosion model is used to evaluate the ballistic limit thickness of multi-layer Kevlar K29 and K49 fabrics when impacted by sub-millimetre steel projectile at a velocity of around 6 km/s. Four simulations were performed with 0.3 and 0.5 mm diameter projectiles for each fabric using coupled structural–thermal analysis. In a different multi-layer configuration, an aluminium projectile of 5.56 mm diameter was made to impact a 6-layer Kevlar KM2 fabric at two different velocities of 3.2 and 6 km/s. The model is also found to be significant in accurate prediction of its damage characteristics such as tensile cracking and fraying/unravelling of yarns. The transverse wave velocity is found to be dependent on thickness, i.e. number of layers. The above simulations results are in good agreement and have been benchmarked with the available experiments in the literature.

Keywords Hypervelocity impact · Multi-layer Kevlar · Residual velocity · Ballistic limit thickness

Nomenclature

V_p	Impact velocity
E_1, E_2	Longitudinal Young's Modulus
E_3	Transverse Young's Modulus
E_l	Longitudinal Young's Modulus of fabric yarn

P. Sivakumar (✉) · J. P. Tripathi
Department of Mechanical Engineering, Thapar Institute of Engineering and Technology, Patiala, India
e-mail: psivakumar_be18@thapar.edu

R. Chaturvedi
UR Rao Satellite Centre (URSC), ISRO, Bengaluru, India

R. Velmurugan et al. (eds.), *Dynamic Behavior of Soft and Hard Materials Volume 1*, Springer Proceedings in Materials 34, https://doi.org/10.1007/978-981-99-6030-9_5

ρ_l, ρ_g	Local (meso) and global (macro) density
a	Projectile acceleration
M_p	Projectile mass
V_{pi}	Projectile velocity just before contact with ith layer
V_n	Projectile velocity between layer n and $n + 1$
n	Layer number
r_l	Longitudinal wave radius
r_t	Transverse wave radius
S	Yarn cross-section
T_i	Tension in each layer
b	Stress transmission factor
D	Projectile diameter
d	Yarn diameter
h	Distance between layers

1 Introduction

The growing number of space debris poses a threat of colliding with currently about 3000 functioning spacecrafts, which are essential to many ground-based operations. Due to the lack of observational capabilities of ground radars, smaller debris of size less than 10 cm cannot be tracked. Therefore, impact of these smaller debris moving at an average speed of 7.5 km/s with the functional spacecrafts could be inevitable and catastrophic [1]. Many investigations have undertaken to develop a shielding configuration, such as whipple, stuffed whipple, and multi-shock shields, mainly to protect the human-occupied spacecrafts. High modulus and high strength fabric such as Kevlar are used as multi-layer intermediate wall between an aluminium bumper wall and rear wall.

A woven fabric consists of many yarns in warp and weft, and each yarn is made up of hundreds of fibres. Many authors have numerically developed a yarn-level geometric model (mesoscale) of dry fabric to illustrate a complete yarn to yarn interactions. But this model tends to be computationally expensive. Duan and co-workers [2] developed a global approach (macroscopic) by using a homogenised 3D elements and predicted ballistic response of fabric, which turns out to be much faster in terms of computation time. So, this work will use a macroscopic approach to perform ballistic impact simulations on Kevlar. The most common methods to solve numerical simulation of hypervelocity impact are Lagrange method (erosion model), meshless method (smooth particle hydrodynamics), etc. In a Lagrange method, the FE elements of projectile and target erode after reaching failure criteria and cannot form an effective cloud debris. It would also result in large element distortion, large mass, and energy loss. But it has high computational efficiency. In the meshless SPH method, the debris cloud is accurate, but computationally more expensive than erosion model. In combined FE-SPH method, the FE elements after failure get converted into SPH particle. By this approach, mass and energy are conserved, and mechanical properties of FE elements are inherited by SPH particles but takes higher computation time than erosion model and SPH model [3]. The focus of this work is to validate the ballistic impact response of macroscale dry Kevlar fabric using erosion model for different range of velocities. Under hypervelocity condition, validating the ballistic limit thickness is an important criterion to measure the performance of the target. Unlike in low or high velocity, the projectile does not stay intact to measure the residual velocity as it would be fragmented into pieces. Therefore, the ballistic limit thickness of multi-layer K29 and K49 fabric is validated using structural–thermal analysis.

2 Modelling Perforation of Multi-layer Fabric Under Impact

Based on the contact with projectile, there are two types of yarns in the fabric. The primary yarns are directly below the projectile and come under contact with it during impact. The remaining yarns in the fabric are referred to as secondary yarns. Only the primary yarns provide the force to resist the perforation of projectile which would induce higher strain in them compared to the secondary yarns. These primary yarns fail under tension when it reaches the ultimate failure strain in tension [4].

The governing equations for multi-layer fabric perforation are given below [4]:

$$v_n = v_{n-1} - a_n(t_n - t_{n-1}) \tag{1}$$

$$a_n = \frac{2}{M_\text{p}} \sum_{i=1}^{n} T_i \frac{(n-i+1)h}{\sqrt{r_{ti}^2 + [(n-i+1)h]^2}} \tag{2}$$

$$T_i = \frac{2D_\text{p}}{d} \cdot E \cdot S \left\{ \frac{\left(\frac{D_\text{p}}{2}\right) + \sqrt{\left\{r_{ti} - \left(\frac{D_\text{p}}{2}\right)\right\}^2 + z_i^2} - r_{ti}}{b^{r_{ti}/d} - 1} \right\} \left\{ \frac{\ln b}{d} \right\} \tag{3}$$

$$h = v_{n-1}(t_n - t_{n-1}) - \frac{1}{2} a_n (t_n - t_{n-1})^2 \tag{4}$$

The areal density ratio is given by the mass of fabric in contact with projectile to the mass of projectile,

$$\Gamma_0 =_p^m / M_\text{p} \tag{5}$$

Using simple momentum conservation, the velocity of the projectile immediately after impact on first layer is given by:

$$V_{m1} = \frac{V_p}{1 + \Gamma_0} \tag{6}$$

In Eqs. (1)–(4), the impact velocity is replaced by V_{mi}. Assuming that the projectile has a velocity of V_{p2} just before contact with second layer, the velocity of projectile just on impact with second layer is given by

$$V_{m2} = \frac{1 + \Gamma_0}{1 + 2\Gamma_0} V_{p2} \tag{7}$$

Similarly, the velocity of projectile just after impact with ith layer can be given by:

$$V_{mi} = \frac{1 + (i - 1)\Gamma_0}{1 + i\Gamma_0} V_{pi} \tag{8}$$

The velocity of transverse wave induced due to impact is given by [4],

$$u = c^{1/3}\left(\frac{V_p}{\sqrt{2}}\right)^{2/3} \tag{9}$$

3 Estimation of Properties for Macroscale Model of Fabric

In the macroscopic model developed by Duan and co-workers [2], the woven fabric is modelled as homogenised continuum and transversely isotropic material. The conversion of orthotropic material properties of woven fabric to 3D solid fabric is done using density approximation, which was initially formulated for multi-scale plain weave fabric. Two main conditions should satisfy for this model; one is to maintain the areal density of mesoscale and macroscale model. The second is the impedance matching to create a smooth transition of stress wave between two region's interfaces. This technique has also been implemented for converting complete mesoscale to macroscale model [5].

The material density of macroscale model is calculated as the ratio of areal density to fabric thickness. The thickness of the 3D fabric model is taken equal to that of the woven fabric. Since porosity between the yarns is not considered, the calculated density of macroscale model would be less than the woven fabric. As defined in paper [2], the magnitude of young's modulus along warp and wept (longitudinal) direction is kept same and is calculated as below:

$$E_1 = E_2 = \frac{\rho_l}{\rho_g} E_l \tag{10}$$

Table 1 Kevlar properties

Kevlar type	Fibre properties			Fabric properties	
	Density (kg/m^3)	Tensile modulus (GPa)	Tensile tenacity (MPa)	Areal density (kg/m^2)	Thickness (mm)
K29	1440	70.5	2920	0.319	0.43
K49	1450	112.4	3000	0.217	0.33
KM2	1440	62	3400	0.18	0.23

The young's modulus along the transverse and shear modulus in all three directions is kept in two and three orders magnitude smaller than the young's modulus along longitudinal direction and poison's ratio in all three directions to be zero, as reported in paper [5]. The failure stress of the macroscale model is specified to be half of the tensile tenacity of yarn, as reported by Yang and co-workers [6]. Properties of Kevlar produced by Dupont are listed in Table 1, as reported in the paper [7].

4 Setting-Up of Simulation Model

A total of three sets of simulations were performed in this work. The first set of simulations performed were based on the impact experiments performed by Rao and co-workers [2]. In this work, a single layer of Kevlar KM2 fabric of dimensions 50.8 mm × 50.8 mm was impacted by a rigid steel projectile of diameter 5.35 mm at speed of 60.6 and 245 m/s. The erosion model was performed for simulation with both impact velocities. The second set of simulations, an aluminium projectile of diameter 5.56 mm diameter was made to impact a 6-layer Kevlar KM2 fabric at a velocity of 3.2 km/s. Cha and co-workers [8] used this multi-layer Kevlar as an intermediate layer in a conventional whipple shield in their work. Using the same multi-layer configuration, this work also includes simulation with an impact velocity of 6 km/s. In the last set of simulations, four simulations were performed using structural–thermal analysis for multi-layer Kevlar K29 and K49 impacted by a sub-millimetre steel projectile. The details of these four simulations are listed in Table 2. All four edges of the fabric are constrained to be fixed.

Table 2 Multi-layer Kevlar simulation data [7]

Fibre	Projectile diameter (mm)	Impact velocity (km/s)	Single layer thickness (mm)	No. of layers	Experimental ballistic limit thickness
K29	0.3	5.89	0.43	10	3.44 mm (8th layer)
K49	0.3	6.14	0.33	12	3.3 mm (10th layer)
K29	0.5	6.27	0.43	15	5.59 mm (13th layer)
K49	0.5	6.13	0.33	17	4.95 mm (15th layer)

The projectile material was defined using the Johnson Cook Plasticity and Damage Criterion. The Kevlar fabrics are defined using orthotropic elastic constants and failure defined using maximum effective failure strain. The interaction between elements of target and projectile was modelled using a general contact.

5 Comparison and Validation with Experimental Data

To validate the first set of simulations with single layer Kevlar KM2 fabric, the residual velocity of the projectile obtained from the simulation is compared with the experimental test result. The projectile impacted with initial velocity of 60.6 m/s bounced back without perforation similar to experimental test as shown in Fig. 1a. The projectile impacted with 245 m/s resulted in a residual velocity of 207.1 m/s experimentally [2]. The residual velocity obtained in the simulation using erosion model is 216 m/s, and the error in the residual velocity is 4.29% compared to experimental data, see Fig. 1b.

An aluminium projectile of 5.56 mm diameter impacting a 6-layer Kevlar KM2 fabric at a velocity of 3.2 km/s resulted in a residual velocity of 2.989 km/s [8]. The projectile and target modelled with erosion model resulted in a residual velocity of 3.06 km/s, as shown in Fig. 2a. The error in residual velocity for this model is 2.37%. Table 3 shows comparison of experimental data with erosion model for two different impact configurations.

The impact of aluminium projectile at an impact velocity of 6 km/s on the 6-layer Kevlar fabric resulted in a complete erosion of the projectile as shown in Fig. 2b. The complete erosion of the projectile makes it difficult to determine its residual velocity, and hence, it is important to validate the simulation of multi-layer Kevlar fabric using ballistic limit thickness. As reported by Lee and co-workers [9], ballistic impact on a 4-layer Kevlar fabric exhibited extensive yarn pullout and well-observed cross-pattern stretching and fabric breakage. Similar damage characteristics were observed in the simulation with impact velocity of 6 km/s as shown in Fig. 4, which shows damage in 6 layers of Kevlar fabric. Two modes of failure are observed during the simulation, one is initial failure caused due to perforation, and the hole diameter

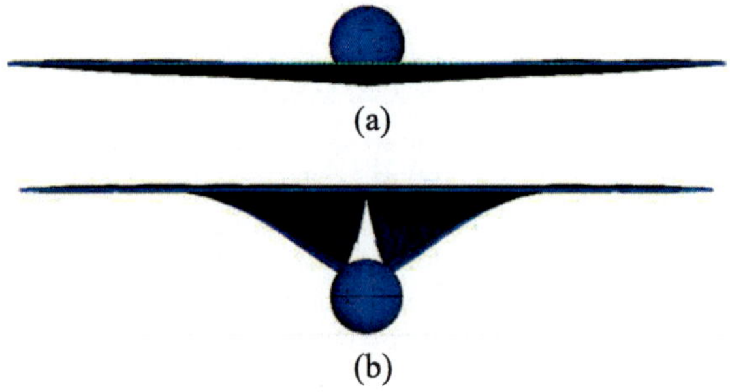

Fig. 1 Single-layer KM2: **a** $V_p = 60.6$ m/s, **b** $V_p = 245$ m/s

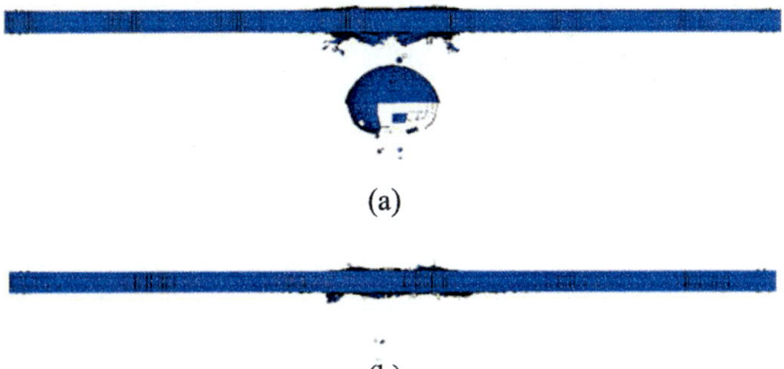

(a)

(b)

Fig. 2 Multi-layer KM2: **a** $V_p = 3.2$ km/s; **b** $V_p = 6$ km/s

Table 3 Experimental and numerical results of erosion model

No. of layers	1	6
Initial velocity	245 m/s	3.2 km/s
Experimental residual velocity	207.1 m/s	2.989 km/s
Simulation		
Model type	Erosion	Erosion
Residual velocity	216 m/s	3.06km/s
Error (%)	4.29	2.37
Computation time (h)	1.36	0.51

is expected to be identical to that of the projectile diameter. The other is the tensile cracking caused due to impact waves moving in all the four directions from the impact area as seen in Fig. 4. The cross-pattern damage occurs when the impact wave originating from the impact zone meets with the impact wave reflected from the edges. Figure 3 shows that the mass loss in each layer of fabric just after perforation is measured to less than the mass loss due to transverse/tensile cracking.

In the third set of simulations, four simulation were performed as mentioned in Table 2. When a 0.3 mm steel projectile impacts a multi-layer K29 and K49 Kevlar fabric, the ballistic limit thickness turns out to be 3.44 mm (8th layer) and 3.3 mm (10th layer), respectively. Similarly, when a 0.5 mm steel projectile impacts a multi-layer K29 and K49 Kevlar fabric, the ballistic limit thickness turns out to be 4.73 mm (11th layer) and 4.62 mm (14th layer), respectively. Comparing with the experimental data in Table 2, the ballistic limit thickness for the first two simulations with 0.3 mm diameter projectile matches exactly with the experiment. The other two simulations with 0.5 mm diameter projectile impacting the multi-layer Kevlar are slightly varying, but well within the margin of error define by the Ref. [8]. The ballistic limit thickness is compared with the experimental data using standard error bars, as shown in Fig. 5.

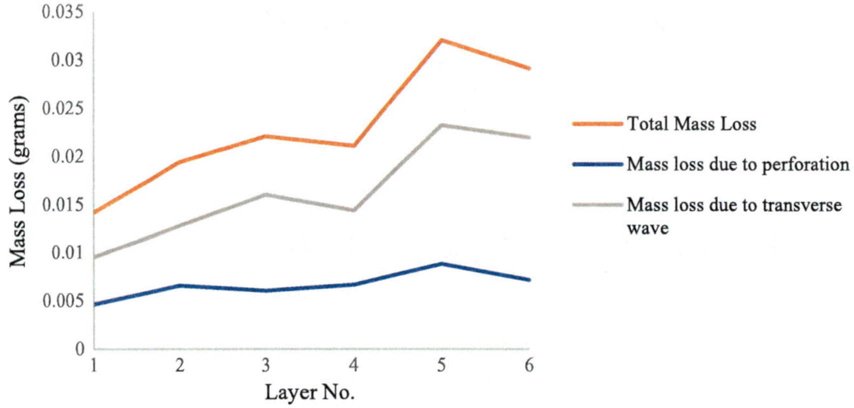

Fig. 3 Mass loss observation in all layers of fabric for impact velocity of 6 km/s

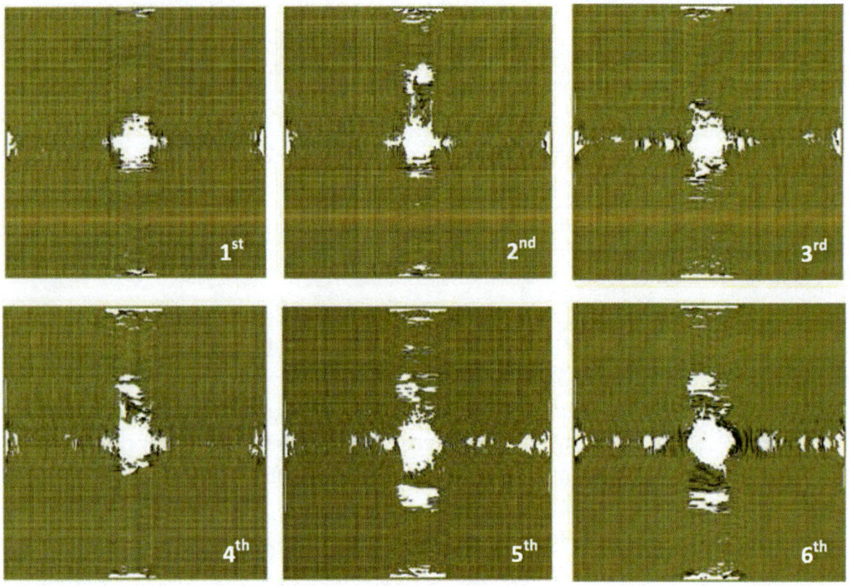

Fig. 4 Damage analysis of all 6 layers of dry Kevlar fabric

Figure 3 shows variation of mass loss in all six layers of fabric. To explain this behaviour, the transverse wave velocity in each layer is observed and analysed. Figure 6 shows the transverse wave velocity of fabric layer 1 and 6 which indicates that the transverse wave originating from the impact zone decreases as it moves towards the edges and vice versa. The transverse wave reflected back from the edges meets with the incoming wave from the impact centre creating a tensile phenomenon, causing the fabric to damage or crack. It is also noticed that the transverse wave

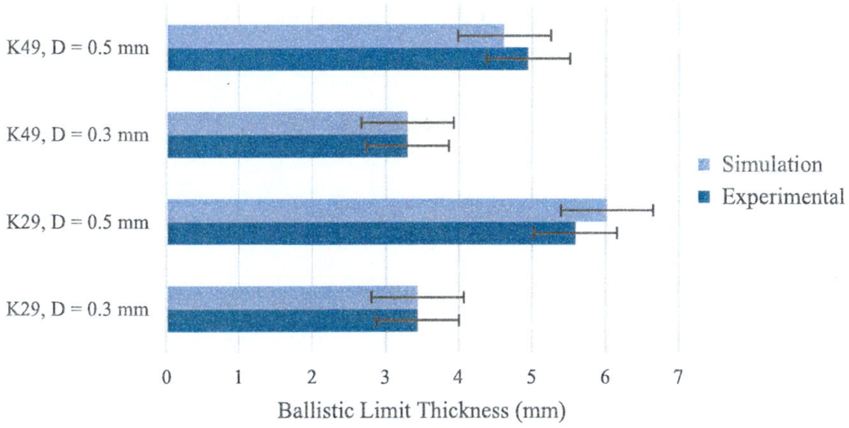

Fig. 5 Comparison of ballistic limit thickness with experimental data

velocity differs in each layer. It increases with each layer, but also decreases as it moves from the impact zone to edges and vice versa. As expected, there is a delay noticed in wave formation in each layer as well as time taken for it to reach to edges. As the velocity of wave increases with each layer, the time taken for it to reach edges and reflect back to the impact zone also decreases, as seen in Fig. 7.

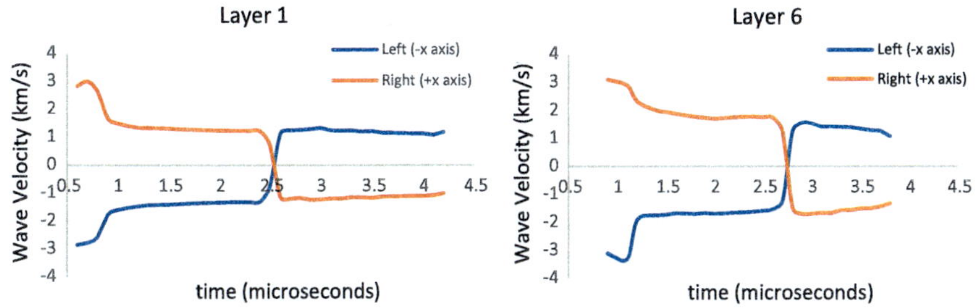

Fig. 6 Wave velocity vs time for layer 1 and layer 6

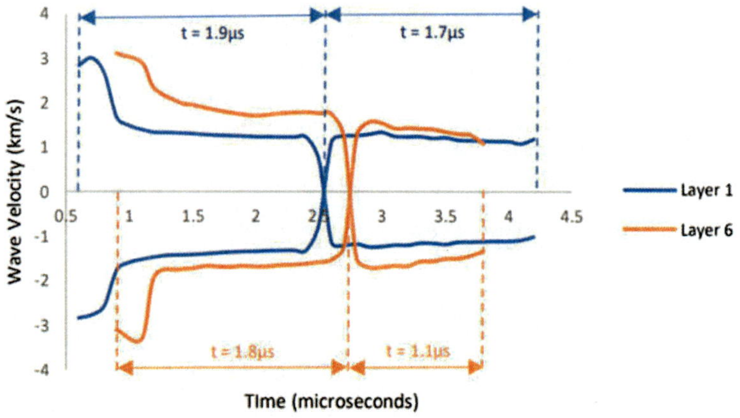

Fig. 7 Comparison of wave velocity between layer 1 and 6

6 Conclusion

In this work, ballistic impact analysis on multi-layer dry Kevlar fabrics is studied when impacted by a small size spherical debris. The Kevlar fabric is represented as a macroscale mechanical model and a transversely isotropic material. Since the macroscale properties of the Kevlar were not readily available in the literature, the yarn properties (mesoscale) were converted into macroscale using density approximation method, which has been developed and tested for ballistic impact conditions in literature. Simulations were performed using erosion model, in which FE elements gets deleted after failure.

The residual velocity and ballistic limit thickness are important criteria for validation of ballistic performance of Kevlar fabrics under low/high velocity and hyper velocity conditions, respectively. So, both criterions were considered for comparison and validation of simulations in this work and found to be in good agreement with results obtained from experimental data taken from literature. Four simulations: two with single layer and the other two with multi-layer Kevlar KM2 fabric were validated using residual velocity. The multi-layer Kevlar K29 and K49 is also impacted by a sub-millimetre steel projectile to validate the ballistic limit thickness using structural–thermal analysis. The damage characteristics is also being accurately predicted for ballistic impact on multi-layer Kevlar fabric. The damage caused in terms of mass loss due to transverse waves is observed to be ~ 50–60% more than the mass loss due to perforation. This variation of damage in each layer can be explained by variation of transverse wave velocity in each layer. Simulations show that the transverse wave velocity increases with each layer and but also decreases from the impact zone to edges and vice versa. The existing equation of transverse wave velocity only considers initial projectile velocity and speed of sound as factors. Through this work, the transverse wave velocity is also found to be dependent on different layers of fabric. Hence, the modification to include the factors of thickness, number of layers, and variation of velocity within and with each layer, is necessary.

References

1. Space Situational Assessment 2021, ISRO. https://www.isro.gov.in/update/24-mar-2022/space-situational-assessment-2021
2. Rao MP, Nilakantan G, Keefe M, Powers BM, Bogetti TA (2009) Global/local modeling of ballistic impact onto woven fabrics. J Compos Mater 43(05). https://doi.org/10.1177/0021998308097684
3. Becker M, Seidl M, Mehl M, Souli M, Legendre JF (2019) Numerical and experimental investigation of SPH, SPG, and FEM for high-velocity impact applications. In: 12 European LS-DYNA conference 2019, Germany
4. Mamivand M, Liaghat GH (2010) A model for ballistic impact on multi-layer fabric targets. Int J Impact Eng. https://doi.org/10.1016/j.ijimpeng.2010.01.003
5. Ha-Minh C, Imad A, Kanit T, Boussu F (2013) Numerical analysis of a ballistic impact on textile fabric. Int J Mech Sci. https://doi.org/10.1016/j.ijmecsci.2013.01.014
6. Yang Y, Liu Y, Xue S, Sun X (2021) Multi-scale finite element modeling of ballistic impact onto woven fabric involving fiber bundles. Compos Struct. https://doi.org/10.1016/j.compstruct.2021.113856
7. Higashide M, Onose N, Hasegawa S (2014) Ballistic limit thickness and weight of high strength fiber fabrics for sub-millimeter steel sphere impact at 6 km/s. Trans JSASS Aerospace Tech Japan. https://doi.org/10.2322/tastj.12.Pr_1
8. Cha J-H, Kim YunHo, Kumar SKS, Choi C, Kim C-G (2019) Ultra-high-molecular-weight polyethylene as a hypervelocity impact shielding material for space structures. Acta Astronaut. https://doi.org/10.1016/j.actaastro.2019.12.008
9. Lee YS, Wetzel ED, Wagner NJ (2003) The ballistic impact characteristics of Kevlar woven fabrics impregnated with a colloidal shear thickening fluid. J Mater Sci. https://doi.org/10.1023/A:1024424200221

FE Investigation on Behaviour of Al-Alloy Tubes Subjected to Axial Impact

Aman Kumar and Vimal Kumar

Abstract For designers, it is a challenging task to design or develop an efficient product and shock absorber which also fulfils the primary requirements in terms of crashworthiness, lightweight, material optimization, and of course economy too. Under impact loading, the thin-walled cylindrical tubes are employed as primary energy-absorbing elements in automotive and some other industries. Therefore, in the present paper, the performance of the aluminium tubes was investigated against their varying thickness (2 and 3 mm) and length (130–170 mm) under impact loading. The effect of variation of length-to-diameter ratio of the tube has been studied. The numerical simulations were carried out using Abaqus Explicit. The metal-plasticity constitutive model was employed for modelling the ductile aluminium alloy. The tubes were fixed at the bottom during the impact, and a known mass was dropped on the other end of the tube with a velocity of 10 m/s. The axial crushing length of the specimen was found to be directly proportional to the length-to-thickness (h/t) ratio. The maximum value of axial crushing length was found to be 151.6 mm for a tube-5 which has 2 mm thickness, 170 mm length, and 50 mm diameter. The initial peak crushing force (IPCF) of the 2 mm cylindrical tube specimens was 35–40% lesser than the 3 mm thick cylindrical specimens. It was found that the specific energy absorption was increased with an increase in the thickness of the tube. The number of folds was also increased under impact with an increase in the tube length.

Keywords Energy absorption · Aluminium · Tube impact · Axial impact · Thin-walled tube · Crashworthiness

---

A. Kumar (✉) · V. Kumar
Department of Civil Engineering, National Institute of Technology Hamirpur, Hamirpur 177005, India
e-mail: araman0101@gmail.com

V. Kumar
e-mail: panchariya.vimal@gmail.com

R. Velmurugan et al. (eds.), *Dynamic Behavior of Soft and Hard Materials Volume 1*, Springer Proceedings in Materials 34, https://doi.org/10.1007/978-981-99-6030-9_6

1 Introduction

For impact energy absorption in different fields of engineering such as safety purposes in fast-moving vehicles, aircraft, bridges, industrial buildings, lifts, and stadium grounds thin-walled elements like tubes are commonly employed because to their high energy absorption, low weight, and ease of fabrication. Thin-walled tubes disperse impact energy by deforming plastically, reducing vehicle damage [1]. Extensive research work there has been done to study the effect of different cross-sections on the impact energy absorption capacity, initial peak force, mean crush force, and specific energy absorption of the tubes are some of the important crash-worthy parameters to be considered while designing the energy absorbers. The research performed by Alexander [2] on thin-walled tubes revealed that material property, the configuration of the cross-section, wall thickness, boundary condition, and other factors affected the deformation response and energy absorption capacity. Hsu and Jones [3] discussed that when aluminium alloys were used in the structure of a vehicle, weight reduction to the extent of 25–35% was achieved compared to conventional steel structures. Aluminium is one of the most preferred choices due to its excellent energy absorption capacity and considerable lightweight [4]. The deformation behaviour and energy absorption capacity of thin-walled tubes with diverse sections such as cylindrical, conical, square, prismatic, hexagonal, double hat, and frusta were studied [4–7]. Out of these sections, cylindrical and square-shaped tubes have attracted the attention of researchers in recent decades. The purpose of this research work is to study the crashworthiness of aluminium alloy 6061-T6 O-rolled cylindrical tubes. The cylindrical tubes were impacted by a 75 kg steel mass. The impact energy was kept constant for all the simulations. Two thicknesses of tubes, i.e. 2 and 3 mm, were used in the present study. The effects of variation of height-to-diameter ratio of the tube were studied. The effect of length-to-thickness (h/t) ratio was investigated in this study which is a new study parameter in the study of thin wall tubular impact absorption studies.

2 Finite Element Modelling

The FE modelling of the crushing problem was carried out in Abaqus software to numerically simulate the crushing of the cylindrical tubes having 2 and 3 mm thicknesses. The material considered for tubes was AA 6061- O for which the properties were given in the study Zehram et al. [8]. The metal-plasticity constitutive model was employed for modelling the ductile aluminium alloy. The diameter of the tube was kept constant for all simulations, whereas the lengths of the tube were taken as 130, 150, and 170 mm. The finite element model of one of the tubes under axial impact is shown in Fig. 1. The impactor was modelled as a rigid plate of mass of 75 kg. The bottom surface of the tube was fixed against rotation and translation. The contact

Fig. 1 Finite element model of the cylindrical aluminium tube

Table 1 Details of tubes

Tube No	Thickness (mm)	Length (mm)	Diameter (mm)	H/t ratio
Tube-1	2	130	50	65
Tube-2	3	130	50	43.33
Tube-3	2	150	50	75
Tube-4	3	150	50	50
Tube-5	2	170	50	85
Tube-6	3	170	50	56.67

between the plate and tube was modelled as general contact with a friction coefficient of 0.2. The size of the 4-node shell elements used for the tube was 0.75 mm × 0.5 mm. The details of the different tubes are provided in Table 1.

3 Results and Discussion

In this study, the influence of forming parameters, i.e. length and thickness on the deformation behaviour of cylindrical tubes, was investigated through numerical simulations. The progressive crushing of cylindrical aluminium tube is shown in Fig. 2 for Tube-1 and Tube-2, Fig. 3 for Tube-3 and Tube-4, and Fig. 4 for Tube-5 and Tube-6, respectively, at 5 different time instances during impact, i.e. 0, 4, 8, 16, and 30 ms. It was seen that tubes under axial impact caused progressive deformation behaviour. The final axial crushing length of the tubes was found to be directly proportional to the length-to-thickness (h/t) ratio for 2 mm thick tubes. On the other hand, the value of axial crushing length was observed to be almost the same for all tubes with 3 mm thicknesses. The development of folds occurred progressively with radial inward and outward in a cyclic manner. It was perceived that the number of folds increased with the increase in length/thickness (h/t) ratio.

The comparison between force–displacement curves is shown in Fig. 5(a, b), respectively, for tube thicknesses 2 and 3 mm. The impact force reaches its peak value when outward radial displacement was at a limiting value. The initial force

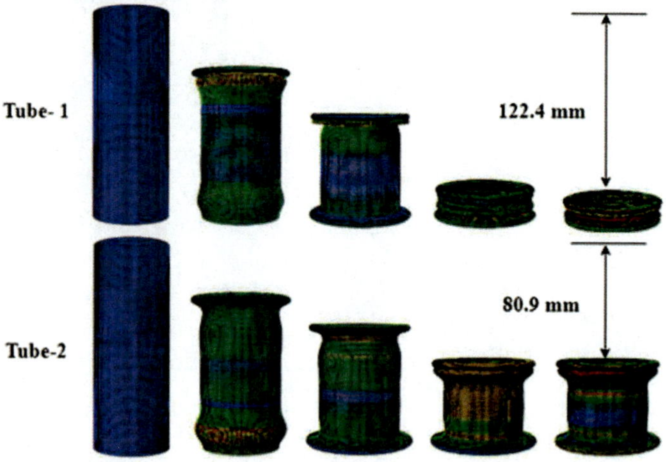

Fig. 2 Axial crushing length and deformation history of Tube-1 and Tube-2

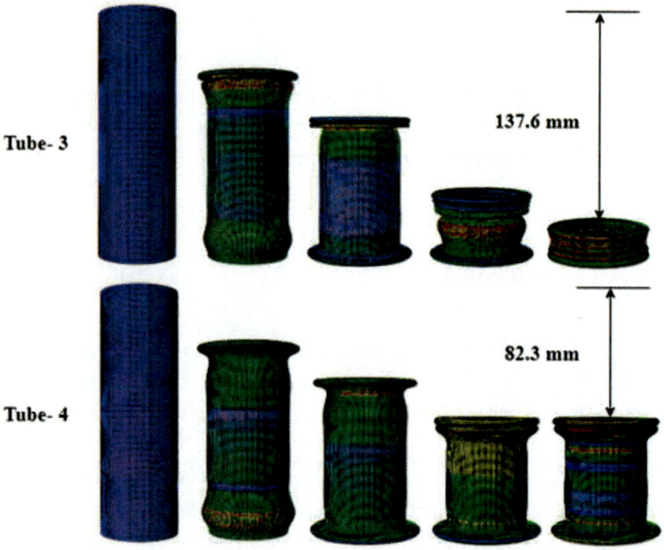

Fig. 3 Axial crushing length and deformation history of Tube-3 and Tube-4

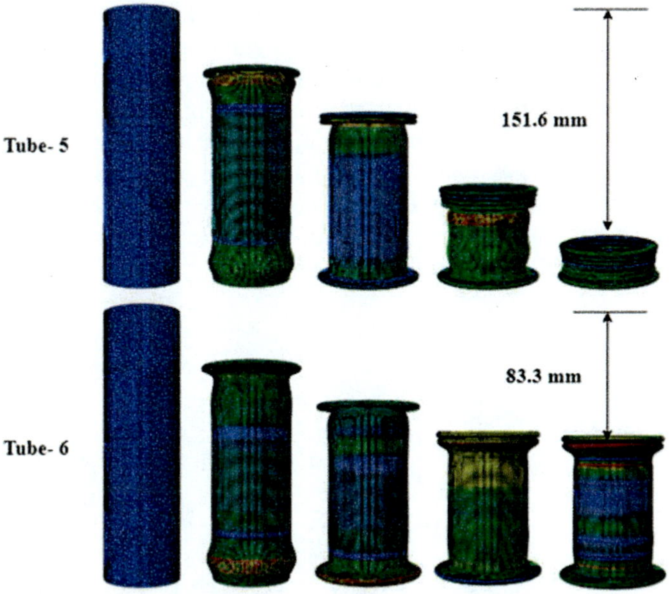

Fig. 4 Axial crushing length and deformation history of 170-50-2 and 170-50-3 al 6061-O tube specimens

starts decreasing as the next inward fold was developed. The initial peak force is the force required to initiate deformation and to start the energy dissipation. But it should be within the acceptable level for occupant safety in a vehicle. The reason for the higher impact force during impact loading was the inertia effect and the contact friction effect.

Figure 6 shows a comparison of the initial peak force for cylindrical tubes with different thicknesses and lengths. The results specified that the initial peak force for

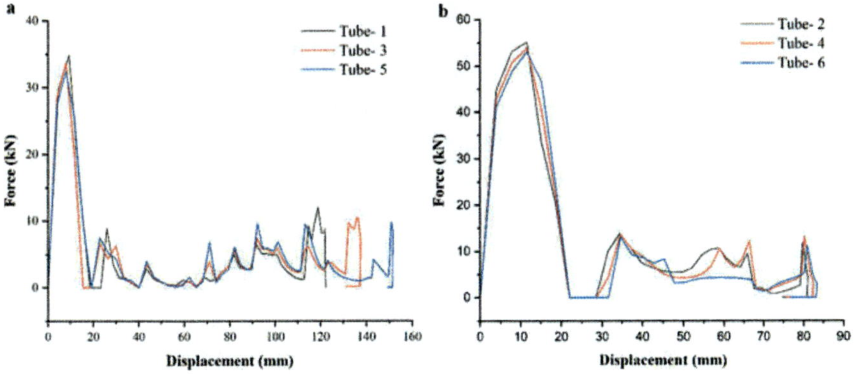

Fig. 5 Force vs displacement curve **a** 2 mm and **b** 3 mm thick tube specimens

2 mm thick tubes was 35–40% lesser than that for 3 mm thick tubes under the same condition. The initial peak crushing force remains almost the same for the different tubes of 3 mm thickness. The impact energy absorption capacity of cylindrical tubes obtained was compared, and the comparison is shown in Fig. 7. From the comparison, it was observed that the tube specimens having 3 mm thickness gained/dissipated comparatively higher energy under crushing. It was observed that increasing the thickness of the tube increased the energy absorption capacity. Also, it was found that tube-5 dissipates energy smoothly by taking more time than other tubes and it has the lowest values of initial peak crushing force.

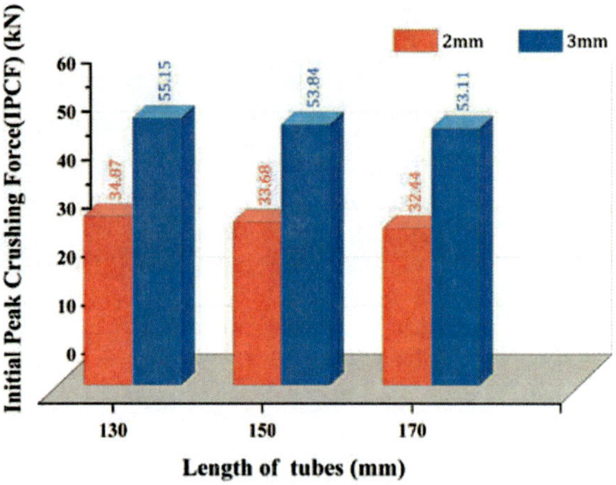

Fig. 6 Initial peak crushing force for various tube specimens

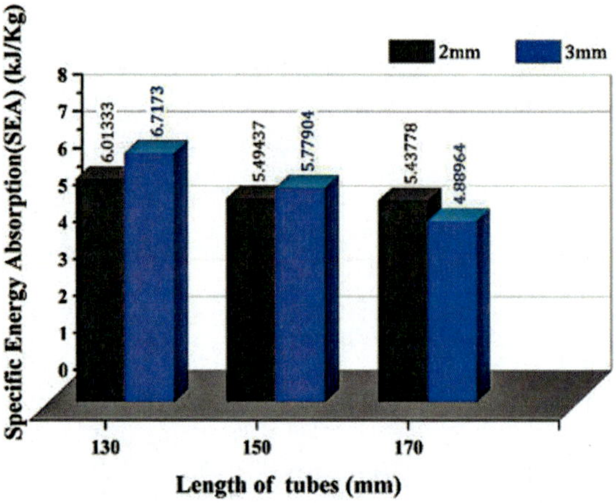

Fig. 7 Specific energy absorption for various tube specimens

4 Conclusion

The present study used numerical simulations to investigate the crashworthiness of aluminium alloy 6061-O-rolled tubes with thicknesses of 2 and 3 mm. The lengths of the tubes ranged from 130 to 170 mm. A 75 kg steel load impacted the cylindrical tubes. An effective impact energy-absorbing (EA) system is one that has a low initial peak crushing force, delays the transmission of impact-induced decelerations into the passenger's section, and has uniform crushing force over the crushing length [9]. In general, the specimen's axial crushing length was determined to be exactly related to the length-to-thickness (h/t) ratio. The maximum value of axial crushing length was found to be 151.6 mm for 170-50-2 tube specimens. The initial peak crushing force (IPCF) of the 2 mm cylindrical tube specimens was 35–40% lesser compared to 3 mm thick cylindrical tube specimens. The length-to-thickness (h/t) ratio has no effect on specific energy absorption. Tube-5 with a thickness of 2 mm, a length of 170 mm, and a diameter of 50 mm dissipates energy more slowly than other tubes and has the lowest initial peak crushing force values. It may be recommended for use as energy absorbers in crash-worthy applications.

References

1. Alkhatib SE, Tarlochan F, Hashem A, Sassi S (2018) Collapse behavior of thin-walled corrugated tapered tubes under oblique impact. Thin-Walled Struct 122:510–528. https://doi.org/10.1016/j.tws.2017.10.044
2. Alexander JM (1960) An approximate analysis of the collapse of thin cylindrical shells under axial loading. Q J Mech Appl Math 13(1):10–15. https://doi.org/10.1093/qjmam/13.1.10
3. Hsu SS, Jones N (2004) Quasi-static and dynamic axial crushing of thin-walled circular stainless steel, mild steel and aluminium alloy tubes. Int J Crashworthiness 9(2):195–217. https://doi.org/10.1533/ijcr.2004.0282
4. Miller WS et al (2000) Recent development in aluminium alloys for the automotive industry. Mater Sci Eng, A 280(1):37–49. https://doi.org/10.1016/S0921-5093(99)00653-X
5. Abramowicz W, Jones N (1984) Dynamic axial crushing of circular tubes. Int J Impact Eng 2(3):263–281. https://doi.org/10.1016/0734-743X(84)90010-1
6. Liu Q, Ma J, He Z, Hu Z, Hui D (2017) Energy absorption of bio-inspired multi-cell CFRP and aluminum square tubes. Compos B Eng 121:134–144. https://doi.org/10.1016/J.COMPOSITESB.2017.03.034
7. Qiu N, Gao Y, Fang J, Feng Z, Sun G, Li Q (2016) Theoretical prediction and optimization of multi-cell hexagonal tubes under axial crashing. Thin-Walled Struct 102:111–121. https://doi.org/10.1016/J.TWS.2016.01.023
8. Zahran MS, Xue P, Esa MS, Abdelwahab MM (2018) A novel tailor-made technique for enhancing the crashworthiness by multi-stage tubular square tubes. Thin-Walled Struct 122:64–82. https://doi.org/10.1016/j.tws.2017.09.031
9. Praveen Kumar A, Mohamed MN, Jusuf A, Dirgantara T, Gunawan L (2018) Axial crash performance of press-formed open and end-capped cylindrical tubes – a comparative analysis. Thin-Walled Struct 124:468–488. https://doi.org/10.1016/j.tws.2017.12.037

Effect of Spherical and Conical Fragment Shape on Damage of Steel Target Plate in Ballistic Applications

S. S. Hirmukhe, A. Kumaraswamy, and S. V. Nimje

Abstract Recent numerical and experimental studies revealed that projectile shapes such as blunt, hemispherical and conical shapes significantly affect residual velocity of projectile as well as the failure mode and the energy absorption of target plate during the penetration in low velocity impact regime. However, in literature few studies have been undertaken on influence of fragment shapes on target plate failure mode in ordinance to hypervelocity impact regime. Therefore, in this study, finite element investigation of penetration characteristic and failure mode of steel target plate impacted by stainless steel spherical and conical fragments in velocity range of 400–2000 m/s, has been performed using the nonlinear finite element code LS-DYNA which is extensively used for impact analysis. The simulation model employs Johnson–Cook constitutive model coupled with Gruneisen equation of state (EOS) and erosion contact algorithm, to accurately predict the large strain developed in target plate due to impact of different fragment shapes (spherical and conical). The simulation results of spherical fragment shape are corroborated with literature experimental observations. The lower fragment perforation time and lower energy absorption of target plate suggested conical fragment would produce better lethality characteristics than spherical fragment of same and material used in fragment generated warheads.

Keywords Ballistic penetration · Preformed fragments · Fragment generator warhead · Numerical analysis

1 Introduction

Fragment generated warheads (FGW) are extensively employed in air defense applications to damage the distant targets and make them incapable to perform their intended function. FGW consists of preformed fragments with different shapes such

S. S. Hirmukhe · A. Kumaraswamy (✉) · S. V. Nimje
Department of Mechanical Engineering, Defence Institute of Advanced Technology (DU), Girinagar, Pune 411025, India
e-mail: akswamy@diat.ac.in

R. Velmurugan et al. (eds.), *Dynamic Behavior of Soft and Hard Materials Volume 1*, Springer Proceedings in Materials 34, https://doi.org/10.1007/978-981-99-6030-9_7

as cubical, blunt, conical, and spherical. These fragments carried by missile system explode near the target so that fragments strike on the target surface at very high velocity and produce maximum lethality. Dhote and Verma [1] investigated the effect of spherical fragment shape impacting on the thin steel plate in hypervelocity range (2000–4000 m/s) using experimental and smooth particle hydrodynamic (SPH) numerical technique. Further, numerical results estimated using former numerical technique match well with experimental results. But this numerical technique requires high end computer configuration as well as takes more computational time. Recent numerical study suggested that erosion contact algorithm is more effective over SPH technique which reasonably capture the deformation behavior of the target material at high velocities with low end computer configuration and less computation time [2]. However, previous literature studies are mainly focused on either selection numerical technique or understanding effect of spherical fragment shape on damage of the thin target plate. Further, to the authors knowledge few studies have been undertaken on influence of different fragment shapes on the thin target plate failure modes in ordinance to hypervelocity impact regime. Moreover, recent numerical and experimental studies revealed that projectile shapes such as blunt, hemispherical and conical shapes significantly affect residual velocity of projectile as well as the failure mode and the energy absorption of target plate during the penetration in low impact velocity regime (i.e., impact velocity < 500 m/s [3]. Therefore, motivated from this study, three-dimensional finite element simulations are performed to investigate the penetration characteristic and failure mode of steel target plate impacted with stainless steel spherical and conical fragments in velocity range of 400–2000 m/s using Johnson–Cook constitutive model coupled with Gruneisen EOS and erosion contact algorithm in LS-DYNA. It is found that residual velocity and kinetic energy of fragments markedly affected by their shapes. The results suggested that conical fragments may cause the maximum damage to the target than spherical fragment of identical material and mass used in FGW.

2 Constitutive Model

To understand the deformation behavior of steel (Steel 1006) target and stainless steel (SS304) fragment subjected impact, the Johnson–Cook (J-C) elasto-viscoplastic constitutive model coupled with damage failure model is used in this study. This constitutive model has shown ability to capture the effect of isotropic strain hardening, strain rate and softening due to adiabatic heating. The equivalent Von-Mises stress ($\overline{\sigma}$) in J–C model is given as [4]:

$$\overline{\sigma}\left(\epsilon^{p}, \dot{\epsilon}^{p}, T\right) = \left[A + B\left(\epsilon^{p}\right)^{n}\right]\left[1 + C\,\mathrm{In}\left(\frac{\dot{\epsilon}^{p}}{\dot{\epsilon}_{0}}\right)\right]\left[1 - T^{*m}\right] \quad (1)$$

$$T^{*} = \frac{T - T_{0}}{T_{\mathrm{melt}} - T} \quad (2)$$

where A, B, C, n and m are material constants determined by performing different mechanical tests. ϵ^p, $\dot{\epsilon}^p$ and $\dot{\epsilon}_0$ are equivalent plastic strain, rate of equivalent plastic strain and reference plastic strain, respectively. Further, T, T_0 and T_{melt} are current, room and melting temperature, respectively. In addition, Johnson and Cook describe the fracture strain which function of strain rate, stress triaxiality and temperature. Fracture strain function is given as:

$$\epsilon_f\left(\overline{\sigma}^*, \dot{\epsilon}^p, T\right) = \left[D_1 + D_2 e^{D_3\overline{\sigma}^*}\right]\left[1 + D_4 \, In \, \dot{\epsilon}^p\right]\left[1 + D_5 T^{*m}\right]. \quad (3)$$

Here, $D_1 - D_5$ represent material constants obtained by performing different mechanical test. $\overline{\sigma}^*$ is stress triaxiality. The J-C constitutive model, Eqs. (1–3) coupled with Gruneisen EOS and erosion contact algorithm, is implemented in LS-DYNA. Further, it is recommended that eroding contact algorithm must be employed whenever solid elements used in the contact definition which are subjected to erosion (element deletion) due to material failure criterion. The material parameters used in current work are presented in Table 1.

Table 1 Material data used in simulation models [1]

Property	Unit	Fragment: stainless steel (304)	Plate: steel 1006
Density	kg/m^3	7896	7896
Poisson's ratio	–	0.3	0.3
Gruneisen coefficient (γ)	–	1.93	2.17
Hugoniot sound speed (C)	m/s	4570	4570
Hugoniot constant (S_1)	–	1.49	1.49
Young's modulus	GPa	200	213
Shear modulus	GPa	77	81.8
Yield stress (A)	GPa	0.34	0.35
Hardening constant (B)	GPa	1.0	0.275
Hardening constant (n)	–	0.65	0.36
Strain rate constant (c)	–	0.07	0.022
Thermal softening exponent (m)	–	1	1
Melting temperature	K	1673	1811
Room temperature	K	293	300
Damage parameters [8, 9]			
D1	–	0.69	0.05
D2	–	0.0	3.44
D3	–	0.0	– 2.12
D4	–	0.0546	0.002
D5	–	0.0	0.61

3 Simulation Aspects

Figure 1a shows finite element mesh assembly of steel target (Steel 1006) with stainless steel spherical fragment (SS 304), and corresponding assembly of steel target with conical fragment is displayed in Fig. 1b. The height and width of target plate are chosen as 200 mm × 200 mm to ensure the plastic zone is well contained within the impacted region so that boundary effect in simulations could be minimized [5–7]. By following the experimental literature, the target thickness and diameter of spherical fragment are taken as 1 mm and 6 mm, respectively [1]. Further, to investigate influence of fragment shapes on the target plate failure mode, mass of conical fragment is considered same as spherical fragment. The target plate and fragments are discretized using 8-noded hexahedral elements. To reduce simulation time without affecting the numerical accuracy, a highly refined mesh is used in impact region of target plate while relatively coarse mesh is employed away from the impact region (refer Fig. 1c). A proper mesh convergence study is carried out, and minimum element size for target plate in impact region is taken as 0.1 × 0.1 × 0.1 mm^3. Whereas, for both fragment shapes it is considered as 0.3 × 0.3 × 0.3 mm^3. A fixed boundary condition is imposed on boundary nodes of target plate, while initial velocity is applied to the fragment. An eroding surface-to-surface contact is assumed between target and fragment. The average crater diameter is calculated by measuring the diameter at different location of hole (refer Fig. 2) for 1 mm thin target plate impacted with 6 mm spherical diameter at 2000 m/s normal impact velocity. The average crater diameter in present study is found to be 8.7 mm, and it is comparable with corresponding experimental carter diameter of 9.0 mm [1]. The deviation in simulation and experimental crater diameter is less than 10%, validates constitutive model and material parameter value used in current study.

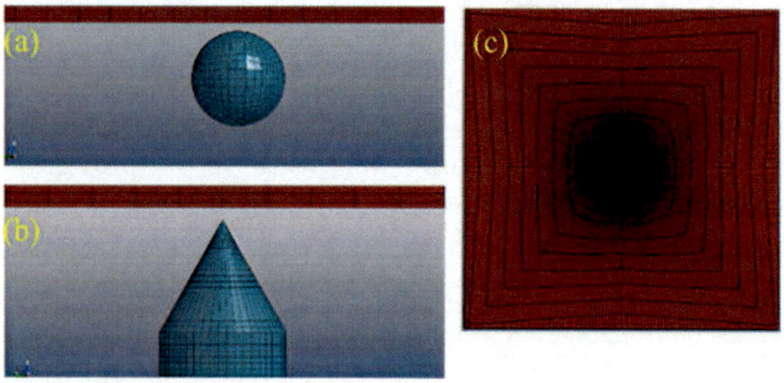

Fig. 1 Finite element mesh assembly of steel target with **a** Spherical fragment, **b** conical fragment. **c** target plate mesh

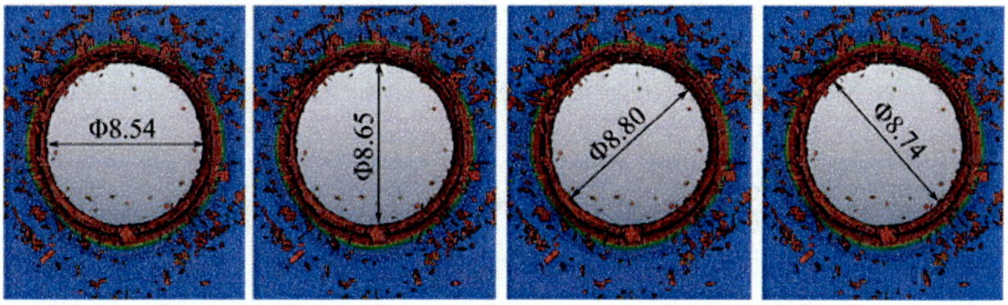

Fig. 2 Simulation crater diameter for Ø6 mm spherical fragment impact on 1 mm thin steel target at 2000 m/s normal impact velocity

4 Results and Discussion

The residual velocity, kinetic energy and fragment propagation time are obtained from the simulation models for SS 304 spherical and conical fragments impacted at normal impact velocities ranging from 400 to 2000 m/s on 1 mm thin Steel 1006 target plates, and it is tabulated in Table 2. From Table 2, it can be noticed that residual velocity and kinetic energy of conical fragment is found to be higher than spherical fragment for all initial impact velocities. Both spherical and conical fragments perforate the 1 mm thin target plate for all impact velocities. The variation of fragment perforation time versus initial impact velocity for spherical and conical fragments is shown in Fig. 3. The perforation time decreases with an increase in fragment initial velocity. The perforation time for a conical fragment is always less than the spherical fragment for all initial impact velocities which suggests that energy absorbed by the target plate due to impact of conical fragment would be lower than spherical fragment. To confirm this observation, the variation of energy absorbed by the target plate against fragment initial impact velocity is shown in Fig. 4 for both fragments (spherical and conical). It is revealed that energy absorption of target plate is more for spherical fragment than conical fragment for all impact velocity. Therefore, it is worth to conclude that conical fragment would cause maximum damage of target plate for given impact velocity. Furthermore, less energy absorbed by the target plate suggests that plastic deformation in the target plate could be less which will be investigated in future research work.

5 Conclusions

In the current study, finite element analysis of penetration characteristics of spherical and conical fragment shapes is investigated using Johnson–Cook material model supplemented with Gruneisen equation of state and erosion contact algorithm in LS-DYNA for 1 mm thin steel plate at various normal impact velocities. The constitutive model is validated with experimental data available in literature by considering the

Table 2 Simulation results for 1 mm thin target plate

Fragment velocity (m/s)	Fragment shape	Residual velocity of fragment (m/s)	Residual kinetic energy of fragment (m/s)	Fragment perforation time (μs)	Perforation in target
2000	Spherical	1733	1324	2	Yes
	Conical	1870	1553	1.2	Yes
1600	Spherical	1381	841	4	Yes
	Conical	1487	982	3	Yes
1200	Spherical	1012	451	6	Yes
	Conical	1100	540	4	Yes
800	Spherical	644	183	8	Yes
	Conical	708	223	5	Yes
400	Spherical	267	31.5	20	Yes
	Conical	308	42.5	12	Yes

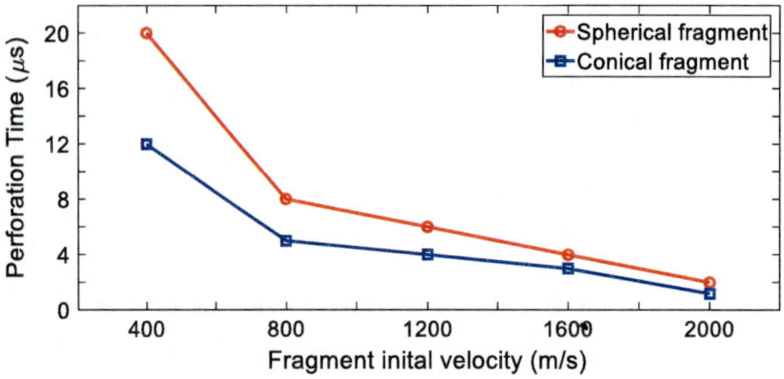

Fig. 3 Fragment perforation time versus fragment initial velocity

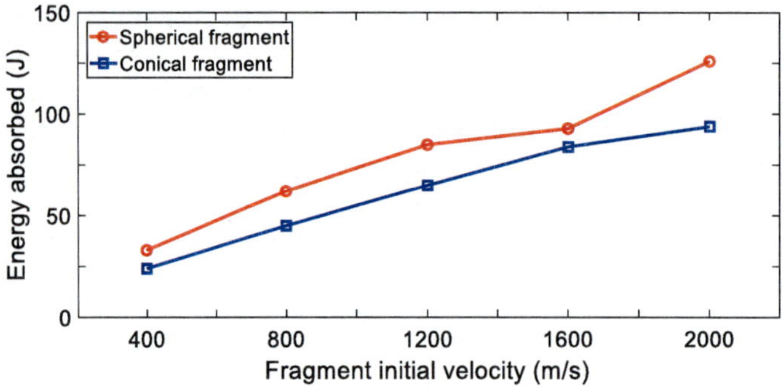

Fig. 4 Variation of target plate energy absorbed versus fragment initial velocity

spherical fragment shape. From present work, it is found that the fragment residual velocity and kinetic energy enhances with increase in fragment impact velocity. While fragment perforation time decreases with increase in impact velocity. The lower fragment perforation time and lower energy absorption of target plate indicate high penetration capability of the fragment. For a given impact condition, the target plate with conical fragment has shown lower fragment perforation time and energy absorption of target plate than spherical fragment suggesting conical fragment caused maximum lethality of target than spherical fragment. Thus, a conical fragment used in FGW can cause maximum damage of target plate as compared to spherical fragment of same mass and material.

Acknowledgements Authors would like to thank Dr. K. D. Dhote (Scientist F), Mr. P. N. Verma (Scientist E), Director ARDE, Pune, and ARMREB panel members for their valuable inputs and suggestions in ARMREB project (Project No. ARMREB/ADMB/2021/246). Also, Authors would like to thank Vice Chancellar, DIAT (DU), Pune for providing computational facilities.

References

1. Dhote KD, Verma PN (2018) Investigation of hole formation by steel sphere impacting on thin plate at hypervelocity. Thin-Walled Struct 126:38–47. https://doi.org/10.1016/j.tws.2017.05.012
2. Mulabagal P, Kumaraswamy A, Saxena A (2021) Numerical investigation of penetration characteristics of preformed spherical fragments at hyper velocity. J Eng Res Special Issue Oct (2021). https://doi.org/10.36909/jer.ICCEMME.15597
3. Borvik T, Hopperstad OS, Berstad T, Langseth M (2002) Perforation of 12 mm thick steel plates by 20 mm diameter projectile with flat, hemispherical and conical noses Part II : numerical simulations. Int J Impact Eng 27:37–64. https://doi.org/10.1016/S0734-743X(01)00035-5
4. Johnson GR, Cook WH (1983) A constitutive model and data for metals subjected to large strains, high strain rates and high temperatures. In: Proceedings of the 7th international symposium on ballistics 21, pp 541–547
5. Hirmukhe SS, Prasad KE, Singh I (2020) Investigation of pressure sensitive plastic flow in nanoglasses from finite element simulations. Scripta Mater 180:45–50. https://doi.org/10.1016/j.scriptamat.2020.01.022
6. Hirmukhe SS, Sharma A, Nandam SH, Hahn H, Prasad KE, Singh I (2022) Investigation of softening induced indentation size effect in nanoglass and metallic glass. J Non-Cryst Solids 577:121316. https://doi.org/10.1016/j.jnoncrysol.2021.121316
7. Hirmukhe SS, Joshi AT, Singh I (2022) Mixed mode (I & II) fracture behavior in nanoglass and metallic glass. J Non-Cryst Solids 580:121390. https://doi.org/10.1016/j.jnoncrysol.2021.121390
8. Karim MR, Fatt MS (2005) Impact of Boeing 767 aircraft into the world trade center. J Eng Mech 131:1066–1072. https://doi.org/10.1061/(ASCE)0733-9399(2005)131:10(1066)
9. Frontan J, Zhang Y, Dao M, Liu J, Galvez F, Jerusalem A (2012) Ballistic performance of nanocrystalline and nanotwinned ultrafine steel. Acta Mater 60:1353–1367. https://doi.org/10.1016/j.actamat.2011.11.029

Numerical Analysis of Penetration Characteristics of Conical Fragments in Ballistic Applications

Pradeep Mulabagal, A. Kumaraswamy, and S. S. Hirmukhe

Abstract Numerical analysis using explicit code LS-DYNA has been carried out to study the penetration characteristics of a spherical fragment of Ø6mm diameter and conical fragments with apex angles 40°, 60°, 80° and 100° of identical mass made of Stainless Steel (SS 304). The spherical and conical fragments impact on Steel 1006 target plates of 6, 3 and 1 mm thickness at 500 to 2000 m/s impact velocity range in the simulation models. The simulation model employed for the spherical fragment of Ø6 mm diameter in the present study has been validated with experimental data available in the literature. The validated simulation model has been extended to analyse the penetration characteristics of conical fragments with apex angles 40° to 100°. Johnson–Cook (J-C) model, a strain rate-dependent plasticity model supplemented with the Gruneisen equation of state (EoS) and an erosion contact algorithm, have been used in the simulation models to capture the penetration characteristics of the spherical and conical fragments at high velocities. The penetration characteristics have been determined from simulation models in terms of residual velocity and residual kinetic energy of the fragment. The conical fragment with 40° apex angle indicated better penetration characteristics than other fragment shapes considered in the current study.

Keywords Ballistic penetration · Preformed fragments · Fragment generator warhead · Numerical analysis

1 Introduction

Surface to air type missile systems predominantly use Fragment Generator Warheads (FGW) to destroy distant targets. FGW carry preformed fragments of various shapes, viz. spherical, conical, cubical, blunt etc., which detonate upon reaching proximity of a target and cause damage to target due to impact of fragments at high velocities.

P. Mulabagal · A. Kumaraswamy (✉) · S. S. Hirmukhe
Department of Mechanical Engineering, Defence Institute of Advanced Technology (DU), Girinagar, Pune 411025, India
e-mail: akswamy@diat.ac.in

R. Velmurugan et al. (eds.), *Dynamic Behavior of Soft and Hard Materials Volume 1*, Springer Proceedings in Materials 34, https://doi.org/10.1007/978-981-99-6030-9_8

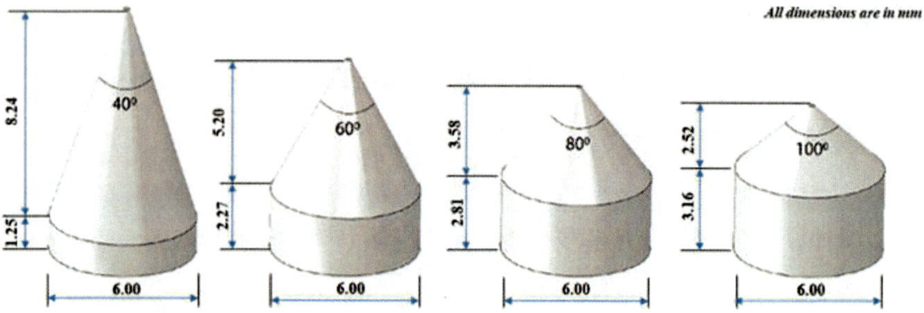

Fig. 1 Dimensions of conical fragments with apex angles 40°, 60°, 80° and 100°

Experimental, empirical and numerical study for the characterization of a hole formed on a steel target due to the impact of spherical steel fragments in ultra-ordnance to hypervelocity range (2–4 km/s) has been reported in the literature [1, 2]. The numerical study reported in the literature has been carried out using a nonlinear hydro-code, viz. smoothed particle hydrodynamics (SPH) in ANSYS Autodyn-3D software. However, the effectiveness of an erosion contact algorithm over SPH hydro-code in predicting the target material behaviour due to impact of spherical fragments at ordnance to hypervelocity range (1–5 km/s) has been reported in the literature [3].

Johnson–Cook (J-C) constitutive material model [4] supplemented with Gruneisen EoS is widely used in explicit codes to simulate the high strain rate deformation behaviour and hydrodynamic response of materials under shock loading. Recently, Mulabagal et al. [3] performed a numerical analysis with spherical fragments, and the simulation model results were well in line with experimental results [1, 2]. However, in previous studies [1–3], the influence of conical shape fragments with different apex angles on the deformation behaviour of the target is not studied. Therefore, this paper deals with a numerical analysis of penetration characteristics of conical fragments with apex angles 40°, 60°, 80° and 100° of identical mass and material as that of Ø6mm SS 304 spherical fragments studied in Mulabagal et al. [3]. Further, the validated simulation model for Ø6mm spherical fragments [3] has been used in the present study to carry out the numerical analysis. The fragments impact on the target plate of 6, 3 and 1 mm thickness in the impact velocity range from 500 to 2000 m/s. The simulation results are presented in terms of residual velocity and residual kinetic energy. Figure 1 depicts the conical fragment shapes studied in this paper.

2 Simulation Model

The meshed conical fragments employed in LS Dyna simulation models are shown in Fig. 2. The meshed target plate and assembly of conical fragments impacting the target plate are represented in Figs. 3a, b, respectively. A validated simulation model used for spherical fragments in Mulabagal et al. [3] is employed in the present study. The target plate and conical fragments are discretized using 8-noded hexahedral elements. A mesh convergence study has been carried out to optimize the simulation time without affecting the numerical accuracy. A highly refined mesh is employed in the impact region of the target plate, while relatively a coarser mesh is used away from the impact region. An optimal element size of 0.2 mm is considered in the thickness direction for all target plates. Further, to minimize the boundary effect, the dimensions of the target plate are considered sufficiently larger [5–7]. Explicit code LS-DYNA suggests using an eroding contact algorithm in the contact definition for solid elements subjected to erosion (element deletion). Therefore, eroding node-to-surface contact algorithm and default element formation have been employed in the simulation model. J-C constitutive material model coupled with Gruneisen EoS is used in the simulation model to obtain the deformation behaviour of both target and fragments. The material properties of the target (Steel 1006) and fragment (SS304) used in the constitutive material model and EoS are listed in Table 1 [2, 8].

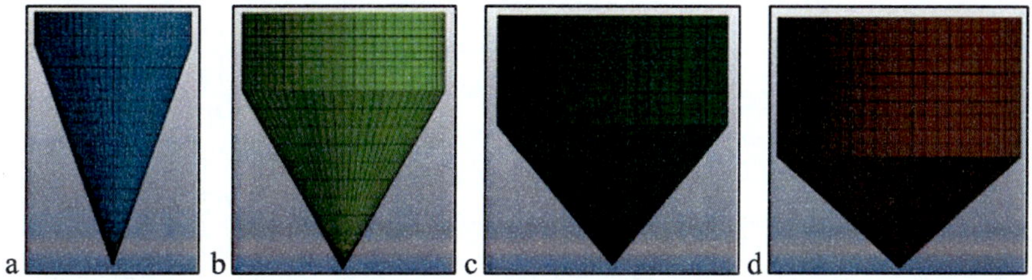

Fig. 2 Meshed conical fragments with apex angles **a** 40°, **b** 60°, **c** 80°, **d** 100°

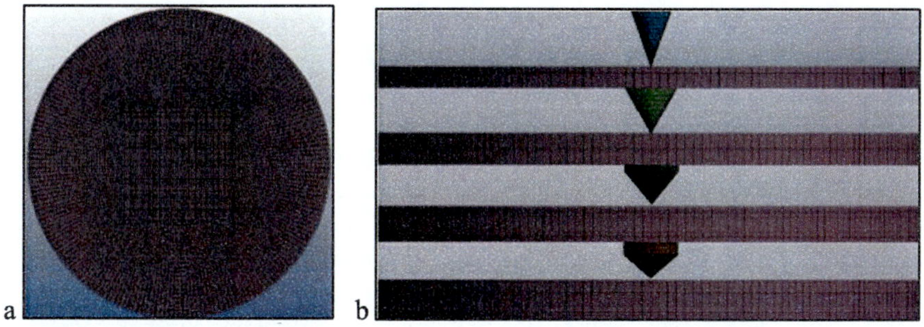

Fig. 3 **a** Meshed target plate; **b** conical projectiles impacting on the target plate

Table 1 Material properties of fragment and target material used in the simulation model

Property	Unit	Fragment: stainless steel (304)	Plate: Steel 1006
Density	kg/m^3	7896	7896
Poisson's ratio	–	0.3	0.3
Gruneisen coefficient (γ)	–	1.93	2.17
Hugoniot sound speed (C)	m/s	4570	4569
Hugoniot constant (S_1)	–	1.49	1.49
Young modulus	GPa	210	210
Shear modulus	GPa	77	81.8
Yield stress (A)	MPa	310	350
Hardening constant (B)	MPa	1000	275
Hardening constant (n)	–	0.65	0.36
Strain rate constant (c)	–	0.07	0.022
Thermal softening exponent (m)	–	1	1
Melting temperature	K	1673	1673
Room temperature	K	300	300
Sound speed	m/s	5406	5406
Damage parameter (D_1)	–	0.05	-0.8
Damage parameter (D_2)	–	3.44	2.1
Damage parameter (D_3)	–	– 2.12	0.5
Damage parameter (D_4)	–	0.002	0.0002
Damage parameter (D_5)	–	0.61	0.61

3 Simulation Results

4 Discussion

The residual velocity and residual kinetic energy of the fragment obtained from simulation models for spherical and conical fragments impacting on 6, 3 and 1 mm thick target plates are presented in Tables 2, 3 and 4, respectively. The variation of fragment residual velocity for various fragment shapes and target thicknesses is shown in Fig. 4. From Fig. 4a, it can be seen that fragment residual velocity increases with an increase in fragment impact velocity for all fragment shapes. Further, conical fragments indicate higher residual velocity than spherical fragments for a given fragment impact velocity. A similar trend is observed for other target thicknesses (Refer to Fig. 4b, c). It is also observed that all fragments could not perforate the 6 and 3 mm thick target plates at an impact velocity of 500 m/s (Refer to Fig. 4a, b). However, at 500 m/s impact velocity, all fragments perforated through a 1 mm thick plate (Refer to Fig. 4c).

Table 2 Simulation results for a 6 mm thick target plate

Fragment impact velocity (m/s)	Initial kinetic energy of fragment (J)	Fragment shape	Residual kinetic energy of fragment (J)	Residual velocity of fragment (m/s)	Perforation in target
2000	1780.00	Cone 40°	668.04	928.37	Yes
		Cone 60°	507.9	694.86	Yes
		Cone 80°	567.7	752.01	Yes
		Cone 100°	665.88	862.93	Yes
		Sphere	38.41	80.346	Yes
1500	1001.25	Cone 40°	410.45	743.68	Yes
		Cone 60°	166.42	438.15	Yes
		Cone 80°	144.20	449.35	Yes
		Cone 100°	152.19	465.78	Yes
		Sphere	–	–	No
1000	445.23	Cone 40°	75.38	316.45	Yes
		Cone 60°	20.16	180.23	Yes
		Cone 80°	–	–	No
		Cone 100°	–	–	No
		Sphere	–	–	No
500	111.25	No perforation for all fragments			

At lower impact velocities, a fragment undergoes elastic deformation during penetration through target plates. Therefore, erosion of fragment elements is negligible at lower impact velocities. As the impact velocity increases, plastic deformation of the fragment occurs which leads to higher erosion of fragment elements. Therefore, to determine the lethality of the fragment, the residual kinetic energy would be a preferable parameter over residual velocity owing to the presence of a mass component in kinetic energy. The variation of residual kinetic energy with the impact velocity of the fragment for different target thicknesses considered in the present study is shown in Fig. 5. Interestingly, from Fig. 5, it is observed that the fragment residual kinetic energy also increases with an increase in impact velocity for all fragment shapes.

Further, for a given impact velocity, all conical fragments have shown higher residual kinetic energy than spherical fragments for impact velocities 1000 m/s and above, indicating the former is more lethal than the latter. In other words, conical fragments would cause more damage to the target due to higher residual kinetic energy at higher impact velocities. Subsequently, to identify the most lethal conical fragment shape, plots of the residual kinetic energy of conical fragments of different apex angles corresponding to target thicknesses at 2000, 1500, 1000 and 500 m/s impact velocity are shown in Fig. 6a–d, respectively. Figure 6 indicates that the residual kinetic energy decreases with a decrease in impact velocity and an increase in cone apex angle. The conical fragment with an apex angle 40° indicates higher

Table 3 Simulation results for a 3 mm thick target plate

Fragment impact velocity (m/s)	Initial kinetic energy of fragment (J)	Fragment shape	Residual kinetic energy of fragment (J)	Residual velocity of fragment (m/s)	Perforation in target
2000	1780.00	Cone 40°	1181.60	1392.51	Yes
		Cone 60°	1269.40	1481.42	Yes
		Cone 80°	1012.00	1239.24	Yes
		Cone 100°	1255.00	1462.90	Yes
		Sphere	244.42	349.59	Yes
1500	1001.25	Cone 40°	628.47	1045.04	Yes
		Cone 60°	474.16	872.64	Yes
		Cone 80°	438.26	858.77	Yes
		Cone 100°	399.60	831.58	Yes
		Sphere	103.86	248.36	Yes
1000	445.23	Cone 40°	185.17	573.03	Yes
		Cone 60°	150.77	531.86	Yes
		Cone 80°	137.60	524.18	Yes
		Cone 100°	120.72	495.83	Yes
		Sphere	69.088	316.02	Yes
500	111.25	No perforation for all fragments			

residual kinetic energy than all other fragment shapes considered in the present study. Therefore, it is worth concluding that a conical fragment with an apex angle of 40° causes maximum damage to all targets under the conditions considered in the present study.

5 Conclusions

Numerical analysis of penetration characteristics of spherical and conical fragment shapes with different apex angles has been investigated using the Johnson–Cook material model supplemented with the Gruneisen equation of state and erosion contact algorithm in LS-DYNA for various target thicknesses and impact velocities. The simulation model validated with experimental data available in the literature for spherical fragments has been extended to conical fragments in the current study. The current study found that the residual velocity and residual kinetic energy of the fragment increase with an increase in impact velocity. High residual kinetic energy indicates high penetration capability of the fragment. Therefore, all conical fragments possess high residual kinetic energy than the spherical fragment at an impact velocity of 1000 m/s and above indicating the former exhibits better lethality than

Table 4 Simulation results for a 1 mm thick target plate

Fragment impact velocity (m/s)	Initial kinetic energy of fragment (J)	Fragment shape	Residual kinetic energy of fragment (J)	Residual velocity of fragment (m/s)	Perforation in target
2000	1780.00	Cone 40°	1712.9	1931.1	Yes
		Cone 60°	1723.6	1936.8	Yes
		Cone 80°	1456.6	1682.9	Yes
		Cone 100°	1441.6	1648.4	Yes
		Sphere	358.30	460.58	Yes
1500	1001.25	Cone 40°	851.33	1328.4	Yes
		Cone 60°	807.29	1278.8	Yes
		Cone 80°	760.55	1235.2	Yes
		Cone 100°	772.82	1257.1	Yes
		Sphere	232.00	433.6	Yes
1000	445.23	Cone 40°	343.37	861.08	Yes
		Cone 60°	307.12	813.32	Yes
		Cone 80°	314.39	815.90	Yes
		Cone 100°	291.23	790.67	Yes
		Sphere	230.88	653.07	Yes
500	111.25	Cone 40°	70.22	395.81	Yes
		Cone 60°	56.19	353.10	Yes
		Cone 80°	42.37	305.24	Yes
		Cone 100°	40.80	300.14	Yes
		Sphere	64.22	378.98	Yes

the latter. Further, the conical fragment with an apex angle of 40° indicates higher residual kinetic energy than all other fragment shapes for a given impact velocity and target plate thickness considered in the present study. Thus, a conical fragment with 40° apex angle could be an ideal replacement for a spherical fragment of same mass and material which can be used in FGW to cause maximum damage to the target.

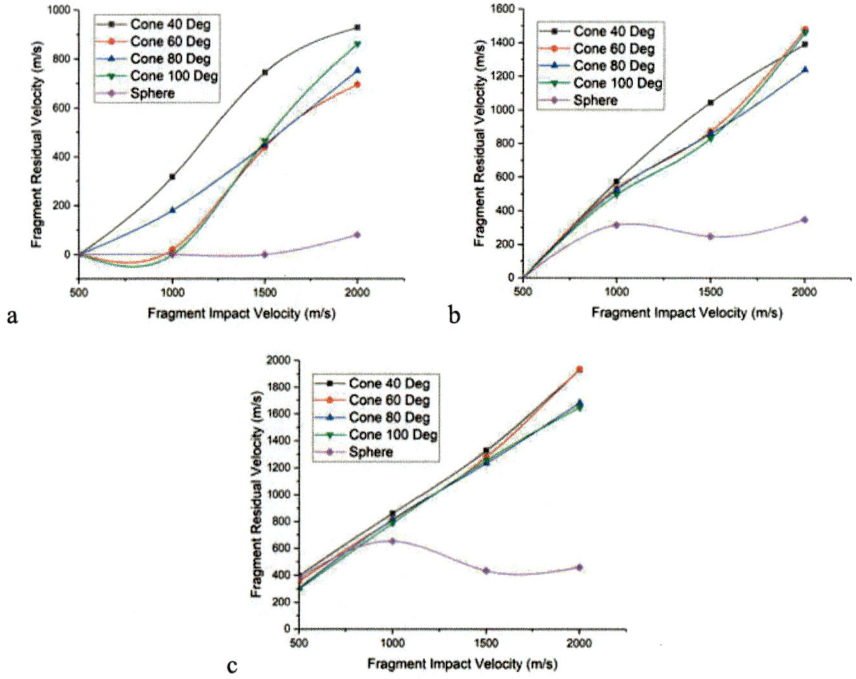

Fig. 4 Residual velocity versus impact velocity of the fragment for target plate of thickness **a** 6 mm, **b** 3 mm, **c** 1 mm

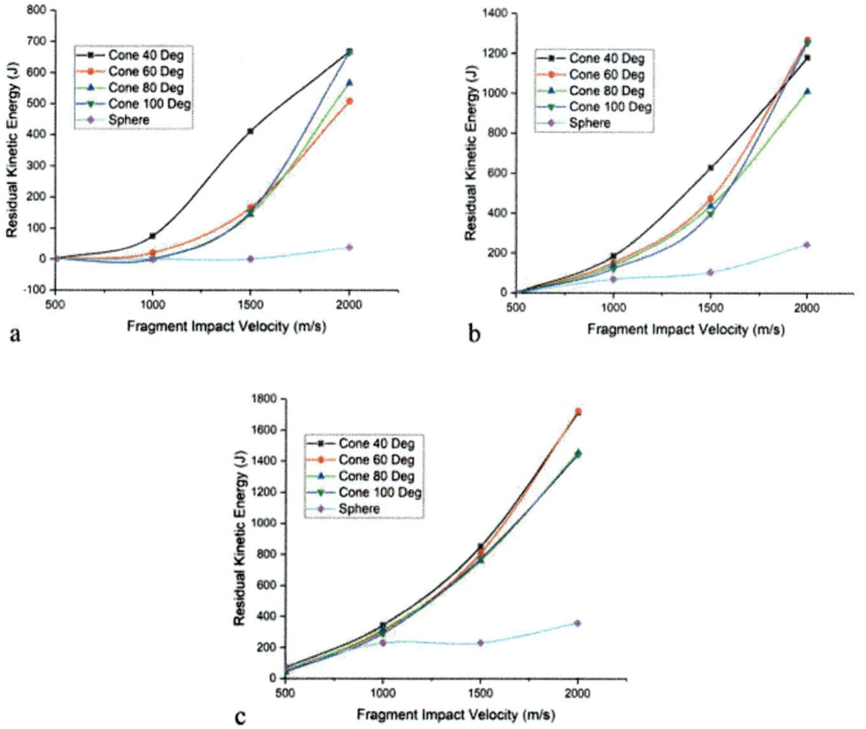

Fig. 5 Residual kinetic energy versus impact velocity of the fragment for target plate of thickness **a** 6 mm, **b** 3 mm, **c** 1 mm

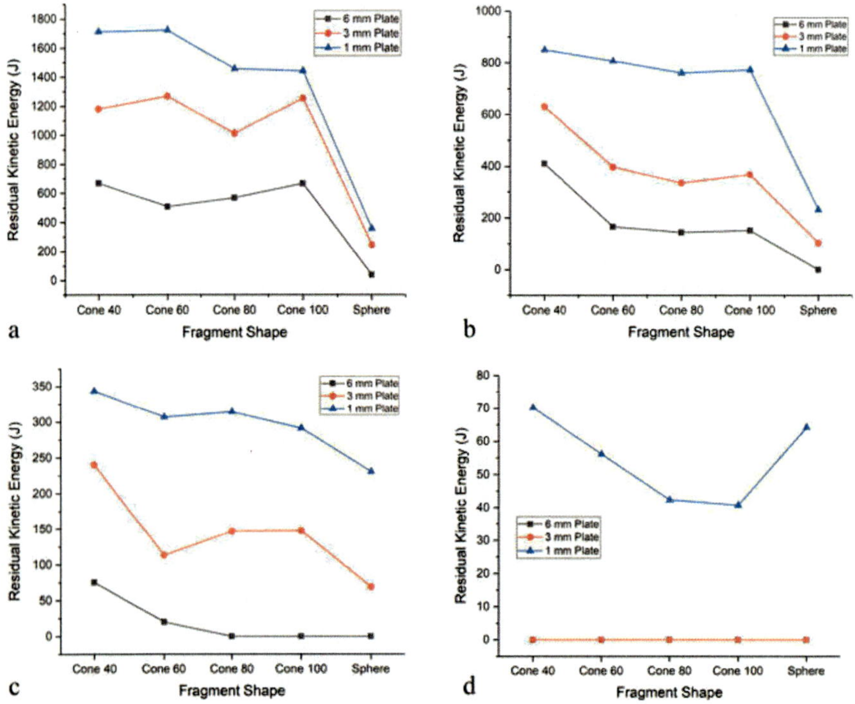

Fig. 6 Variation of residual kinetic energy of conical fragments at impact velocity **a** 2000 m/s, **b** 1500 m/s, **c** 1000 m/s and **d** 500 m/s for 6, 3 and 1 mm thick target plates

References

1. Dhote KD, Verma PN (2017) Investigation of hole size formation by steel sphere impacting on thin plate at hypervelocity. Procedia Eng 173:323–330
2. Dhote KD, Verma PN (2018) Investigation of hole formation by steel sphere impacting on thin plate at hypervelocity. Thin-Walled Struct 126:38–47
3. Mulabagal P, Kumaraswamy A, Saxena A (2021) Numerical investigation of penetration characteristics of preformed spherical fragments at hyper velocity. J Eng Res Special Issue Oct
4. Johnson GR, Cook WH (1983) A constitutive model and data for metals subjected to large strains, high strain rates and high temperatures. In: Proceedings of the 7th international symposium on ballistics, 21, pp 541–547
5. Hirmukhe SS, Prasad KE, Singh I (2020) Investigation of pressure sensitive plastic flow in nano glasses from finite element simulations. Scripta Mater 180:45–50
6. Hirmukhe SS, Sharma A, Nandam SH, Hahn H, Prasad KE, Singh I (2022) Investigation of softening induced indentation size effect in nanoglass and metallic glass. J Non-Cryst Solids 577:121316
7. Hirmukhe SS, Joshi AT, Singh I (2022) Mixed mode (I & II) fracture behavior in nanoglass and metallic glass. J Non-Cryst Solids 580:121390
8. Gintalas M, Ziliukas A, Ainsworth RA (2014) New equation for the plastic correction factor for J-integral determination from test results of three-point-bend specimens. Mater Perform Charact 3(3):125–143

Approach for Ensuring Fuel Tank Crashworthiness Using LS-Dyna Simulation

Kalinga Gulbarga, Gururaj Bankapur, Amit Kumar, A. T. Rao, and A. Sakthivel

Abstract Helicopters are widely used in military and civil fields owing to its unique vertical take-off and landing, excellent hover performance and low speed characteristics. During the operative life of helicopter, it is possible that it could incur into emergency landing or crashes could generate serious injuries and even death of occupants may be due to fuel tank failures. Hence, the crashworthiness is crucial concern for the design of the helicopter fuel tank and to improve the survivability of helicopter with its surrounding structures under crash situations. FAR 29.952 provides the certification regulations to ensure safe and crashworthy fuel system. The drop test involves 50 ft drop of fuel tank with its surrounding fuselage structure filled with 80% water inside the fuel tank. The acceptance criteria are to ensure that there is no leakage of fuel from the fuel tank after ground impact. In case of leakage of fuel, redesigning and repetition of drop test will incur high cost and time delay for certification of fuel tank. In this paper, a simulation model using LS-DYNA has been developed and validated through drop test and describes the detailed FE idealization and contact definition to accurately predict drop test simulation. The smooth particle hydrodynamics (SPH) technique has been used to model fluid structure interaction (FSI) coupling. It is observed that good correlation between simulation and drop test results. Further, the validated FE model is utilized for design modifications without repeating experimental tests to achieve airworthiness certification.

Keywords Helicopter · Fuel tank · Crashworthiness · LSDyna · SPH

K. Gulbarga (✉) · G. Bankapur · A. Kumar · A. T. Rao
RWRDC, HAL, Bangalore 560017, India
e-mail: kgulbarga@gmail.com

A. Sakthivel
RCMA (Helicopters), Bangalore 560017, India

R. Velmurugan et al. (eds.), *Dynamic Behavior of Soft and Hard Materials Volume 1*, Springer Proceedings in Materials 34, https://doi.org/10.1007/978-981-99-6030-9_9

1 Introduction

Rotorcraft crash resistant fuel system is to enhance the survivability of the occupants during emergency landing after impacting to the ground. As a part of certification process, a comprehensive crash-resistant fuel system (CRFS) test has to be carried out in compliance with FAR 29.952, Advisory Circular AC 29-2C [1]. The conventional materials used modern helicopters are replaced by lightweight composite material like sandwich panels, monolithic frames and composites corrugated/sine wave panels to enhance the energy absorption capabilities. The impact behaviour of airframe structure is to ensure which may cause potential damage to fuel tank by piercing sharp edges into the fuel tank fitted inside the airframe structure or loss of structural integrity after impacting to ground [2, 3]. Hughes et al. [4] investigated the energy absorption of airframe structure after impacting to ground which could be achieved by providing proper trigger mechanisms like kinks in the stiffeners/stingers which allows controlled collapse of the structure instead of simple buckling mode of failure.

The crash-resistant fuel system is drop tested by specially designed cage made of steel structure. Necessary precautions are taken as the fuel tank with structure will not slide during the drop test till it impacts to ground. The drop test is expensive and time consuming as they involve fabrication of steel cage structure and joining mechanism to crane hook, fabrication/assembly of airframe structure and manufacture of flexible fuel tank. In case of leakage in the fuel tank after the drop test, design and development schedules will be affected and also adverse effects of time and costs due to redesigning and repeating of drop tests.

Hence, the fuel tank criticality upon drop test has to be identified before the actual test in order to avoid damages to fuel tank. The fuel tank drop analysis simulation involves a complex fluid and structure interactions (FSI) which is highly nonlinear in nature. LS-DYNA nonlinear finite element explicit code is the promising software with capability to model partially filled fuel tanks using smooth particle hydrodynamics (SPH) technology [5]. It has user-friendly pre- and post-processor which necessities to complete the analysis in reasonable time.

Anghileri et al. [6] investigated the fuel tank crashworthiness from experimental and numerical point of view. The simulations of sloshing behaviour of fuel inside the tank during crash events are assessed and compared with experimental and numerical results. Kim et al. [7] used commercially available nonlinear transient finite element (FE) analysis code LS-DYNA to simulate dynamic response of fuel tank installed inside the rotorcraft. Shahram Shokouhfar et al. [8] simulated the automotive tire rubber with multiple layered material moulded using LS-DYNA Part-Composite FE modelling [9, 10] approach to simulate dynamic behaviour of tire. In this present paper we investigated the failure mode of fuel tank after drop impact, and detailed finite element simulation using LS-DYNA was performed by defining the accurate fluid and structure interaction cards, contact definitions between the parts, velocity boundary conditions and correlating test and simulation results, improving the design based on simulation results to achieve airworthiness certification considering above literature inputs.

2 Fuel Tank Drop Test

The main aim of the drop test is to achieve the certification of the fuel tank system. Based on the drop test results, LS-DYNA FE model is developed to validate the test results and further to identify the design modifications.

2.1 Specimen Description

The test specimen consists of fuel tank bladder filled with 80% normal capacity of fuel tank volume and surrounding airframe structure. The crash-resistant fuel tank material is adhesively bonded with nylon, rubber and composite Kevlar layer. The nylon/rubber acts as fuel barrier with leak proof, and Kevlar layer improves crashworthiness. The surrounding airframe structure consists of major components like transverse composite sandwich frames, sandwich bottom shells and stiffened longitudinal keel beams with stiffeners\stringers. The fuel tank is connected to airframe structure at keel beam side cutout shown in Figs. 1 and 2. All the structural components are connected by rivets, and fuel tank side cutout to keel beam is connected using bolts and washers. The rigid ground surface is constructed and ensured there is no yielding of ground surface during drop test.

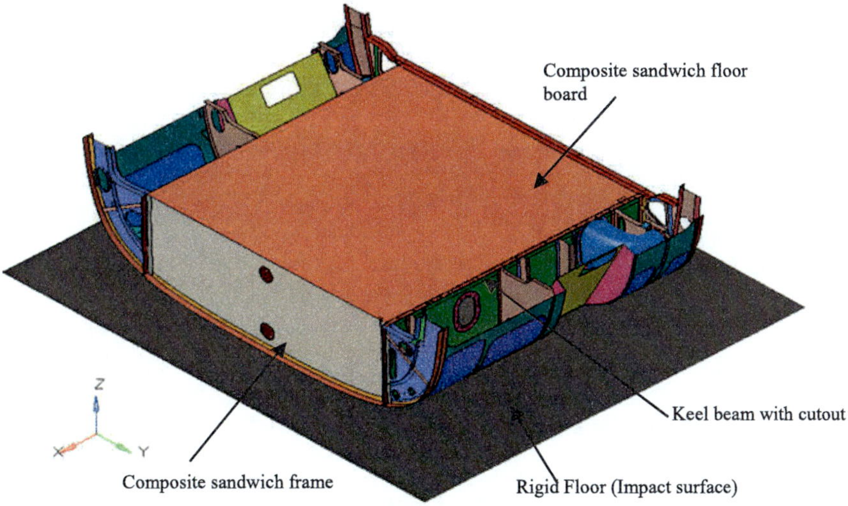

Fig. 1 Airframe structure enclosing fuel tank

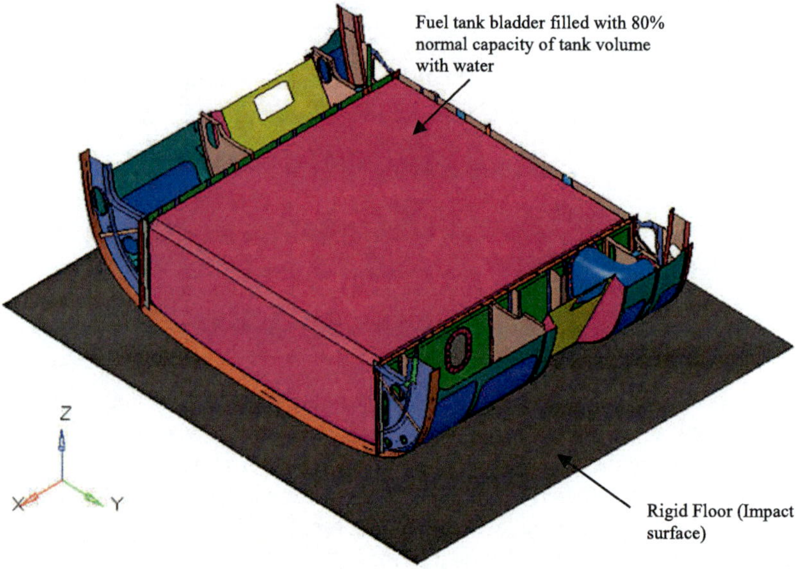

Fig. 2 Fuel tank with surrounding airframe

2.2 *Drop Test Requirements*

The drop test to be performed for each tank, or the most critical tank with surrounding structure as per civil requirements FAR/CS 29.952, and drop test parameters are as follows:

(1) The drop height at least 50 ft (15.25 m)
(2) The drop impact surface is non-deforming
(3) The tanks filled with water to 80% of the normal capacity
(4) The tanks enclosed in a surrounding structure representative of the installation
(5) The tank must drop freely and impact in a horizontal position ± 10°
(6) After the drop test, there must be no leakage.

2.3 *Test Setup Arrangement*

The tank test specimen is placed inside the carriage cage and was raised to height of 50 ft (15.25 m) using crane. Straps and foam blocks are tied to the test specimen to avoid sliding of the specimen from cage while raising to specified height and also to provide safety to personals on ground. An electrically and remotely operated quick release mechanism, capable of handling 10 ton, is used for dropping the object. Two high-speed wide-angle cameras was used for capturing high-speed events and to determine the angle of impact in both pitch and roll angles at impact instant, and further cameras are also used to record drop events, provide displacements, velocity and acceleration time history data during impact. The cage with quick release mechanism is shown in Fig. 3. The cage with fuel tank assembly when raised to height of 50 ft is held by 4 tight ropes such that complete assembly will not slide or tilt at any instant during ground impact. The wind velocity should be noted as it not affects the test setup to tilt during drop. Leak test of fuel tank is performed before the start of the test. The water is filled to 80% of capacity of fuel tank and mixed with potassium permanganate solution in order to identity the leak in case of fuel tank damage.

Fig. 3 Test setup showing carriage cage with quick release mechanism

Fig. 4 Damage at bottom cutout region

2.4 *Post-test Observations*

The data captured in high-speed camera was post processed to obtain displacement, velocity and acceleration time history plots; the velocity of impact was 17.20 m/s (theoretical estimate is 17.30 m/s). It is confirmed that angle of impact along pitch was 0.63° and roll was 1.82°, thus meeting airworthiness requirements. There was damage noticed in the fuel tank specimen near bottom cutout region and found leakage of fluid. The cutout connects the fuel tank bladder to airframe structure via keel beam using bolted joint. The fuel tank undergone large deformation on impacting to ground, since the gap between cutout edge and airframe bottom surface of tank is small, it caused the rigid cutout connection at bottom edge to pull the fuel tank and structure interface (keel beam) which led to tearing of bottom edge cutout along with fuel tank bladder. The actual cutout damage is shown in Fig. 4.

3 Validation of Test Results

To capture drop test results, a finite element model was developed to simulate actual test condition, and fuel tank drop analysis was performed using commercial explicit finite element LS-Dyna code. Once the numerical model is validated, it helps the designers to modify the structure and study its effects numerically.

3.1 *Finite Element Modelling of Fuel Tank Drop Test*

The airframe structure consisting of metallic and composite structure like keel beam, cutout, floor board, side shells, bottom shells, stiffeners/stringers has been modelled with 2D shell elements. The flexible fuel tank bladder is also modelled with 2D shell elements. The cutout connection at keel beam and fuel tank interface is modelled with 1D beam elements to idealize bolted connections. The keel beam to floor board

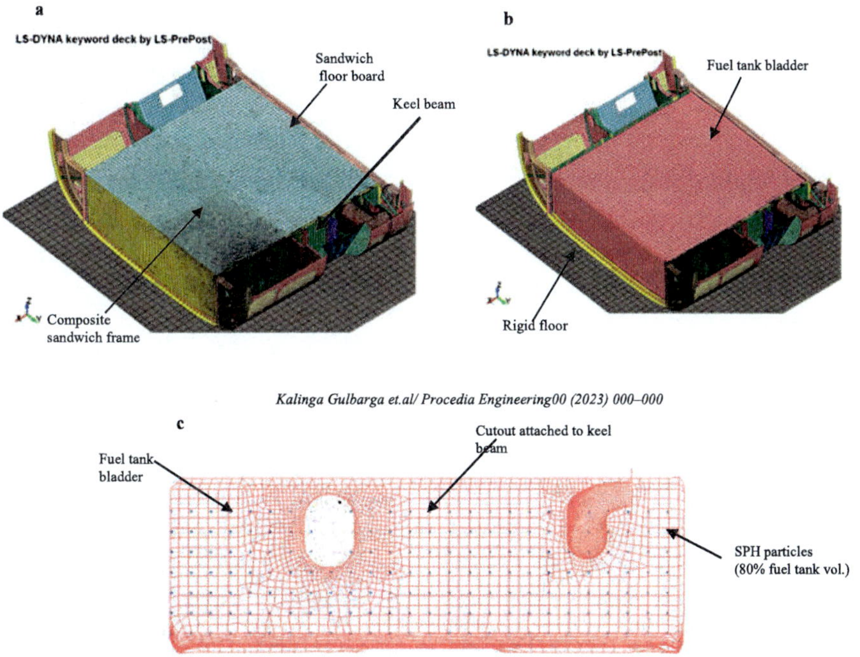

Fig. 5 **a** FE model of fuel tank with airframe structure; **b** FE model of fuel tank bladder with airframe structure; **c** FE model of fuel tank bladder with SPH particles

connection is idealized as 1D elements to represent riveted and screw/anchornut connections. The FE details with cutout connection are shown in Fig. 5.

The fluid(water) inside the fuel tank is modelled with SPH elements by assigning suitable density of water. SPH is mainly used for analysis of fluid structure interaction, it is mesh free method, and there are no nodal or elemental connectivity between them. There are collection of points to represent the fluid domain called as particles. Each SPH particle is formulated based on conservation of mass, moment and energy. The SPH uses smoothening length option to smoothen the domain using kernel function.

3.2 LS-DYNA Material Models for Fuel Tank Drop Assembly

The material models of composite sandwich panels like floor board, side shells, bottom shells *MAT_ENHANCED_COMPOSITE_DAMAGE and *MAT_COMPOSITE_DAMAGE were used. The suitable composite sandwich and its layup schemes were called *PART_COMPOSITE card. The material model for metallic parts including bolts, rivets were defined using *MAT_PIECEWISE_LINEAR_PLASTICITY. The material model for fluid(water) was modelled using *MAT_

NULL with *EOS_LINEAR_POLYNOMIAL is employed. The rigid floor is defined using *MAT_RIGID model. The material for fuel tank bladder consisting of nylon, rubber and crashworthy layer is modelled with *MAT_COMPOSITE_DAMAGE, and together it was called in *PART_COMPOSITE card. The finite-element model consisted of 65,144 nodes and 65,629 elements. The computational time was approximately 37 h on a DELL work station for 35 ms impact simulation.

3.3 Contact Definitions and Velocity Boundary Condition for Fuel Tank Drop Assembly

The fuel tank drop analysis is a highly nonlinear transient behaviour because of the interaction between fluid and structure associated with sloshing of water inside the tank. An suitable contact cards are defined between parts for predict accurate simulations. The contact between fluid and tank bladder is defined using*CONTACT_AUTOMATIC_NODES_TO_SURFACE. The contact between flue tank bladder and airframe structure is defined using *CONTACT_AUTOMATIC_SURFACE_TO_SURFACE. The contact between airframe bottom structure and hard rigid bottom is defined using CONTACT_AUTOMATIC_SURFACE_TO_SURFACE, and contact between SPH particles to hard rigid surface is defined using *CONTACT_AUTOMATIC_NODES_TO_SURFACE. The initial velocity to entire assembly is defined using *INITIAL_VELOCITY.

3.4 Finite Element Idealization of Side Cutout and Fuel Tank Bladder Interface

The finite element idealization of side cutout connection and fuel tank bladder is technically complex as it is the weak region consisting of fuel tank bladder which bonded to blanking plate. This weak region could fail due insufficient interface bond strength. The weak zone is modelled by using spot-weld rigid link feature in LS-DYNA with failure criteria. The failure force of weld element is defined equivalent to bond strength at that location. The spot-weld rigid link modelling idealization between the nodes is shown in Fig. 6.

4 Simulation Results

The LS-DYNA simulation results shown that crash behaviour of the drop test and analysis were similar. The large deformation of airframe structure was seen in the vicinity of the cutout region, and it is found bottom cutout tear out along with the fuel

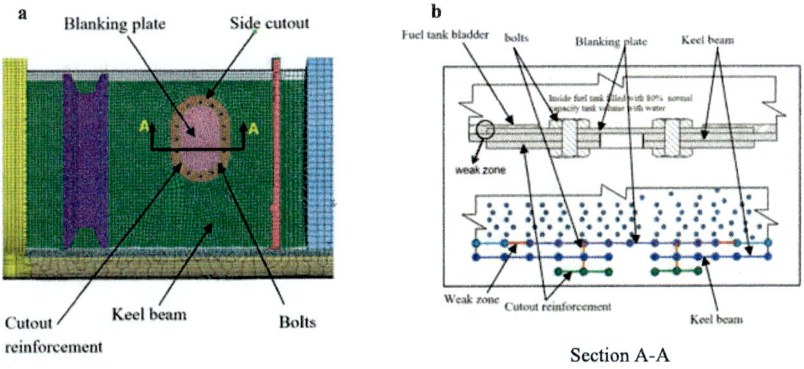

Fig. 6 **a** FE idealization of side cutout with keel beam; **b** idealisation keel beam with tank bladder using spot-weld elements

tank bladder connection. The fuel tank bladder connection to airframe structure is via keel beam, and rest of the bladder retains inside the tank through surface contact. After the impact to ground, large forces develop due to fuel sloshing inside the tank which lead to failure near the weak zone due to pulling of the tank bladder connection. Similar failure is observed in the LS-DYNA simulation results which are shown in Fig. 7. The surrounding structure parts like floor board, bottom shell, keel beam, transverse frames and side shells are deformed and found intact. The floor board connections to keel beam given way at few places and no major structural failure. There is no puncturing of fuel tank bladder from these local structural deformations.

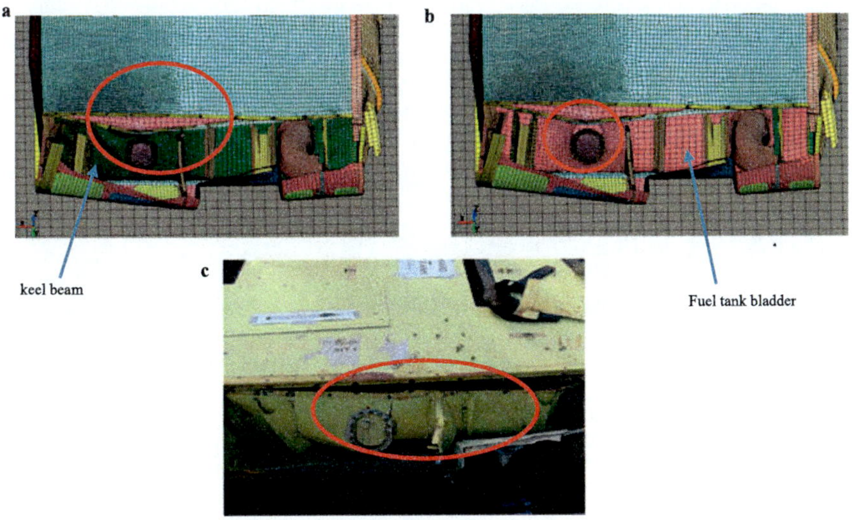

Fig. 7 **a** Analysis keel beam failure; **b** fuel tank bladder failure; **c** actual drop test

5 Design Modifications

Based on the correlation of test results and simulation results, the design modification was carried out. The side cutout was moved to noncritical zone, and tank bladder connection to keel beam was removed. The detail design modification of FE model is shown in Fig. 8. The other parts of the airframe structure were not disturbed, and further same LS-DYNA FE simulation procedure was followed as mentioned in previous section.

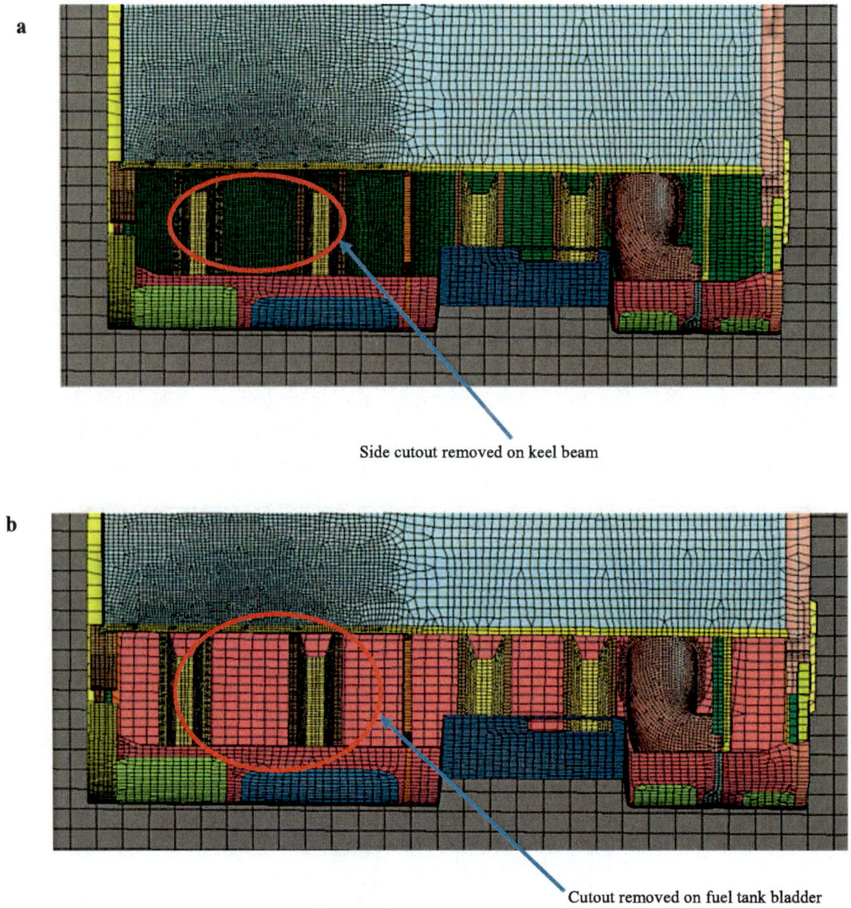

Fig. 8 **a** FE model of design modification by removing side cutout on keel beam; **b** FE model showing cutout removed on fuel tank bladder

5.1 Simulation Results for Modified Fuel Tank Assembly

The improved design modification of LS-DYNA simulation results shown that airframe structure remained intact and found no failure in fuel tank bladder, and hence, no leakage is observed. The fuel tank is intact, and it remained inside the airframe keel beam structure. There is tearing of keel beam structure due to deformation after impact refer Fig. 9. Based on the design modification, the dynamic behaviour of the fuel tank assembly has changed, and it has more uniform deformation compared to previous one.

6 Conclusions

Rotorcraft fuel tank crashworthiness is crucial concern for the design of the helicopter fuel tank and to improve the survivability of helicopter and ensure safety of occupants upon emergency landing of rotorcraft. The drop test has been carried out as per FAR 29.952 which provides certification regulations to ensure safety and crashworthy of fuel system. The fuel tank drop test is expensive and difficult to perform. In the present work, an alternative nonlinear transient FE LS-DYNA software was used to simulate fuel tank drop test as per regulation standards using powerful SPH technique available within the software.

The developed numerical model predicted fuel tank connection failure and deformation patterns of the surrounding structure. The test results correlated with simulation prediction and found validated.

Design modifications of drop test simulations were performed. The simulation results shown no leakage and tank puncture from the deformed airframe structure upon impact to ground. The LS-DYNA simulation results helped to achieve the regulation requirement of fuel tank crashworthiness. Further, it reduced design and development cycle time, cost of repeating actual drop tests.

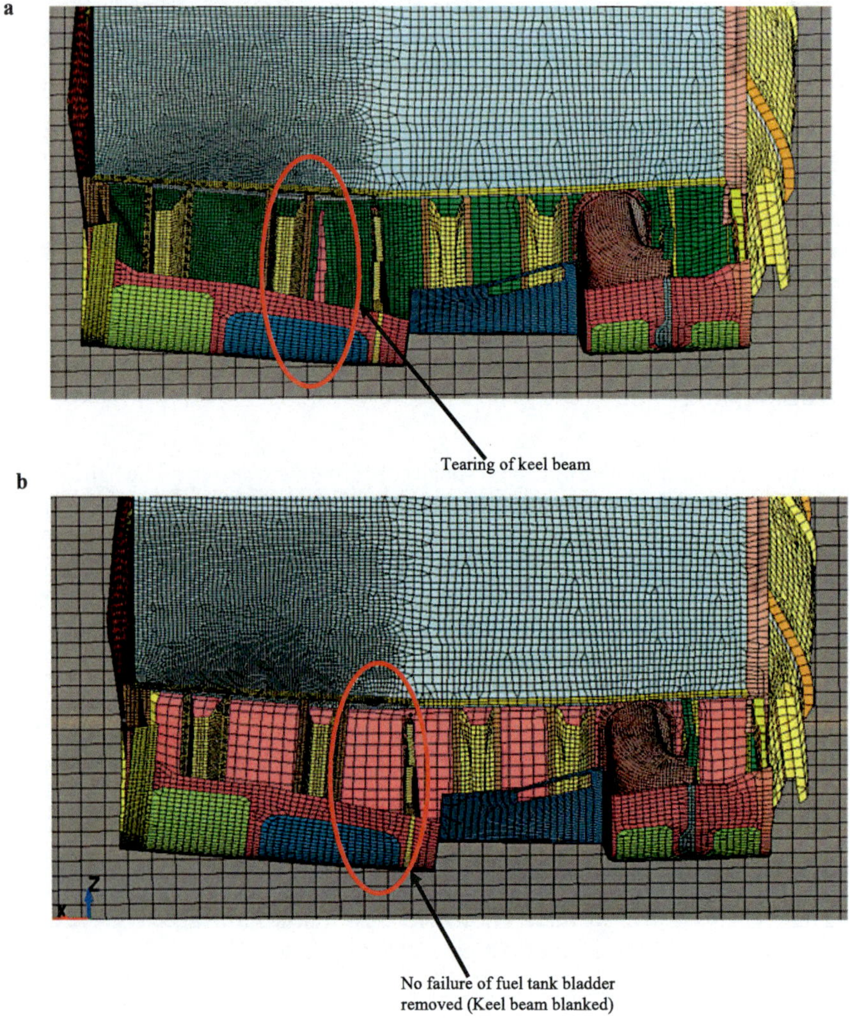

Fig. 9 **a** Deformation of modified fuel tank with airframe keel beam structure; **b** Deformation of fuel tank bladder

References

1. Federal Aviation Regulations, Part 29—Airworthiness Standards Transport Category Rotorcraft
2. Riccio A, Saputo S, Sellitto A, Russo A, Di Palma L. An insight on the crashworthiness behavior of full scale composite fuselage section at different impact angles. Aerospace 2019.
3. Di Palma L, Di Caprio F, Chiariello A, Ignarra M, Russo S, Ricco A, Caputo F (2020) Vertical drop test of composite fuselage section of a regional aircraft. AIAA J 58:474–487

4. Hughes K, Vignjevic R (2007) J Campbell Experimental observations of an 8 m/s drop test of a metallic helicopter underfloor structure onto a hard surface: part 1. J Aerospace engg 221:661–678
5. Cristillo D, Di Caprio F, Pezzella C, Paciello C, Magistro S, Di Palma L, Belardo M (2022) On numerical models for cube drop test of bladder fuel tank for aeronautical applications. J. compos. sci. 6:99
6. Anghileri M, Castelletti LML, Tirelli M (2005) Fluid–structure interaction of water filled tanks during the impact with the ground. Int J Impact Eng 31(3):235–254
7. Kim HG, Kim S (2014) Numerical simulation of crash impact test for fuel cell group of rotorcraft. Int J Crashworthiness 19:639–652
8. Shahram Shokaouhfar, Subash Rakheja and Moustafa.EI-Gindy , Verification of the Part Composite approach for modeling the multilayered structure of a rolling truck tire. In the 10th European LS-Dyna Conference (Germany), 2015.
9. LS-DYNA Keyword User's Manual (2012) Volume I, Version 971, Livermore Software Technology Corporation, California
10. LS-DYNA Keyword User's Manual (2012) Volume II, Version 971, Livermore Software Technology Corporation, California

Effect of Projectile Nose Angle on Resistance of Aluminium Target Against Oblique Impact

Aman Kumar and Vimal Kumar

Abstract In today's world, when it comes to the design of things like bulletproof jackets, armoured vehicles, bunkers, and the like, one of the primary concerns is safety. This is especially true in the aggressive technology sector, which includes the defence industry. In this chapter, a numerical investigation was carried out to study the ballistic performance of an aluminium plate (Al-7075 T6) when subjected to an oblique impact by a projectile with a conical shape. For modelling and simulating the current issue, the Abaqus finite-element code was utilized. In the current investigation, the nose angles of the cylindro-conical projectiles that were taken into consideration were 90° and 150°, respectively. For this particular study, it was assumed that the length, width, and thickness of the target plate, as well as the radius and mass of the projectile, would remain unchanged. The obliquity angles for impact were 15°, 30° and 45°. The residual velocity of the 150° nose angle projectile was higher than 90 nose angle projectiles at 15° obliquity. The residual velocity was decreased with increase in obliquity angle. At the ballistic limit (i.e. 750 m/s), the projectile's energy dissipation was at its highest, which was 93% for a nose angle of 90° and 86% for a nose angle of 150°. The loss of energy was observed to be increasing with increasing obliquity. In addition, a longer amount of time was necessary to achieve the minimum value of kinetic energy when the obliquity was set at 45° rather than 15°. The plate offered higher reaction forces for projectiles with nose angles of 150° compared to those with nose angles of 90°. Additionally, the reaction forces of the plate decreased with an increase in the angle of obliquity. The size of the hole caused by a projectile with a nose angle of 90° was greater than that caused by a projectile with a nose angle of 150°. Petalling and plugging were both contributing factors in the failure of the plate, which was caused by both types of projectiles. The shape of the hole changed from circular to ovular with increasing obliquity.

A. Kumar (✉) · V. Kumar
Department of Civil Engineering, National Institute of Technology Hamirpur, Hamirpur 177005, India
e-mail: araman0101@gmail.com

V. Kumar
e-mail: panchariya.vimal@gmail.com

R. Velmurugan et al. (eds.), *Dynamic Behavior of Soft and Hard Materials Volume 1*, Springer Proceedings in Materials 34, https://doi.org/10.1007/978-981-99-6030-9_10

Keywords High-velocity impact · Conical projectile · Ballistic limit · Energy dissipation · Aluminium plate

1 Introduction

Researchers conducted a comprehensive review of a variety of aluminium alloys, reporting their machinability characteristics, properties, and comparative analysis [1, 2]. Aluminium 7075 is an Al–Zn–Mg–Cu-based commercial alloy commonly used for structural components in the aerospace and automotive industries. This is due to the alloy's high strength-to-weight ratio, reasonable machinability, and excellent fatigue strength. These are susceptible to damage from other objects like high-velocity projectiles or missiles. Therefore, it is essential to determine the ballistic resistance of aluminium alloys to improve their structure. The ballistic performance and failure mode of thin ductile targets are highly dependent on several parameters, including impact velocity, target and projectile material, projectile shape, and impact angle or incidence angle.

The shape of the projectile has a significant impact on the target's failure pattern [3, 4]. The behaviour of a ductile target under oblique impact have studied in the past [5, 6]. The angle of obliquity (α), which is defined as the angle subtended by the velocity vector and the normal to the surface of the target (Fig. 1), is a crucial parameter that influences the ballistic resistance of the target [7]. It was discovered that the velocity drop increases with increasing impact obliquity and that the oblique incidence influences the shape of perforation. While most numerical and experimental studies focus on the determination of failure mode and ballistic limit and residual velocities, this study focuses on the effect of the nose angle of the conical projectile to predict the stressing of plate and dissipation of energy in various forms. AA-7075 T-6 was considered for the 6 mm thick target plate with uniform thickness. Impact velocities ranged from 800 m/s to 1500 m/s at incidence angles of 15°, 30° and 45°.

2 Constitutive Modelling

In this study, AA 7075-T6 was considered for the plate, while AISI 4340 steel was used for the projectile. The plate's modulus of elasticity and Poisson's ratio are 71.7 GPa and 0.33, respectively. Similarly, the elasticity modulus and Poisson's ratio of the projectile are 205 GPa and 0.33, respectively. The Johnson–Cook elasto-viscoplastic material model was used in numerical simulations to study the impact behaviour of both the target and the projectile [8]. This model incorporates the effects of yielding, plastic flow, isotropic strain hardening, strain rate hardening, and softening linear thermo-elasticity resulting from adiabatic heating and damage. The equivalent von-Mises stress ($\bar{\sigma}$) of the Johnson–Cook model is denoted by the

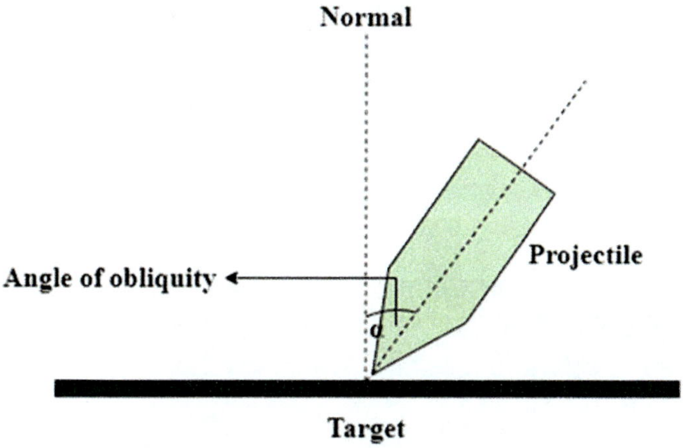

Fig. 1 The angle of obliquity

following Eq. (1):

$$\overline{\sigma} = \left[A + B\left(\overline{\varepsilon}^{pl}\right)^n\right]\left[1 + C\ln\frac{\dot{\overline{\varepsilon}}^{pl}}{\dot{\varepsilon}_0}\right]\left[1 - \hat{T}\right] \tag{1}$$

where $\overline{\varepsilon}^{pl}$ is the equivalent plastic strain; A, B, n, C, and m are the specific material parameters; $\frac{\dot{\overline{\varepsilon}}^{pl}}{\dot{\varepsilon}_0}$ is the dimensionless plastic strain rate, where $\dot{\overline{\varepsilon}}^{pl}$ is the equivalent plastic strain rate; and $\dot{\varepsilon}_0$ is reference strain rate. The homologous temperature $\hat{T}$ is the non-dimensional temperature defined in Eq. (2):

$$\hat{T} = \frac{(T - T_0)}{(T_{melt} - T_0)}, \quad T_0 \le T \le T_{melt} \tag{2}$$

where T is the absolute temperature, T_0 is the room temperature, and T_{melt} is the melting temperature. The fracture model proposed by [9] considers the effect of stress triaxiality, strain rate and temperature on the equivalent fracture strain, shown in Eq. (3).

$$\overline{\varepsilon}_f^{pl} = \left[D_1 + D_2\exp(D_3\frac{\sigma_m}{\overline{\sigma}})\right]\left[1 + D_4\ln\left(\frac{\dot{\overline{\varepsilon}}^{pl}}{\dot{\varepsilon}_0}\right)\right]\left[1 + D_5\hat{T}\right] \tag{3}$$

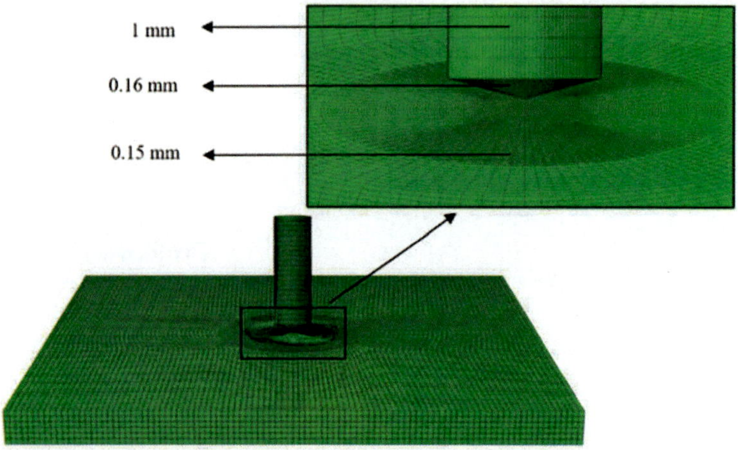

Fig. 2 Details of the projectile and target plate meshing

3 Finite-Element Modelling

Abaqus CAE was used to create the target and projectile finite-element models. The size of the target plate was 100 mm × 100 mm × 6 mm. The projectiles were cylindro-conical in shape with two different cone angles of 90° and 150°. For all simulations, the projectile's mass (7.85 gm) and diameter (7.62 mm) were kept constant. A three-dimensional deformable body was used to model the target and projectile. Figure 2 depicts a typical projectile and target finite-element model. The penalty method was used to model the contact between the projectile and the plate, with a constant coefficient of friction of 0.2. To optimize the mesh, the geometry of the plate was divided into two parts: a circular central part of 30 mm diameter (more than double the diameter of the projectile, allowing to begin the process of crack propagation accurately without significant effects on the energy balance) and an exterior part that complements the structure. C3D8R elements were used to mesh the central portion of the plate (Fig. 2).

4 Results and Discussions

4.1 Ballistic Limit

The residual versus impact velocity plots are shown in Fig. 3a and b. Figure 3a depicts the effect of projectile nose angle on the residual velocity at 15° obliquity. The ballistic limit for both projectiles was determined to be 750 m/s. At impact velocities above the ballistic limit, it was observed that the residual velocity was

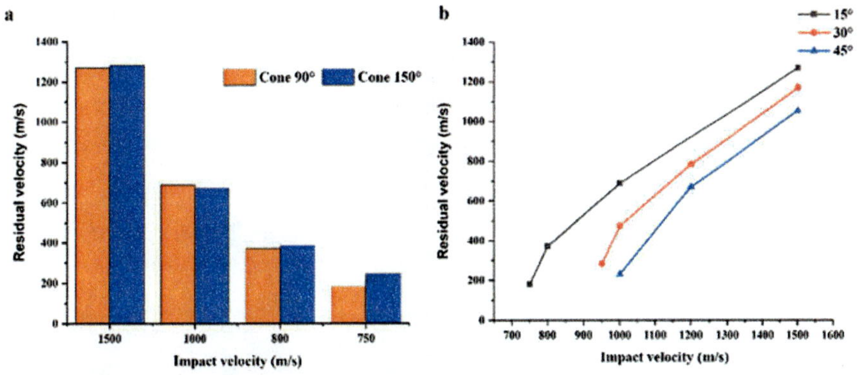

Fig. 3 **a** Effect of nose angle and **b** effect of obliquity on projectile residual velocity

almost identical. Figure 3b depicts the effect of the angle of obliquity for the 90° nose angle projectile. It was reported that the ballistic limit at, 30, and 45° was observed to be 750 m/s, 950 m/s, and 1000 m/s, respectively. In general, the residual velocity decreased as the angle of obliquity increased.

4.2 *Energy Dissipation*

The kinetic energy of the projectile is converted into many different forms of energy during the process of dissipation. These forms of energy include plastic deformation energy, elastic deformation energy, and frictional dissipation, among others. However, it was discovered that the majority of the kinetic energy of the projectile was lost in the form of plastic deformation of the target plate. In Fig. 4, we see a comparison of the percentage of energy lost by projectiles when they hit a plate at three different impact velocities at a typical obliquity angle of 15°. This occurs while the projectiles are penetrating the plate. At the ballistic limit (i.e., 750 m/s), it was observed that the projectile's energy dissipation was at its highest, which was 93% for a nose angle of 90° and 86% for a nose angle of 150°.

Figure 5. depicts the observed relationship between the amount of energy lost and the passage of time for a variety of obliquity angles. It was found that there was a correlation between the angle of obliquity and the amount of energy that was lost. The loss of energy was observed to be increasing with increasing obliquity. In addition, it was found that a longer amount of time was necessary to achieve the minimum value of kinetic energy when the obliquity was set at 45° rather than 15°.

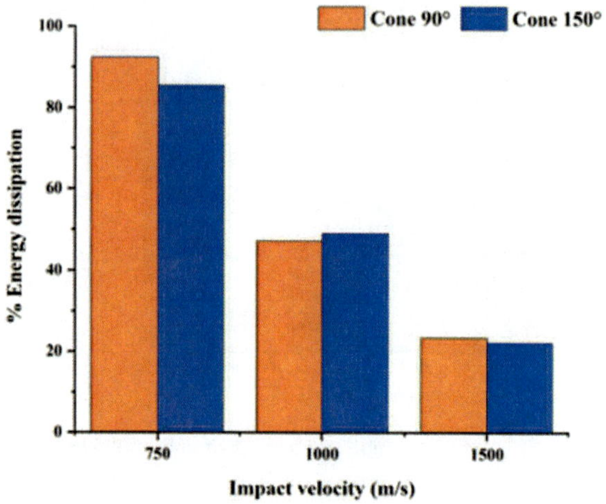

Fig. 4 Comparison of percent energy dissipation for conical projectiles with nose angles of 90° and 150°

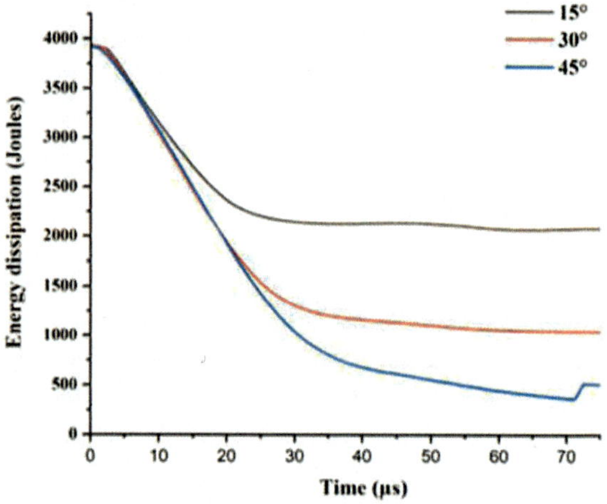

Fig. 5 Energy dissipation for a conical projectile with a 90° nose angle at various obliquity angles

4.3 Reaction Forces

Figure 6a and b shows the reaction forces that are given by the target plate. The reaction forces indicate the stresses that are being produced in the aluminium plate

It was reported that the plate offered higher reaction forces for projectiles with nose angles of 150° compared to those with nose angles of 90°. Additionally, it was observed that the reaction forces of the plate decreased with an increase in the angle of obliquity.

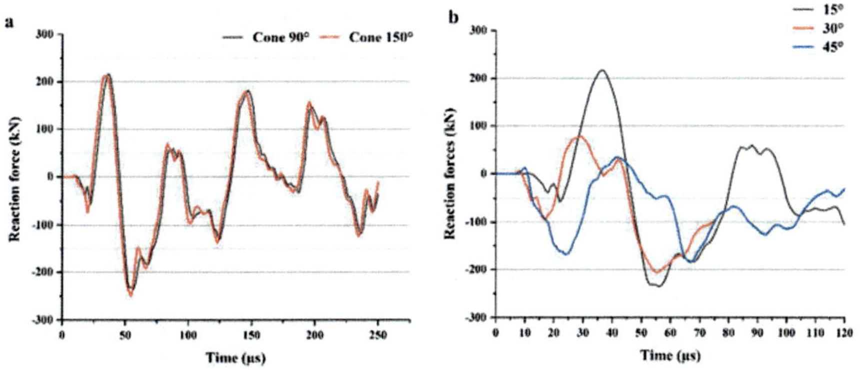

Fig. 6 The reaction force of the plate at **a** varying nose angles and **b** varying angles of obliquity for 90° nose angle projectile

4.4 *Damage Pattern*

Figure 7 shows, the damage profile for the plate that was caused by two separate projectiles with varied nose angles. It was evaluated at three distinct impact velocities, 750 m/s (the ballistic limit), 1000 m/s, and 1500 m/s. It was found that the angle of the projectile's nose had a significant influence on the damage profile of the plate. It was discovered that the size of the hole caused by a projectile with a nose angle of 90° was greater than that caused by a projectile with a nose angle of 150°. Petalling and plugging were both contributing factors in the failure of the plate, which was caused by both types of projectiles. Figure 8 demonstrates the damage profile given to the plate by a projectile with a 90° conical nose at different obliquity angles. It was observed that the shape of the hole changed from circular to ovular with increasing obliquity. It was reported that the angle of impact at the exit of the projectile changed more with decreasing velocity. Also, the tendency for the plate to fail due to petalling increased with increasing obliquity. The reason for this may be the increased frictional or contact forces.

5 Conclusions

In this work, numerical simulations were carried out to determine the numerous parameters like ballistic limit, residual velocity, and damage profile of the aluminium alloy plate. The plate was struck by a conical projectile of two distinct nose angles, that is, 90° and 150°. The influence of the obliquity angle was also studied in this work. The impact velocity of the projectile varied from 700 to 1500 m/s. The ballistic limit of both projectiles were found 750 m/s at 15° obliquity. The residual velocity of 150 nose angle projectile was higher than 90° nose angle projectile. The residual velocity was decreased with an increase in obliquity angle. At the ballistic limit

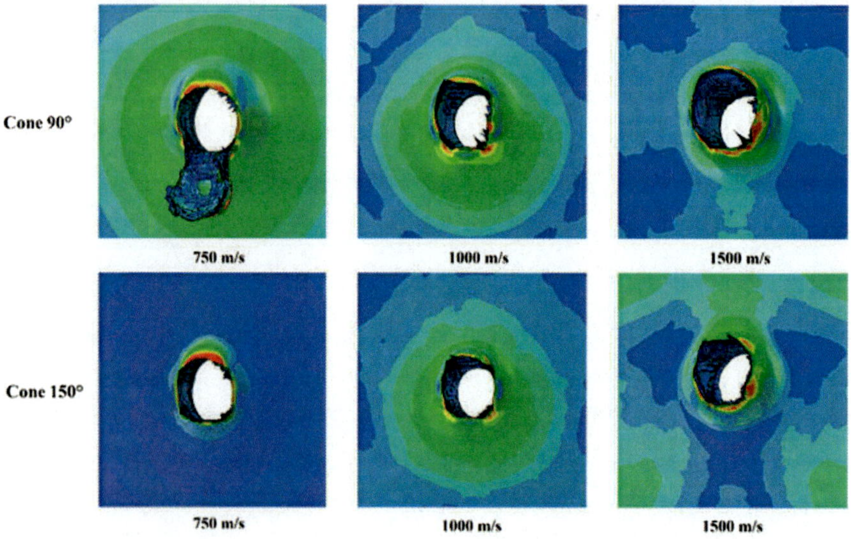

Fig. 7 Damage profile for the plate inflicted by 90° and 150° nose projectiles

(i.e. 750 m/s), the projectile's energy dissipation was at its highest, which was 93% for a nose angle of 90° and 86% for a nose angle of 150°. The loss of energy was observed to be increasing with increasing obliquity. In addition, a longer amount of time was necessary to achieve the minimum value of kinetic energy when the obliquity was set at 45° rather than 15°. The plate offered higher reaction forces for projectiles with nose angles of 150° compared to those with nose angles of 90°. Additionally, the reaction forces of the plate decreased with an increase in the angle of obliquity. The size of the hole caused by a projectile with a nose angle of 90° was greater than that caused by a projectile with a nose angle of 150°. Petalling and plugging were both contributing factors in the failure of the plate, which was caused by both types of projectiles. The shape of the hole changed from circular to ovular with increasing obliquity. The angle of impact at the exit of the projectile changed more with decreasing velocity. Also, the tendency for the plate to fail due to petalling increased with increasing obliquity. The reason for this may be the increased frictional or contact forces.

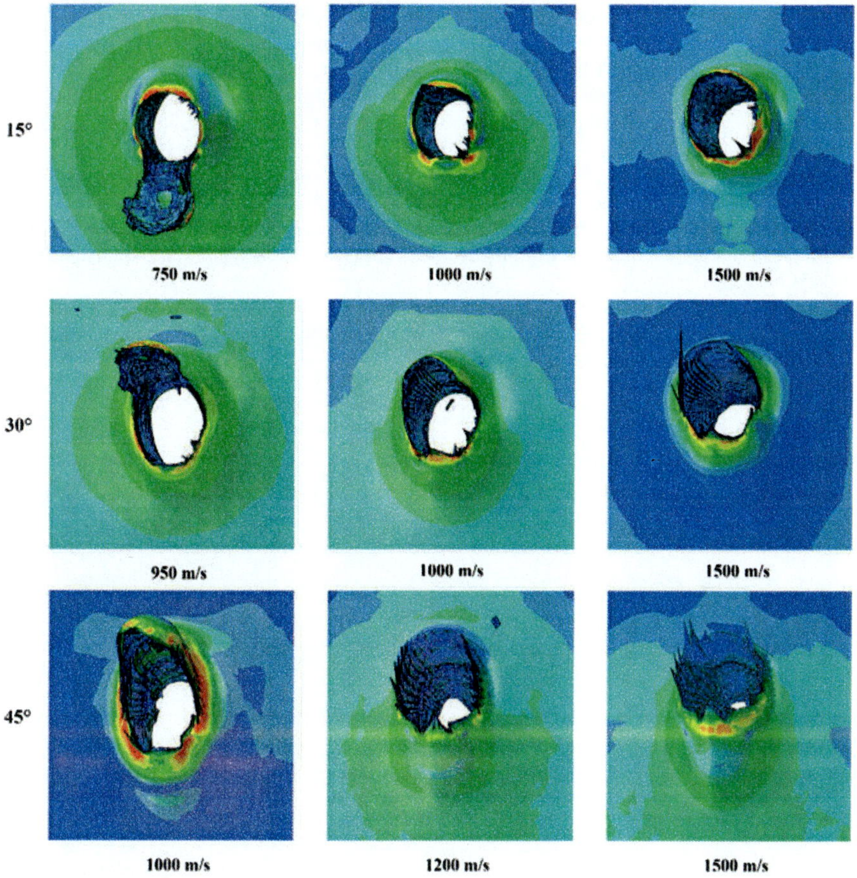

Fig. 8 Damage profile for the plate caused by a projectile with a 90° conical nose at various obliquity angles

References

1. Santos MC, Machado AR, Sales WF, Barrozo MAS, Ezugwu EO (2016) Machining of aluminum alloys: a review. The Int J Adv Manuf Technol 86(9):3067–3080. https://doi.org/10.1007/s00170-016-8431-9
2. Mali RA, Agrahari MD, Gupta TVK (2020) FE based simulation and experimental validation of forces in dry turning of aluminium 7075. Mater Today: Proc 27:2319–2323. https://doi.org/10.1016/J.MATPR.2019.09.120
3. Dey S, Børvik T, Hopperstad OS, Leinum JR, Langseth M (2004) The effect of target strength on the perforation of steel plates using three different projectile nose shapes. Int J Impact Eng 30(8–9):1005–1038. https://doi.org/10.1016/J.IJIMPENG.2004.06.004
4. Deng Y, Wu H, Zhang Y, Huang X, Xiao X, Lv Y (2022) Experimental and numerical study on the ballistic resistance of 6061–T651 aluminum alloy thin plates struck by different nose shapes of projectiles. Int J Impact Eng 160:104083. https://doi.org/10.1016/J.IJIMPENG.2021.104083

5. Gupta PK, Iqbal MA, Mohammad Z, Baqi A, Gupta NK (2018) Energy absorption in thin metallic targets subjected to oblique projectile impact: a numerical study. Thin-Walled Struct 126:58–67. https://doi.org/10.1016/J.TWS.2017.08.005
6. Iqbal MA, Gupta G, Gupta NK (2010) 3D numerical simulations of ductile targets subjected to oblique impact by sharp nosed projectiles. Int J Solids Struct 47(2):224–237. https://doi.org/10.1016/j.ijsolstr.2009.09.032
7. Backman ME, Goldsmith W (1978) The mechanics of penetration of projectiles into targets. Int J Eng Sci 16(1):1–99. https://doi.org/10.1016/0020-7225(78)90002-2
8. Johnson GR (1983) A constitutive model and data for materials subjected to large strains, high strain rates, and high temperatures. In: Proceedings 7th information symposium, Ballistics, pp 541–547
9. Johnson GR, Cook WH (1985) Fracture characteristics of three metals subjected to various strains, strain rates, temperatures and pressures. Eng Fract Mech 21(1):31–48. https://doi.org/10.1016/0013-7944(85)90052-9

Numerical Analysis of Windowed Origami-Ending Crash Tubes

Haris Farooq and Manoj Kumar

Abstract Crash tubes are structural components that reduce damage to passengers and cargo in a low-speed collision. They are placed at the front of the vehicle and designed to dissipate the kinetic energy and reduce impact force during a crash. This paper presents a family of novel Origami-Ending crash Tubes (OETs) with circular, square and rectangular windows. Validation, modelling and numerical simulations are performed using ABAQUS EXPLICIT. Dynamically loading the crash tubes showed that most of the tubes collapsed under mixed-mode consisting of both symmetric and extensional modes. Theoretically, employing origami at the ends reduces the initial impact forces and having windows increases the specific energy absorbed. We demonstrated that for OETs, a considerable reduction in initial peak force is observed. Also, an increase in Specific Energy Absorbed (SEA) is noticed compared to the conventional crash box. Other comparative studies were performed based on crashworthiness parameters like mean force, specific energy absorbed and initial peak force. Substantial gains over the crashworthiness of the structure make OETs worthy for engineering applications.

Keywords Windowed origami-ending tube · Axial crushing · Numerical simulation · Crashworthiness · Deformation mode

1 Introduction

Vehicular design involves various challenges like fuel economy, comfort, looks, ease of control and engine performance. One of the important challenges in vehicle design is passenger and cargo safety during an impact. Various systems are installed in a vehicle for safe riding. Structural components like crash tubes are designed and integrated with the vehicle frame in a specific way to protect cargo and avoid severe injuries to passengers during the impact. Crash tubes are thin shells specifically

H. Farooq · M. Kumar (✉)
Department of Mechanical Engineering, Dr. B. R. Ambedkar National Institute of Technology, Jalandharo, India
e-mail: kumarm@nitj.ac.in

R. Velmurugan et al. (eds.), *Dynamic Behavior of Soft and Hard Materials Volume 1*, Springer Proceedings in Materials 34, https://doi.org/10.1007/978-981-99-6030-9_11

designed to achieve maximum energy absorption and minimize impact forces. They experience large forces and are generally involved in the propagation phase of a collision. In this phase, the chances of damage to the occupants are very high as they experience significant deceleration. Hence, the crash tube design is crucial for a vehicle's better safety rating.

Crushing of these thin-walled structures has been a vast topic of research for many decades now. Wierzbicki and Abramowicz [1] gave a theoretical basis to determine the mean force in the case of simple cross-sections. Using super folding theory and the concept of effective crushing distance, [2–4] proposed various modes of deformation in the case of square and circular sections and gave formulae to calculate the mean force in each case. Authors [5–8] proposed various modifications in the theoretical model so that the experimental results match well with the proposed model.

Simple square and circular crash tubes transmit high impact forces or in general, have poor crashworthiness characteristics. For these reasons, various changes have been made in the design of these sections so that the overall crashworthiness can be improved. Authors [9, 10] utilized the concept of corrugation and grooves to improve the crashworthiness of these tubes.

Windowed sections have better crashworthiness when the mass of the tube is a critical factor in the design process. These windows induce stress concentrations in the tube so that a desirable mode of collapse can be triggered. Authors[11–13] studied the effect of introducing cut-outs in different tubes under various loading conditions and observed significant improvements in specific energy absorption. Though the specific energy absorption of the tube improves, the overall energy absorption degrades since the material is removed. To overcome this shortage, origami structures have been utilized.

Origami is defined as the art of paper folding. Since no cutting is done to get to different shapes, this property has made origami very popular in modern engineering. Origami engineering provides one of the efficient means to alter the stiffness of the tubes. Since stiffness is a function of variables like cross-sectional area, length, and Young's modulus, we can easily manipulate these variables and design structures of required stiffness by providing slight deformations or bends at tube edges. A large number of studies have been performed on the usage of origami engineering to improve the crashworthiness of the crash tube. Authors [14–17] carried out detailed research on origami-based tubes. They introduced simple origami patterns in conventional shells to reduce initial impact deceleration and increase.

mean force. Wang et al. [18] also proposed an Origami-Ending Tube (OET) resembling a conventional column with tapered ends. These OETs performed better than origami crash tubes absorbing 99% more energy and reducing the initial peak by 40% in the optimal case. Wang et al. [19] took their research further by applying the concept of OETs to long columns. They found that their collapse modes were highly dependent on imperfections, and tubes deformed in symmetric corner and mixed-mode.

2 Crashworthiness Parameters

Crashworthiness is defined as the ability of the automobile to dissipate or absorb kinetic energy during a collision. The basic idea behind crashworthiness design is to give optimum stiffness to the structure so that the damage is minimized. There are numerous ways in which we can judge the performance of the crash tube hence its crashworthiness. Following are some of the parameters which are more meaningful to this study.

2.1 *Initial Peak Force* ($\mathbf{F_{max}}$)

It is the maximum force experienced in the initial phase of the collision. It arises because of the high initial stiffness of the tube.

2.2 *Energy Absorbed* ($\mathbf{E_a}$)

It is defined as the total energy absorbed in all phases of a collision It is effective absorption of impact energy and is the most crucial characteristic in a crash impact event. Mathematically, it is defined as follows

$$E_a = \int_0^{\delta} F ds \tag{1}$$

δ is total crush-displacement, F is crush-force at any instant. The area under the curve represents the total energy absorbed during a collision.

2.3 *Mean Force* ($\mathbf{F_m}$)

It is the mean force experienced by a vehicle during the span of impact. Mathematically, it is defined as the ratio of the total energy absorbed to the energy total crush-displacement.

$$F_m = \frac{E_a}{\delta} \tag{2}$$

2.4 Specific Energy Absorption (SEA)

This term is essential when the component's weight is considered during the design process. SEA is defined as the ratio of the total energy absorbed to the mass of the crushed portion of the crash tube. Mathematically, it is defined as:

$$SEA = \frac{E_a}{\rho A \delta} \tag{3}$$

where ρ is the density of the tube, and A is the cross-sectional area.

3 Numerical Analysis

Numerical analysis, modelling and simulations of dynamic loading of the conventional and windowed origami-ending tubes to determine their failure modes, deformed shapes and different crashworthiness parameters was performed using Abaqus Explicit. Numerical studies of conventional tubes were performed to validate the model and comparative analysis. Present studies deal with numerical analysis of windowed Origami-Ending Tubes (OET) consisting of simple OET, circular OET, square OET and rectangular OET. The geometries of all the conventional and origami tubes are described in the next section.

All the tubes considered have a square cross-section. The tube's width, length and thickness are 35 mm and 90 mm and 2 mm, respectively. Simple OET was made by introducing a Triangular Origami Pattern (TOP) on one end of the conventional tube. Windows are introduced in simple OET to give a windowed origami-ending tube. One pair of opposite faces in each tube, except the simple tube, contains windows. The geometries of the conventional and proposed tubes are shown in Figs. 1 and 2, respectively. All the tubes are made of aluminium alloy AA6060-T4. Fig. shows the stress–strain curve of the alloy. Other properties of the alloy are shown in the Table 1. The strain-rate effect has been ignored as aluminium is not strain-rate sensitive at normal temperatures. The crushing of tubes was simulated by placing the tube between rigid plates. The top plate was allowed to move in the axial direction while the bottom plate was fixed. The tubes analysed have two axes of symmetry. Hence, symmetry of the model was utilized, and quarter modelling of all the tubes was performed. The endplates are assumed rigid, and to ensure full crushing of the tube, the analysis was carried out by giving a velocity of 15 m/s and a mass of 500 kg to the top plate.

All the tubes were crushed up to 70 mm in length. Since the tube is made of shell features, it was meshed using S4R shell elements. The global size of the shell element was chosen as 1 mm for validation and 0.8 mm for all other cases. To simulate the contact behaviour between tube and plates and for tube self-contact, general contact was assumed. The interaction property chosen was frictional contact, which takes

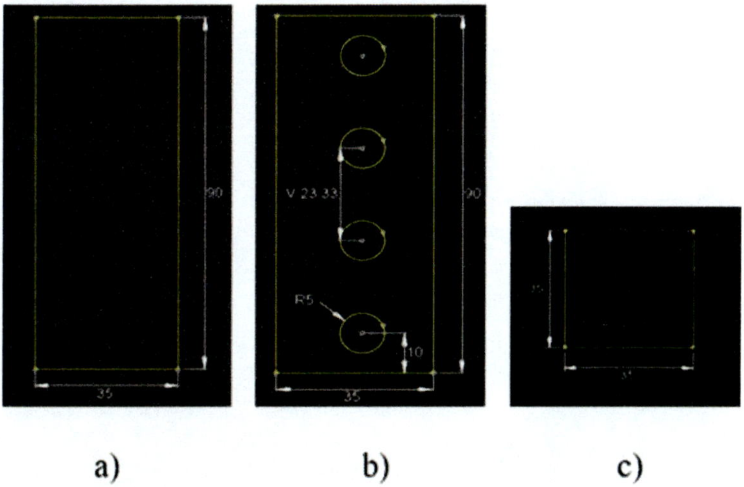

Fig. 1 **a** Simple tube **b** Simple circular tube **c** Top view

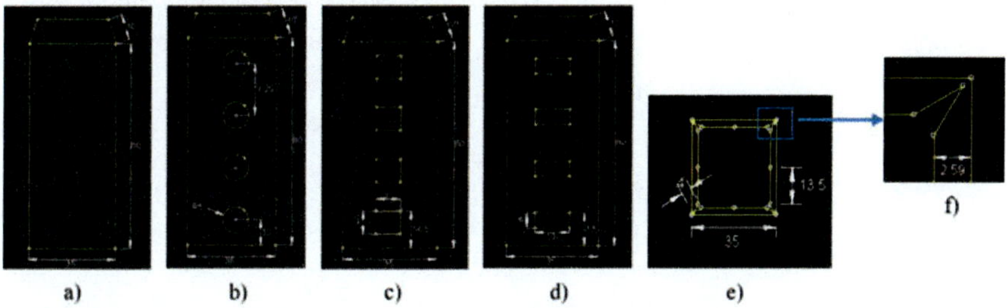

Fig. 2 **a** Simple OET **b** Circular OET **c** Square OET **d** Rectangle OET **e** Top view **f** Detailed view of corner

Table 1 AA6060-T4 material properties

Property	Value
Density(kg/m^3)	2700
Young's modulus (MPa)	68
Poisson's Ratio	0.33

care of the relative tangential motion or finite sliding. The frictional coefficient was taken as 0.15.

When a crash tube is subjected to loading, be it dynamic or quasi-static, the tube starts to crush in the form of folds. The energy dissipated by the tube depends upon the shape of each fold or lobe. Based on the shape of the fold, we have the following modes of deformation as shown in Fig. 3.

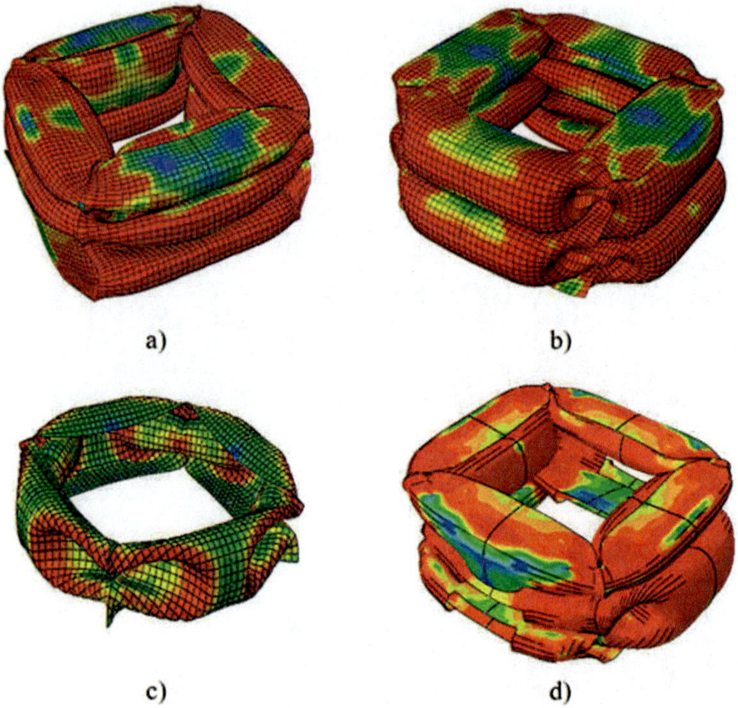

Fig. 3 Modes of deformation a Extensional mode b Symmetric mode c Diamond mode ud Mixed mode

To account for the unavoidable manufacturing fault, the tube was given a tiny flaw by projecting the first four buckling modes onto the tube side. The magnitude of each buckling mode was constant at 0.05 mm.

4 Validation

Numerical simulation of the simple conventional tube was performed to validate the FE model by comparing the findings with those proposed in the paper by Nikkhah et al. [12]. A simple conventional tube with the exact dimensions as described earlier was crushed. Figure 4 shows the force–displacement plots and bar graphs of the experimental and numerical findings and Fig. 5 shows that the loading arrangement. Initial peak force and SEA were the two crashworthiness parameters used for comparison. The error in the numerical value of initial peak force and SEA is around 3.5% and 5%, respectively, compared to experimental data as shown in Fig. 6. Deformed shapes of simple conventional tubes obtained from both simulation and experimental

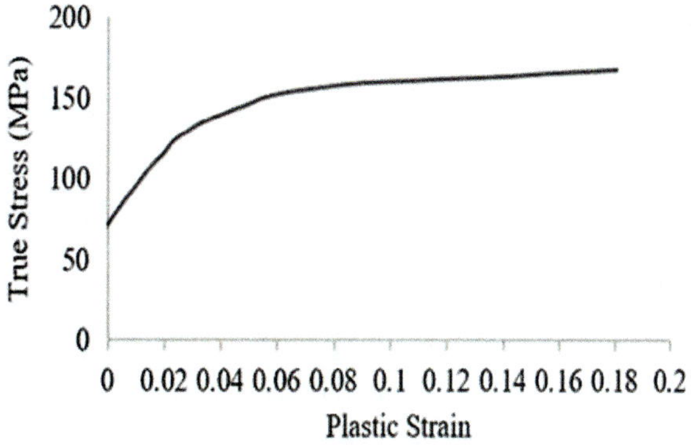

Fig. 4 True stress-plastic strain curve of AA6060-T4

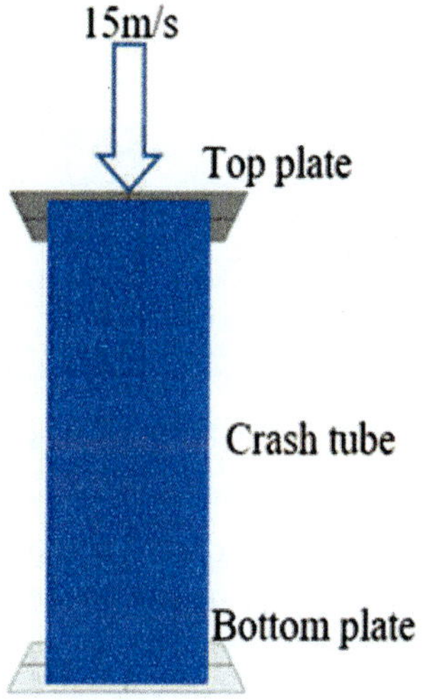

Fig. 5 Loading arrangement

results were compared, as shown in Fig. 7, implying that the numerical and experimental results are in good agreement. So, the finite-element modelling approach is acceptable for analysing all the other cases.

5 Results and Discussion

5.1 Crushing of Conventional Tubes

The simple tube is analysed first as it is the reference to evaluate the crashworthiness of all the proposed tubes. Figure 8a represents the progressive crushing of the simple tube. The tube collapses in symmetric mode with opposite pairs of sides collapsing

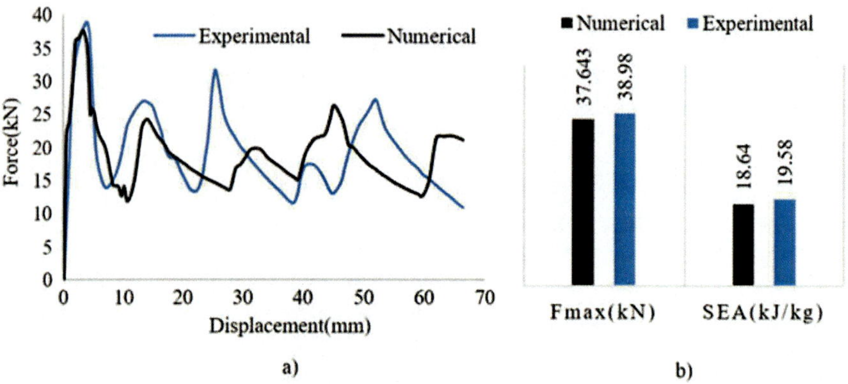

Fig. 6 **a** Force–displacement plots of simple tube **b** Numerical and experimental values of Fmax and SEA for simple tube

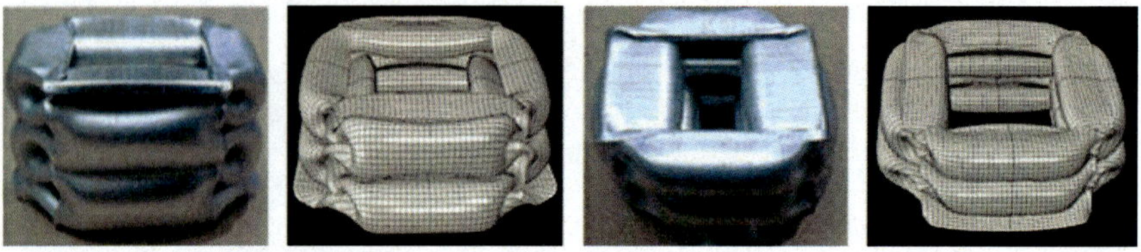

Fig. 7 Deformed shapes of simple tube obtained by experiments [13] and simulations

either inward or outwards, resulting in the formation of three lobes in the longitudinal direction. The crushing of the simple circular-windowed tube (Fig. 8b) is similar to that of the simple tube as both deform in symmetric mode with the formation of three lobes. This behaviour can also be judged directly from the force–displacement plot, as shown in Fig. 9a. Figure 10 shows that due to the presence of windows in the simple tube, the initial peak force is reduced from 35.257 to 31.391kN. The mean force of simple circular tube is marginally less than the simple tube, decreasing from 17.58 to 17.035kN. But, due to the lower weight of the windowed tube, its SEA is slightly higher than the simple tube, increasing from 18.086kJ/kg to 18.445kJ/kg.

5.2 *Crushing of OETs*

Figure 8c shows the progressive crushing process of simple OET. Due to tapered ends, the formation of lobes starts from the top in contrast to that of the simple tube. The tube collapses under local buckling mode or extensional mode, with all the sides collapsing in the same direction, either inwards or outwards. Due to an origami-ending pattern in this tube, the initial impact forces are reduced by almost 37% compared to a simple tube, as illustrated in Figs. 9b and 10. Since simple OET

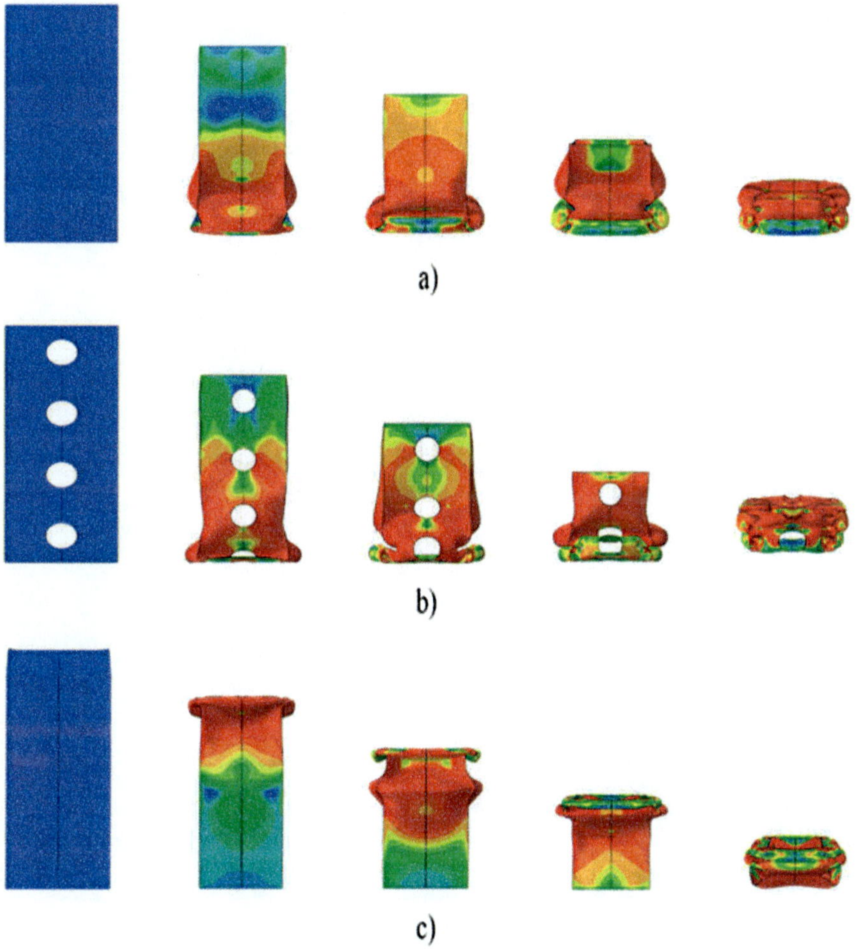

Fig. 8 The deformation processes of conventional and proposed tubes; **a** Simple tube **b** Circular simple tube **c** Simple OET **d** Circular OET **e** Square OET **f** Rectangle OET

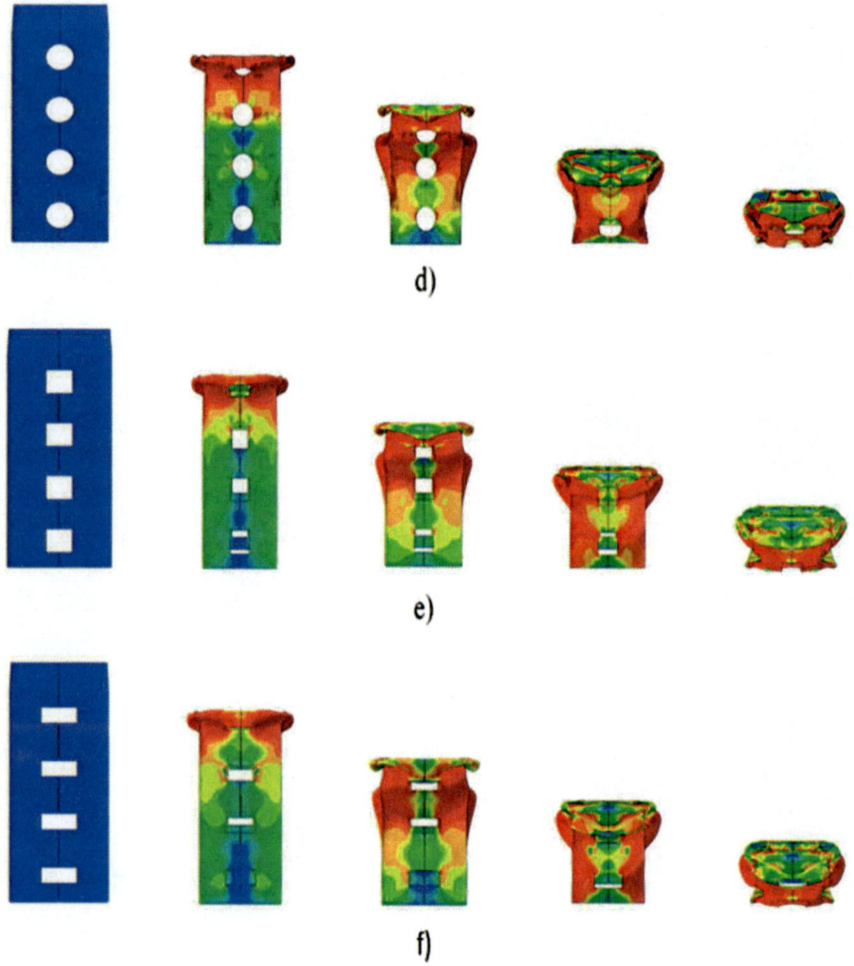

Fig. 8 (continued)

deforms in extensional mode rather than symmetric mode, its mean force is higher than the simple tube.

Also, as the weight of both tubes is the same and the mean force of simple OET is higher, the SEA of simple OET is more significant than the simple tube. The circular OET deforms in mixed mode, as shown in Fig. 8d. Mixed-mode consists of symmetric and extensional modes of deformation. Initially, circular OET collapses under extensional mode. Later the mode changes to symmetric, with one pair of opposite sides moving inwards and the other pair moving outwards. The initial peak force in circular OET is 16% higher than in simple OET but 10% lower than in the simple tube, as shown in Fig. 10.

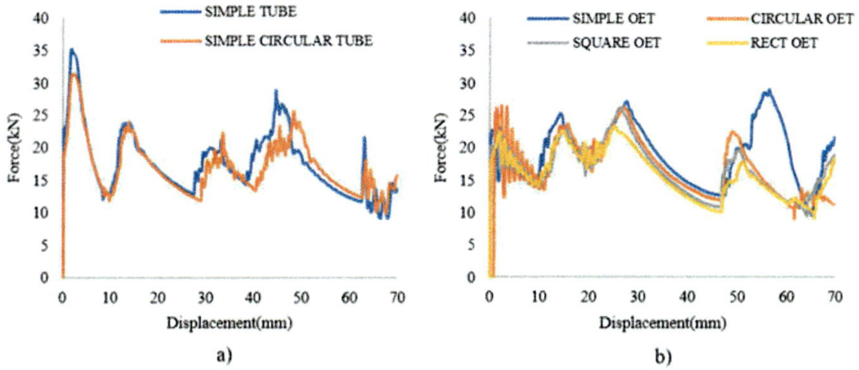

Fig. 9 Force–displacement plots for **a** Conventional tubes **b** Windowed OETs

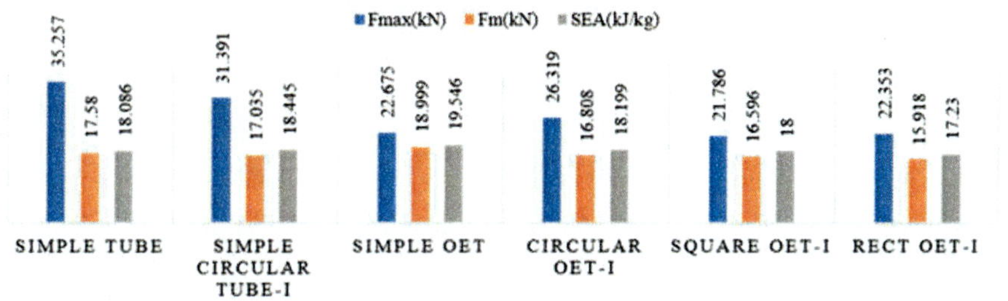

Fig. 10 Comparison of each tube based on initial peak force (Fmax), mean force (Fm) and specific energy absorption (SEA)

In fact, among all the OETs initial peak is found maximum in circular OETs as shown in Fig. 9b. Similarly, the mean force and SEA in simple OET is higher than the simple tube. The deformation pattern or mode observed in both the square and rectangular OETs are also mixed mode, initially deforming in extensional mode and later changing to symmetric mode. Compared to simple OET, the mass and SEA of square OET are around 5 and 8% less, respectively. The peak force observed in the square OET is 21.786 kN, the lowest among all the tubes, and for rectangular OET, it is 22.353 okN, a little higher than square OET. The transition from the square to the rectangular shape of the window also decreases mean force, hence energy absorption capacity. It can be seen from Fig. 10 that rectangular OET has the lowest SEA among all tubes, and for square OET it is a little higher.

6 Conclusions

The crashworthiness of windowed OETs was explored in this research. In the axial direction, the tubes were subjected to impact loading conditions. The FE code ABAQUS EXPLICIT was used to accomplish the numerical modelling. A crash study of a simple conventional tube was used to validate the model, and the results were in good agreement with experimental observations. The model was then used to undertake analysis on a variety of windowed OETs. When all of the data were taken into account, it was discovered that simple OET had the best crashworthiness characteristics of all the tubes, despite the fact that the initial peak was higher than square OET. In comparison to the other two windowed sections, the first peak of circular OET was found to be quite high, while the performance of rectangular OET was practically identical to square OET. Square OET has demonstrated to be more crashworthy among the windowed parts. The benefits of utilising windows and an origami-ending pattern in the tube were confirmed by the findings.

References

1. Wierzbicki T, Abramowicz W (1983) On the crushing mechanics of thin-walled structures. J Appl Mech 50:727–734
2. Abramowicz W, Jones N (1984) Dynamic axial crushing of square tubes. Int J Impact Eng 2:179–208
3. Abramowicz W, Jones N (1984) Dynamic axial crushing of circular tubes. Int J Impact Eng 2:263–281
4. Abramowicz W, Jones N (1986) Dynamic progressive buckling of circular and square tubes. Int J Impact Eng 4:243–270
5. Mamalis AG, Manolakos DE, Baldoukas AK, Viegelahn GL (1991) Energy dissipation and associated failure modes when axially loading polygonal thin-walled cylinders. Thin-Walled Structo 12(1):17–34. https://doi.org/10.1016/0263-8231(91)90024-D
6. Singace AA (1999) Axial crushing analysis of tubes deforming in the multi-lobe mode. Int J Mech Sci 41:865–890
7. Reddy TY, Reid SR (1986) Axial splitting of circular metal tubes. Int J Mech Sci 28:111–131
8. Chen W, Wierzbicki T (2001) Relative merits of single-cell, multi-cell and foam-filled thin-walled structures in energy absorption. Thin-Walled Struct 39:287–306
9. Singace AA, El-Sobky H (1997) Behaviour of axially crushed corrugated tubes. Int J Mech Sci 39:249–268
10. Daneshi GH, Hosseinipour SJ (2002) Grooves effect on crashworthiness characteristics of thin-walled tubes under axial compression. Mater Des 23:611–617
11. Song J, Guo F (2013) A comparative study on the windowed and multi-cell square tubes under axial and oblique loading. Thin-Walled Struct 66:9–14
12. Song J (2013) Numerical simulation on windowed tubes subjected to oblique impact loading and a new method for the design of obliquely loaded tubes. Int J Impact Eng 54:192–205
13. Nikkhah H, Guo F, Chew Y, Bai J, Song J, Wang P (2017) The effect of different shapes of holes on the crushing characteristics of aluminium square windowed tubes under dynamic axial loading. Thin-Walled Struct 119:412–420
14. Ma J, You Z (2014) Energy absorption of thin-walled square tubes with a pre-folded origami pattern—Part I: geometry and numerical simulation. J Appl Mech 81

15. Zhou C, Zhou Y, Wang B (2017) Crashworthiness design for trapezoid origami crash boxes. Thin-Walled Struct 117:257–267
16. Ma J, Dai H, Shi M, Yuan L, Chen Y, You Z (2019) Quasi-static axial crushing of hexagonal origami crash boxes as energy absorption devices. Mech Sci 10:133–143
17. Yuan L, Shi H, Ma J, You Z (2019) Quasi-static impact of origami crash boxes with various profiles. Thin-Walled Struct 141:435–446
18. Ming S, Song Z, Li T, Du K, Zhou C, Wang B (2020) The energy absorption of thin-walled tubes designed by origami approach applied to the ends. Mater Des 192:108725
19. Ming S, Song Z, Zhou C, Li T, Du K, Xu S, Wang B (2021) The energy absorption of long origami-ending tubes with geometrical imperfections. Thin-Walled Struct 161:107415

Ballistic Impact Simulation of Alumina Using Smoothed Particle Hydrodynamics (SPH) Method

Nikhil Andraskar, Gaurav Tiwari, and Manmohan Dass Goel

Abstract This study investigated the ballistic performance of alumina 99.5% plate of 5 mm thickness against an Ogival projectile through numerical simulation. Due to the dynamic nature of the impact process, both the target and the projectile dispersed into several fragments. Carrying out the numerical simulation of the impact process using the conventional meshing technique can prove a difficult task. In this chapter, Smoothed Particle Hydrodynamics (SPH) method was used in ANSYS-LS-DYNA® code to simulate the fragmentation and perforation of ceramic alumina (Al_2O_3) by steel 4340 projectile. Simulated results using SPH method are obtained for residual velocity and residual kinetic energy and compared with experimental studies and Finite-Element Model FEM. The alumina tile of 100 × 100 mm and thickness of 5 mm was hit with the projectile with velocities in the range of 94–275 m/s. A cylindrical projectile with an ogive nose of 10.9 mm in diameter and 52 mm in length was used. The simulation using SPH correctly predicted the fragmentation of both target and projectile and the values of residual velocity and kinetic energy were closer to the experimental results with all remaining parameters being constant. There was only minimal damage to the projectile tip as no backing plate was provided. Also, the smoothing length feature of the SPH has been explored and as the smoothing length parameter was increased, the predicted residual velocity of the projectile is found to be reduced by a small margin.

Keywords Ceramic · Ballistic impact · SPH · LS-DYNA

N. Andraskar (✉) · G. Tiwari
Department of Mechanical Engineering, Visvesvaraya National Institute of Technology, Nagpur 440010, India
e-mail: andraskarnikhil@gmail.com

M. D. Goel
Department of Applied Mechanics, Visvesvaraya National Institute of Technology, Nagpur 440010, India

R. Velmurugan et al. (eds.), *Dynamic Behavior of Soft and Hard Materials Volume 1*, Springer Proceedings in Materials 34, https://doi.org/10.1007/978-981-99-6030-9_12

1 Introduction

Ceramics possess high strength and hardness and low density, making them an excellent option to use in protective structures like vehicles and body armor. When a projectile impacts a ceramic plate at high velocity, compressive waves are generated in the ceramic which travel back from the free surface in the form of tensile stress waves. Ceramics have high compressive strength but very low strength in tension. Hence, the tensile stress waves cause cracks to form in the ceramics resulting in fragmentation at higher kinetic energy of the projectile, both the target and projectile undergo fracture and fragmentation. A number of studies have been conducted to evaluate the performance of ceramic structures against projectiles travelling at high velocity. Wilkins [1–5] in his earlier studies found that the blunt projectile perforated the ceramic–metal target at a lower velocity than a sharp projectile because the blunt projectile makes contact with the target plate at higher contact area thereby causing more damage that a sharp projectile impact. Further studies with ceramic bi-layer having metallic plate at the back and ceramic at the front have been performed [6] which evaluated that the backing metal layer provides the necessary tensile strength to the ceramics. Khan et al. [7] performed experimental as well as numerical study on the monolithic independent ceramic plate to study about the fragmentation behavior of ceramic at different impact velocities. The size of the fragments reduced as the impact velocity was increased.

Cottrell et al. [8] used meshless particle method along with adaptive remeshing and material erosion. Only the boron carbide elements were converted into non-deformable particles. Bresciani et al. [9] used adaptive mesh to Smoothed Particle Hydrodynamics (SPH) particles conversion and cohesive law which was used for modeling the projectile with prefragmented parts. The ceramic tile model changed from Finite-Element Model (FEM) into SPH once their failure criteria was met, but this model showed poor prediction of residual mass. Han et al. [10] used Meshless Local Petrov-Galarkin (MLPG) approach to build a three-dimensional code that defined contact, impact and penetration phenomena.

SPH method has been used to study the high velocity impact on brittle materials by [11] and [12] while Rabczuk and Eibl [13] studied about concrete fragmentation using Normalization Smoothed Particle Hydrodynamics (NSPH) and Moving Least Square Approach (MLSPH). Conventional SPH method is generally more suited to hypervelocity problems where the material upon impact behaves like a fluid and sometimes may not be suited for ballistic impact problems [14]. Particle contact algorithms were developed to solve this problem [15].

In this work, ANSYS-LS-DYNA® analysis software is used for performing the simulations using SPH solver. The ceramic alumina tile of 5 mm thickness was impacted by a steel 4340 projectile. Both the target plate and the projectile were modeled using SPH method. Residual velocity and residual kinetic energy of the projectile were predicted. The simulation results predicted were compared with the experimental results and FEM simulation results reported by Khan et al. [7] to demonstrate the effectiveness of SPH method over FEM technique. Also, the effect of smoothing length on the results were studied.

2 Numerical Modeling

As discussed in the previous section, SPH is a meshless method which uses kernel function to interpolate the physical parameters of each particle based on its relation with the neighboring particles.

2.1 SPH Model Formulation

SPH model for ceramic tiles as well as projectile is shown in Fig. 1. The model was established using ANSYS-LS-DYNA® analysis software. The square ceramic alumina plate (alumina 99.5%) of size 100 × 100 × 5 mm and the dimension of the projectile is as shown in Fig. 1. The projectile has total length of 52.6 mm and was made of steel 4340 material. The diameter of the projectile was 10.9 mm in the middle cylindrical portion. The impact velocity was in the range 94 m/s to 275 m/s. The model configuration was based on the experimental work performed by Khan et al. [7]. Both projectile and ceramic is modeled using SPH method and as it is capable of handling large deformation and fragmentation, without element erosion criteria normally required in FEM simulations. The particle spacing in both target and projectile is kept same. The smoothing length defined by parameter CLSH is kept to default value of 1.2 at the start and then varied to study its effect on the predicted values. The smoothing length of the particles is constrained using HMIN and HMAX parameters set at 0.2 and 2.0 respectively. IAVIS parameter defines the artificial viscosity which is set to default value of 0 in CONTROL_SPH card. The contact between projectile and ceramic plate is defined by using DEFINE_SPH_TO_SPH_COUPLING keyword which is generally preferred for impact problems [16]. For this method, parameter CONT in CONTROL_SPH card is set to 1.

2.2 Constitutive Modeling

The ceramic material was modeled using Johnson–Holmquist (JH-2) material model whereas the steel 4340 projectile was modeled using Johnson–Cook (JC) material model. JH-2 Model predicts the captures the behavior of brittle ceramic as observed in the literatures.

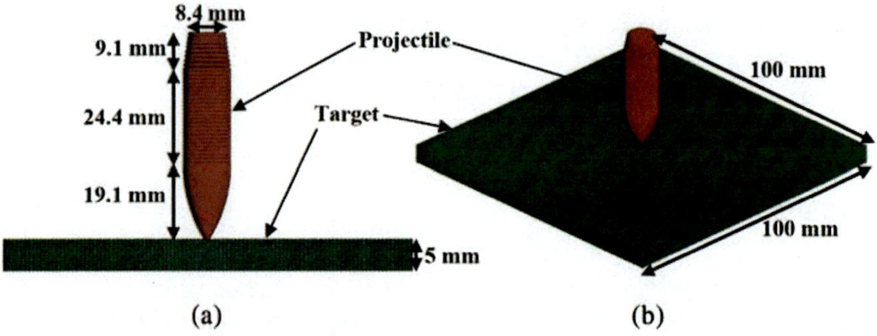

Fig. 1 SPH model along with plate and projectile dimensions **a** side view **b** isometric view

2.3 Constitutive Model for Ceramic

JH-2 model [17] is generally preferred to define the behavior of ceramic materials and has models for intact strength, fracture strength, pressure and bulking effect. The normalized strength is given by

$$\sigma^* = \sigma_i{}^* - D\left(\sigma_i{}^* - \sigma_f{}^*\right) \tag{1}$$

where $\sigma_i{}^*$ is the normalized intact equivalent stress, and $\sigma_f{}^*$ is the normalized fracture stress, and D is the damage ($0 \leq D \leq 1.0$)

The normalized intact equivalent stress is given in Eq. 2 and the normalized fractured equivalent stress in Eq. 3

$$\sigma_i{}^* = A\left(P^* + T^*\right)^N\left(1 + Cln\dot{\varepsilon}^*\right) \tag{2}$$

$$\sigma_f{}^* = B\left(P^*\right)^M\left(1 + Cln\dot{\varepsilon}^*\right) \tag{3}$$

where A, B, C, M and N are material constants and the normalized pressure $P^* = P/P_{HEL}$ is the actual pressure and P_{HEL} is the pressure at Hugoniot Elastic Limit (HEL). The HEL is the net compressive stress corresponding to the uniaxial strain (shock wave) exceeding the elastic limit of the material. The normalized maximum tensile hydrostatic pressure is $T^* = T/P_{HEL}$ where T is the maximum tensile hydrostatic pressure the material can withstand and the dimensionless strain rate, $\dot{\varepsilon}^*=\dot{\varepsilon}/\dot{\varepsilon}_0$ where $\dot{\varepsilon}$ is actual equivalent strain rate and $\dot{\varepsilon}_0$ is reference strain rate considered as 1 s^{-1}.The hydrostatic pressure before and after the damage was expressed through Eq. (4) and (5) respectively;

$$P = K_1\mu + K_2\mu^2 + K_3\mu^3 \tag{4}$$

$$P = K_1\mu + K_2\mu^2 + K_3\mu^3 + \Delta P \tag{5}$$

where K_1 is a bulk modulus and K_2, K_3 are the pressure constants. The volumetric strain $\mu = \rho/\rho_0 - 1$ where ρ is current density and ρ_0 is the initial density of the material. ΔP is pressure increment due to bulking of the material due to accumulation of damage.

The damage criterion for the fracture model is given by

$$D = \sum \frac{\Delta \varepsilon_p}{\varepsilon_f} \tag{6}$$

where $\Delta\varepsilon_p$ is the increment in equivalent plastic strain and ε_f is the plastic strain to fracture. The plastic strain to fracture of the material is described as;

$$\varepsilon_f = D_1^*(P^* + T^*)^{D_2^*} \tag{7}$$

where D_1^* and D_2^* are the damage constants. The material parameters are as mentioned in Table 1. β is the fraction of elastic energy loss converted to hydrostatic energy.

Constitutive model for steel projectile

Johnson–Cook material model is based on von-Mises yield criterion. It incorporates the effects of thermo-elasticity, strain rate hardening and plastic flow and predicts the behavior of metals at high strain rates, high temperatures and pressures [18, 19] The equivalent von-Mises stress is given by following equation.

$$\sigma = \left[A^* + B^*(\varepsilon)^n\right]\left[1 + C^* ln\left(\frac{\varepsilon}{\varepsilon_0}\right)\right]\left[1 + \dot{T}^m\right] \tag{8}$$

where ε is the equivalent plastic strain, A^*, B^*, C^*, m and n are material parameters. $\dot{T}$ is further defined as

$$\dot{T} = \begin{cases} 0 & \text{for } T < T_0 \\ \dfrac{T - T_0}{T_{melt} - T_0} & \text{for } T_0 \le T \le T_{melt} \\ 1 & \text{for } T > T_{melt} \end{cases} \tag{9}$$

Table 1 Material model constants for Al_2O_3 (JH-2 material model) [7]

Strength constant								
A	*B*	*C*	*M*	*N*	HEL(GPa)	Reference strain rate EPSI	P_{HEL}(GPa)	*T* (GPa)
0.93	0.31	0	0.6	0.6	2.79	1	1.46	0.2
Damage parameters			Equation of state				Density ρ (kg/m^3)	Shear modulus *G* (GPa)
D_1^*	D_2^*		K_1 (GPa)	K_2 (GPa)	K_3 (GPa)	β		
0.005	0.1		130	0	0	1	3700	90.16

Table 2 Material model constants for Steel 4340 (Johnson–Cook material model) [7]

Shear modulus	Johnson–Cook strength model constant				
G (GPa)	A^* (GPa)	B^* (GPa)	C^*	m	n
77	0.95	0.725	0.15	0.625	0.375
Johnson–cook fracture model constant					Density
D_1	D_2	D_3	D_4	D_5	ρ(kg/m^3)
−0.8	2.1	−0.5	0.002	0.61	7770
Polynomial EOS		Elastic Modulus	Melting temp T_M	Reference temp T_r	Effective strain rate
Bulk modulus K_1 (GPa)	Specific Heat C_r (J/kgK)	E (GPa)			EPSO(s^{-1})
233	477	201	1790 K	300 K	1

The damage parameter D which defines the fracture model is given as

$$D = \sum \frac{\Delta\varepsilon}{\varepsilon_f} \tag{10}$$

where $\Delta\varepsilon$ denotes the increment in equivalent plastic strain and ε_f denotes failure strain which is given by,

$$\varepsilon_f = \left[D_1 + D_2 exp\left(D_3 \frac{\sigma_m}{\sigma}\right)\right]\left[1 + D_4 ln\left(\frac{\varepsilon}{\varepsilon_0}\right)\right]\left[1 + D_5 \dot{T}\right] \tag{11}$$

where σ_m is the mean stress and σ is the equivalent von-Mises stress. D_1, D_2, D_3, D_4 and D_5 are material damage parameters. EPSO is the effective plastic strain rate. The material parameters for steel 4340 are mentioned in Table 2 as taken from [7].

3 Simulation Results and Discussion

The simulation results are presented in this section and are compared with experimental to verify the SPH method for residual velocity and residual kinetic energy prediction. Also, the comparison is made with FEM model to check the accuracy of SPH model against FEM model.

The fragmentation of the ceramic plate from side on and from the back side is shown in Fig. 2. The ceramic breaks into number of fragments as the projectile perforate the plate. There is a slight damage to the tip of the projectile due to impact with ceramic plate. The fragmentation pattern was similar to one presented in [7]. A ceramic ejecta is formed at the start which then developed into a conical shape as the projectile travels further.

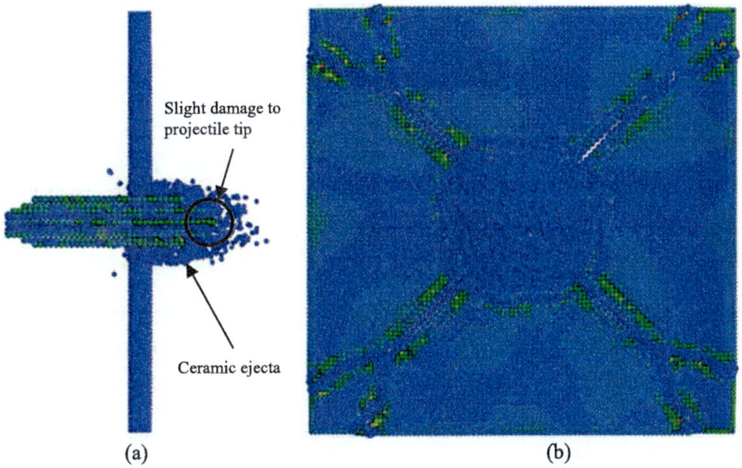

Fig. 2 Fragmentation view in SPH simulation from **a** side on and **b** back side of the ceramic plate at velocity of 275 m/s

3.1 Residual Velocity and Residual Kinetic Energy

Numerical simulation of ogive nosed hardened steel projectile of 10.9 mm dia and 30 gm mass impacted over a square alumina tile (Al_2O_3 99.5%) of size 100 × 100 mm and thickness of 5 mm was carried out on ANSYS-LS-DYNA®. The ceramic plate and the projectile both were modeled using SPH method. The numerical results were validated against the experimental results in [7] as shown in Table 3 and graphically represented in Fig. 3. Close correlation was observed between the results obtained in ANSYS-LSDYNA® with experimental and numerical results from [7]. The simulations were performed with impact velocity in the range of 94–275 m/s.

Table 3 Comparison between numerical and experimental results

Impact velocity (m/s)	Residual velocity (m/s)			Kinetic energy absorbed (Joules)		
	Experimental [7]	FEM simulation [7] (% Error)	SPH simulation	Experimental [7]	FEM simulation	SPH simulation
275	239	212 (11%)	242 (1%)	276.81	460.215	255.8
264	215	201 (6%)	228 (6%)	351.88	439.425	308.57
249	182	190 (4%)	197 (8%)	432.54	388.515	394.4
215	153	151 (1%)	171 (11%)	340.4	351.36	265.14
179	116	126 (8%)	129 (11%)	277.5	242.475	231.89
112	70	45 (35%)	61 (12%)	114.7	157.785	132.75
94	58	23 (60%)	41 (29%)	82.19	124.605	107.83

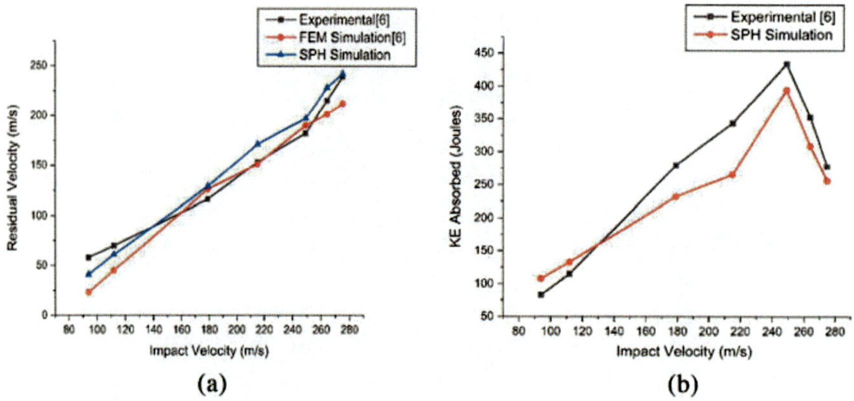

Fig. 3 Graph for **a** residual velocity and **b** KE absorbed using experiments and numerical simulation results

The velocity of the rigid body measured after the perforation of the target plate was taken as the residual velocity of the projectile. Kinetic energy absorbed by the target was taken as the difference between the kinetic energy of the projectile measured before the impact and kinetic energy of the projectile measured after the projectile passes through the target. The residual velocity obtained via SPH method was compared with experimental and FEM simulation results mentioned in [7] and as shown in Fig. 3a, The SPH method shown to have predicted closer to the experimental values. The graph for kinetic energy absorbed by the target plate was also plotted in comparison with experimental and SPH simulation results. The nature of the graph matched the one obtained for experimental results but the variation in one of the cases was 22%.

3.2 Effect of Smoothing Length Parameter

The smoothing length described by parameter CLSH in ANSYS-LS-DYNA® has a default value of 1.2 and its range is defined between 0.2 and 2. The CLSH parameter was varied from 1.2 to 1.5 to study its effect on the residual velocity of the projectile. As shown in Table 4, the effect of smoothing length on the predicted residual velocity for impact velocity of 275 m/s showed a slight decrease in the residual velocity as the smoothing length increased. As the CLSH value was increased, the damage and strain become less localized and increase in artificial viscosity was observed. Also, it was found that the computation time also increased when CLSH parameter was varied from 1.2 to 1.5 as mentioned in Table 4.

Table 4 Effect of smoothing length on residual velocity

Impact velocity (m/s)	Residual velocity (m/s) at different CLSH values			
	1.2	1.3	1.4	1.5
275	242	238	241	231
249	197	196	196	191
215	171	169	170	166
179	129	128	128	123

4 Conclusions

Numerical simulations using SPH method were performed to study the performance of Al_2O_3 ceramic tile against ogive nose shaped projectile. The simulated results were compared with the experimental results. The effect of smoothing length parameter on the residual velocity prediction was also explored. The following conclusions were drawn.

1. SPH was found to be a feasible method to study behavior projectile and target during high velocity impact process. The bulge formation and fragmentation of target plate was properly reproduced along with projectile damage.
2. Good prediction accuracy was observed for residual velocity using SPH method with experimental results against that of FEM simulations. The kinetic energy values were slightly over predicted in few cases.
3. The smoothing length parameter CLSH also had an effect on the predicted residual velocity values which were shown to decrease as the CLSH value was increased.

References

1. Wilkins M (1967) Second progress report of light armor program, in, Lawrence Livermore National Lab.(LLNL), Livermore, CA (United States)
2. Wilkins M (1977) Use of boron compounds in lightweight armor. In: Boron and refractory borides, Springer, pp 633–648
3. Wilkins M, Honodel C, Sawle D (1967) Approach to the study of light armor. In: California University, Livermore. Lawrence Radiation Lab
4. Wilkins ML (1978) Mechanics of penetration and perforation. Int J Eng Sci 793–807
5. Anderson C (2007) A review of computational ceramic armor modeling, advances in ceramic armor II. In: Frank LP (ed) Proceedings of the 30th international conference and explosion on advanced ceramics and composites, pp 1–18
6. Hetherington JG (1992) The optimization of two component composite armours. Int J Impact Eng 409–414
7. Khan MK, Iqbal MA, Bratov V, Morozov NF, Gupta NK (2020) An investigation of the ballistic performance of independent ceramic target. Thin-Walled Struct 154:106784

8. Cottrell MG, Yu J, Owen DRJ (2003) The adaptive and erosive numerical modelling of confined boron carbide subjected to large-scale dynamic loadings with element conversion to undeformable meshless particles. Int J Impact Eng 1017–1035
9. Bresciani LM, Manes A, Romano TA, Iavarone P, Giglio M (2016) Numerical modelling to reproduce fragmentation of a tungsten heavy alloy projectile impacting a ceramic tile: adaptive solid mesh to the SPH technique and the cohesive law. Int J Impact Eng 3–13
10. Han Z-D, Liu H-T, Rajendran A-M, Atluri S-N (2006) The applications of meshless local petrov-galerkin (MLPG) approaches in high-speed impact, penetration and perforation problems. Comput Model Eng Sci 119–128
11. Clegg RA, Hayhurst CJ (1999) Numerical modelling of the compressive and tensile response of glass and ceramic under high pressure dynamic loading. pp L3.01
12. Hedayati E, Vahedi M (2017) Numerical investigation of penetration in ceramic/aluminum targets using smoothed particle hydrodynamics method and presenting a modified analytical model. Comput Model Eng Sci 295–323
13. Rabczuk T, Eibl J (2003) Simulation of high velocity concrete fragmentation using SPH/MLSPH. Int J Numer Methods in Eng 1421–1444
14. Johnson GR, Stryk RA, Beissel SR (1996) SPH for high velocity impact computations. In: Computer methods in applied mechanics and engineering, pp 347–373
15. Seo S, Min O (2006) Axisymmetric SPH simulation of elasto-plastic contact in the low velocity impact. Comput Phys Commun 583–603
16. Xiao Y, Wu H, Ping X (2020) On the simulation of fragmentation during the process of ceramic tile impacted by blunt projectile with SPH method in LS-DYNA. Comput Model Eng Sci 923–954
17. Johnson GR, Holmquist TJ (1994) An improved computational constitutive model for brittle materials. In: AIP conference proceedings, pp 981–984
18. Johnson GR, Cook WH (1985) Fracture characteristics of three metals subjected to various strains, strain rates, temperatures and pressures. Eng Fract Mech 31–48
19. Johnson GR (1983) A constitutive model and data for materials subjected to large strains, high strain rates, and high temperatures. In: Proceedings 7th information symposium Ballistics, pp 541–547

Comparative Study of Single Tunnel and T- Intersection Subjected to Impact Load

Shipra Sinha, K. Seshagiri Rao, and Tanusree Chakraborty

Abstract Intersection is the region where two or more openings share common points. These are constructed in several subsurface structures including road tunnels, railway tunnels, mines and hydel projects. In road tunnels, cross passages are constructed at regular intervals to facilitate access to the egress tunnel. The junction of cross passage with the main tunnel is the critical part of the tunnel. Tunnels are subjected to impact load when projectile hits the surface. The purpose of this study is to understand the influence of intersection of tunnels by comparing it with single tunnel constructed in rockmass. This is accomplished by comparing the displacements observed at the tunnel crown in both configurations, that is, single and T-section, when subjected to different low-velocity and high-velocity impacts at the ground surface. Tunnel intersections fall in the ambit of three-dimensional problem owing to the geometrical aspect and hence, the problem is attempted in this study through numerical simulation using FEM-based explicit module of ABAQUS. The tunnels are circular in shape of diameter 5 m and intersect each other at the same level from the ground surface at cover depth / diameter ratio of 1. The location of impact is above the centre of the tunnel crown. The study focusses on solving the combination of non-linear problem of impact and criticality of the geometry. The results provide an idea about the magnitude and extent of damage incurred by the roof in single and T-section. The displacements at the crown in the single tunnel are found to be less as compared to that in the T-section but the deformed area is more in the former. This eventually develops an insight on ways to improve the roof support system.

Keywords Tunnel intersection · Impact load · Numerical modelling · FEM

S. Sinha (✉) · K. S. Rao · T. Chakraborty
Department of Civil Engineering, Indian Institute of Technology, Delhi, New Delhi 110016, India
e-mail: shipra.sinhaiitd@gmail.com

R. Velmurugan et al. (eds.), *Dynamic Behavior of Soft and Hard Materials Volume 1*, Springer Proceedings in Materials 34, https://doi.org/10.1007/978-981-99-6030-9_13

1 Introduction

Underground tunnels have long been in use for transportation but a possibility for human dwellings below the grounds seems not too far. This imminent structural revolution will necessitate a lot of utility tunnels in addition to the main openings and hence, intersections are going to be very common in the whole underground system.

Intersection is the region where two or more openings share common points. These are constructed in several subsurface structures including road tunnels, railway tunnels, mines and hydel projects. In road tunnels, cross passages are constructed at regular intervals to facilitate access to the egress tunnel. The junction of cross passage with the main tunnel is the critical part of the tunnel. Tunnel intersections fall under 3D problems. The 3D nature is due to geometrical considerations [1]. State of stress is three dimensional due to superimposition of the stress distribution around individual openings. The possibility of carefully constructed tunnel being subjected to any unexpected dynamic load such as projectiles hitting the tunnel, airplane crash or accidental boulder impact cannot be ruled out.

Impact of two objects is associated with wave propagation, deformation in the form of cracks, fragments, chipping, spalling, distortion, etc. [2]. The different phenomena associated with impact depends on several factors. These include characteristics of impactor as well as target along with striking conditions [3]. Impact velocity is one of several factors defining the striking condition. Before understanding these, it is imperative to first know the types of impact. Impact can be broadly classified into high-velocity and low-velocity impacts. Classification of impacts in terms of velocity ranges are presented in Table 1. Examples of high-velocity impact are impact due to rigid projectile and hydrodynamic shocks. Examples of low-velocity impact are impact due to boulders and airplane crash [4].

Apart from impact velocity governing the response generated in an impact, important characteristics of impactor that play a role as well include its weight, geometry and size. The shape of the impactor tip has a marked effect on both peak impact force as well as duration of the first impact impulse [6]. Talking about the role of target, especially rock targets, for low-velocity impact, rock penetration depends on its shear strength where internal angle of friction is more important than cohesive strength. For hypervelocity impact, penetration is dependent on the material density [7]. Hypervelocity impacts are practically explosions where the phase of the target may change, such as solid to fluid. Hypervelocity impact studies concern those in

Table 1 Classification of impact based on velocity [5]

Type of impact	Velocity
Low velocity	< 10 m/s
Intermediate velocity	10–50 m/s
High (Ballistic)	50–1000 m/s
Hypervelocity	2 km/s – 5 km/s

the field of planetary objects. For structures such as tunnels, impact by objects up to high velocity can be expected.

There are different methods to assess response of tunnel under impact load. These are: large-scale field tests, small-scale physical modelling in laboratories and numerical simulations [8–10]. Physical modelling comprises the scaling down of the expected real phenomenon in terms of size of the model as well as impact energy. Model studies can be destructive or non-destructive; destructive. Models can be prepared either from natural rocks or by equivalent material modelling. The problem with in situ tests is that these tests are expensive and difficult to conduct. Numerical simulation results seem to be the easiest option to anticipate the type of response any structure would present in the event of any condition/load of interest.

The purpose of this study is to gain a behavioural insight of different sections of the tunnel constructed in weak rockmass by means of numerical modelling in the event of an impact load. The impactor hits the tunnel top surface with velocities of 5 and 200 m/s to simulate different types of velocity impacts.

2 Numerical Modelling

The problem in this study is attempted through numerical simulation using FEM-based explicit module of ABAQUS [11]. Two different configurations of single tunnel and T—intersection are modelled. T- intersection is the region where two tunnels intersect at an angle of 90° to each other. Both the intersecting tunnels and single tunnel are circular in shape with diameter of 5 m, at cover-depth to diameter ratio of 1, that is, located at a depth of 5 m from the surface. The placement of tunnel below top surface presents the case of shallow tunnel. Both models are subjected to low-velocity and high-velocity impact. For the generation of impact load, an impactor is modelled with a hemispherical tip of radius 2.5 m. The geometric details of tunnel model and impactor are presented in Fig. 1. The lateral extent of the model is selected in accordance with the zone of influence for single tunnel which is 2.5 times of the radius of the tunnel [8, 12, 13]. The tunnel model is of the cuboidal shape with dimensions of 30 m width, 30 m height and 35 m length.

The surrounding material around tunnels is chosen to be weathered quartzite. The material is modelled using Mohr–Coulomb plasticity model. The material properties of weathered quartzite are shown in Table 2. The dilation angle is taken one-eighth of the friction angle [14].

The impactor is modelled as discrete rigid part with weight of 700 kg. The weight is allocated to the impactor by attaching a reference point to it and adding weight to it using 'Inertia' feature. The location of impact load is on the top surface at the centre of the model, that is, above the centre of the tunnel crown. Two types of impact: low-velocity (5 m/s) and high-velocity impact (200 m/s) is studied. The step time for low velocity and high velocity is kept 0.14 s and 0.013 s, respectively. Thus, five different models are analysed. The tunnel model is fixed at the base with impactor having degree of freedom in the vertical direction only. The parts are meshed accordingly,

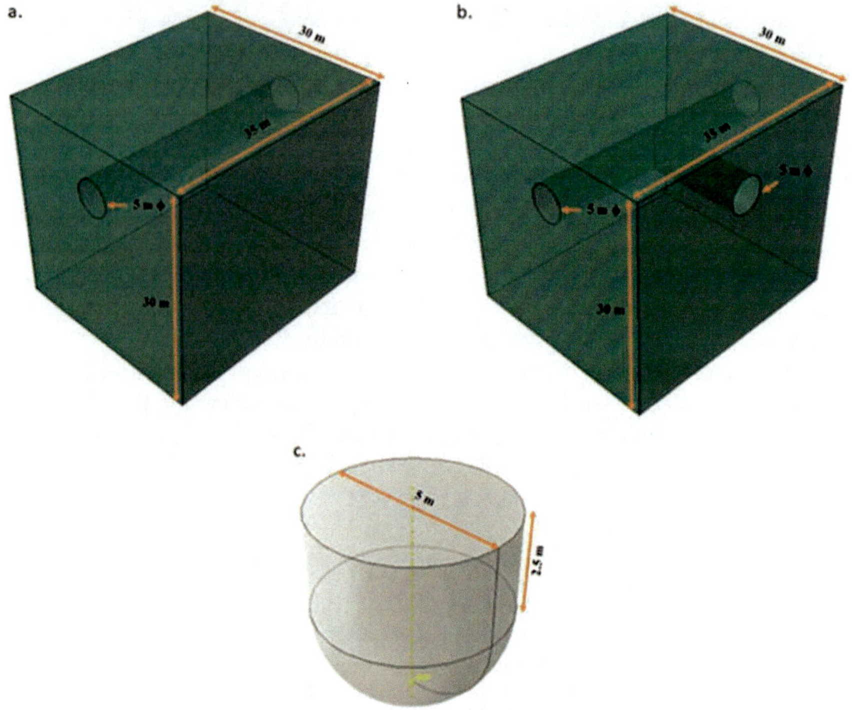

Fig. 1 Geometric details **a** Single tunnel; **b** T- intersection; **c** Impactor

Table 2 Material properties for weathered quartzite [15]

Property	Value	Property	Value
Density, ρ (kg/m^3)	2420	Cohesion, c (MPa)	15.75
UCS, σ_c (MPa)	32.20	Friction angle, φ (°)	49.49
Elastic modulus, E_{t50} (GPa)	12.18	Dilation angle, ψ (°)	6
Poisson's ratio, ν	0.21		

the details of which are provided in The meshing is done such as to minimize the penetration of master nodes of rigid impactor into the top surface of tunnel model which is the slave surface.

Table 3 and shown in Fig. 2.

The meshing is done such as to minimize the penetration of master nodes of rigid impactor into the top surface of tunnel model which is the slave surface.

The interaction between the impactor and the tunnel is formulated using General Contact (Explicit) keeping tangential behaviour as 'Frictionless' and normal behaviour set to 'Hard contact'.

Table 3 Properties of meshed model

Part	Element type	No. of elements	
		Single	T- section
Tunnel	C3D8R	95,340	199,652
Impactor	R3D4	21,756	390

a. b.

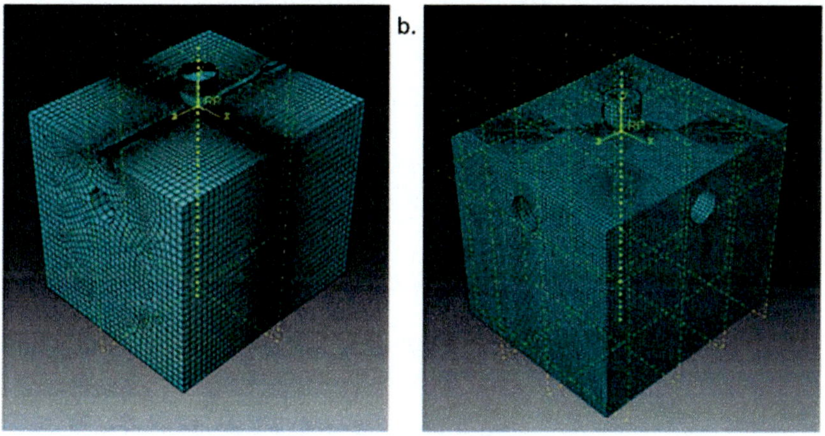

Fig. 2 Meshed models of **a** Single Tunnel; **b** T- intersection

3 Results and Discussion

The prime concern in any tunnel project is to manage deformation of the tunnel. Hence, the results of analyses present in Table 4 presents displacements and width of deformed zones. The visual representation of data in Table 4 is depicted through Fig. 3. The red line on the tunnel cross section is the path along which the displacement values are plotted.

Table 4 Values of peak displacement and deformed zone width

Case	Peak displacement at crown (m)	Deformed zone width along the tunnel axis (m)
Low Velocity_Single 1.0	0.093	12.25
Low Velocity_Single 0.5	0.494	8.4
Low Velocity_T 0.5	0.466	7.88
High Velocity_T 0.5	3.008	13.99
High Velocity Single 1.0	1.393	18.68

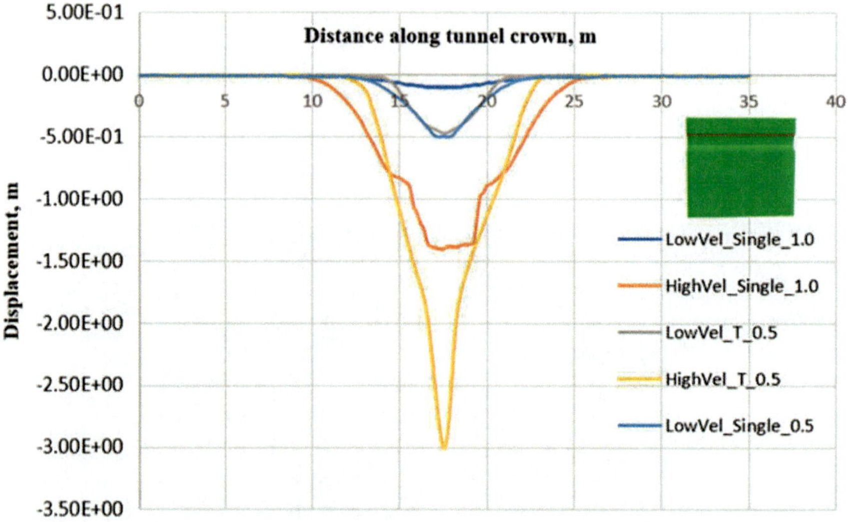

Fig. 3 Displacement plots of tunnel crown in longitudinal direction

3.1 Effect of Varying Configuration on Crown Displacement

Firstly, the analysis results of single and T- section are compared for low velocity impact, that is, keeping impact velocity equal to 5 m/s. As, it is evident from Fig. 4, the tunnel crown at centre of the model seems visibly deformed in T- section. Also, from Table 4, it is found that the deformed zone has less width in T-section compared to the single tunnel.

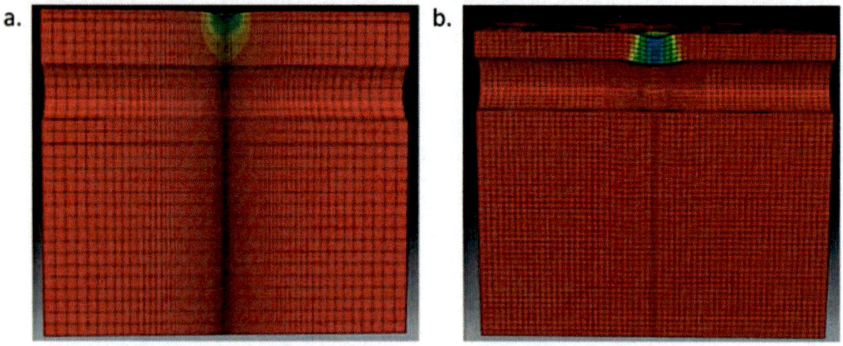

Fig. 4 Displacement contours at low velocity (5 m/s) impact for **a** Single Tunnel (C/D = 1); **b** T-intersection (C/D = 0.5)

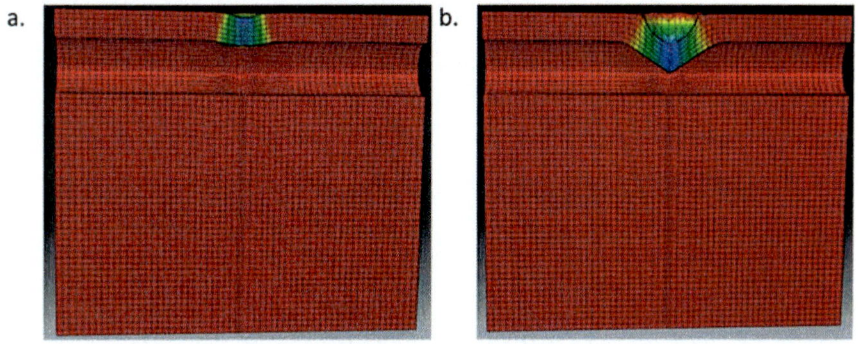

Fig. 5 Displacement contours for T- intersection **a** Low-velocity (5 m/s) impact; **b** High-velocity (200 m/s) impact

3.2 Effect of Low- and High-Velocity Impact on T- Section

The effect of impact in different velocity ranges varies in terms of localization of deformation. Figure 5 clearly demonstrates the localized displacement of tunnel crown in high-velocity impact. Also, the magnitude of displacement is 6.43 times more in high-velocity impact as compared to low-velocity impact. The deformed zone as mentioned in Table 4 is comparatively more in high-velocity case.

4 Conclusion

The study of both types of tunnel configurations: single and T- intersection with the aid of FEM-based numerical modelling leads to the following conclusions:

- There is a marked difference in peak displacement value at tunnel crown between the two configurations. This points out the considerable influence of geometry. Lack of sidewall support at one side and wider roof span to be supported in intersections makes them critical section of any tunnel system.
- Though the displacement values observed is more in case of low cover depth intersection but the length of deformed zone is more in single tunnel case. This suggests that the roof support in case of intersection need not to be overly reinforced in longitudinal direction rather should be optimized as per the deformed zone in both the intersecting tunnels.
- The displacement values observed is comparable for same C/D ratio in case of intersection as well as single tunnel but the length of deformed zone is more in single tunnel case.
- The width of damage zone across the tunnel axis is more in T-intersection.
- In the event of high-velocity impact, damage incurred in tunnels seems to be irreparable beyond measures.

- High-velocity impact seems to generate more localized damage.

The study aims at contributing towards preliminary understanding of a critical section of a tunnel system with future scope of improvements in tunnel support system design in accordance to the analyses' results.

References

1. Gerçek H (1986) Stability considerations for underground excavation intersections. Min Sci Technol 4(1):49–57. https://doi.org/10.1016/S0167-9031(86)90194-5
2. Li QM et al (2005) Local impact effects of hard missiles on concrete targets. Int J Impact Eng 32(1–4):224–284. https://doi.org/10.1016/j.ijimpeng.2005.04.005
3. Zukas JA (1990) High velocity impact dynamics. Wiley-Interscience
4. Kœchlin P, Potapov S (2009) Classification of soft and hard impacts—application to aircraft crash. Nucl Eng Des 239(4):613–618. https://doi.org/10.1016/j.nucengdes.2008.10.016
5. Vaidya UK (2011) Impact response of laminated and sandwich composites. In: Impact engineering of composite structures, Springer, Vienna, pp 97–191. https://doi.org/10.1007/978-3-7091-0523-8_4
6. Li H, Chen W, Hao H (2019) Influence of drop weight geometry and interlayer on impact behavior of RC beams. Int J Impact Eng 131:222–237. https://doi.org/10.1016/j.ijimpeng.2019.04.028
7. Heuze FE (1990) An overview of projectile penetration into geological materials, with emphasis on rocks. Int J Rock Mech Mining Sci Geomech Abstracts 27(1):1–14. https://doi.org/10.1016/0148-9062(90)90003-K
8. Mishra S, Rao KS, Gupta NK, Kumar A (2018) Damage to shallow tunnels in different geomaterials under static and dynamic loading. Thin-Walled Struct 126:138–149. 10/1016/j.tws.2017.11.051
9. Zhou L, Zhu Z, Dong Y, Fan Y, Zhou Q, Deng S (2019) The influence of impacting orientations on the failure modes of cracked tunnel. Int J Impact Eng 125:134–142. https://doi.org/10.1016/j.ijimpeng.2018.11.010
10. Dhamne RR, Mishra S, Kumar A, Rao KS (2021) Deformation behavior of D-shaped shallow tunnels under dynamic loading conditions. Structures 33:3973–3983. https://doi.org/10.1016/j.istruc.2021.06.097
11. ABAQUS/CAE 2017 (2019) Dassault Systemes
12. Obert L, Duvall WI, Merrill RH (1960) Design of underground openings in competent rock (No. Bulletin 587). United States. Government Printing Office
13. Brown ET (1980) In: Underground excavations in rock. CRC Press
14. Hoek E, Brown ET (1997) Practical estimates of rock mass strength. Int J Rock Mech Mining Sci 34(8):1165–1186. https://doi.org/10.1016/S1365-1609(97)80069-X
15. Gupta AS (1997) Engineering behaviour and classification of weathered rocks (Doctoral dissertation, IIT Delhi)

Ballistic Response of Composite Helmet

Gaurav Tiwari and S. L. S. Hamsi

Abstract The present study has been carried out in order to investigate the ballistic performance of both CFRP (Carbon Fibre-Reinforced Polymer) and Kevlar composite helmet when impacted against ogive projectile and spherical projectile within the subordnance velocity regimes. The shell thickness of the helmet was considered to be 7.5 mm. Creo software was used for modelling the helmet and LS-PrePost finite-element code was used for performing the simulations of the 3D model of helmet when it is impacted by the projectiles. The ballistic limits of both the composites helmets were obtained by varying the incident velocities of the projectiles. Further, the effect of shell thickness of the helmet, effect of the different composites, effect of the impact locations on performance and the effect of the projectile shapes on the performance parameters like maximum inside surface deformation, ballistic limit, dent region diameter and damage pattern were evaluated.

Keywords CFRP · Ballistic limit · Kevlar · Ogive projectile · Energy absorption

1 Introduction

Personal armours also referred as body armours are one of the prominent equipment used by the solider. Some of the body armours that are utilized by front head soldiers are ballistic helmets and bullet proof vests. The prime factor that should be considered while designing the helmet is the reduction of weight along with comfortability and also to enhance the strength of the helmet so that the soldiers are protected from the injuries. However, in recent years, composite materials has been grabbing attention in the development of helmets under ballistics due to its advanced mechanical properties like high strength, specific stiffness, resistance to corrosion, etc. In this context, we are mainly focusing on Kevlar and CFRP composite. Yet, the complex response of composite materials linked with high costs and limited amount of data

G. Tiwari (✉) · S. L. S. Hamsi
Department of Mechanical Engineering, Visvesvaraya National Institute of Technology, Nagpur 440010, India
e-mail: gauraviitdelhi@gmail.com

R. Velmurugan et al. (eds.), *Dynamic Behavior of Soft and Hard Materials Volume 1*, Springer Proceedings in Materials 34, https://doi.org/10.1007/978-981-99-6030-9_14

from ballistic testing has led to experimental studies of ballistic helmet becomes expensive and time-consuming. Therefore, finite-element analysis can be used as a method to characterize the response of composite ballistic helmet and to obtain information on parameters affecting impact phenomena. Kulkarni et al. [1] analysed both ballistic and blast performance of the US army helmet and also investigated about the threats which are caused in the warfare. The traumatic brain injury caused to the head was studied. It was found that, when the blast waves pass through the skull, the traumatic brain injuries occurred which directly led to the death. Jamroziak et al. [2] performed the studies to scrutinize the ballistic behaviour of the helmet experimentally when the projectile impacts on the helmet at different velocities. Goldsmith et al. [3] conduced numerical and analytical analysis on woven carbon laminates to find the ballistic resistance of the plates against cylindroid—conical shape of projectile. They concluded that, as the thickness of the laminate increased the ballistic resistance towards the projectile also got increased and the mechanism due to which the damage occurred were the crack propagation, delamination of the fibres and the enlargement of the hole. Velmurugan et al. [4] analysed the performance of the laminate made of CFRP composite while varying the orientation of the ply. Totally, three different orientations are accommodated by changing the thickness from 2 to 4 mm when impacted by a spherical ball. It was found that as the thickness of the laminate increased, the residual velocities for the corresponding impact velocities decreased rapidly. Rodríguez et al. [5] determined the ballistic response of the Kevlar laminate by conducting numerical analysis when it was impacted by the cylindrical projectile. Four projectiles namely conical-flat, conical–conical, flat–flat and flat-conical were used in the study. It was concluded that the helmet had highest ballistic limit when impacted against conical-conical and had local damage area. Tham et al. [6] performed both experiments and numerical simulations on US army helmet against spherical, FSP, FMJ bullet impacts. It was found that the ballistic limit was 610 m/s against a FSP impact. After comparing the ballistic limit of both laminate and helmet, they concluded that the helmet had higher ballistic limit than the laminate. Magnus and Kleiven [7] mainly focused on the effect of load level on human head by varying the helmet shell stiffness and impact angle direction of the projectile. Shell thickness of 9.5 mm was used. It was shown that as the stress was increasing with the increase in contact between the skull and the helmet shell and also the pressure was high for compliant shell. Therefore, to get the optimum results the shell should not be very stiff or it should not be very compliant. It was concluded that the stress increased with increase in the impact angle and impact angle of 45° causes high strain in brain.

Numerous researches are going on throughout the centuries to design the helmet according to the military standards. During the literature survey, it was noticed that mostly FE simulations were performed on the laminates and only few papers conveyed the analysis on the helmets and its ballistic resistance to the ammunitions. The majority of the studies on helmets were conducted using Kevlar 29 and no other composites were used in the study. From the literature, it was seen that the major lacuna is that there are only limited studies present on the effect of helmet's shell thickness on the performance of Kevlar when impacted by a projectile. Therefore, it was decided to numerically investigate the ballistic response of CFRP and Kevlar helmets against spherical projectile and ogival projectile impacts. The maximum inside surface deformation, ballistic limit, dent region diameter and the damage pattern within the Kevlar and CFRP helmet was studied within subordnance regime of velocity.

2 Numerical Modelling

In the present study, numerical simulations were performed to determine the ballistic performance of Kevlar and CFRP helmet when a spherical projectile and ogive nosed projectile impacts the helmet on the rear side. The simulations were performed using LS-PrePost software. The element size of 7 × 7 mm was considered for the helmet shell. Proper care was taken to keep the aspect ratio of elements unity. An overall of 38,567 elements were utilized in this model through the thickness. Damage material model was considered for studying the composite helmet and the properties of both CFRP and Kevlar are taken from literature [8, 12]. Helmet was considered as a shell element and assigned with a thickness of 7.5 mm; same thickness value has been assigned at all the four nodes present on the helmet. Both the spherical and ogive projectiles were modelled as a rigid body and the material properties of steel were assigned to the projectiles. The diameter and length of ogive nosed projectile was 19.4 and 50.8 mm, respectively, whereas for spherical projectile diameter was 14.2 mm. The modulus of elasticity of steel was considered as 210 GPa and density was 7850 Kg/m^3. The projectiles were given an element size of 2 × 2 mm for both projectiles. The meshed model of both the projectiles and helmet are shown in the Fig. 1. AUTOMATIC_SURFACE_TO_SURFACE Contact type interaction property has been appraised between helmet and projectile. For Kevlar composite, the value of static friction is 0.3 and the value of dynamic friction coefficient is 0.28 [8]. For helmet made of CFRP composite, the static friction coefficient is 0.35 and dynamic friction coefficient was 0.135 [9]. Figure 1 shows the assembly of both the helmet and the projectiles. The rim of the helmet, that is, the outer edge of the helmet is fixed in this study by using the BOUNDARY_SPC_NODE. So, all the translational and rotational movements in all axis are restricted. Material properties for CFRP and Kevlar has shown in Table 1.

As the outer layer of the helmet was modelled from Fibre-Reinforced Composite (FRC) the failure material model used was Chang-Chang composite damage model. Matrix cracking and the fibre fracture are induced by the tensile failure. In addition to this, composite failure criteria was also used to anticipate the failure in fibre shear out, compression of matrix, matrix cracking and fibre rupture. Tensile failure of composite is predominantly used as an element detection criterion as the matrix and the fibre of helmet holds on the compressive forces and also resists the compressive forces after failing in compression.

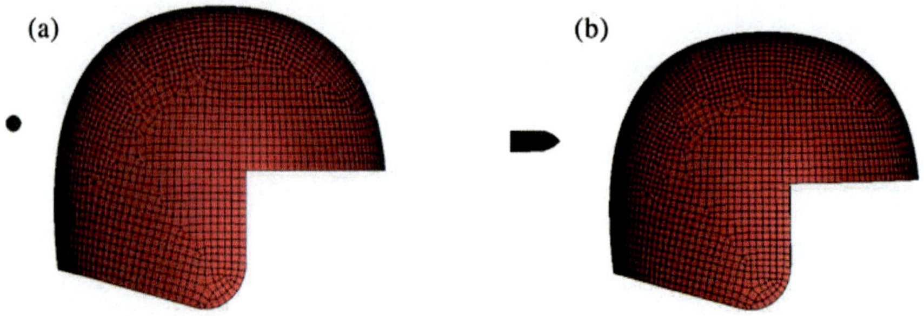

Fig. 1 Finite-element model **a** With spherical ball, **b** With ogive projectile

Table 1 Material properties of Kevlar and CFRP

Material property	Symbol	CFRP [12]	Kevlar [8]
Coefficient of elasticity in fiber direction (a), (GPa)	E_a	132	18.51
Coefficient of elasticity in transverse direction (b), (GPa)	E_b	7.8	18.51
Coefficient of elasticity in transverse direction (c), (GPa)	E_c	7.8	6
Poisson's ratio, ac plane	ϑ_{ab}	0.33	0.33
Poisson's ratio, bc plane	ϑ_{ac}	0.35	0.33
Poisson's ratio, ab plane	ϑ_{bc}	0.33	0.25
Shear modulus, ac plane, (GPa)	G_{ab}	4.81	2.5
Shear modulus, ab plane, (GPa)	G_{ac}	4.81	0.77
Shear modulus, bc plane, (GPa)	G_{bc}	3.76	2.5
Tensile strength in fiber direction (a), (GPa)	Xa^T	2.07	0.555
Tensile strength in transverse direction (b), (GPa)	Ya^T	0.061	0.55
Compressive strength in transverse direction (b), (GPa)	Ya^T	0.15	1.086
Shear strength, ab plane, (GPa)	Sa^L	1.3	0.588
Shear strength, ac plane, (GPa)	Sb^T	0.11	1.06
Shear strength, bc plane, (GPa)	Sc^T	0.11	1.06
Density (Kg/m^3)	ρ	1600	1230

3 Results and Discussion

3.1 Validation of Numerical Model

Firstly, the projectile impacts at two locations, in the front and other is in the right side at a velocity of 205 m/s and 220 m/s, respectively. The dent region diameter and inside surface deformation were validated with the results obtained from Tham's [6] experiments and the numerical results from B. T. Long [10] and Zhihua Cai [8]. Table 2 shows the validation of dent diameter and deformation (in bracket difference

% value of corresponding model from numerical results has been shown). Secondly, the Kevlar helmet was impacted at disparate locations with varying initial velocities of a spherical projectile. The residual velocities attained from the numerical simulations were correlated with the experimental results obtained from Rubio [11]. The validations based on residual velocities are given in Table 3. It was found that the numerical results from the present model and the results from the experiments performed in literature for residual velocities were in good correlation. The impact velocities of the projectile considered for validation were higher than V50 velocity regime. This proved that the helmet numerical model was capable of predicting the ballistic performance of both Kevlar and CFRP under different projectiles shape impact.

Four thicknesses 5.5 to 8.5 were taken and the respective parameters were compared. All the impacts were done against the spherical projectile. From the results, it was found that the dent region diameters were 42 mm, 37 mm, 34 mm and 30 mm for 5.5 mm, 6.5 mm 7.5 mm and 8.5 mm shell thickness, respectively. Values of mass increased with increase in the thickness, whereas the values of back face deformation decreased with increase in thickness. It was found that after thickness of 8.5 mm the mass value reached very high which is not suitable for wearing.

Table 2 Validation of FE model for CFRP material

Impact direction	Front 205 m/s				Side 220 m/s			
Comparisons	Tham's model [6]	Lee's model [9]	Cai's model [8]	Results obtained	Tham's model [6]	Lee's model [9]	Cai's model [8]	Results obtained
Dent region (diameter) (mm)	42 (23.5%)	32 (5.8%)	28 (17.6%)	34	42 (16.6%)	32 (11.1%)	32 (11.1%)	36
Maximum inside surface deformation (mm)	15 (56%)	6.4 (33%)	6.5 (32%)	9.58	18.7 (71.7%)	10.4 (4.5%)	9.6 (11.8%)	10.89

Table 3 Validation of FE model of Kevlar material

Impact location	Impact velocity (m/sec)	Residual velocity (m/sec)		Difference (%)
		Experimental (Rubio's model [11])	Current numerical results	
Frontal	641	242	235.786	2.63
Left	662	307	302.39	1.52
Rear	671	408	398.4	2.40
Left	679	408	401.58	1.59

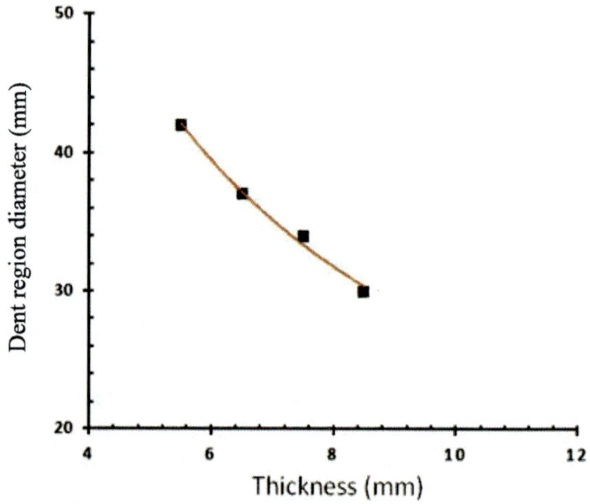

Fig. 2 Variation of performance parameters with change in thickness dent region diameter

3.2 Ballistic *Performance of CFRP Helmet*

See Fig. 2.

3.3 Effect of the Projectile Shapes on Ballistic Performance of CFRP Helmet

3.3.1 Ballistic Limit

The mostly used parameter to assess the impact performance of the target body is the ballistic limit. Numerical simulations were carried on 7.5 mm thick CFRP helmet with two different projectiles namely spherical and ogival. Both projectiles were impacted on the rear side of the helmet with varying velocities, see Table 4. The ballistic limit for both spherical projectile and the ogive nosed projectile were found out. Based on the numerical simulations, it was found that the ballistic limit for ogive projectile and spherical projectile were 143.5 and 155 m/s, respectively. For the ogive projectile, the bullet partially penetrates at a velocity of 138 m/s and completely penetrates at a velocity of 145 m/s. The ballistic limit of the CFRP helmet impacted against spherical projectile was 7.41% higher than impacted by the ogive nosed projectile. Though the ballistic limit was nearer, it was seen that the impact of different projectiles could affect the ballistic limit. It was noted that when the ogive projectile impacts on the helmet the sharp edge of projectile strikes the helmet shell which led to the failure of the composite at a lower ballistic limit when compared to spherical ball. Figure 3 shows the comparison of ballistic limit for both the projectiles.

Table 4 Numerical results for residual velocity against ogival and spherical nosed projectile

	Spherical projectile		Ogival projectile	
Impact direction	Impact Velocity (m/sec)	Residual Velocity (m/sec)	Impact Velocity (m/sec)	Residual Velocity (m/sec)
Back side	250	239.236	250	242.346
	200	161.727	200	177.062
	180	137.596	180	149.287
	170	105.95	170	131.076
	165	76.33	160	99.325
	160	0	155	71.708
			145	0

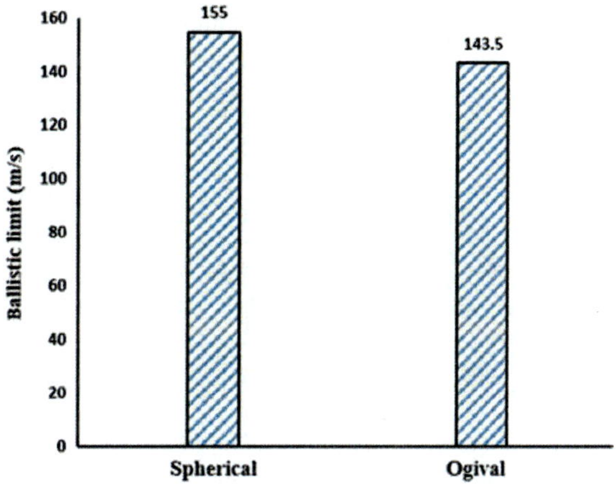

Fig. 3 Ballistic limit comparison for different projectiles against CFRP helmet

3.3.2 Damage Assessment

The damage on the outer and inner surface of the CFRP helmet against different projectiles, under the complete perforation obtained from the simulation results are shown in Fig. 4 against spherical projectile and Fig. 5 against ogive nosed projectile. Damage assessment has been done of both the fibres by their extreme deformation in the direction of projectile motion. During numerical simulation when stiffness of the element becomes zero, it eroded from the parent model. By measuring the distance of eroded element from the point of contact, damage assessment was calculated. When the CFRP helmet was impacted by both the projectiles, the major damage area was seen at the inner face. Most importantly delamination is phenomenon due to which the composite fails and it is most likely to observed on the back side. As the ogive projectile has relatively sharp nose, the rupture was seen through the tensile failure of the composite. The ruptured fibre remained in contact with helmet by creating illusion that the bullet did not penetrate the helmet. When compared to the ogive projectile impact, higher damages were present when impacted by a spherical projectile. When spherical projectile impacts on the helmet, then more stretching of fibre was observed when compared to ogive projectile impact. Central delamination of the inner face took place on the back side of the helmet almost from top to bottom which is half of the helmet height. Intense cracks and delamination were observed on both inner and outer side of the helmet. The damages and delamination are more for the spherical projectile impact because of the high energy absorption. Cross failure pattern was seen on the helmet due to the directional straining of composite from fibre rupture. Circular localized failure was observed on the outer side of the helmet. Centralized damage was observed on the outer side of the helmet when impacted by ogive projectile. All most complete delamination of the fibre was observed on the inner surface of the helmet which was mainly due to the fibre rupture. Localized fibre rupture in the impact region is the main phenomenon when impacted by ogive projectile and the energy absorption increased when there is more stretching of the fibre before the failure. As the projectile penetrates the helmet, inter laminar shear stresses were developed which can be seen in the form of the delamination. It was seen that the ogive projectile was more effective penetrator.

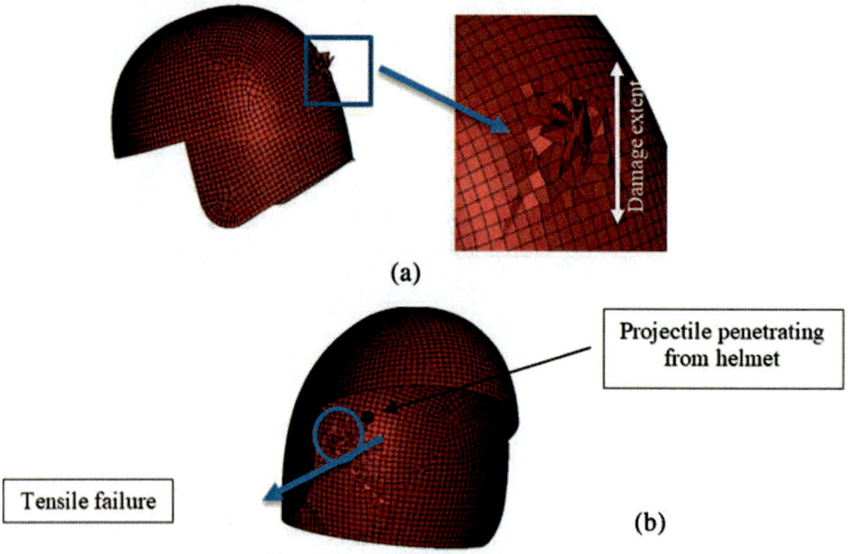

Fig. 4 Damage assessment of CFRP helmet when impacted by spherical projectile **a** Outside view at impact location, **b** Inside view at impact location

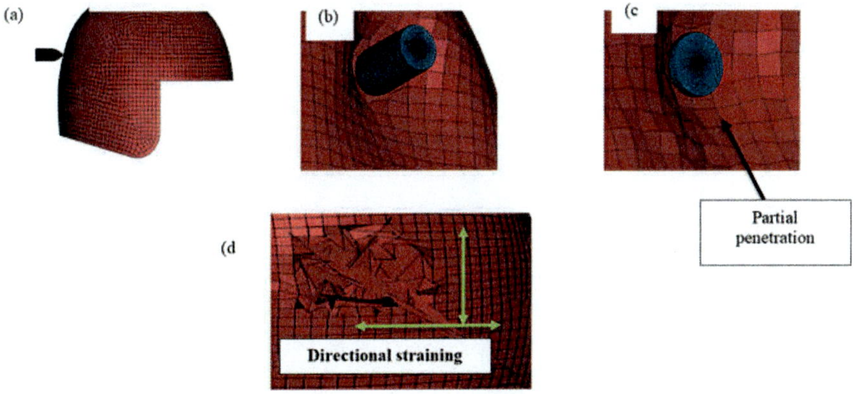

Fig. 5 Damage assessment of CFRP helmet when impacted by ogive projectile **a** at time $t = 1$ ms, **b** $t = 3$ ms, **c** $t = 4$ ms, **d** $t = 6$ ms

3.4 Ballistic Performance of Kevlar Helmet

Figure 6 shows the residual velocity values of the helmet against both the projectiles. The ballistic limit for both spherical projectile and the ogive nosed projectile were acquired for a Kevlar helmet. The projectiles were only impacted on the back face of the helmet as on the back side more severe injuries. For an ogival impact, partial penetration of the projectile was seen at a velocity 525 m/s where the residual velocity was zero. The projectile completely penetrated at a velocity of 593 m/s. This indicated that the ballistic limit could be between 525 and 593 m/s. It was perceived that the projectile partially penetrated up to 551 m/s and after that the projectile completely penetrated.

For a spherical projectile impact, partial penetration of the projectile was seen at a velocity 537.8 m/s where the residual velocity became zero. The projectile completely penetrated at a velocity of 604 m/s. So, the ballistic limit would be in between 537.8 and 604 m/s. The limit was found to be 570.9 m/s. Finally, from the numerical analysis the ballistic limit was concluded as 551.5 m/s for ogive projectile and 570.9 m/s for a spherical projectile. The ballistic limit of the helmet made of Kevlar impacted against spherical projectile was 3.39% higher than impacted by the ogive nosed projectile. Though the ballistic limit was nearer, it was seen that different projectiles had larger impact on the ballistic limit and the reason behind lower ballistic limit against ogive projectile was due to the sharp nose of it. Damage assessment is almost similar for both type of helmets.

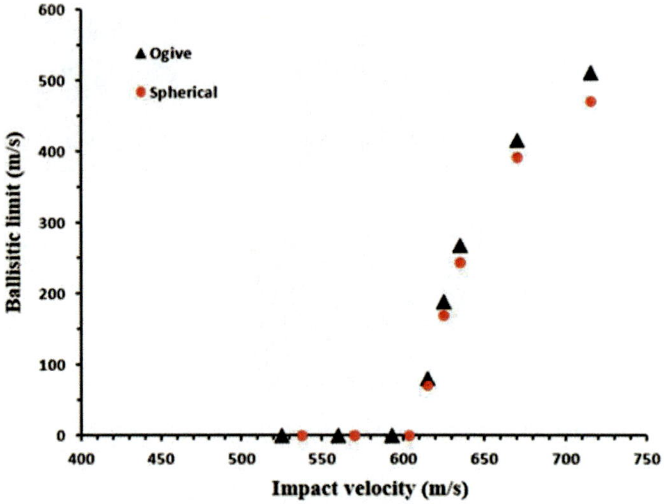

Fig. 6 Variation of residual velocity with respect to impact velocity

4 Conclusion

Ballistic performance of composite helmet with a 7.5 mm shell thickness against ogive and spherical projectile was evaluated numerically by varying the helmet material. Two composites namely, CFRP and Kevlar were used. Numerical simulations were performed by using LS-PrePost software. The ballistic performance of the helmet was evaluated under the basis of ballistic limit, damage pattern, inside surface deformation and dent region diameter. The shell thickness of the helmet significantly affected the ballistic performance of the composite helmet. The dent region diameter and the back face deformation decreased with increase in the thickness. After a thickness of 8.5 mm the mass was considered to be very high. The composite material of helmet remarkably affected the ballistic performance of the helmet against both projectiles. The ballistic limit of a CFRP helmets were 143.5 m/s and 155 m/s against ogival and spherical projectile respectively. The energy absorbed against spherical projectile was 6.4% higher than the energy that was absorbed against ogive projectile. The ballistic limit of a helmet made of Kevlar composite were 551.5 and 570.9 m/s against ogival and spherical projectile respectively. As the sharpness of projectile's tip increases, the ballistic limit decreased for both the composites. When the helmet was impacted by ogive projectile; localized rupture was noticed, whereas for a spherical projectile stretching of fibre and delamination was perceived. From the comparative study, it was found that Kevlar helmet offered more ballistic resistance when compared to the CFRP on the basis of the ballistic limit values.

References

1. Kulkarni S, Gao X-L, Horner S, Zheng J, David NV (2012) Ballistic helmets: their design, materials, and performance against traumatic brain injury. Composite Struct 101. 313–331. https://doi.org/10.1016/j.compstruct.2013.02.014
2. Jamroziak K, Bajkowski M, Bocian M, Polak S, Magier M, Kosobudzki M, Stepien R (2019) Ballistic head protection in the light of injury criteria in the case of the Wz.93 combat helmet. Appl Sci 9:2702. https://doi.org/10.3390/app9132702
3. Goldsmith W, Dharan H, Chang H (1995) Quasi-static and ballistic perforation of carbon fiber laminates. Int J Solids and Struct 32:89–103. https://doi.org/10.1016/0020-7683(94)00109-A
4. Velmurugan R, Naresh K, Shankar K (2017) Influence of fibre orientation and thickness on the response of CFRP composites subjected to high velocity impact loading. Adv Mater Process Technol 4.1–12. https://doi.org/10.1080/2374068X.2017.1410688
5. Rodríguez Millán M, Moreno C, Marco M, Santiuste C, Miguelez H (2015) Numerical analysis of the ballistic behaviour of Kevlar(R) composite under impact of double-nosed stepped cylindrical projectiles. J Reinforced Plastics and Compos 35. https://doi.org/10.1177/0731684415608004
6. Tham CY, Tan VBC, Lee H (2008) Ballistic impact of a KEVLAR((R)) helmet: experiment and simulations. Int J Impact Eng 35:304–318. https://doi.org/10.1016/j.ijimpeng.2007.03.008
7. Aare M, Kleiven S (2007) Evaluation of head response to ballistic helmet impacts using the finite element method. Int J Impact Eng 34:596–608. https://doi.org/10.1016/j.ijimpeng.2005.08.001
8. Cai Z, Li Z, Dong J, Mao Z, Wang L, Xian C (2016) A study on protective performance of bullet-proof helmet under impact loading. J Vibroeng 18. https://doi.org/10.21595/jve.2015.16497
9. Lee H, Gong, Wei S (2010) Finite element analysis for the evaluation of protective functions of helmets against ballistic impact. In: Computer methods in biomechanics and biomedical engineering, vol 13. pp 537–50. https://doi.org/10.1080/10255840903337848
10. Van Hoof J, Deutekom MJ, Worswick MJ, Bolduc M (1999) Experimental and numerical analysis of the ballistic impact response of composite helmet materials. In: 18th International symposium on Ballistics, San Antonio, TX
11. Rubio Diaz I, Rodríguez Millán M, Marco M, Olmedo A, Loya (2019) Ballistic performance of aramid composite combat helmet for protection against small projectiles. Composite Struct 226:111153. https://doi.org/10.1016/j.compstruct.2019.111153
12. Gharde N, Tiwari G, Shrivastava S (2021) Comparative study of impact response of GFRP and CFRP composite plates against blunt nosed projectile. Mater Today: Proc 41:1055–1059

Dynamic Indentation of Prismatic Li-Ion Battery Cells

Krishna Prasath Logakannan, Feng Zhu, David Sypeck, Jie Deng, and Sangyeon Kim

Abstract Safety design of the Electric Vehicles (EVs) involves protecting the battery system during a crash. A battery system consists of one or more battery packs. Each battery pack is the assembly of battery modules consisting of multiple cells connected in series or parallel. It is crucial to understand the failure behaviour of an individual cell to improve its safety in the crash accidents. The cell studied in this chapter consists of four layers of jellyrolls enclosed in an aluminium case. Jellyrolls are highly strain-rate dependent and exhibit non-linear response due to the presence of liquid electrolyte. Compression tests were carried out on the jellyroll material at the strain rates ranging from 9×10^{-4}/s to 657/s to characterize its dynamic behaviour. Based on the compression data, Finite Element (FE) model was established to predict the indentation damage of the cells subject to flat and hemispherical punches. Experimental results from our previous work were utilized to validate the model in terms of force–displacement curve and deformation mode.

Keywords Lithium-ion batteries · Impact · Dynamic indentation · Finite-element model

K. P. Logakannan · F. Zhu (✉)
Hopkins Extreme Materials Institute, Johns Hopkins University, Baltimore, MD, USA
e-mail: fzhu8@jhu.edu

F. Zhu
Department of Mechanical Engineering, Johns Hopkins University, Baltimore, MD, USA

Ralph S. O'Connor Sustainable Energy Institute, Johns Hopkins University, Baltimore, MD, USA

D. Sypeck
Department of Aerospace Engineering, Embry-Riddle Aeronautical University, Daytona Beach, FL, USA

J. Deng · S. Kim
Department of High Voltage Battery System, Ford Motor Company, Dearborn, MI, USA

R. Velmurugan et al. (eds.), *Dynamic Behavior of Soft and Hard Materials Volume 1*, Springer Proceedings in Materials 34, https://doi.org/10.1007/978-981-99-6030-9_15

1 Introduction

Li-ion batteries are being used as the power source for Electric Vehicles (EV) [1]. Currently, there are several types of Li-ion cell in use, namely pouch cell, prismatic cell, cylindrical cell, and elliptical cell, out of which, the most commonly used types on EVs are pouch and prismatic cells. Understanding the behaviour of these cells subject to impact loading is critical for improving EVs' crash safety, as we know material properties may change at high strain rate [2, 3]. Up to date, most of the work in the open literature is focused on pouch cells [4–11]. Prismatic cells are much less studied although they are also popular. A prismatic cell is formed by stacking jellyrolls and enclosed by a metal case. Zhu et al. [12] conducted tests and numerical simulations to study quasi-static indentation response of prismatic cells and jellyrolls. Another study [13] tested the dynamic indentation behaviour of similar prismatic cells. However, no numerical simulations on the dynamic loading conditions have been reported. FE modelling can be performed through two different ways, namely a detailed model [14] and a simplified model [12]. In the detailed model [14], all the components of jellyrolls are described in detail and an accurate prediction can be made at high computational cost. On the other hand, in simplified models [12], the jellyrolls are modelled as a homogeneous foam-type material which saves computational time and allows investigating the basic trend rather than detailed failure of each individual component. Jellyrolls are known to exhibit strain rate-dependent properties [15–17]. Unlike conventional solid materials, presence of liquid electrolyte influences the strength of jellyroll in a non-linear manner. A recent study showed the effect of liquid on the behaviour of jellyroll using pore-fluid model [18]. In this work, to develop the numerical models for prismatic cells, we first performed compression tests of jellyroll at a wide range of strain rates. Stress–strain curves were then used as the input for the jellyroll material modelling. Based on the jellyroll model, the dynamic indentation for cells was simulated. The FE model predictions were validated using our test data [19].

2 Battery Structure and Experimental Procedure

2.1 Battery Cell

Prismatic battery cells used in this work were obtained from a commercial battery module. The capacity of the cell is 63 Ah and rated voltage of 3.7 V. More information about the battery specification, extraction of cell, and jellyroll can be found in our previous work [13, 19]. All the tests were carried out on discharged cells and jellyrolls to prevent any potential fire hazard. Image of the cell together with the dimension is shown in Fig. 1a.

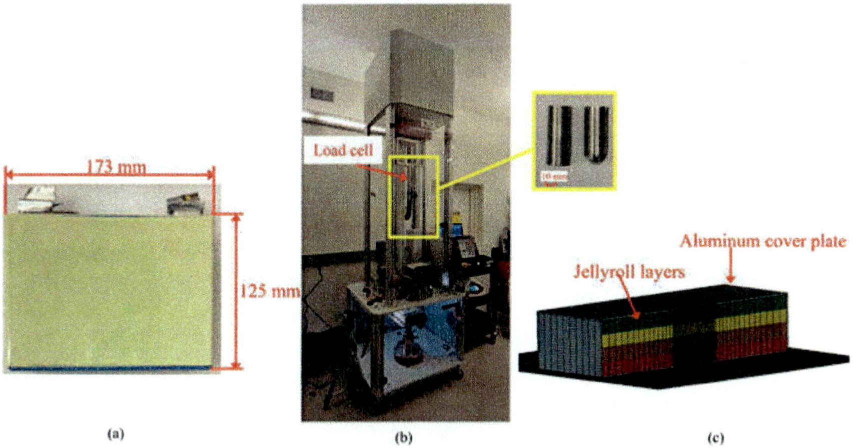

Fig. 1 **a** Picture of the battery cell (out of plane dimension is 45 mm); **b** Dynamic indentation test setup for the cells with the two indenters used; and **c** FE model showing the four layer jellyrolls

2.2 *Jellyroll Compression Tests*

Jellyrolls extracted from the cells were subjected to the uniaxial compressions in the thickness direction at various strain rates ranging from 9×10^{-4}/s to 657/s. Compression tests for strain rate tests up to 9/s were conducted in the constant speed mode using an MTS machine (Criterion$^{(R)}$ Series 40) equipped with a 100 kN load cell. Higher strain rate compressive tests were conducted in the impact mode with an INSTRON drop tower CEAST 9350 setup with the load cell capacity of 90 kN. More details regarding the strain rates and test setup are described in Table 1. Impact mode tests were conducted at the initial impact speed of 1.5 m/s, 4.5 m/s, and 7.5 m/s, corresponding to the strain rates of 132/s, 412/s, and 654/s, respectively. Additional masses of 60 and 6 kg were added for compression tests at 1.5 m/s and 4.5 m/s, respectively, to increase the impact energy to rupture the jellyroll layers. For all strain rates, jellyrolls were cut into approximately 25 mm × 25 mm using a machine saw. The thickness of the jellyroll was approximately 11 mm. The samples were tested within an hour of cutting as the electrolyte tends to alter the properties of jellyroll under prolonged exposure to atmospheric conditions.

2.3 *Dynamic Cell Indentation Tests*

The dynamic cell indentation tests were carried out using two different indenters, namely, a flat punch and a hemispherical punch, which were mounted on the aforementioned drop tower system. The test setup and punches are shown in Fig. 1b. The machine is equipped with a damping mechanism which prevents multiple impacts

Table 1 Details of the strain rates and facilities in the jellyroll compression tests

S. No.	Strain rate (/s)	Machine and capacity
1	9×10^{-4}	MTS and 100 kN
2	9×10^{-3}	MTS and 100 kN
3	9×10^{-2}	MTS and 100 kN
4	9×10^{-1}	MTS and 100 kN
5	9	MTS and 100 kN
6	132	Instron drop tower and 90 kN
7	412	Instron drop tower and 90 kN
8	657	Instron drop tower and 90 kN

on the sample. The diameters of the punches were 1 inch. More details regarding the cell tests can be found in our previous work [19].

3 Finite-Element Model

Finite-element simulations were conducted to predict the cell deformation and impact force measured in the tests using LS-DYNA explicit solver. In order to reduce the computational time, only one half of the cell was modelled with the symmetry boundary applied at the cut section as shown in Fig. 1c. In this work, only the highest velocity (7.5 m/s) case was simulated. The models for all other speeds were all similar. The indenters were modelled as a rigid shell and set an initial velocity of 7.5 m/s. Density of the rigid material was adjusted to match the mass of indenter plus the drop tower. The cell was placed on a fully fixed rigid plate. ERODING_SURFACE_TO_SURFACE contact was applied between aluminium outer casing and jellyroll stack. Jellyroll was simplified and modelled as a porous medium using MAT_MODIFIED_CRUSHABLE_FOAM (MAT_163) material law, which is an extension of MAT_CRUSHABLE_FOAM model (MAT_63). MAT_163 allows users to define strain rate dependent stress–strain curves of foam. The test data obtained in Sect. 2.2 were used as the model input. Density, elastic modulus, Poisson's ratio, and tensile stress cut off values were 2.6×10^{-6} kg/mm^3, 0.5 GPa, 0.01, and 0.01 GPa, respectively. Failure of the foam was applied using MAT_ADD_EROSION with criteria of EPSSH = 0.1 (maximum shear strain). Foam failure was not modelled for flat punch model since no failure was observed in experiments. Aluminium was modelled as MAT_JOHNSON_COOK_ORTHOTROPIC (MAT_099) and the corresponding material constants are shown in Table 2.

Table 2 Material properties used for cover plate

Density (kg/mm^3)	Elastic modulus (GPa)	Poisson's ratio	Johnson–Cook model parameters			
			A (GPa)	B (GPa)	n	C
2.7 × 10^{-6}	52	0.33	0.163	0.156	0.32	0.015

4 Results and Discussion

4.1 *Strain Rate Effect and Failure Mode of Jellyrolls Observed in the Tests*

Figure 2 shows the stress–strain curves obtained from the compression tests of jellyrolls at various strain rates. It can be seen that the curves at each strain rate show a repeatable pattern. In all the cases, the initial peak stress corresponds to the failure of the jellyroll. Figure 3a and b shows the comparison of the typical stress–strain curves at various strain rates obtained through constant speed and impact modes, respectively. For constant speed compressions, the initial peak stress decreases drastically for 0.9/s and 9/s. This trend is due to the presence of liquid electrolyte and its permeability. As discussed by Kisters et al. [18], the presence of liquid can alter the peak stress and slope during deformation. They also reported that highly permeable liquid did not change the initial slope much as speed increases, however, low permeable liquid increased the slope as speed increases. In this case, the peak stress decreases with relatively no change in slope, which indicates the liquid is highly permeable.

For impact mode tests, curves have similar trends for all the strain rates without significant change in peak stress which could be due to strain rate effect as jellyroll tends to become stronger at higher strain rates [20], and this would limit the effect of liquid electrolyte. Figure 4 shows the failure modes of the jellyroll at constant speed mode and impact mode. For both cases similar periodic V-shaped fracture pattern can be observed.

4.2 *Finite Element Modelling Results for Cells Indentation*

Figures 5 and 6 compare the force—displacement curves and deformation obtained from experiments and finite-element simulations. Experimental curves and deformation modes shown in Figs. 5b and 6b have been reported in [19], and they are used here for comparison purpose. For flat punch (Fig. 5a), it can be seen that the force increases till the displacement of around 3 mm and starts to drop. The drop in force corresponds to the failure of the cover plate as noticed in the experiments and FE model. The oscillation in experimental force is due to the stress waves generated in the impactor and force transducer during the impact. The cover plate is ruptured along the circumference of the flat end of the punch. Similar deformation was observed in both

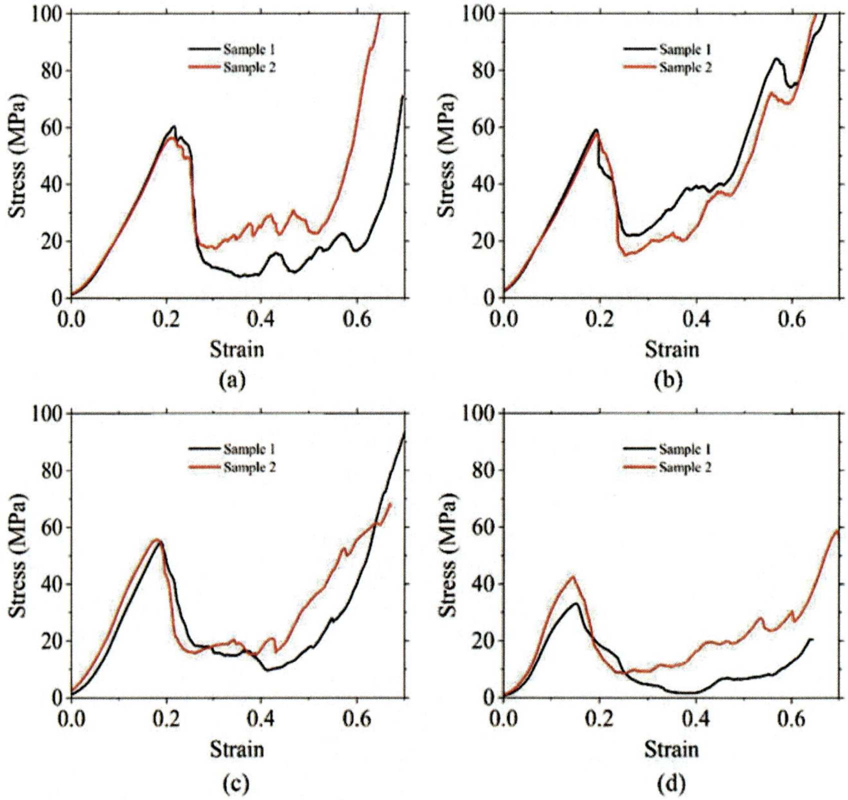

Fig. 2 Stress–strain curves of jellyroll obtained through compression tests at various strain rates: **a** 9×10^{-4} /s; **b** 9×10^{-3} /s; **c** 9×10^{-2} /s; **d** 9×10^{-1} /s; **e** 9 /s; **f** 132 /s; **g** 412 /s; and **h** 657 /s

experiments and FE model (Fig. 5b and c). No failure was observed in the jellyrolls. For hemispherical punch (Fig. 6a), the force increases to the peak at the displacement of 6 mm and then begins to drop. The deformation is compared in Fig. 6b and c. The aluminium cover plate is ruptured due to tensile tearing, and subsequent compression of the first layer of jellyroll. The hemispherical punch can produce a deeper indentation than the flat punch does. This could be due to the higher contact area for flat punch and therefore more energy is needed to penetrate the aluminium cover. The voltage changes measured during the tests cannot be simulated with an FE model, so they are not shown in this chapter.

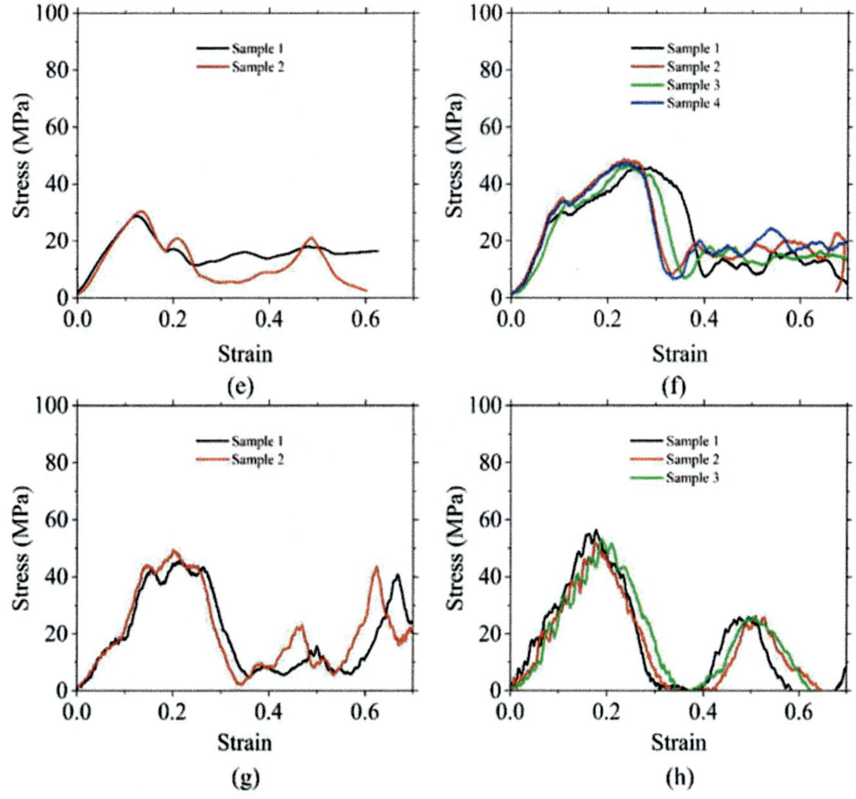

Fig. 2 (continued)

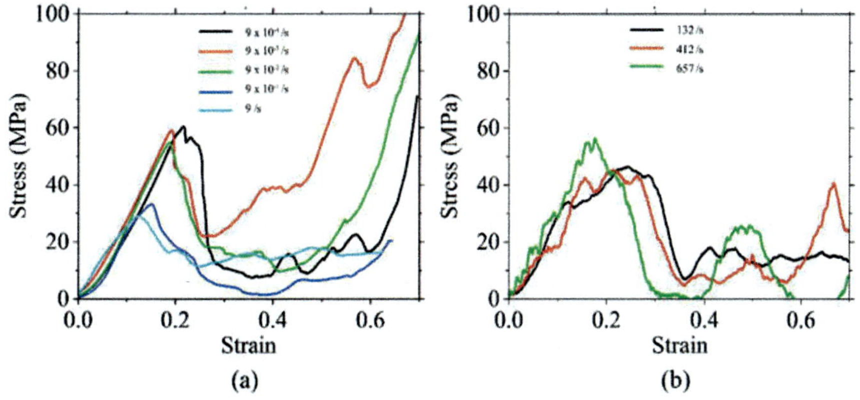

Fig. 3 Typical stress–strain curves obtained in the jelly compression tests at various strain rates; **a** Constant speed mode; and **b** Impact mode

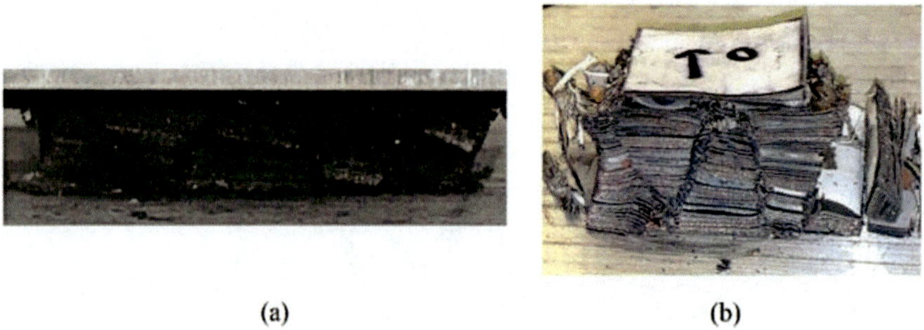

Fig. 4 Failure modes of jellyroll in the compression tests **a** Constant speed mode; and **b** Impact mode

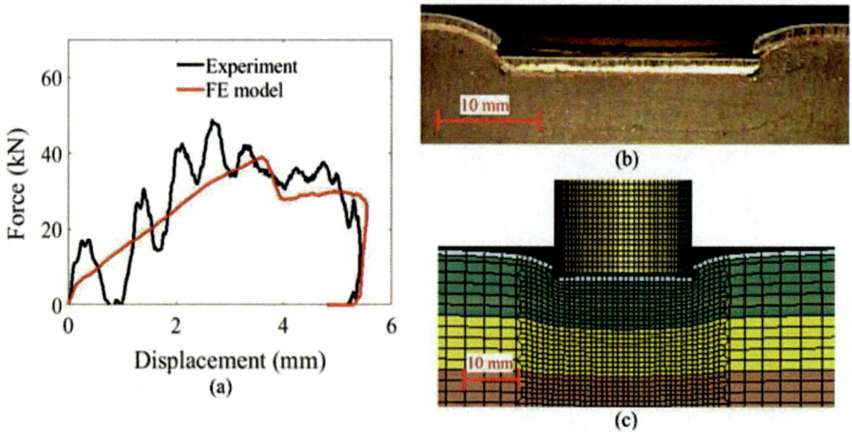

Fig. 5 Comparison of test and simulation results for cell indentation at 7.5 m/s using a flat punch: **a** Force—displacement curves; **b** Deformation observed in the experiment; and **c** Deformation predicted by the FE model

5 Conclusions

Jellyrolls in the prismatic Li-ion battery cells were subjected to compressions at various strain rates. The results show that the initial peak stress decreases when the strain rates are in the range from 9×10^{-4} to 9 /s. However, at higher strain rates, the difference is not significant. Fracture of jellyroll exhibits a repeated V pattern, and the mode can be observed for all strain rates. Results of the compression tests were used as the input of jellyroll numerical model. Jellyroll and aluminium housing structure models were then integrated to form the cell model to simulate the two punch indentation tests at 7.5 m/s. The model simulated responses were compared with the cell test data for validation. The FE model was able to predict the deformation

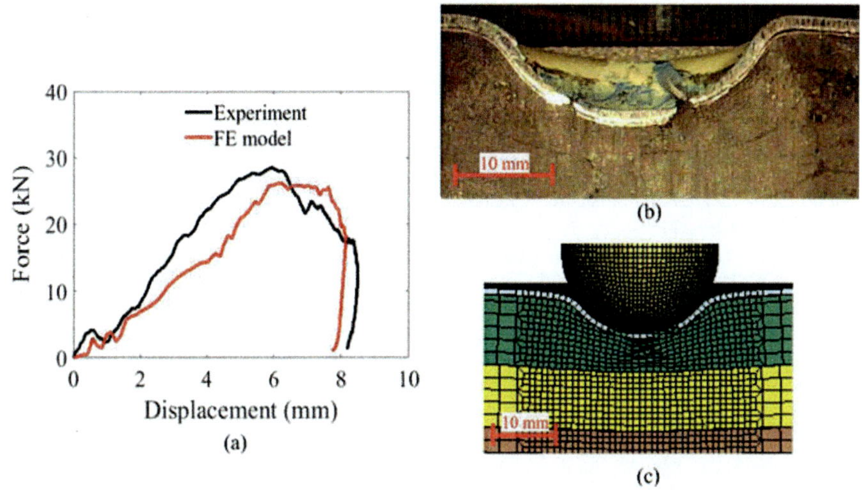

Fig. 6 Comparison of test and simulation results for cell indentation at 7.5 m/s using a hemispherical punch: **a** Force–displacement curves; **b** Deformation observed in the experiment; and **c** Deformation predicted by the FE model

and forces well. As an ongoing work, this validated cell model with be combined with the other components to simulate the crush tests at the battery module level.

Acknowledgements This work is funded by Ford Motor Company through a University research program and Hopkins Extreme Materials Institute (HEMI) through a Seed Grant. The financial support is gratefully acknowledged.

References

1. Masias A, Marcicki J, Paxton WA (2021) Opportunities and challenges of lithium ion batteries in automotive applications. ACS Energy Lett 6:621–630. https://doi.org/10.1021/acsenergylett.0c02584
2. Naresh K, Shankar K, Rao BS, Velmurugan R (2016) Effect of high strain rate on glass/carbon/hybrid fiber reinforced epoxy laminated composites. Compos B 100:125–135. https://doi.org/10.1016/j.compositesb.2016.06.007
3. Velmurugan R, Naresh K, Shankar K (2018) Influence of fibre orientation and thickness on the response of CFRP Composites subjected to high velocity impact loading. Adv Mater Process Technol 4:120–131. https://doi.org/10.1080/2374068X.2017.1410688
4. Sahraei E, Hill R, Wierzbicki T (2012) Calibration and finite element simulation of pouch lithium-ion batteries for mechanical integrity. J Power Sources 201:307–321. https://doi.org/10.1016/j.jpowsour.2011.10.094
5. Sahraei E, Meier J, Wierzbicki T (2014) Characterizing and modeling mechanical properties and onset of short circuit for three types of lithium-ion pouch cells. J Power Sources 247:503–516. https://doi.org/10.1016/j.jpowsour.2013.08.056

6. Chung SH, Tancogne-Dejean T, Zhu J, Luo H, Wierzbicki T (2018) Failure in lithium-ion batteries under transverse indentation loading. J Power Sour 389:148–159. https://doi.org/10.1016/j.jpowsour.2018.04.003
7. Li W, Xia Y, Chen G, Sahraei E (2018) Comparative study of mechanical-electrical-thermal responses of pouch, cylindrical, and prismatic lithium-ion cells under mechanical abuse. Sci China Technol Sci 61:1472–1482. https://doi.org/10.1007/s11431-017-9296-0
8. Zhu F, Du X, Lei J, Audisio L, Sypeck D (2021) Experimental study on the crushing behaviour of lithium-ion battery modules. Int J Crashworthiness 26:598–607. https://doi.org/10.1080/13588265.2020.1766397
9. Kisters T, Sahraei E, Wierzbicki T (2017) Dynamic impact tests on lithium-ion cells. Int J Impact Eng 108:205–216. https://doi.org/10.1016/j.ijimpeng.2017.04.025
10. Pan Z, Li W, Xia Y (2020) Experiments and 3D detailed modeling for a pouch battery cell under impact loading. J Energy Storage 27:101016. https://doi.org/10.1016/j.est.2019.101016
11. Tancogne-Dejean T, Grolleau V, Mohr D (2022) Strain rate dependent plasticity of lithium-ion pouch cells: experiments and simulations. Int J Impact Eng 159:104048. https://doi.org/10.1016/j.ijimpeng.2021.104048
12. Zhu F, Zhou R, Sypeck D, Deng J, Bae C (2022) Failure behavior of prismatic li-ion battery cells under abuse loading condition—a combined experimental and computational study. J Energy Storage 48:103969. https://doi.org/10.1016/j.est.2022.103969
13. Chen X, Wang T, Zhang Y, Ji H, Ji Y, Yuan Q (2019) Dynamic mechanical behavior of prismatic lithium-ion battery upon impact. Int J Energy Res er.4774. https://doi.org/10.1002/er.4774
14. Zhang C, Santhanagopalan S, Sprague MA, Pesaran AA (2015) Coupled mechanical-electrical-thermal modeling for short-circuit prediction in a lithium-ion cell under mechanical abuse. J Power Sources 290:102–113. https://doi.org/10.1016/j.jpowsour.2015.04.162
15. Kermani G, Sahraei E (2019) Dynamic impact response of lithium-ion batteries, constitutive properties and failure model. RSC Adv 9:2464–2473. https://doi.org/10.1039/C8RA08898E
16. Zhu J, Luo H, Li W, Gao T, Xia Y, Wierzbicki T (2019) Mechanism of strengthening of battery resistance under dynamic loading. Int J Impact Eng 131:78–84. https://doi.org/10.1016/j.ijimpeng.2019.05.003
17. Zhou M, Hu L, Chen S, Zhao X (2021) Different mechanical-electrochemical coupled failure mechanism and safety evaluation of lithium-ion pouch cells under dynamic and quasi-static mechanical abuse. J Power Sources 497:229897. https://doi.org/10.1016/j.jpowsour.2021.229897
18. Kisters T, Gilaki M, Nau S, Sahraei E (2022) Modeling of dynamic mechanical response of li-ion cells with homogenized electrolyte-solid interactions. J. Energy Storage 49:104069. https://doi.org/10.1016/j.est.2022.104069
19. Logakannan KP, Zhu F, Sypeck D, Deng J, Kim S (2022) Impact response of prismatic li-ion battery jellyrolls and cells. Int J Impact Eng 170:104352. https://doi.org/10.1016/j.ijimpeng.2022.104352
20. Xu J, Wang L, Guan J, Yin S (2016) Coupled effect of strain rate and solvent on dynamic mechanical behaviors of separators in lithium ion batteries. Mater Des 95:319–328. https://doi.org/10.1016/j.matdes.2016.01.082

Effect of Adding Expanded Polypropylene and Polyurethane Foams on Crashworthiness Response of the Aluminium-Reinforced Honeycomb Structure

Vaibhav Saptal, G. Tiwari, and T. Thomas

Abstract This chapter aimed to find the effects of adding Expanded Polypropylene and Polyurethane foams on the crashworthiness response of reinforced aluminium honeycomb structures. Cell size and cell wall thickness were taken as 20 mm and 0.11 mm respectively for a honeycomb of node length 25 mm. Expanded Polypropylene (EPP) foams of densities of 20, 40, 60 kg/m^3 and Polyurethane foams (PU) of densities 16, 29 and 42 kg/m^3 were added to either side of the reinforced sheet. Finite-element simulations for foam-filled reinforced honeycombs were performed under quasi-static loading conditions using finite element code LS-DYNA. The effect of varying foam-filling percentages in reinforced honeycomb on crashworthiness parameters was also studied. The effect of varying PU foam density by keeping EPP foam density constant showed an increase in crashworthiness parameters peak load, mean load and specific energy absorption but crush force efficiency showed fluctuating behaviour in all the cases. Crashworthiness of both distinct type of structure showed superior performance as foam density increases.

Keywords Honeycomb core · Reinforced honeycomb · Quasi-static crushing

1 Introduction

One of the primary goals of the aerospace and automotive industries is to design lightweight and superior energy-absorbing crash-worthy structures to ensure occupant's safety. In recent years, honeycombs and foams got the attention of researchers due to their lightweight and energy absorption capabilities. Meran et al. [1] numerically studied the effect of varying foil thickness, cell size, cell expanding angle and impact velocity on the crashworthiness parameters of aluminium honeycomb under out-of-plane stress conditions. Results showed that as foil thickness and cell

V. Saptal (✉) · G. Tiwari · T. Thomas
Department of Mechanical Engineering, VNIT Nagpur, Nagpur 440010, India
e-mail: vaibhavsaptal9@gmail.com

R. Velmurugan et al. (eds.), *Dynamic Behavior of Soft and Hard Materials Volume 1*,
Springer Proceedings in Materials 34, https://doi.org/10.1007/978-981-99-6030-9_16

size increased; crush force efficiency and total energy absorption decreased. Crashworthiness parameters were unaffected by impact mass and velocity. Yin et al. [2] used a new approach for estimating the mean crushing stress of honeycomb structures with varying cell specifications under axial loading based on the super folding element theory and had a better agreement with the numerical simulation results. Tao et al. [3] studied the effect of strain rate on aluminium honeycombs to perform the crushing with quasi-static and dynamic loading and found that the rate-dependent shock theory yields more accurate predictions. The strain rate effect made a considerable contribution to the dynamic enhancement of metallic honeycombs across a wide velocity range. Pietras et al. [4] conducted static and low-velocity compression testing on polyurethane foam-filled aluminium honeycombs. The strength and energy absorption of the honeycomb structure was increased by more than 41% with the addition of 0.02 per cent reduced graphene oxide flakes. Due to the interaction effect, the foam-filled honeycomb showed better properties (up to 61%) than the bare honeycomb. The in situ foam-filling method was better than the ex-situ foam-filling method and the composition of foam greatly influences the collapse mechanism of the honeycomb. Nia and Sadeghi [5] conducted studies to determine the impact response of foam filling on the plastic behaviour and mechanical characteristics of honeycomb. Experimentation showed that foam filling enhanced the mean crushing strength and energy absorption capacity of panels up to 300 per cent. Moreover, the sum of the mean crush strengths of bare honeycomb and foam alone is lesser than the mean crush strength of foam-filled panels. Zhou et al. [6] studied the lateral crushing response of concrete foam-filled auxetic honeycomb made of aluminium under low velocity and quasi-static loading conditions. It was found that as foam density increases; modes of failure shift from crushing failure to failure under shear with high peak load. Zhang et al. [7] performed static and dynamic impact tests on expanded polypropylene foam-filled aluminium honeycomb and found higher peak and mean force but Specific Energy Absorption (SEA) was decreased. The initial peak strength, mean strength and SEA of filled specimens increased with increasing impact velocity. Compared to the static compression test, the initial peak strength in the dynamic impact test rose, however, the mean strength and SEA dropped. When different types of filling were compared, it was discovered that single-cell filling was an excellent choice for improving load resistance while utilising the least amount of filler material. Mahmoudabadi and Sadighi [8] theoretically and experimentally studied the aluminium honeycomb's crushing behaviour under quasi-static and low-velocity impact loading conditions. A novel static model was developed for determining the crushing strength and modes of failure of honeycomb structure by employing the energy method. Khan et al. [9] used Digital Image Correlation and performed experiments to crush aluminium honeycomb both axially and laterally and discovered that the honeycomb's crushing strength in the axial direction is greater than in the lateral direction. The ribbon direction strength was two times stronger than the transverse direction in lateral crushing. The deformation nucleated from the shear band at 45 degrees in the sample, according to the results of Digital Image Correlation. The deformation was restricted to the shear band's areas and the local plastic strain in the core was substantially larger for a given global strain level. Aminanda et al. [10] crushed honeycomb specimens made of paper, Nomex and aluminium under quasi-static loading conditions and found that all the honeycombs showed similar kinds of force–displacement curves. For Nomex honeycombs indented by various indenters, an analytical model was developed and compared results with experiments. Both analytical and experimental results were found close.

From the literature study, it is observed that reinforced honeycombs got less attention than conventional honeycombs. Very few studies are available on foam filled reinforced honeycomb structures. In this chapter, finite-element analysis for quasi-static crushing of foam-filled reinforced honeycomb in out of plane direction was performed using finite element code LS-DYNA. The effect of the addition of expanded polypropylene (by varying densities of 20, 40 and 60 kg/m^3) and polyurethane (by varying densities of 16, 29 and 42 kg/m^3) foams together on reinforced honeycomb was studied. The effect of varying foam filling percentages on crashworthiness parameters such as peak load, mean load, specific energy absorption and crushing efficiency were calculated and analysed.

2 Material Modelling

2.1 *Mat 024 (Piecewise_linear_plasticity)*

MAT 024 (PIECEWISE_LINEAR_PLASTICITY) was used to simulate the honeycomb material's behaviour i.e., aluminium under numerical simulation [11, 12]. This material model is a nonlinear constitutive model based on Von Mises yield criteria, which considers the nonlinear behaviour of aluminium until a fracture takes place [13].

The yielding function for the model is defined as follows [14],

$$\phi = \frac{1}{2} S_{\mathrm{mn}} \times S_{\mathrm{mn}} - \frac{S_y^2}{3} \leq 0 \tag{1}$$

where

$$S_y = \beta \times \left[S_0 + f_h\left(\varepsilon_{\mathrm{eff}}^p\right) \right] \tag{2}$$

ϕ = Yield function, S_y = Yield Surface's current radius, S_o = Flow stress.

$f_h\left(\varepsilon_{\mathrm{eff}}^p\right)$ = function for hardening, β = Strain rate effect parameter, S_{mn} = Deviatoric stress

Elasticity modulus, mass density, Poisson's ratio and yield strength for aluminium were taken as 69 GPa, 2.72 kg/m^3, 0.3 and 98.147 MPa respectively. To determine the strain rate effect parameter two options were available, the first one was the Cowper-Symonds model and the second one was the use of the stress–strain curve obtained

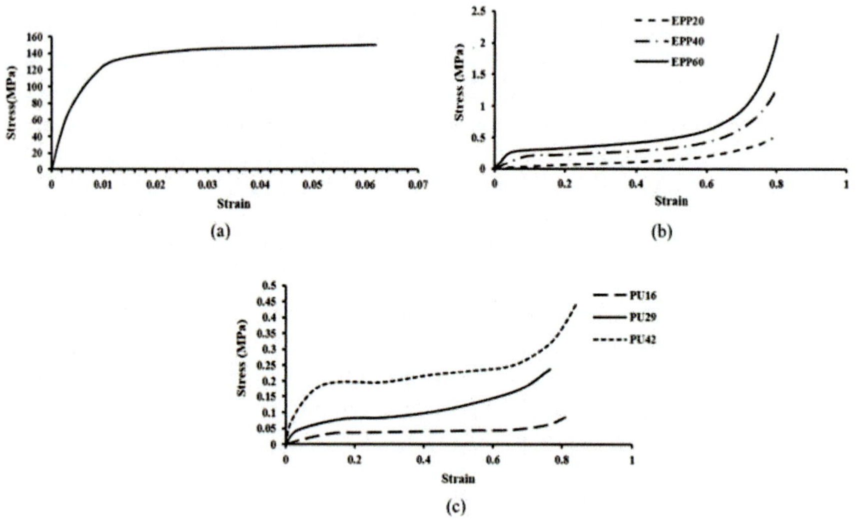

Fig. 1 Stress vs Strain curves for **a** Aluminium **b** Expanded polypropylene **c** Polyurethane

from the uniaxial tension test [14]. The stress–strain curve used in the simulation for aluminium was shown in Fig. 1a.

2.2 *Mat 057 (Low_density_foam)*

The material model of Mat 057 (LOW_DENSITY_FOAM) was used to simulate the expanded polypropylene and polyurethane foams behaviour under numerical simulation [15]. This foam model was used to simulate highly compressible low-density foams [14, 16]. It depicts compressive behaviour, which includes hysteresis during unloading [14]. Under the application of tensile stress, the material shows linear behaviour until it breaks. The low-density foam model was basically a Maxwell fluid consisting of a series of springs and dampers [14]. The modulus of elasticity and decay constant were used to characterise this in the input. For expanded polypropylene foams having densities of 20, 40 and 60 kg/m^3, Young's modulus values were taken as 0.667, 3.395 and 5.084 MPa, respectively. For polyurethane foams having densities of 16, 29 and 42 kg/m^3, Young's modulus values were taken as 0.2974, 1.2764 and 1.8405 MPa, respectively. As foam will not fail in tension in this application, the tension cut-off stress is left at the default value. The form of the unloading curve is controlled by two non-dimensional parameters, that is, the hysteric unloading factor and the shape factor, both of which are set to 0.101 and 25 respectively [16]. The damping coefficient of 0.5 is used to increase stability [16]. The default value of 0.0 is used for both Young's relaxation modulus and the decay constant. Stress–Strain curves were shown in Fig. 1.

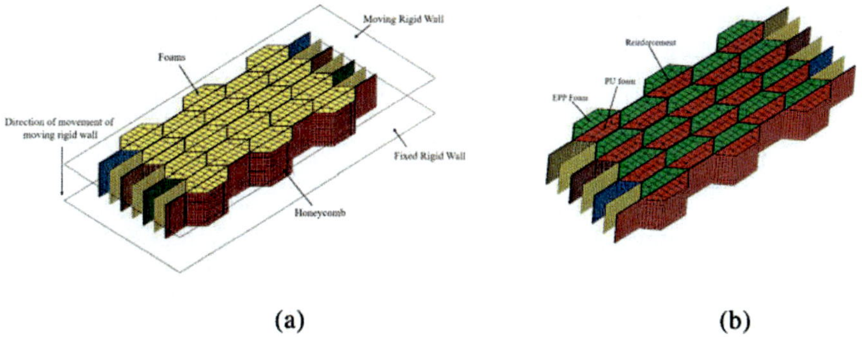

Fig. 2 **a** Typical boundary conditions for all honeycomb structures **b** Foam-filled reinforced honeycomb with two different foams together

3 Numerical Modelling Details

The honeycomb of node length 25 mm, cell size 20 mm and wall thickness 0.11 mm were used in the analysis. LS-DYNA was used to model the crushing behaviour of the aluminium honeycomb and foam. Santosa et al. [17] given concepts for quasi-static simulations, which states that the ratio of total kinetic energy to total internal energy must be very small throughout the simulation and the load–displacement curve should not be dependent on applied velocity when the rate of loading was increased to speed up the analysis were used for analysis. The completely integrated shell elements were used to model honeycomb structures [11, 12]. AUTOMATIC_ SURFACE_TO_SURFACE_ TIEBREAK contact was provided between two layers of honeycomb core with adhesive properties of normal stress and shear stress 45 MPa and 40 MPa, respectively [12]. The coefficient of friction between the surfaces was given as 0.3[12]. Between the aluminium honeycomb and expanded polypropylene foam AUTOMATIC_SURFACE_TO_SURFACE_CONTACT was used with a friction value of 0.2 [16]. The honeycomb model was put between two rigid planes. The top moving plane moved at 0.5 mm/s, whereas the bottom plane was fixed in all directions [12]. Typical Boundary conditions for all structures were shown in Fig. 2a.

4 Result and Discussion

4.1 Model Validation

Lu et al. [18] performed an experiment to crush bare honeycomb and Expanded Polypropylene foam-filled conventional honeycomb in the out of plane direction. Commercially available honeycomb structures having cell wall thickness, node length and cell width of 0.062 mm, 20 mm and 6.2 mm, respectively, were used by them for experiments. They performed quasistatic crushing tests on conventional honeycomb structures and Expanded Polypropylene foam-filled honeycomb structures (of varying densities 20, 40 and 60 kg/m^3) using a uniaxial tension testing machine. To validate the numerical model, conventional honeycombs and foam-filled honeycombs of the same dimensions and mechanical properties were used by Lu et al. [18] were used into LS-DYNA explicit FEA software. As shown in Fig. 3, the results obtained through the numerical model for both conventional honeycomb and EPP foam (of density 20 kg/m^3) filled honeycomb were in accordance with the experimental results presented by Lu et al. [18]. Therefore, the numerical model can be used in further studies (Table 1).

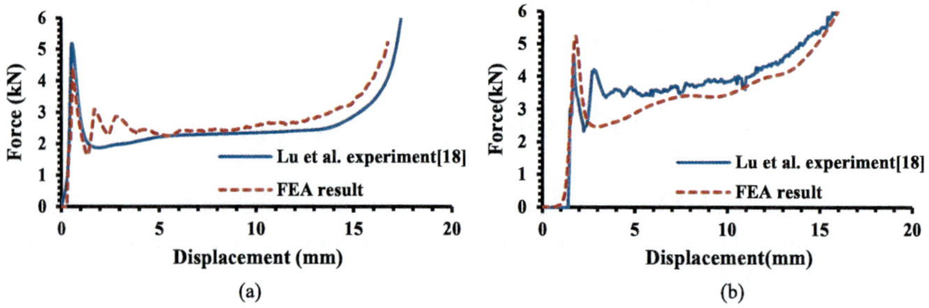

Fig. 3 Model validation **a** conventional honeycomb **b** EPP foam (of density 20 kg/m^3) filled honeycomb

Table 1 Model validation

	Conventional honeycomb		Foam-filled honeycomb	
Parameters	Lu et al. experiment [18]	FEA result	Lu et al. experiment [18]	FEA result
Peak load (kN)	5.14	4.42	5.33	4.66
Mean load (kN)	2.34	2.61	3.69	3.91
Crushing efficiency	45.56%	59.19%	69.32%	83.95%
Energy absorption (kN-mm)	34.9	38.2	52.1	54.3
Maximum crush length (mm)	14.8	14.6	14.1	13.88

4.2 *Effect of Filling Two Different Foams Together on Crashworthiness Response of Reinforced Honeycomb Structure*

The reinforced honeycomb structure provides the feature to add two different foams into the same structure as shown in Fig. 2b. With the help of this feature, properties of two foams can be used to enhance the crashworthiness performance of the honeycomb structure. So, for reinforced honeycomb structure, keeping expanded polypropylene foam of densities of 20, 40, 60 kg/m^3 on one side of the reinforced sheet and varying polyurethane foam of densities of 16, 29 and 42 kg/m^3 on another side of the reinforced sheet different cases were formed and analysed. The load–displacement curve showed similar behaviour to any crashworthy structure during its crushing. It showed three regions: elastic region, plateau stress region and densification region.

4.2.1 Keeping EPP Foam Density 20 kg/m^3 and Varying PU Foam Densities 16, 29, 42 kg/m^3 in Foam-Filled Reinforced Honeycomb

Effect of filing Expanded Polypropylene foam of density 20 kg/m^3 and varying Polyurethane foam of densities 16, 29, 42 kg/m^3 on the other side of the honeycomb structure showed the following variations in crashworthiness properties. As the density of polyurethane foam increases peak load increases. Peak load showed a maximum value of 37.8 kN for EPP20_PU42 (see Fig. 4b). Mean load also showed similar behaviour as peak load. As the density of polyurethane foam increases mean load also increases as shown in Fig. 4b and it showed a maximum value of 19.89 kN for EPP20_PU42. Crush force efficiency showed a sudden increase when PU foam density changed from 16 to 29 kg/m^3, but after that, it again decreased when the density of polyurethane foam changed from 29 to 42 kg/m^3. Specific energy absorption also increased as the density of foam increased from 16 to 42 kg/m^3.

4.2.2 Keeping EPP Foam Density 40 kg/m^3 and Varying PU Foam Densities 16, 29, 42 kg/m^3 in Foam-Filled Reinforced Honeycomb

Effect of filing Expanded Polypropylene foam of density 40 kg/m^3 and varying Polyurethane foam of densities 16, 29, 42 kg/m^3 on another side of the honeycomb structure showed the following variations in crashworthiness properties. As the density of polyurethane foam increases peak load increases. Peak load showed a maximum value of 38.5 kN for EPP40_PU42 (see Fig. 4d). Mean load also shows similar behaviour as peak load. As the density of polyurethane foam increases, mean load also increases as shown in Fig. 4d. It shows a maximum value of 19.89 kN for EPP20_PU29. Crush force efficiency showed a sudden increase when PU foam density changed from 16 to 29 kg/m^3, but after that, it again decreased when the

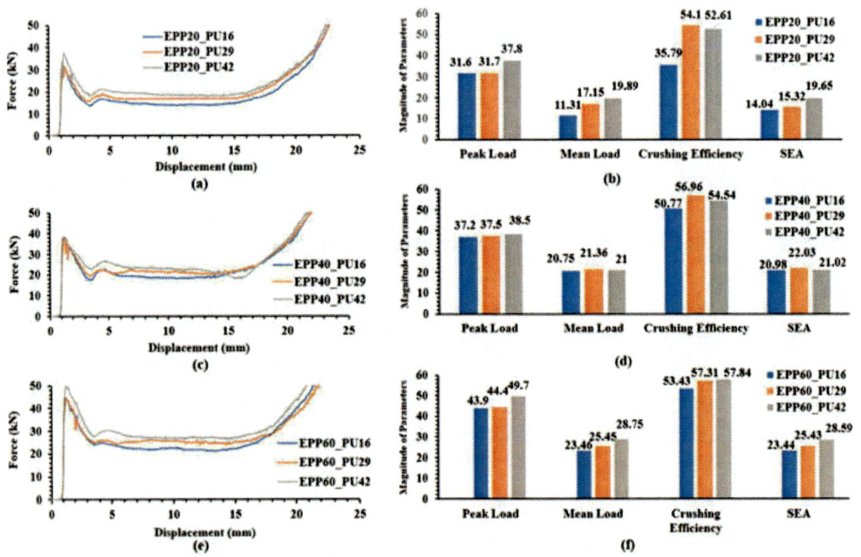

Fig. 4 Effect of filling two different foams together on crashworthiness response of reinforced honeycomb structure

density of polyurethane foam changed from 29 to 42 kg/m^3. Specific energy absorption also increased as the density of foam increased from 16 to 29 kg/m^3 but it showed decrement when the density of foam increased from 29 to 42 kg/m^3.

4.2.3 Keeping EPP Foam Density 60 kg/m^3 and Varying PU Foam Densities 16,29,42 kg/m^3 in Foam-Filled Reinforced Honeycomb

Effect of filing Expanded Polypropylene foam of density 60 kg/m^3 and varying Polyurethane foam of densities 16, 29, 42 kg/m^3 on the other side reinforced sheet showed the following variations in crashworthiness properties. As the density of polyurethane foam increases, the peak load showed an increment. Peak load maximum value of 49.7 kN for EPP60_PU42. Mean load also shows similar behaviour as peak load as the density of polyurethane foam increases mean load also increases as shown the in figure and shows a maximum value of 28.75 kN for EPP60_PU42. Crush force efficiency showed an increase when PU foam density changed from 16 to 42 kg/m^3. Specific energy absorption also increased as the density of foam increased from 16 to 42 kg/m^3. SEA variation with density was shown in Fig. 4f.

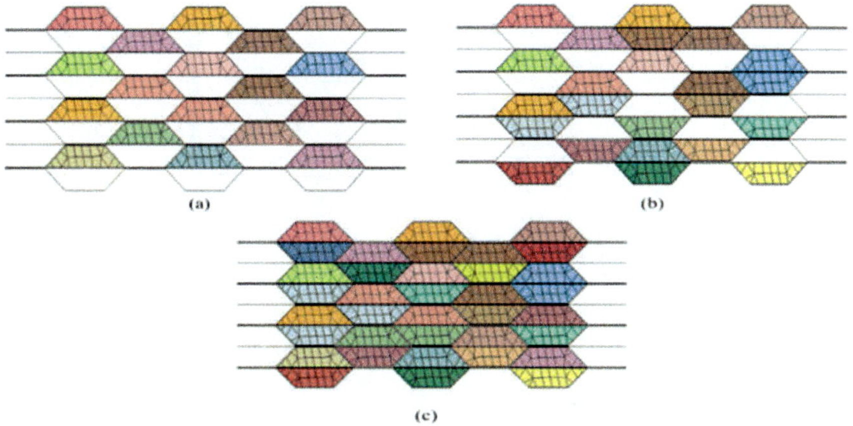

Fig. 5 Typical arrangement of honeycomb structure for **a** 50% foam filling **b** 66% foam filling **c** 100% foam filing

4.3 *Effect of Different Types of Arrangements of Foam Filling in Foam-Filled Honeycomb Structures*

In order to analyse the effect of the arrangement of foams on the crashworthiness properties of foam-filled reinforced honeycomb structure, three arrangements with 50% foam filling, 66% foam filling and 100% foam filling were taken for the study. The arrangement of foams was shown in Fig. 5. While arranging, care has been taken that the effect addition of foams will distribute uniformly throughout the section. Two types of foams named expanded polypropylene and polyurethane were used for the analysis. Here also, the Load displacement curve showed similar behaviour to any crashworthy structure during its crushing.

4.3.1 Expanded Polypropylene Foams

As shown in Fig. 5, 50% foam filling, 66% foam filling and 100% foam filling were done for EPP foam. As the foam filling percentage increases peak load increases. The mean load also showed an increment with an increase in density. Crush force efficiency increases when the foam filling percentage increases from 50 to 66%, but it again decreases when the foam filling percentage increases from 66 to 100%. Specific energy absorption increases with an increase in foam filling percentage. Variation in parameters was shown in Fig. 6a and b.

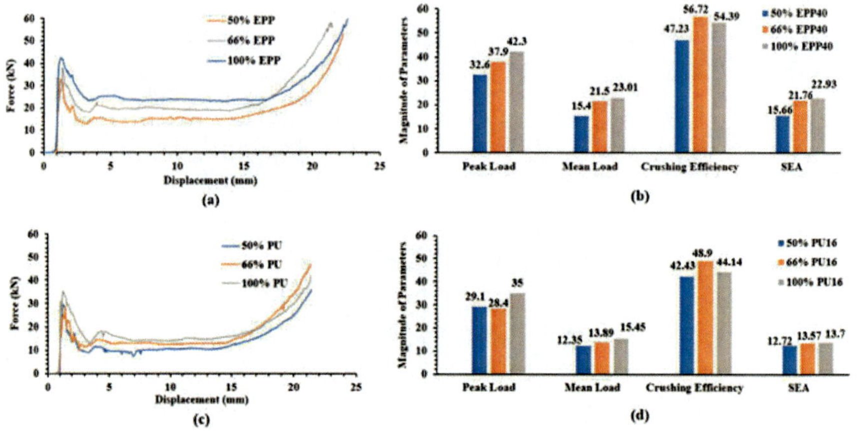

Fig. 6 Effect of different types of arrangements of foam filling in foam-filled honeycomb structures

4.3.2 Polyurethane Foams

PU foams showed similar trends in response as EPP foams. As shown in Fig. 5, 50% foam filling, 66% foam filling and 100% foam filling were done for expanded PU foam. As the foam filling percentage filling increases peak load increases. The mean load also showed an increment with an increase in density. Crush force efficiency increases when the foam filling percentage increases from 50 to 66%, it again decreases when the foam filling percentage increases from 66 to 100%. Specific energy absorption increases with an increase in foam filling percentage. Variation in parameters was shown in Fig. 6 c,d.

5 Conclusion

In this study, foam-filled reinforced honeycomb structures were crushed axially under a quasi-static loading condition using finite element simulation. The effect of filling expanded polypropylene foam (varying densities 20, 40, 60 kg/m^3) and polyurethane foam (varying densities 16, 29, 42 kg/m^3) in combinations on the crashworthiness of reinforced honeycomb structures was studied. Also, the effect of varying foam filling percentages of honeycomb cell (55%, 66%, 100%) on crashworthiness parameters was studied. From the study, it was concluded that as the density of PU foam filled into the reinforced honeycomb increased by keeping EPP foam density constant, crashworthiness parameters peak load, mean load and specific energy absorption was increased. In both EPP and PU foam-filled reinforced honeycombs, as foam filling percentage increased, parameters such as peak load, mean load and specific energy absorption also increased.

References

1. Partovi Meran A, Toprak T, Muğan A (2014) Numerical and experimental study of crashworthiness parameters of honeycomb structures. Thin-Walled Struct 78:87–94. https://doi.org/10.1016/j.tws.2013.12.012
2. Yin H, Wen G (2011) Theoretical prediction and numerical simulation of honeycomb structures with various cell specifications under axial loading. Int J Mech Mater Des 7(4):253–263. https://doi.org/10.1007/s10999-011-9163-5
3. Tao Y, Chen M, Chen H, Pei Y, Fang D (2015) Strain rate effect on the out-of-plane dynamic compressive behavior of metallic honeycombs: experiment and theory. Compos Struct 132:644–651. https://doi.org/10.1016/j.compstruct.2015.06.015
4. Pietras D, Linul E, Sadowski T, Rusinek A (2020) Out-of-plane crushing response of aluminum honeycombs in-situ filled with graphene-reinforced polyurethane foam. Composite Struct 249. https://doi.org/10.1016/j.compstruct.2020.112548
5. Alavi Nia A, Sadeghi MZ (2010) The effects of foam filling on compressive response of hexagonal cell aluminum honeycombs under axial loading-experimental study. Mater Design 31(3):1216–1230. https://doi.org/10.1016/j.matdes.2009.09.030
6. Zhou H, Jia K, Wang X, Xiong MX, Wang Y (2020) Experimental and numerical investigation of low velocity impact response of foam concrete filled auxetic honeycombs. Thin-Walled Struct 154. https://doi.org/10.1016/j.tws.2020.106898
7. Zhang Y, Liu Q, He Z, Zong Z, Fang J (2019) Dynamic impact response of aluminum honeycombs filled with expanded polypropylene foam. Compos B Eng 156:17–27. https://doi.org/10.1016/j.compositesb.2018.08.043
8. Zarei Mahmoudabadi M, Sadighi M (2011) A theoretical and experimental study on metal hexagonal honeycomb crushing under quasi-static and low velocity impact loading. Mater Sci Eng A 528(15):4958–4966. https://doi.org/10.1016/j.msea.2011.03.009
9. Khan MK, Baig T, Mirza S (2012) Experimental investigation of in-plane and out-of-plane crushing of aluminum honeycomb. Mater Sci Eng A 539:135–142. https://doi.org/10.1016/j.msea.2012.01.070
10. Aminanda Y, Castanié B, Barrau JJ, Thevenet P (2005) Experimental analysis and modeling of the crushing of honeycomb cores. Appl Compos Mater 12(3–4):213–227. https://doi.org/10.1007/s10443-005-1125-3
11. Thomas T, Tiwari G (2019) Energy absorption and in-plane crushing behavior of aluminium reinforced honeycomb. Vacuum 166:364–369. https://doi.org/10.1016/j.vacuum.2018.10.057
12. Tiwari G, Thomas T, Khandelwal RP (2018) Influence of reinforcement in the honeycomb structures under axial compressive load. Thin-Walled Struct 126:238–245. https://doi.org/10.1016/j.tws.2017.06.010
13. Liu B, Villavicencio R, Guedes Soares C (2013) Experimental and numerical plastic response and failure of pre-notched transversely impacted beams. Int J Mech Sci 77:314–332. https://doi.org/10.1016/j.ijmecsci.2013.09.032
14. Livermore Software Technology Corporation (2003) LS-DYNA Theory Manual Version 970
15. Ramaswamy K, Patham B, Savic V, Tripathy B (2017) Stable and accurate LS-DYNA simulations with foam material models: optimization of finite element model parameters. SAE Int J Mater Manuf 10(2):226–233. https://doi.org/10.4271/2017-01-1338
16. Tay YY, Lim CS, Lankarani HM (2014) A finite element analysis of high-energy absorption cellular materials in enhancing passive safety of road vehicles in side-impact accidents. Int J Crashworthiness 19(3):288–300. https://doi.org/10.1080/13588265.2014.893789
17. Santosa SP, Wierzbicki T, Hanssen AG, Langseth M (2000) Experimental and numerical studies of foam-filled sections
18. Liu Q et al (2017) Axial and lateral crushing responses of aluminum honeycombs filled with EPP foam. Compos B Eng 130:236–247. https://doi.org/10.1016/j.compositesb.2017.0

Assessment of Crash Survivability of Typical Helicopter Using LS-DYNA Simulation

Errababu Kunchala, C. Krishna Prasad, and A. Sakthivel

Abstract Contemporary helicopters have a demanding role to play both in military and in civil operations. The helicopter fatality rates are higher than fixed wing aircraft during crashes, mainly due to inherently risky operations close to ground, bad weather conditions and also non-availability of ejection seats. Since crash impact velocities are much lower for helicopters compared with fixed wing, there is a considerable scope for improving helicopter crash survivability as per MIL-STD-1290A (AV). Typically, during crash, the impact energy is judiciously distributed among various energy absorbing structural members namely landing gear, bottom structure and seat. The primary design goal for crashworthiness is to limit the impact forces transmitted to the occupants, and to maintain structural integrity of the fuselage to ensure a minimum safe occupant volume. The rotor, transmission and engine should not disintegrate during crash and cause injury to the occupants. In this chapter, full-scale airframe finite-element model of a typical helicopter was developed and simulated using HyperMesh and LS-DYNA solver. It is considered that seat is certified to withstand 51 g for military applications. Considering the constant landing gear energy, the balance energy absorbed by the bottom structure decides the crash survivability of helicopter. Iterative analysis was carried out to evaluate the conventional bottom structure to ensure uniform load distribution with high-energy absorption. But the requisite crash survivability was not achieved. Hence, the bottom structure was modified by incorporating the additional crashworthy features like kinks, lightening holes and other trigger mechanisms and subsequently iterative analysis was carried out. The results were found encouraging. Graphs were plotted for global energies, velocity and accelerations responses. The energy absorbed by modifying the structure has increased and 'g' loads experienced by pilot and co-pilot floor were reduced.

E. Kunchala (✉) · C. Krishna Prasad
RWRDC, HAL, Bangalore 560017, India
e-mail: errababu.k@gmail.com

A. Sakthivel
RCMA (Helicopters), Bangalore 560017, India

R. Velmurugan et al. (eds.), *Dynamic Behavior of Soft and Hard Materials Volume 1*, Springer Proceedings in Materials 34, https://doi.org/10.1007/978-981-99-6030-9_17

Keywords Crashworthy · Crash survivability · Airframe · Helicopter · Impact · Ls-Dyna

Nomenclature

MIL-STD	Military Standard
FEM	Finite-Element method
C.G	Centre of Gravity
LRU	Line Replaceable Units
CAD	Computer Aided Design
1D	One Dimensional
2D	Two Dimensional

1 Introduction

Crashworthiness is the ability of rotorcraft to maintain a protective space for the occupants inside the airframe cabin throughout the crash impact sequence and to eliminate injuries and fatalities by limit the magnitude and duration of accelerations [1]. To provide the occupant protection to maximum extent in the crash events, a harmonious systems approach to crash resistance must be followed. There are three main key energy absorption contributors in rotorcraft that are landing gear assembly, airframe structure and occupant seats which are designed to work together in order to absorb the rotorcraft's kinetic energy and slowdown the occupants to rest without injurious. An example of system approach is depicted in Fig. 1 [2]. The landing gear absorbs energy to reduce the impact velocity of fuselage through oleo collapse or permanent deformation of skid tubes. The structure absorbs the energy by controlled collapsing at frame and sheet stringer constructions. The usage of composite materials like laminated and sandwich constrictions with triggered mechanisms helps to directional collapse of airframe structure and limit the 'g's to the occupants [3]. The seat and restraint system are designed to restrain the occupant within the protective shell in the event of crash and to provide additional energy absorbing stroke to limiting occupant loads [3, 4].

The prototype full-scale rotorcraft crash testing, redesigning and performing the test with different impact attitudes has significant effect on cost, test setup and development time of rotorcraft. Developing crash simulation methods will significantly reduce development cost and time.

In order to investigate the structural dynamic behaviour during crash events, an commercially non-linear finite-element transient dynamics codes like LSDYNA [5], MSC Dytran [6] and PAM-CRASH [7] were available. Recently significant contributions in the field of rotorcraft crashworthiness research has been made in order

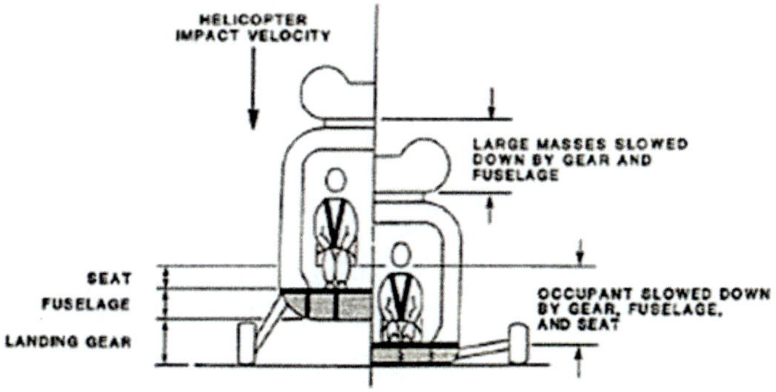

Fig. 1 Schematic illustrating a systems approach to crashworthy design [2]

Fig. 2 Typical subfloor structure [14]

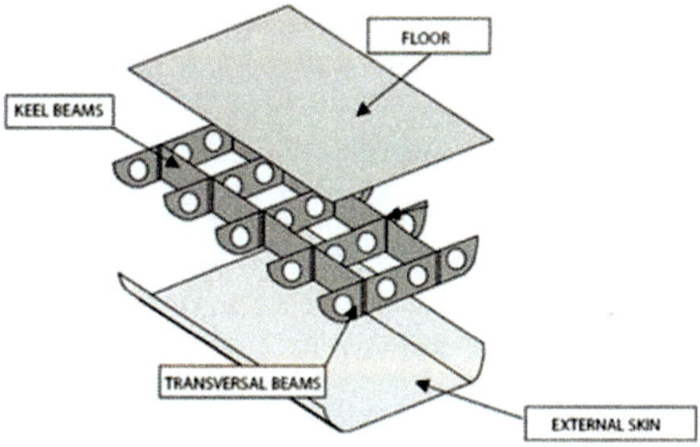

to attenuate the energy in the crash events by effective design of subfloor structure [8–13].

In this paper, the rotorcraft fuselage subfloor structure was emphasized to maximise the energy absorption characteristics. The major subfloor structure consists of keelbeams, transverse beams/frames, floor board and bottom panel skin as shown in Fig. 2 [14]. The full-scale finite element model of typical rotorcraft structure was developed using LS-DYNA FE software to idealise the subfloor structure [15, 16].

USARTL-TR-79-22 U.S. Army Research and Technology Laboratories

2 Model Development

The non-linear explicit finite-element analysis code LS-DYNA was used to perform rotorcraft crash analysis. The full-scale rotorcraft FE model was developed using the CAD model and FE pre-processing was performed in Altair HyperMesh and LS-DYNA software. The fuselage, keel beams, transverse beams/frames, floor board and composite sandwich panels were idealized using 1D, 2D elements with appropriate element types, properties and materials. The seat with dummies, heavy mass items like transmission system, main rotor, tail rotor, engine and LRU's (Line Replacement Unit) were not modelled but were idealized using lumped mass elements assigning mass values at their respective C.G locations [15, 17]. The material properties and allowable are referred from material handbook data [18]. The FE model units followed were length in mm, weight in Tons and time in seconds. The analysis solution time step depends on smallest element dimension; therefore, the smallest elements lead more run time and hence Mass scaling was used to control the minimum time step. Mass scaling adds artificial mass to smaller dimensions elements and the solution run time can be reduced.

The metallic properties were defined using the MAT24 card with failure strain (MAT_ PIECEWISE_LINEAR_PLASTICITY) available in LS-DYNA. The composite properties were defined as PART_COMPOSITE with MAT55 (MAT_ ENHANCED_COMPOSITE_DAMAGE). The structural connections like rivets and bolts idealised with 1D beam elements using MAT_SPOTWELD_DAMAGE_ FAILURE. The rigid impact surface was idealised using MAT_20 (MAT_RIGID).

An automatic single surface CONTACT_AUTOMATIC_SINGLE_SURFACE was defined for each structural member in order to avoid self-penetration. The contact between bottom surface of rotorcraft to ground was defined as CONTACT_ AUTOMATIC_NODES_TO_SURFACE [9]. The contact between frames to impact surface was defined using CONTACT_ AUTOMATIC_NODES_TO_SURFACE and the contact between fasteners to structural members were defined using CONTACT_ SPOTWELD.

2.1 Analysis

The FEM of the full-scale typical rotorcraft is as shown in Fig. 3, which consists of 10,70,554 nodes and 10,28,742 elements including 8,32,450 quadrilateral shell elements,18,692 triangular shell elements, 10,216 beam elements and 157 mass points. As a quality checks on model fidelity, the total weight, mass distribution and C.G location [19] of the full-scale FEM is compared with the actual weight of typical helicopter structure. A crash landing may occur in different conditions of attitude and predominant attitude is with no roll, pitch, or yaw as per crash statistical data survey [9, 20]. It is assumed that the impact occurs with no roll, pitch, or yaw. The survivability of rotorcraft is defined in MIL-STD-1290 & USARTL-TR-79–22 [3, 4] which is statistical distribution of percentile crash survivability vs sink speed of rotorcraft as shown in Fig. 4.

The targeted sink speed of rotorcraft was estimated from percentile crash survivability vs sink speed graph as shown in Fig. 4 and calculated the fuselage contact velocity by subtracting landing gear energy from energy balance equation. At this condition, both the fuselage bottom surface and landing gear are in contact with floor. Both the landing gear and fuselage continue to absorb energy during fuselage crushing. The simulation was carried out in this condition. An initial velocity was

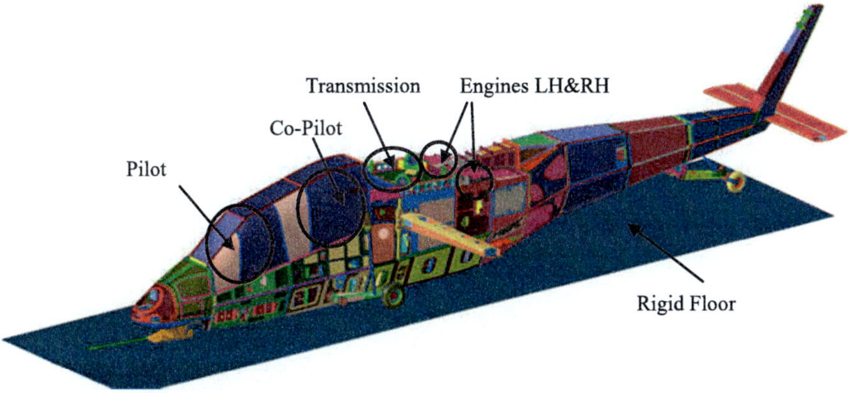

Fig. 3 Finite-element crash model of typical rotorcraft

Fig. 4 Vertical velocity changes for survivable rotary-wing aircraft accidents [3]

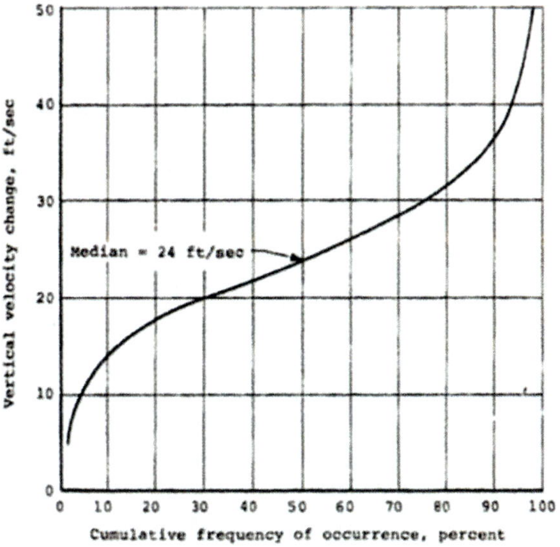

assigned to all nodes in the model except rigid floor and impacted on the rigid floor with no roll, pitch, or yaw attitudes [15]. The nodes of rigid floor were constrained in translational and rotational degree of freedom. The normal time step of the simulation is 2.7e-7 s which requires couple of days for completing the analysis for single core machine. Hence, mass scaling of one micro second time step was used to reduce the analysis simulation time. The percentage of added mass is less than 1% of total mass due to mass scaling. The model was executed in LS-DYNA 971R6.0 for duration of 0.05 s (50 ms) of simulation time, which required 50 h of CPU time.

From the above global energy plots (Fig. 5), the total energy of the system is maintaining constant throughout the analysis. The kinetic energy of the helicopter is

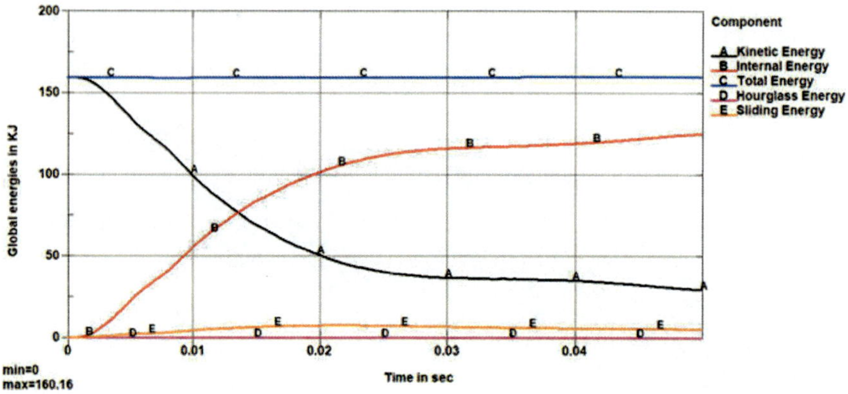

Fig. 5 Global energies plot

decreasing and internal energy of the structure is increasing at the same time and the hourglass energy is showing zero. A general thumb rule is that the hourglass energy should be less than 10% of the total energy of the solution [21]. The behaviour of FE model is as per the expected physical behaviour. The structural acceleration pulse recorded from a crash is composed of a large spectrum of frequencies superimposed together. The structure has many components, each with its own fundamental mode of oscillation, plus many harmonics. In crash dynamics, one is often concerned with the magnitude and duration of the low-frequency (Fundamental) acceleration pulse that will be input into the passenger [19].

From the acceleration and velocity responses (Figs. 6 and 7), the node numbers 6794 and 6795 represents co-pilot and pilot floor, respectively. The co-pilot floor touches the ground first followed by pilot floor due to orientation of helicopter and pilot floor is rebounding more than co-pilot floor. The co-pilot floor having maximum of 91 g at eight milliseconds followed by pilot floor maximum of 89 g at 15 ms. However, 91 and 89 g is the absolute peak of the high-frequency oscillatory response, but not basic fundamental pulse. The pilot and co-pilot floor velocity goes to zero at a time of 0.017 s and by approximately 0.034 and 0.029 s. It has gone positive to approximately 2540 and 1210 mm/sec for pilot and co-pilot floors respectively which is the rebound velocity. Thus, the total velocity change including rebound is 9860 and 8530 mm/sec. The duration of the fundamental pulse is about 0.034 and 0.029 s. Thus, the fundamental frequency is about 1/T or 30 and 35 Hz. The maximum acceleration of the fundamental pulse of 79 g on the pilot floor and 73 g on the co-pilot floor which were obtained by simply computing the maximum slope of the velocity curves. The raw acceleration data of pilot and co-pilot floor were filtered with various SAE filters and 30 and 45 Hz filters data was very close to the value calculated from the slope of the velocity curve as shown in Fig. 6. The raw acceleration and filtered 30 and 45 Hz acceleration data were integrated to obtain velocity response. The velocity

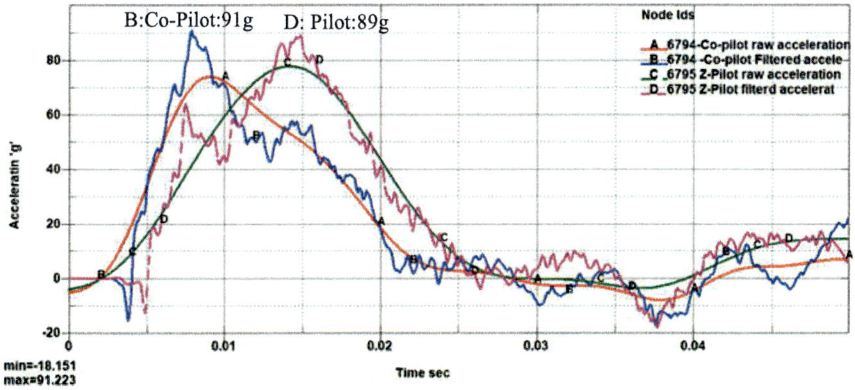

Fig. 6 Acceleration raw data and filtered data of pilot and co- pilot floor

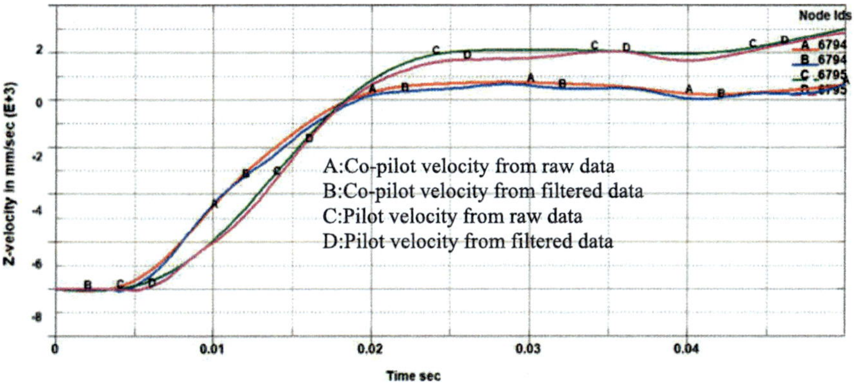

Fig. 7 Comparison of the velocity responses obtained from integrating the raw data and integrating the filtered acceleration data

response obtained by integrating the 30 and 45 Hz filtered accelerations follows the velocity response obtained from the raw acceleration data quite well [19] as shown in Fig. 7.

2.2 Additional Crashworthy Features and Analysis

The fundamental peak acceleration observed in the analysis was 79 g on the pilot floor and 73 g on the co-pilot floor which are higher than the maximum allowable floor acceleration of 51g's [22]. Due to these high decelerations at floor level will cause fatal injury to pilot and co-pilots. Hence the bottom structure of fuselage was improved by providing additional crashworthy features like cut outs in keel beams,

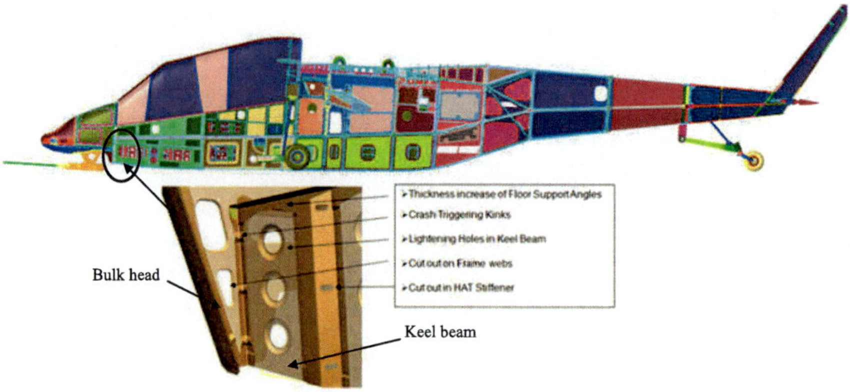

Fig. 8 Typical structural modifications

weak links in vertical stiffeners and one of the machined frame bottom is converted into sheet metal with stringer construction and added vertical stiffeners below fuel tanks area [9]. The modifications are shown in Fig. 8.

The updated FE model was run and extracted acceleration responses for pilot, co-pilot and large masses. Figure 9 depicts the raw accelerations data in the pilot and co-pilot floors. The co-pilot floor having maximum of 75g at 10 ms followed by pilot floor maximum of 67g at 14 ms. The maximum acceleration of the fundamental pulse of 54g on the pilot and co-pilot floors which were obtained by simply computing the maximum slope of the velocity curves as shown in Fig. 10.The maximum raw accelerations of large masses engine LH, engine RH and transmission system are 38g, 50g and 41g respectively for short durations as shown in Fig. 11.The structural deformations of typical rotorcraft at various time intervals of 0, 10, 25 and 50 ms were illustrated in Fig. 12. It shows that sub-floor structure is stroking and absorbing energy.

3 Results and Discussions

The comparison of accelerations of pilot and co-pilot floor before and after structural modifications was illustrated in Fig. 13. The maximum peak raw decelerations of pilot and co-pilot floor were reduced from 89 to 67g and 91g to 75g, respectively. The fundamental decelerations of pilot and co-pilot floor which were derived from maximum slope of velocity responses are reduced from 79 to 54g and 73g to 54g, respectively, after incorporating structural modifications. The velocity response of pilot and co-pilot floors are illustrated in Fig. 14. The maximum slope of velocity response of the modified structure is less than the unmodified structure. The total internal energy of structure before and after structural improvements is 125 and 143 kJ as shown in Fig. 15.

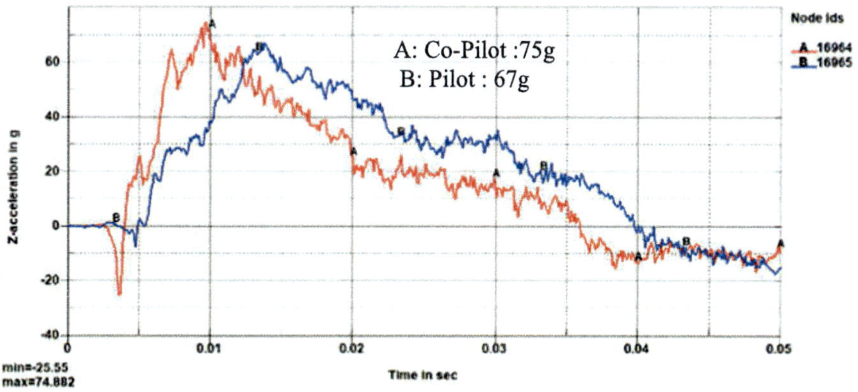

Fig. 9 Pilot and co-pilot floor raw acceleration responses

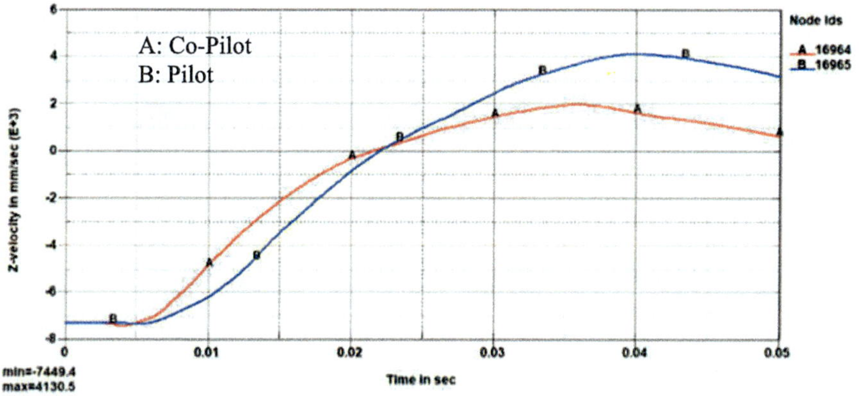

Fig. 10 Pilot and co-pilot floor velocity response

4 Concluding Remarks

A full-scale three-dimensional finite element model of a typical helicopter airframe was developed and executed crash simulation in nonlinear explicit transient code LS-DYNA. The full-scale rotorcraft was impacted on rigid ground and structural 'g' responses was analysed along with energy balance. The fundamental accelerations responses 'g' loads experienced by occupants reduced by 32% and energy absorbed by structure is increased by 13%. The average 'g' loads for large masses are within allowable limits of 20 g and cabin volume reduction is less than 15% with incorporation of crashworthy features in bottom structure. Hence due to reduction of 'g' loads in the occupants, the fatality rates of the occupants will reduce and there by percentile crash survivability of rotorcraft will enhance.

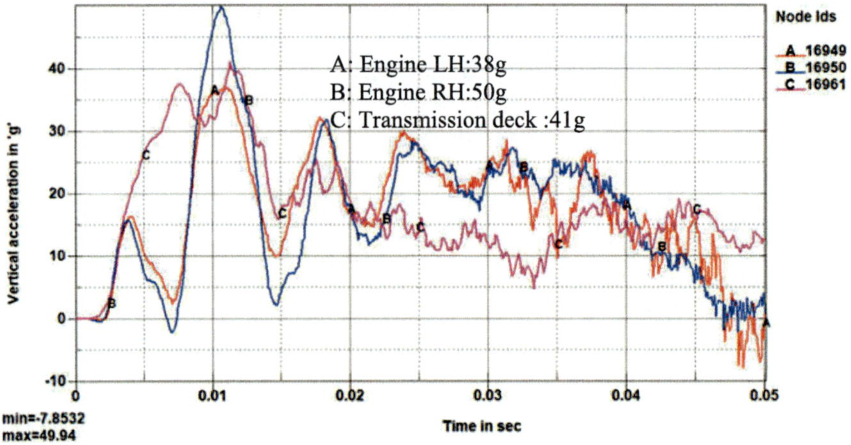

Fig. 11 Large masses raw acceleration responses

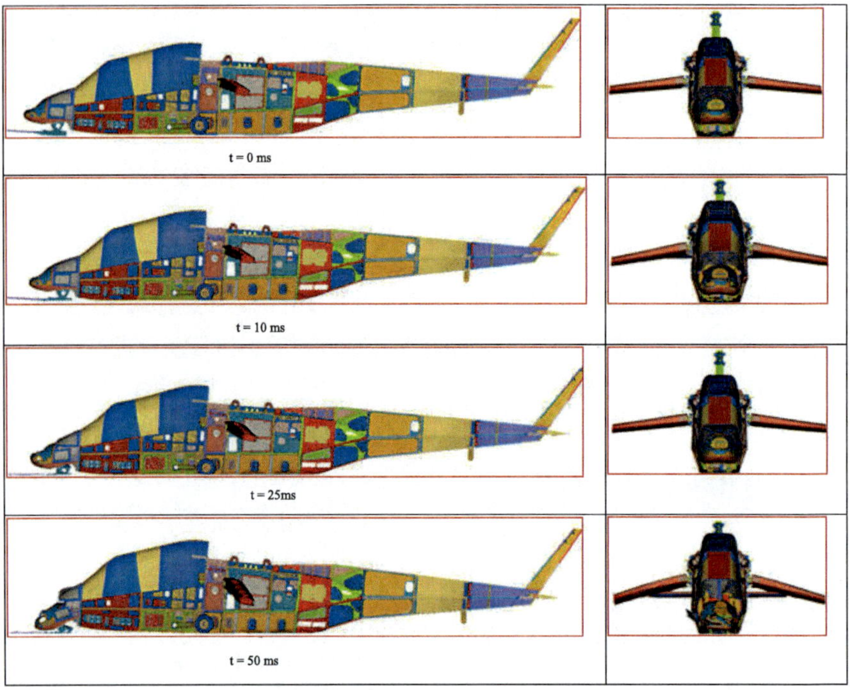

Fig. 12 Structural deformations of typical rotorcraft at various time intervals

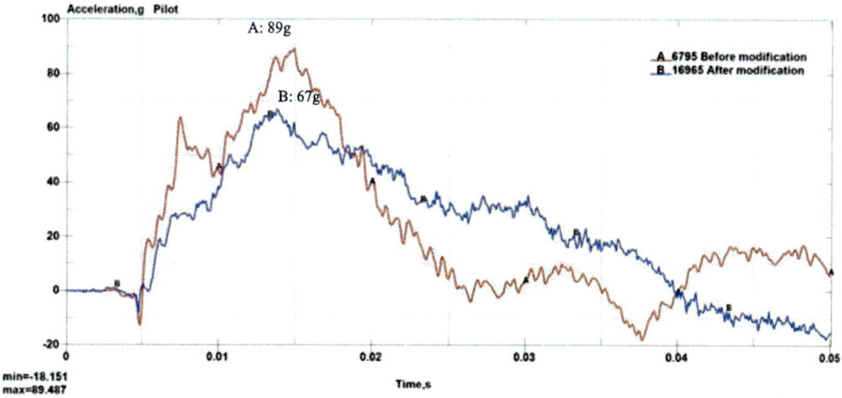

(a) Pilot Floor

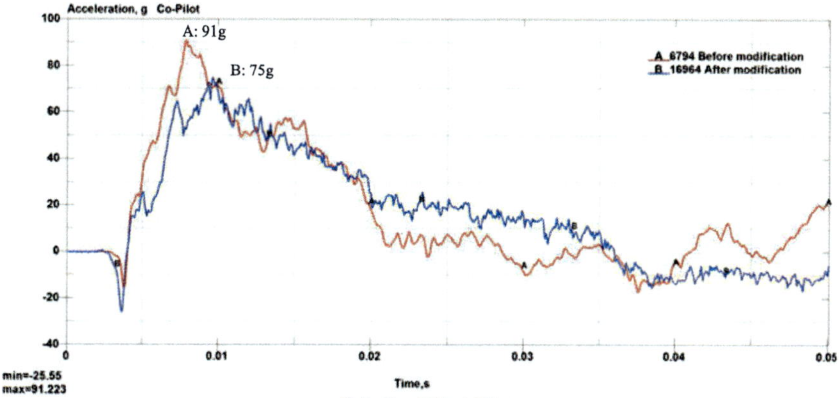

(b) Co-Pilot Floor

Fig. 13 Comparison of vertical raw acceleration responses before and after structural modifications

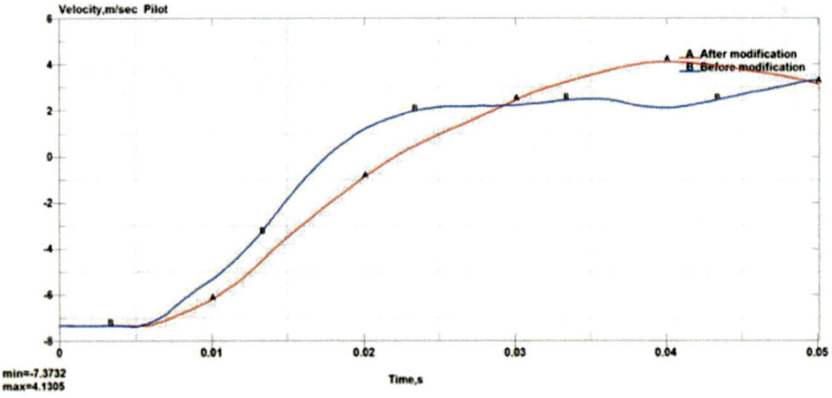

(a) Pilot floor

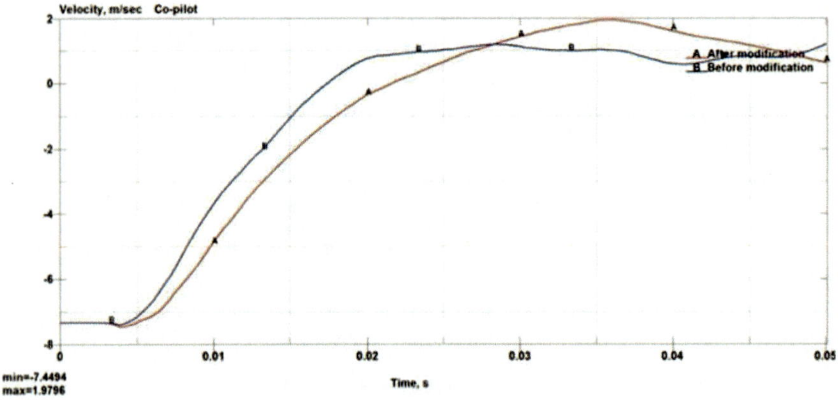

(b) Co-Pilot floor

Fig. 14 Comparison of vertical velocity responses before and after structural modifications

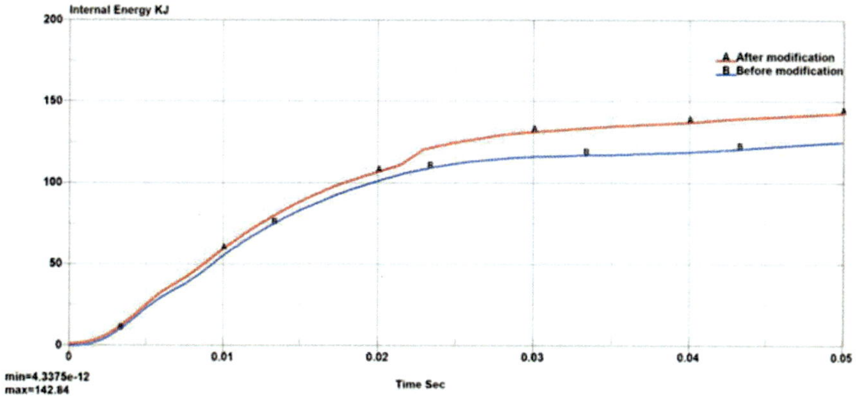

Fig. 15 Comparison of Internal energy before and after structural modifications

References

1. Jones LE (2002) Overview of the NASA systems approach to crashworthiness program.NASA Langley Research Center, Hampton, VA, June 11–13
2. Aircraft Crash Survival Design Guide (1989) Volumes 1, USAAVSCOM TR 89-D-22A, December 1989
3. Aircraft Crash Survival Design Guide (1980) Volumes 2, USARTL TR- 79--22B, December 1980
4. Military Standard, MIL-STD-1290A (AV) (1988) Light fixed and rotary-wing aircraft crash resistance, Department of Defence, Washington, DC, September
5. LS-DYNA Keyword User's manual volume-1,2, February 2012
6. MSC.DYTRAN User's Manual, MSC.Software Co., Los Angeles, CA, 2002. 53
7. PAM-CRASH User's Manual, Engineering Systems International, F-94588, 20 Rue Saarinen, Silic
8. Boitnott RL (1989) Crashworthy design of helicopter composite airframe structures. U. S. Army Aviation Research and Technology Activity Langley Research Center Hampton, Virginia 23665–5225, U.S.A.
9. Engleder A, Markmiller J, Mueller R (2015) The evoluation of airbus helicopters crashworthy composite airframes for transport helicopters. AIRBUS HELICOPTERS Deutschland GmbH, 41st European Rotorcraft Forum 2015
10. Cacchione B, Airoldi A, Astori P (2011) A hybrid modelling technique for the energy absorption capabilities of a crashworthy helicopter structure
11. De!etombe E, Delsart D, Johnson AF, Kohlgrtiber D (1998) Composite helicopter structural crashworthiness. ONERA, Institut de Mecanique des F\uides de Lille, Lille 59045, France DLR, Institut fUr Bauweisen und Konstruktionsforschung, Stuttgart D-70503, Gennany, 24 European Rotorcraft Forum Marseilles, France -September 1998
12. Bisagni C, Mirandola C (2015) Experimental and numerical investigation of crash behavior of composite helicopter cruciform elements. Dipartimento di Ingegneria Aerospaziale, Politecnico di MilanoVia La Masa 34, 20156 Milano, Italy
13. Gia Votto V, Sala G, Anghileri M (1990) Crash Analysis and Correlation with test of a Composite Helicopter Sub-floor structure. Aerospace Engineering Department Politecnico D1 Milano, Italy, Nineteenth European Rotorcraft Forum
14. Astori P (2018) cabin safety sensitivity to the mechanical parameters of the main crashworthy stages. In: 44th European Rotorcraft Forum, Delft, The Netherlands, 19–20 September, 2018, Politecnico di Milano (Italy) Thiva Ramachandran, Mecaer Aviation Group (Canada)

15. Jackson KE, Fasanella EL (2005) Test-analysis correlation of a crash simulation of a vertical drop test of a commuter-class aircraft. U.S. Army Research Laboratory, Vehicle Technology Directorate NASA Langley Research Center Hampton, VA
16. Jackson KE, Fasanella EL, Lyle KH (2006) Crash certification by analysis—are we there yet. NASA Langley Research Center, Hampton, VA, Presented at the American Helicopter Society 62nd Annual Forum, Phoenix, AZ, May 9–11
17. Fasanella EL, Jackson KE (2004) Crash simulation of a boeing 737 fuselage section vertical drop test. US Army Research Laboratory, Vehicle Technology Directorate, Hampton, VA
18. MIL-HDBK-5J (2003) 31 January 2003
19. Fasanella EL, Jackson KE (1985) Best practices for crash modelling and simulation. NASA and US Army Research Lab, NASA Langley Research Center, October 2002, 2700, 64578 Rungis, Cedex, France
20. Jackson KE (2018) NASA langley research center Advances in rotorcraft crashworthiness: trends leading to improved survivability. J Amer Helicopter Soc 63
21. Full Spectrum Crashworthiness Criteria for Rotorcraft by US Army RDECOM, Dec (2011)
22. MIL-S-58095 (1986) Seat system: crash-resistant, non-ejection, aircrew, general specification for, Department of Defense, Washington, DC, 31 January 1986

Experimental and Numerical Investigation of High Velocity Response of Aramid Honeycomb Core Sandwich Structure

Saurabh Rathod, Gaurav Tiwari, and Nikhil Khaire

Abstract In this research, an experimental and numerical analysis of a flat sandwich structure with an aramid honeycomb core using a cylindrical steel ogive-shaped projectile was performed to study the impact resistance of the structure. The flat structure consists of a circular aramid honeycomb core of 200 mm in diameter, having a cell size ranging from 3.2 to 9.2 mm and a 20 mm thickness, backed by circular aluminum alloy facesheets with varying thicknesses from 0.7 to 2 mm on both sides. Models of sandwich structures with honeycomb cores and aluminum skins were built in three dimensions using ABAQUS/Explicit. The numerical analysis has also been expanded to include the impact of core and facesheet geometries, on the energy absorption properties of sandwich structures. In terms of energy absorption, a positive relationship was identified between cell wall thickness and facesheet thickness. In contrast, the cell length negatively impacted the structure's energy absorption for the ogive projectile. The investigation revealed that with an increase in facesheet thickness from 0.7 to 2 mm, the ballistic limit increased progressively by 23.04, 26.81, and 26.83%, respectively. Conversely, as the cell length increased incrementally from 3.2 to 9.2 mm, the ballistic limit decreased by 3.79, 4.83, and 6.28%, respectively. Similarly, with an increase in cell wall thickness from 0.03 to 0.09 mm, the ballistic limit increased by 4.48, 1.47, and 2.9%, respectively.

Keywords Ballistics response · Sandwich structures · Impact behavior · Energy absorption

---

S. Rathod (✉) · G. Tiwari
Department of Mechanical Engineering, Visvesvaraya National Institute of Technology, Nagpur 440010, India
e-mail: mr.saurabh.rathod@gmail.com

N. Khaire
Department of Mechanical Engineering, Indian Institute of Technology, New Delhi 110016, India

R. Velmurugan et al. (eds.), *Dynamic Behavior of Soft and Hard Materials Volume 1*, Springer Proceedings in Materials 34, https://doi.org/10.1007/978-981-99-6030-9_18

1 Introduction

Sandwich structures are often used and very important in industries like aerospace, defense, marine, and automobiles because of their excellent performance and high energy absorption properties under impact scenarios. Sandwich panels consist of a lightweight core mostly backed by two similar thin facesheets bonded together by adhesive. In recent years, the number of publications in this field has increased due to the growing interest in sandwich panels. In particular, honeycomb has been receiving more attention as the core of the sandwich panel. Gibson and Ashby [1] reviewed the analytical study of the honeycomb behavior of cellular solids, providing a comprehensive description of the mechanics of cellular solids. Based on their analyses of the three stages of perforation, Fatt et al. [2] established an analytical model for aluminum sandwich plates penetration under low velocity impact. In contrast, Feli and Namdari [3] analytically modeled the effects of high velocity on sandwich plates containing composite plates and an aluminum honeycomb core. Recently, many researchers have conducted detailed research to study mechanical behavior of aramid honeycomb cores. Othman et al. [4] researched the failure characteristics and mechanical behavior of Nomex honeycomb structure and found that most of the energy was absorbed by deformation of facesheet than its fracture. Harizi et al. [5] examined Nomex honeycomb cores and present a comprehensive discussion of mechanical behavior and the failure mechanism of sandwich panels in three-point bending test. Numerous studies have been conducted to determine the parameters influencing ballistic resistance and energy absorption for ballistic impact. In the simulation, Foo et al. [6] increased the density of aluminum honeycomb by raising the thickness of the wall and decreasing the size of the cells and discovered that increasing the density of honeycomb sandwich panels can enhance their impact resistance. Khaire et al. [7–10] conducted several numerical and experimental studies on high velocity impact with energy absorption properties using different types of projectiles and discovered that for thicker skin sandwich panels, the effect of core geometries was minimal. Børvik et al. [11] performed an experimental and analytical research of perforation phenomenon of aluminum plates impacted by conical and ogive nose bullets to understand the mechanisms and key characteristics affecting aluminum plates. Rathod et al. [12] evaluated the composite structure's ballistic resistance by considering various parameters, such as depth of penetration, ballistic limit, and target areal density. In addition, Zhang et al. [13] reported that projectile shape determined the failure mechanism and conical and ogive are most effective penetrators. Kolopp et al. [14] shown that Nomex composites are lighter and more resistant to impact than other similar sandwich panels. Based on the reviewed literature, it has been observed that numerous investigations have been conducted on the response of aluminum honeycomb structures to high velocity impact. Also, most of the studies performed on sandwich structure with an aramid honeycomb core are limited to low velocity impact. However, few literatures are available about the impact resistance of an aramid honeycomb core when subjected to high velocity. Therefore, a comprehensive investigation must be conducted on ballistic performance of the aramid honeycomb core. Hence, the current research evaluated impact resistance and the energy absorption of the flat aramid structure using an ogive-nosed projectile. Furthermore, the failure mechanism of an aramid honeycomb sandwich structures has been analyzed and discussed. Finally, the effect of various parameters on residual velocity, energy absorption, and ballistic limit has been investigated. The proposed aramid honeycomb core sandwich structures offer high specific strength and stiffness, good energy absorption, and excellent thermal and acoustic insulation. These properties make them suitable for use in various industries, such as in the aerospace industry, it can be used for aircraft flooring, interior panels, and helicopter rotor blades. In the defense industry, it can be used for military vehicles, armor plating, and ballistic panels. Aramid honeycomb core sandwich structures can also be used in the marine industry for boats and ships, the automotive industry for car body panels, and the construction industry for roofing and cladding. Additionally, they can be used in sporting goods, such as bicycle frames and snowboards, and in wind energy for the construction of wind turbine blades.

2 Experimental Setup

The compressed air gun testing system was used to conduct the ballistic impact experiments. This system consists of a reciprocating compressor that produces pressurized air, a pressure vessel to store that pressure, a barrel serves as guide for the projectile to find target, an actuator valve that releases pressure for the fraction of time, a target mounting plate that holds the sandwich structure, and a catcher box to catch the penetrated projectile, refer Fig. 1. It is capable of propelling projectiles at up to 300 m/s with its compressed air gun. However, a projectile's precise speed is largely determined by its initial input pressure and its mass. Various projectile speeds were used in this experiment, between 35 and 150 m/s. The approach for selecting the impact velocity was based on a step-by-step increment in velocity until the ballistic limit was obtained, which is the velocity at which the projectile penetrates the sandwich structure. The infrared sensors were installed to measure the velocity of projectile. These sensors consist of two sets with each set containing two transmitters and two receivers. The impact velocity of the projectile was measured using the first set, while using the second set, the projectile's residual velocity was measured. The energy absorbed by a flat structure is determined based on these velocities. To study the influence of facesheet thickness, cell wall thickness, and cell length, standard sizes were chosen for the honeycomb core, including cell lengths of 3.2, 5.2, 7.2, and 9.2 mm, and cell wall thicknesses of 0.03, 0.05, 0.07, and 0.09 mm. Facesheet thicknesses ranging from 0.5 to 2 mm in increments of 0.5 mm were examined to determine the effect of facesheet thickness. However, the 0.5 mm facesheet thickness was found to exhibit minimal resistance to impact during preliminary experimentation. Consequently, the next available standard thickness of 0.7 mm was chosen for further experimentation.

Fig. 1 Experimental set up for ballistic impact test

3 Finite Element Modeling

The flat sandwich structures were modeled with ABAQUS/Explicit [15], a commercial finite element solver, for high velocity impact using a FE model, as shown in Fig. 2. It consists of an aramid honeycomb core, an ogive-nosed projectile, and two facesheets with variable thicknesses (*F*) of 0.7, 1, 1.5, and 2 mm. Also, between the front facesheet–core and the rear facesheet–core, a layer of 0.2 mm-thick epoxy resin was used. The flat faceplates were modeled as solid models with a radius of 110 mm. Similarly, the ogive-nosed projectile, which was considered as analytically rigid, was also modeled as a solid model. Aramid honeycomb was modeled as an elastoplastic material, whereas Johnson Cook viscoelastic plastic material was applied to model the aluminum facesheet, refer Table 1 for material properties. For parametric study, honeycomb cores were modeled with different cell lengths (L) from 3.2 to 9.2 mm, and cell wall thicknesses (C) of 0.03, 0.05, 0.07, and 0.09 mm. In the present study, honeycomb cores were modeled using four-nodded shell elements (S4R) with hourglass control and reduced integration, whereas eight-node brick elements (C3D8R) were used to mesh the facesheets and adhesive layers with reduced integration. During meshing, the projectile impact area was meshed more finely (0.2 $\times$ 0.2 mm) to obtain accurate results and capture failure modes, while other areas were meshed coarsely (2 $\times$ 2 mm) to reduce computation time. To restrict all six degrees of freedom, the circumferential edges of the sandwich structure were subjected to an encastre boundary condition, which restricted translation in all three directions (U1, U2, and U3) as well as rotation around all three axes (UR1, UR2, and UR3). Also, surface-based contacts were utilized to investigate the bonding between the facesheets, adhesive layers, and honeycomb core. Specifically, surface-to-surface explicit contact was used to model the contact between the outer

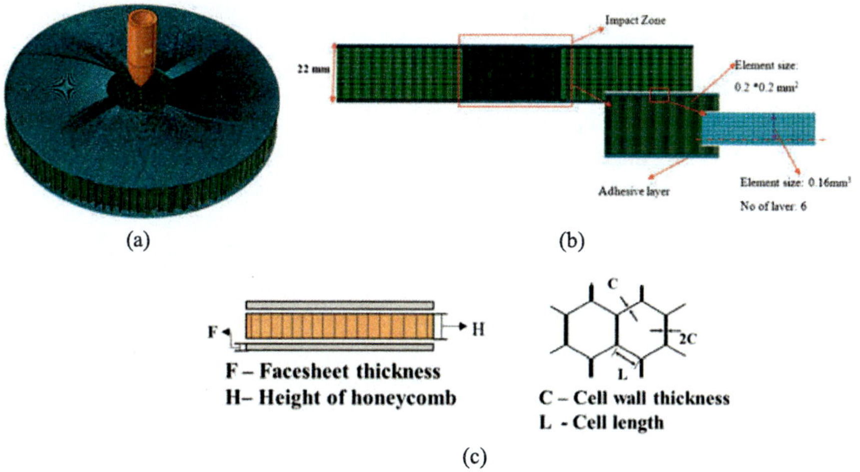

Fig. 2 Numerical model of flat sandwich structure **a** ogive projectile **b** mesh detail **c** nomenclature

Table 1 Material properties of the flat sandwich structure components

Properties	Modulus of elasticity, E (N/mm^2)	Poison's ratio, ν	Density (Kg/m^3)	Yield stress, A (N/mm^2)	B	N	Reference strain rate
Facesheet AL-1100 [16]	65,762	0.3	2700	148.361	345.513	0.183	1
Adhesive layer [17]	2500	0.33	1500	48			
Aramid honeycomb core [18]	2000	0.4	710	30			

Properties	C	M	T_{melt} (K)	T_0 (K)	Specific heat C_p (J/Kg-K)	Inelastic heat fraction,	D_1	D_2	D_3	D_4	D_5
Facesheet AL-1100 [16]	0.001	0.859	893	293	920	0.9	0.071	1.248	−1.14	0.147	0

surface of the projectile and the surfaces of the facesheet and adhesive layer. Similarly, surface-to-surface contact was used to examine the interaction between the honeycomb cores and projectiles.

4 Result Discussion

In this study, a flat aramid sandwich structure with an aramid honeycomb core has been experimentally and numerically tested against an ogive-nosed projectile. By comparing predicted results to the actual results that were obtained by experimentation, the predicted results were validated. Also, the numerical study has been further extended to take into account the effect of core geometries, i.e., cell wall thickness, cell length, and facesheet thickness, on the ballistic limit of sandwich structures. In this section, the ballistic performance, failure modes, and energy absorption characteristics were examined against the ogive-nosed projectile.

4.1 Experimental Validation

At high velocity impact, the impact response of a flat sandwich structure with an aramid honeycomb core was explored against an ogive-nosed projectile. The experimental results of residual velocity were compared with the numerical simulation. Table 2 shows the relationship between residual velocity and impact velocity, which is graphically illustrated in Fig. 3. The value of residual velocity determined by numerical analysis was found to be close to the experimental result. The numerical analyses marginally overpredicted the actual data, with a maximum inaccuracy of 9.40 m/s. Figure 3 shows that the projectile's velocity drop was increased by decreasing initial impact velocity, and the drop was greater near to the ballistic limit velocity. Additionally, the ballistic limit was also calculated and found to be 84.9 m/s for experiment and 81.7 m/s for numerical simulation, since the ballistic limit shows the lowest velocity required for a projectile to completely pierce a given structure. The experimental and numerical values were close to each other with an error of 3.92% for the ogive-nosed projectile.

Table 2 Test data of residual velocity impacted against ogive projectile

Impact velocity (m/s)	Residual velocity (m/s)	
	Experimental	Numerical
144.32	92.02	94.77
122.05	75.45	79.49
102.80	47.56	53.72
90.07	24.85	32.15
86.43	13.77	21.81
83.37	0.00	9.40
80.03	0.00	0.00

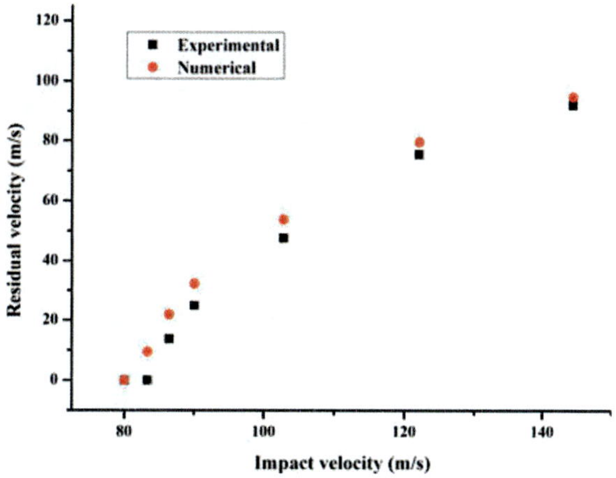

Fig. 3 Residual velocity versus impact velocity of flat sandwich structure

4.2 Failure Modes

An experimental and numerical analysis of penetration mechanisms in flat sandwich structure was evaluated. Figure 4 shows the failure mode of the flat sandwich structure having both facesheet thickness of 1 mm against the ogive-nosed projectile at a velocity of 122.05 m/s. Three phases of the failure process were identified: dishing and damage to the front facesheet, damage to the aramid honeycomb core, and dishing and damage to the rear facesheet. In the first phase, the front facesheet was primarily deformed by local dishing, confirmed by circumferential strain resulting in tensile stretching. Finally, a conical projectile penetrated the front facesheet, creating petals and radial cracks. Consequently, the aramid honeycomb core was deformed in a progressive manner near the top of the honeycomb core, resulting in slight buckling at the top while keeping the rear facesheet undeformed. In the second phase, the core was compressed and penetrated, resulting in energy absorption. The third phase showed a similar failure of dishing as the front facesheet, with the formation of petals. It was also observed that debonding between the rear facesheet and core occurred due to extensive contact force.

Figure 5 shows the complete perforation observed during experimental and numerical analysis against the ogive projectile. It was observed that the numerical model accurately captured the failure modes occurred in experimentation. In the experimentation and numerical simulation, four petals were formed in the front facesheet and rear facesheet. Moreover, when projectile fully penetrated the sheets, thin and sharp tip petals are formed due to thinning of the sheet from root to tip. In addition, it was found that the core collapsed in the folding and buckling of the cell wall, followed by core crushing.

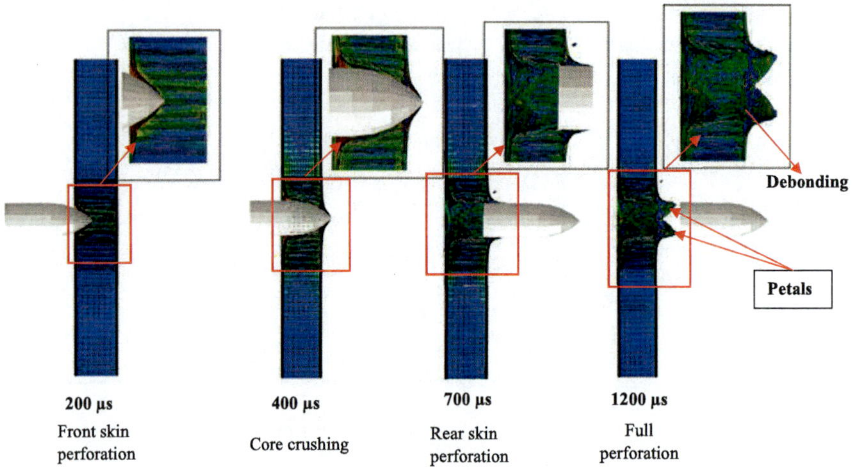

Fig. 4 Perforation progress of flat structure at impact velocity 122 m/s against ogive projectile

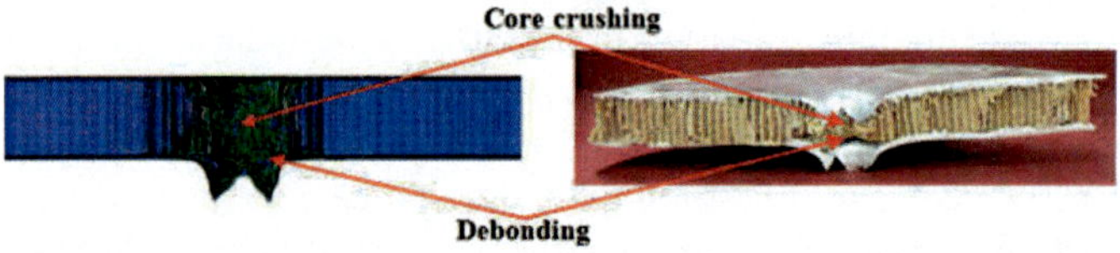

Fig. 5 Numerical and experimental failure mode during a full perforation

4.3 Influence on Ballistic Resistance of Structure

In this section, the influence of the flat sandwich structure's various parameters, i.e., facesheet thickness, cell length, and cell wall thickness, on ballistic resistance is presented. The ballistic resistance was measured in terms of the residual velocity and ballistic limit. Table 3 and Fig. 7 show the influence of the parameters on the ballistic resistance of flat sandwich structures, it was examined that the ballistic limit raised as the facesheet and cell thickness increased; however, increased cell length resulted drop in ballistic limit. The relationship between the initial impact velocity and the residual velocity has been graphically represented in Fig. 6. It was identified that rise in the impact velocity led to increased residual velocity, also the influence of the cell wall thickness was less on the residual velocity. In addition, the velocity drop increased as the facesheet thickness increased and cell length decreased.

Table 3 Influence of geometrical parameters on ballistic limit and energy absorption

Parameter	Influence of facesheet thickness				Influence of cell length				Influence of cell wall thickness			
Notation	F 0.7	F 1.0	F 1.5	F 2.0	L 3.2	L 5.2	L 7.2	L 9.2	C 0.03	C 0.05	C 0.07	C 0.09
Ballistic limit (m/s)	66.4	81.7	103.6	131.4	81.7	78.6	74.8	70.1	78.2	81.7	82.9	85.3
Energy absorption (J)	114.6	173.5	279	448.91	173.5	160.63	145.47	127.76	159.00	173.55	178.68	189.18
Specific energy absorption (J/kg)	262	329	456.9	607	329	345.5	366.5	396.9	340.7	329	299.6	287.1

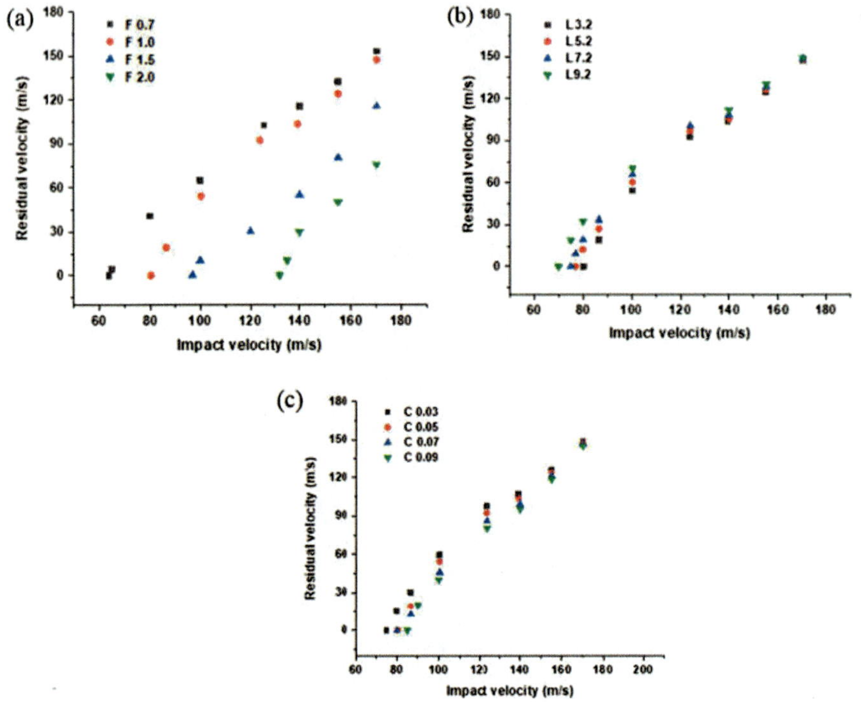

Fig. 6 Influence of **a** facesheet thickness **b** cell length **c** cell wall thickness on the residual velocity for ogive projectile

4.4 Influence on Energy Absorption

Energy absorption is the amount of energy that is required for the projectile to penetrate the structure. It is measured by the difference between the projectile's initial velocity (V_i) and the residual velocity (V_r).

$$\text{EA} = \frac{1}{2} m_p \left(V_i^2 - V_r^2\right) \tag{1}$$

where m_p is mass of projectile (here $m_p = 0.052$ kg).

The evaluated energy absorption for various influential parameters is shown in Table 3 and Fig. 8. Facesheet thickness significantly improves energy absorption, and its value increased as sheet thickness increased. A marginal increase in energy absorption was also observed for the structure with a rise in the cell thickness contrary to the facesheet thickness. In contrast to cell thickness, increasing cell length decreased the energy absorption of the structures.

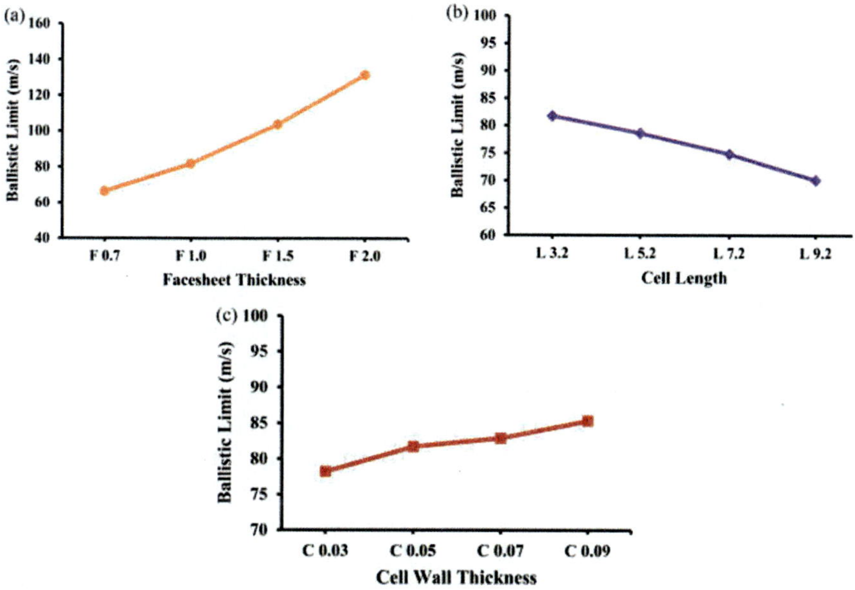

Fig. 7 Influence of variation in geometrical parameters on ballistic limit **a** facesheet thickness **b** cell length **c** cell wall thickness

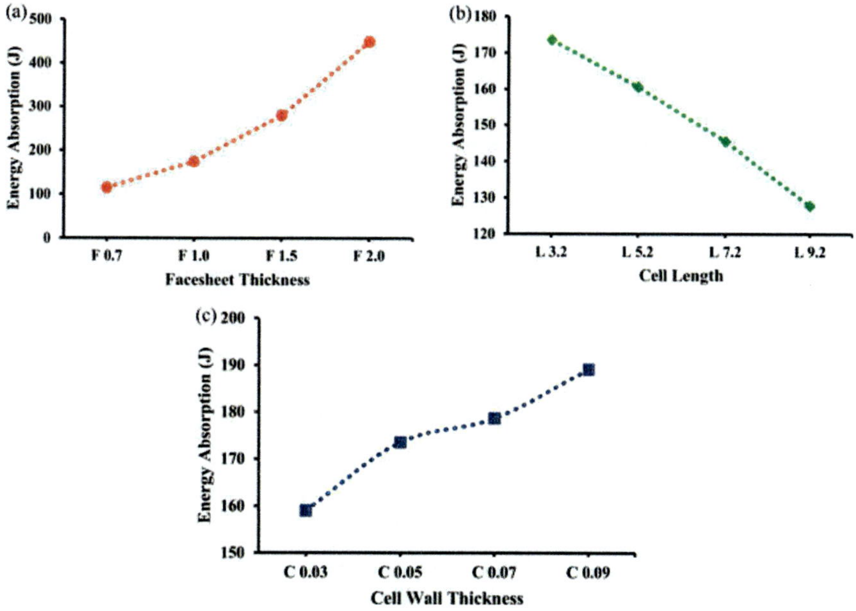

Fig. 8 Influence of variation in geometrical parameters on energy absorption **a** facesheet thickness **b** cell length **c** cell wall thickness

4.5 Influence on Specific Energy Absorption

Specific energy absorption is the energy absorption per unit weight of the sandwich structure.

$$\text{SEA} = \frac{EA}{w_s} \tag{2}$$

where w_s is weight of aramid honeycomb core sandwich structure.

Table 3 and Fig. 9 demonstrate the specific energy absorption values for various influential parameters. The results indicate that facesheet thickness is a significant factor that improves specific energy absorption, with higher values observed for thicker sheets. There was a decrease in specific energy absorption for structures with thicker cell walls, which is contrary to the effect of facesheet thickness. On the other hand, increasing cell length resulted in an increase in specific energy absorption for the structures, in contrast to the effect of cell wall thickness.

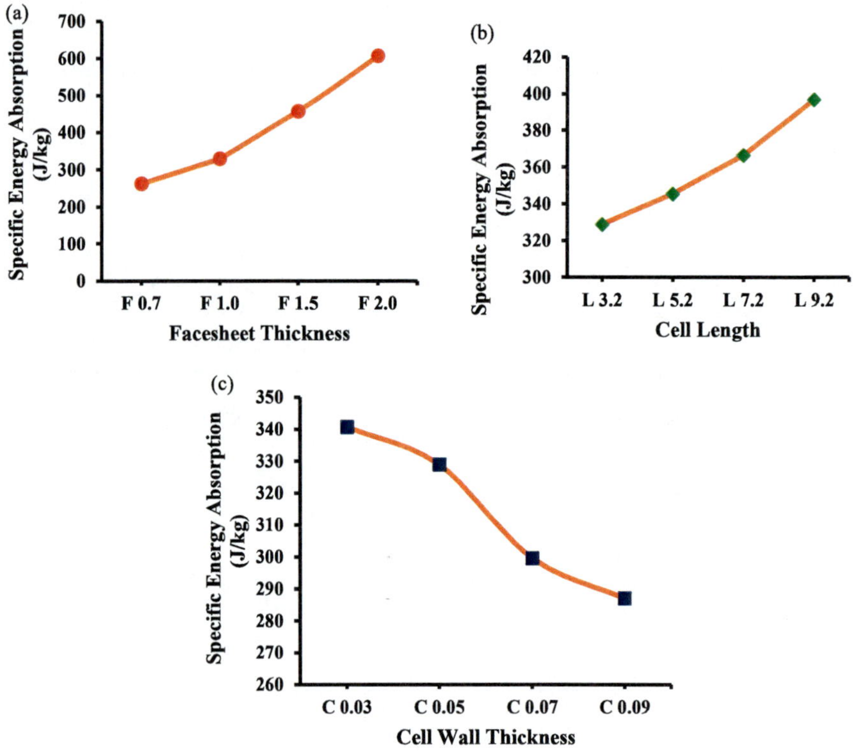

Fig. 9 Influence of variation in geometrical parameters on specific energy absorption **a** facesheet thickness **b** cell length **c** cell wall thickness

5 Conclusion

The objective of this study was to analyze the energy absorption and impact response of a flat sandwich structure with an aramid core and aluminum facesheets against an ogive projectile. Also, a parametric study was conducted to examine the influence of the thickness of the cell wall, the cell length, and the thickness of the facesheets on the flat structure. It was evident that a thicker facesheet increases ballistic performance. Moreover, structures with thicker cell walls and reduced cell lengths had a higher ballistic limit. An increase in cell wall and facesheet thickness resulted in greater energy absorption, whereas with an increase in cell length, energy absorption decreased, i.e., the thickness of the facesheet and cell wall positively affected energy absorption against the ogive projectile, while the cell length had a negative impact. The recommended application of the proposed structure is in aerospace industry for aircraft flooring, interior panels, and helicopter rotor blades; in defense industry for military vehicles, armor plating, and ballistic panels; in marine industry for boats and ships; in the automotive industry for car body panels, etc.

References

1. Gibson LJ, Ashby MF (2014) Cellular solids. Cambridge University Press
2. Hoo FM, Park KS (2000) Perforation of honeycomb sandwich plates by projectiles. Compos Part A Appl Sci Manuf 31(8):889–899
3. Feli S, Namdari PM (2012) An analytical model for composite sandwich panels with honeycomb core subjected to high-velocity impact. Compos B Eng 43(5):2439–2447
4. Othman AR, Barton DC (2008) Failure initiation and propagation characteristics of honeycomb sandwich composites. Compos Struct 85(2):126–138
5. Harizi W, Anjoul J, Acosta Santamaría VA, Aboura Z, Briand V (2021) Mechanical behavior of carbon-reinforced thermoplastic sandwich composites with several core types during three-point bending tests. Compos Struct 262
6. Foo CC, Seah LK, Chai GB (2008) Low-velocity impact failure of aluminium honeycomb sandwich panels. Compos Struct 85(1):20–28
7. Khaire N, Tiwari G, Rathod S, Iqbal MA, Topa A Perforation and energy dissipation behaviour of honeycomb core cylindrical sandwich shell subjected to conical shape projectile at high velocity impact. Thin Walled Struct 171
8. Khaire N, Tiwari G, Iqbal MA (2021) Energy absorption characteristic of sandwich shell structure against conical and hemispherical nose projectile. Compos Struct 258
9. Tiwari G, Khaire N (2022) Ballistic performance and energy dissipation characteristics of cylindrical honeycomb sandwich structure. Int J Impact Eng 160
10. Khaire N, Tiwari G (2020) Ballistic response of hemispherical sandwich shell structure against ogive nosed projectile. Thin-Walled Struct 154
11. Børvik T, Forrestal MJ, Warren TL Perforation of 5083-H116 aluminum armor plates with ogive-nose rods and 7.62 mm APM2 bullets. Exp Mech 50(7):969–78
12. Rathod S, Tiwari G, Chougale D (2021) Ballistic performance of ceramic–metal composite structures. Mater Today Proceed 41:1125–1129
13. Zhang QN, Zhang XW, Lu GX, Ruan D (2016) Ballistic impact behaviors of aluminum alloy sandwich panels with honeycomb cores: an experimental study. J Sandwich Struct Mater 20(7):861–884

14. Kolopp A, Rivallant S, Bouvet C (2013) Experimental study of sandwich structures as armour against medium-velocity impacts. Int J Impact Eng 61:24–35
15. Smith M (2009) ABAQUS/standard user's manual, Version 6.9. Dassault Systèmes Simulia Corp: Providence, RI
16. Gupta NK, Iqbal MA, Sekhon GS (2006) Experimental and numerical studies on the behavior of thin aluminum plates subjected to impact by blunt- and hemispherical-nosed projectiles. Int J Impact Eng 32(12):1921–1944
17. Sun G, Chen D, Wang H, Hazell PJ, Li Q (2018) High-velocity impact behaviour of aluminium honeycomb sandwich panels with different structural configurations. Int J Impact Eng 122:119–136
18. Foo CC, Chai GB, Seah LK (2008) A model to predict low-velocity impact response and damage in sandwich composites. Compos Sci Technol 68(6):1348–1356

An Experimental Approach to Analyze the Effect of Impact Loading on Shallow Tunnels in Weak Rockmass

Swapnil Mishra, Ankesh Kumar, K. Seshagiri Rao, and N. K. Gupta

Abstract The threat to underground structures has emerged as a vital issue and attracts the attention of researchers worldwide. Underground tunnels prove to be an essential mode of transportation for sustainable development. Urban tunnels are primarily located in the upper few kilometers in the ground. Therefore, most of these urban tunnels are situated in weathered rockmasses. Hence, these tunnels are highly susceptible to deformations under low-to-high-frequency dynamic loading. Therefore, it is essential to determine the dynamic response of tunnel lining under different sets of dynamic loading conditions. Thus, in the present study, a new impact testing facility (ITF) of 40 kg capacity was conceptualized, designed, and developed at the Indian Institute of Technology Delhi, India, for the physical modeling of tunnels at certain depth subjected to impact loading. The developed system is multifaceted, and testing is possible for a wide range of drop weights. The small-scale rock tunnel physical models of different cover depths can be tested under a sudden loading stress state, simulating in-situ stress conditions. The system consists of an impact load unit, air compressor unit, eight-channel data acquisition system, and a personal computer to record all load and deformation data. The working of the ITF facility was verified by conducting trial impact tests on several model specimens of rock tunnel blocks with three C/D ratios 1.0, 0.7, 0.5, and surrounding synthetic rockmass. The test results of a few rock tunnel models with surrounding material GM1 and with C/D 1.0 under varying drop energie conditions are discussed in the current study.

S. Mishra
Department of Mining Engineering, Indian Institute of Technology (ISM) Dhanbad, Dhanbad, Jharkhand, India

A. Kumar (✉)
Department of Civil Engineering,
Indian Institute of Technology Palakkad, Palakkad, Kerala, India
e-mail: ankesh@iitpkd.ac.in

K. S. Rao
Department of Civil Engineering, Indian Institute of Technology Delhi, New Delhi, India

N. K. Gupta
Department of Applied Mechanics, Indian Institute of Technology Delhi, New Delhi, India

R. Velmurugan et al. (eds.), *Dynamic Behavior of Soft and Hard Materials Volume 1*,
Springer Proceedings in Materials 34, https://doi.org/10.1007/978-981-99-6030-9_19

Keywords Tunnels · Shallow depth · Rock tunnel models · C/D ratio · Drop energies · Impact testing

Nomenclature

C	Cover depth
D	Diameter of tunnel
m	Mass of drop weight
v	Impact velocity
h	Height of fall
GM1	Synthetic rockmass
ITF	Impact testing facility
PoP	Plaster of Paris
Z	Zone
δ	Deformation

1 Introduction

Underground openings are proven to be vital structures due to their applications for different essential purposes, such as storage of nuclear wastes, transportation as metro tunnels, etc. Recent findings suggest that the underground structures are highly susceptible to damage when subjected to high-strain rate loadings like earthquake, impact, and blast. The behavior of buried structures under impact and blast differs from the above-ground structures. In the case of underground structures, the confinement, overburden, and length—to—diameter ratio play a prominent role in the dynamic response. Therefore, the extensive research is carried out by the researchers to understand the dynamic response of rock subjected to impact loads and their dynamic failure mechanisms [1–5]. It is evident from past experiences that full-scale experimental investigations are difficult to perform; hence, different numerical tools are widely used to study the response of rockmass under impact loading [4, 6–9].

Zhao et al. [10] monitored the effect of blast-induced vibration from the adjacent tunnel on the existing tunnel at Nanjing airport line, China. Chen et al. [11] carried out a structural model experiment on a buried scaled-down reinforced-concrete arch structure subjected to underground explosions. Flathau et al. [12] and Grubaugh [13] studied arch and dome structures through full-scale experiments subjected to nuclear detonation effects. The dynamic response of tunnels subjected to the surface explosion of different TNT charge weights has been reported for a metro tunnel [14]. The actual deformation of tunnel crown due to projectile impact followed by blast loading is reported rarely: most of the researchers have used the modified Split Hopkinson Pressure Bar (SHPB) apparatus for determining the dynamic strength of intact rocks [15–19]. Furthermore, the impact hardness of rocks was determined by different researchers using a drop hammer system [20–22]. Mishra [23] showed that the dynamic response of lining under impact load could be investigated in the laboratory through a physical model testing of rock tunnels under different drop weights.

Many researchers critically observed the material behavior by testing rocks under a high strain rate system, but the behavior of lined and unlined tunnels under low and medium strain rate loading is not addressed to date. Hence, experimental studies are needed to understand the mechanism and effects of impact and blast loads on tunnels. However, the full-scale dynamic impact and blast tests are costly and inherently complex; obtaining such data becomes expensive and challenging. In spite of being the easiest way to formulate deformations inside a tunnel under impact load, physical modeling is found to be rarely used in the studies. Further, scaling-up the impact test results from physical modeling is difficult due to the number of unknowns involved. However, the recent development in numerical modeling has overcome the difficulty associated with scaling-up the physical models.

Therefore, in the present research, a drop weight type system is designed and fabricated, which is capable of conducting impact tests on small-scale rock tunnel models with varying cover depths and surrounding rockmasses. The newly developed impact testing facility (ITF) by the researchers at the Indian Institute of Technology, Delhi, works on the principle of energy conservation, and load is applied through a load hanger with the varying height of fall and weight of discs. The testing is carried out for different impact energy. Further, the experimental results of some of the materials are compared with the FEM-based numerical results, and the crown deformation velocity is extracted from the results. Finally, the numerical simulation is carried out to investigate the effect of impact loads on the prototype models by using crown deformation velocity as the main parameter.

2 Experimental Setup

The types of equipment available so far are not fully capable of conducting drop tests on rock tunnel models, which helps to determine the tunnel lining behavior under varying impact energies and cover depths in model materials of different strengths [24]. Hence, there is a need to develop equipment to overcome these difficulties and limitations. Hence, an ITF has been designed and fabricated, which is capable of conducting tests on different rock tunnel models, with or without lining, under different drop energies. Some trial testing systems were fabricated, and tests were carried out before conceptualizing the final design of the ITF. The size and dimensions of different components have been fixed to meet the requirements of the testing models based on the trial tests and available literature. The different sections of ITF were designed and finally developed in the workshop of M/s Hydraulic Engineering Instruments Co. (HEICO), New Delhi. The schematic view of ITF is shown in Figs. 1 and 2.

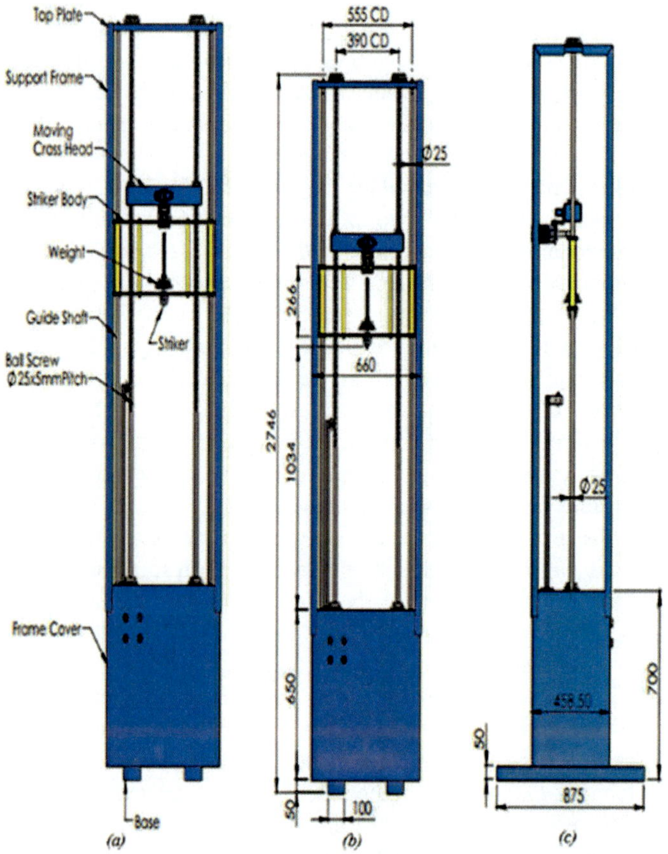

Fig. 1 2D Representation of impact testing facility (ITF) (**a**) nomenclature, (**b**), (**c**) ITF dimensions (all in mm)

Fig. 2 Schematic view of impact testing facility (ITF) and other units

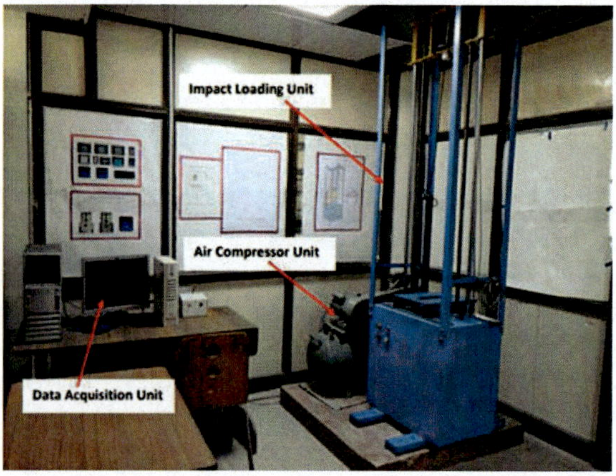

The testing arrangement comprises three significant units (as shown in Fig. 2) (a) impact loading unit, (b) air compressor unit, and (c) data acquisition unit. The dimensions have been decided and fixed to meet the casted models' requirements, whereas the casted models' dimensions are fixed based on the literature and scaling laws. The capacity of the load hanger is achieved by performing a structural analysis of the complete system by using commercially available FEM software. Based on the analysis, the height of fall, sample area, and other components are finalized.

2.1 Impact Loading Unit

The impact loading unit comprises vital segments assembled together to carry out planned experiments on the physical models. The important parts of the unit are a load hanger with the hemispherical striker, pneumatic actuator, limit switch, gear and toolbox, and support system. The working principle of the designed experimental setup (ITF) is based on the law of conservation of energy. The kinetic energy associated with an object of mass 'm' and moving with a velocity 'v' is $\frac{1}{2}mv^2$.

Hence, for applying various magnitudes of energy, one may use different masses and drop height. However, in the case of the facility used, the height of fall 'h' is kept constant (= 1.16 m), making the velocity of fall under gravity equal to $(2gh)^{0.5}$. Therefore, to achieve various impact energy, different masses of falling discs are used. The impact of falling objects on the physical model resembles the case of a projectile (missile) hitting the ground surface. The minimum weight of the load hanger without any loading discs contributes nearly about 3 kg, and the maximum capacity of the drop weight which can be loaded on the hanger is about 30 kg.

The pneumatic actuator is one of the important units of ITF. It is connected with a moving crosshead and is used for gripping the load hanger and drop weight assembly. The assembly can be moved upward up to a maximum height 1.12 m. A pneumatic actuator is used as a switch to release the drop weights from a certain height, due to which the stored potential energy is converted into kinetic energy. The pneumatic actuator has a holding clamp, which is connected to the moving head and a fitting plate. The movement of the holding clamp is controlled by compressed air wires, which are connected to the air compressor unit. The working mechanism of the ITF is controlled by four switches, which are housed inside the blue frame of the ITF. These four switches control the movement of the gear and toolbox. All four switches are mechanically connected to the load hanger for proper functioning and release of impact loads. The toolbox or frame also consists of an electric motor and gearbox along with three sets of driven pulleys.

2.2 Air Compressor Unit

The air compressor is connected to the pneumatic actuator for maintaining the required pressure for holding and releasing the load hanger while testing. A shutter valve is also provided to release the extra pressure if it exceeds the requisite amount.

2.3 Data Acquisition Unit

A data acquisition unit consists of three LVDTs, accelerogram, load cell, HBM data logger and CPU. The HBM make data logger contains 8 channels to record the output signal from LVDTs, accelerogram, and load cell. The displacement in three LVDTs, load, and acceleration w.r.t. time can be stored for every drop weight separately. AC slim line LVDT 005 is used to capture the displacement, which has a full stroke range of ±5 mm. All the LVDTs are well-calibrated before testing. It was noted during calibration that the error was within permissible limits (0.6%). Proper placement of LVDTs in the casted model is one of the essential requirements of the designed experimentation to acquire a clear perspective of the extent of the damaged length of the tunnel liner owing to sudden impact load. Figure 3 shows the chosen locations for the placement of LVDTs, where 'L' is the total length of the tunnel. An accelerometer manufactured by Measurement Specialties (Model 4810A) is used in experimentation to measure the time history of acceleration in impact loading. The signals of the accelerometer are recorded in terms of acceleration time history using a data acquisition system (DAS). The load cell is used to determine the force developed in the hammer during impact. Load cells are calibrated as per the Indian standard IS 1828 (Part I), and the error was found to be within permissible limits.

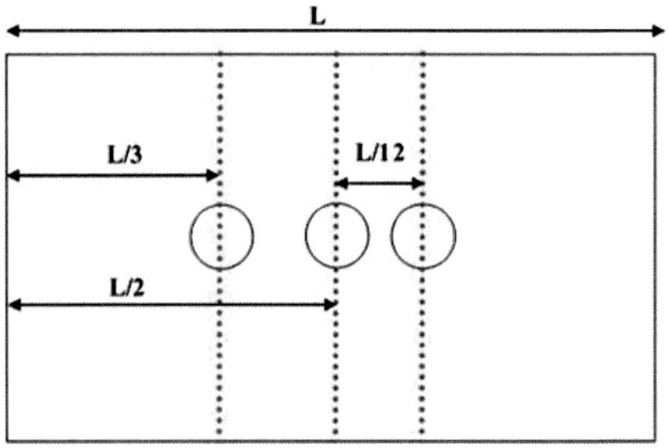

Fig. 3 LVDT locations for the measurement of deformations

3 Model Preparation

In order to evaluate the stability of urban tunnels and their failure under dynamic loading, the small-scale physical tunnel models are created in the laboratory for this study, as shown in Fig. 4. The physical tunnel models are formed by using a Perspex mould with dimensions of 0.35 × 0.30 × 0.30 m and the chosen geomaterial. Further, the circular holes of 0.05 m diameter are created on two parallel plates at a depth of 0.025, 0.035, and 0.050 mm from the top to represent cover depth (Fig. 5). Furthermore, recent research indicates that the variation in material qualities with depth is not substantial in the case of intact rocks at shallow depths. As a result, the changing of material qualities with depth is not considered in this study. However, the material characteristics will vary with the depth of the rockmass.

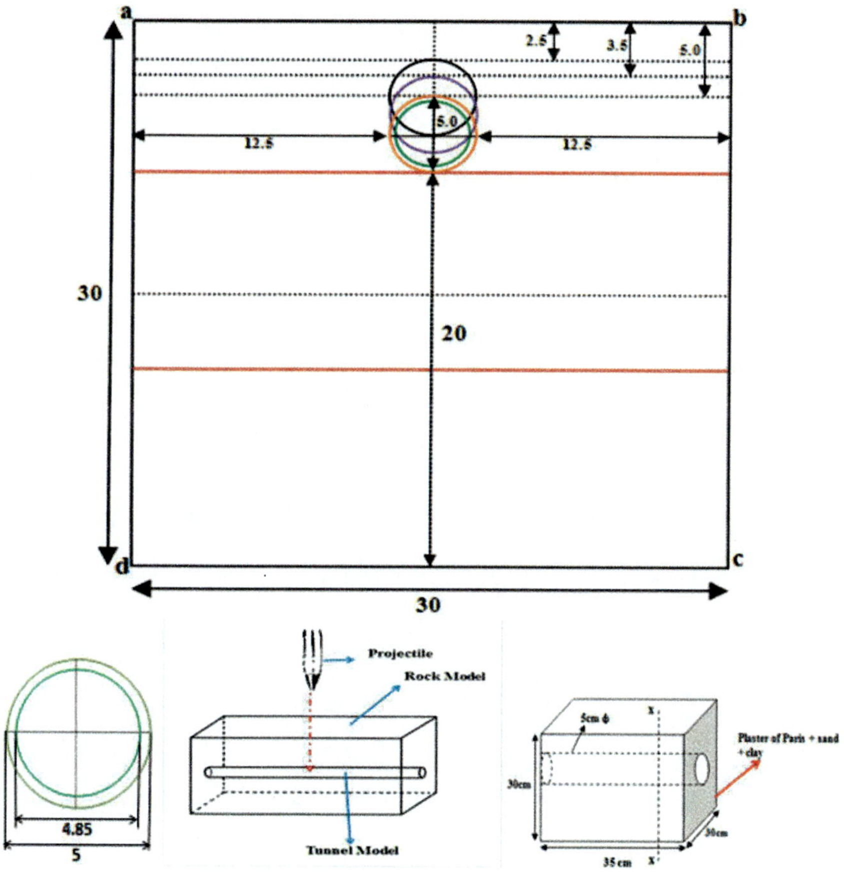

Fig. 4 Geometrical view of model with varying overburden depth along x-x axis; rock tunnel model; lining thickness

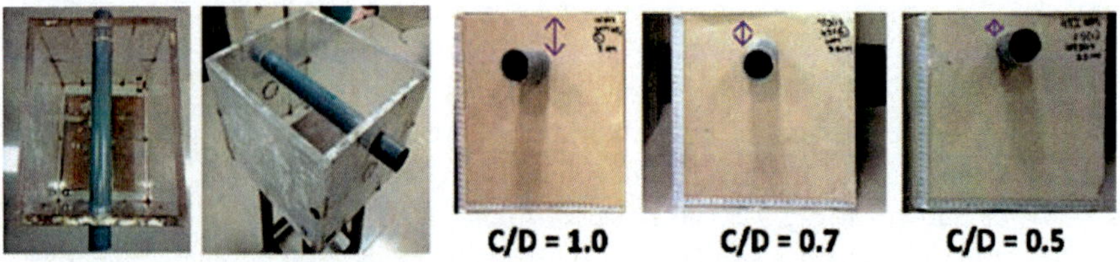

Fig. 5 Perspex mold and prepared rock tunnel model with varying C/D ratio

T-48 Quazigund railway tunnel in Katra, Jammu, was chosen as a prototype, and the properties of the surrounding rockmass and tunnel lining were picked for setting the model parameters. For model studies to replicate in-situ weak and worn rocks at shallow depths, the ratio of the elastic modulus of lining material to the elastic modulus of surrounding rock (Elining / Erock) is kept constant. In the current work, the lining material and thickness employed for physical modeling are chosen based on adequate scaling of the structural interaction of real field conditions. For scaled experiments, PVC hollow tubes are chosen as a lining material. The tunnel lining thickness is determined by the ratio of lining thickness to tunnel diameter for the T-48 railway tunnel project. Because the model's diameter is already established, the lining thickness is calculated using the ratio acquired from the T-48 tunnel. For experimentation, around 180 models are prepared. Figure 6 shows prepared and cured physical models.

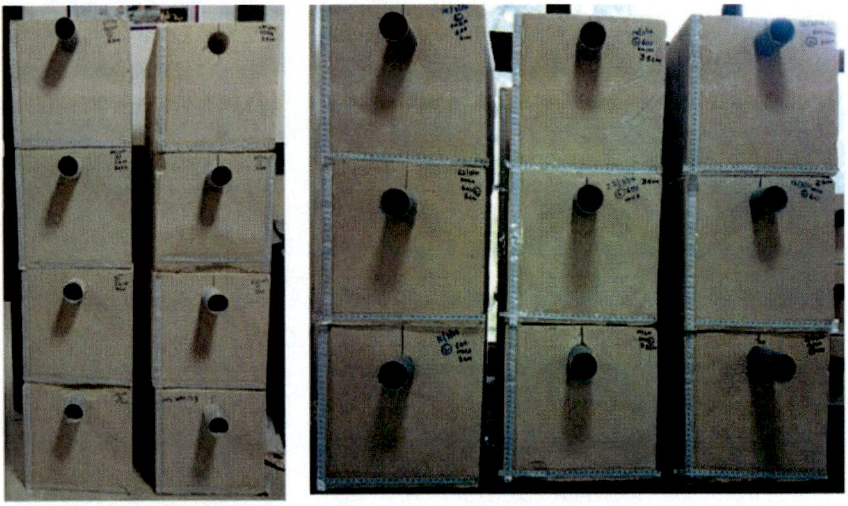

Fig. 6 Casted lined tunnel models

Table 1 Deformation of unlined tunnel under impact load

Z	Distance ($\times 10^{-2}$ m)	C/D = 1.0			C/D = 0.7			C/D = 0.5		
		12.4 kg	15 kg	20 kg	7.4 kg	10 kg	15 kg	3 kg	4.4 kg	5.8 kg
		δ (mm)	δ (mm)	δ (mm)	δ (mm)	δ (mm)	δ (mm)	δ (mm)	δ (mm)	δ (mm)
L1	0	0	0	0	0	0	0	0	0	0
L2	11.67	0	0	0	0	0	0	0	0	0
L3	14.58	−0.013	−0.0218	−0.061	−0.099	−0.174	−0.317	−0.009	−0.051	−0.132
L4	17.5	−0.205	−0.708	−1.52	−0.592	−0.84	−1.278	−0.122	−0.588	−0.767
L3'	20.41	−0.013	−0.0218	−0.061	−0.099	−0.174	−0.317	−0.009	−0.051	−0.132
L2'	23.33	0	0	0	0	0	0	0	0	0
L1'	35	0	0	0	0	0	0	0	0	0

4 Experimental Results

Impact testing is carried out on the rock tunnel models made up of GM1 (synthetic material) material by varying cover depth to diameter ratio (C/D). The dynamic response of both unlined and lined tunnels is analyzed at the crown and floor of the tunnel lining surface experimentally and numerically. In the present work, results are shown presented only for GM1 material for different cover depths.

4.1 Unlined Tunnel

The synthetic rockmass GM1 represents the geomaterial with the highest strength. It is observed that cover depth dominates the deformation response. Table 1 presents the results in terms of the deformations recorded for the case of the tunnel at cover depth of 50, 35, and 25 mm with a diameter of 0.05 m and subjected to three different drop weights. The maximum deformation is recorded at the center of the tunnel, which corresponds to the point of impact. It is found that the deformation decreases as the recording point moves away from the center of the tunnel. The deformations for the C/D ratio of 0.5 case are higher as compared to the cases of the C/D ratio of 1.0 and 0.7.

4.2 Lined Tunnel

At shallow depths, the horizontal stress acting over the tunnel is very high compared to the vertical stress. Lining offers resistance to both radial and tangential deformations. These deformations occurred due to redistribution of locked in-situ stresses in tunnel. When the tunnel is subjected to any impact loading or terrorist attack, external

Table 2 Deformation of Lined Tunnel under Impact Load

Z	Distance (× 10^{-2} m)	C/D = 1.0			C/D = 0.7			C/D = 0.5		
		17.6 kg	20 kg	25 kg	12.4 kg	15 kg	17.6 kg	5.8 kg	10 kg	12.4 kg
		δ (mm)	δ (mm)	δ (mm)	δ (mm)	δ (mm)	δ (mm)	δ (mm)	δ (mm)	δ (mm)
L1	0	0	0	0	0	0	0	0	0	0
L2	11.67	0	0	0	0	0	0	0	0	0
L3	14.58	−0.445	−0.535	−0.700	−0.316	−0.606	−0.694	−0.196	−0.363	−0.568
L4	17.5	−0.952	−1.273	−1.865	−0.787	−1.049	−1.355	−0.552	−1.233	−1.451
L3'	20.41	−0.445	−0.535	−0.700	−0.316	−0.606	−0.694	−0.196	−0.363	−0.568
L2'	23.33	0	0	0	0	0	0	0	0	0
L1'	35	0	0	0	0	0	0	0	0	0

stresses increase and exceed the internal lining pressure, causing overall failure of the structure and surrounding rockmass. In this study of the rock tunnel model, PVC is used as a lining material to replicate in-situ conditions. Various researchers have widely used PVC pipes, wire mesh, and asbestos pipes as lining materials for the physical modeling of tunnels subjected to various loading conditions [25]. Table 2 presents the results in terms of the deformations recorded in the lined tunnel for different C/D ratios of 1.0, 0.35, and 0.25, respectively. For the impact energy due to 17.6 kg of drop weight, the deformation increases by 0.403 mm as cover depth changes from 50 to 35 mm. Similarly, for 12.4 kg drop weight, deformation increases by 0.843 mm when C/D changes from 0.7 to 0.5.

During experimentation, it is observed that the depth of the crater at the surface of models and deformation just below the penetration increases with an increase in drop energy. Some of the deformation profiles for GM1 material for C/D ratio of 1.0, 0.7, and 0.5 are shown in Fig. 7a–c. Tunnels with greater cover depth are safe against the current impact loading condition. However, deformations can also be observed at greater depth if the impact load or drop energy magnitude increases beyond the safe limits. Also, large deformations are encountered in thinner cover depths, while for deep tunnels, all the energy imparted at the point of impact is dissipated in the upper strata of the ground. Hence, deeper tunnels are less influenced because of the cushioning action of the upper strata. The profile of the damage zone for unlined and lined tunnels is shown in Figs. 8 and 9, respectively.

5 Conclusions

A 40-kg impact testing facility (ITF) was developed, and its working was verified with testing on physical models of synthetic rockmass for both lined and unlined tunnels. The newly developed ITF is very useful in simulating weak weathered rockmass behavior under sudden impact loading. The system is highly versatile, and testing is possible by changing different parameters like C/D ratio, surrounding

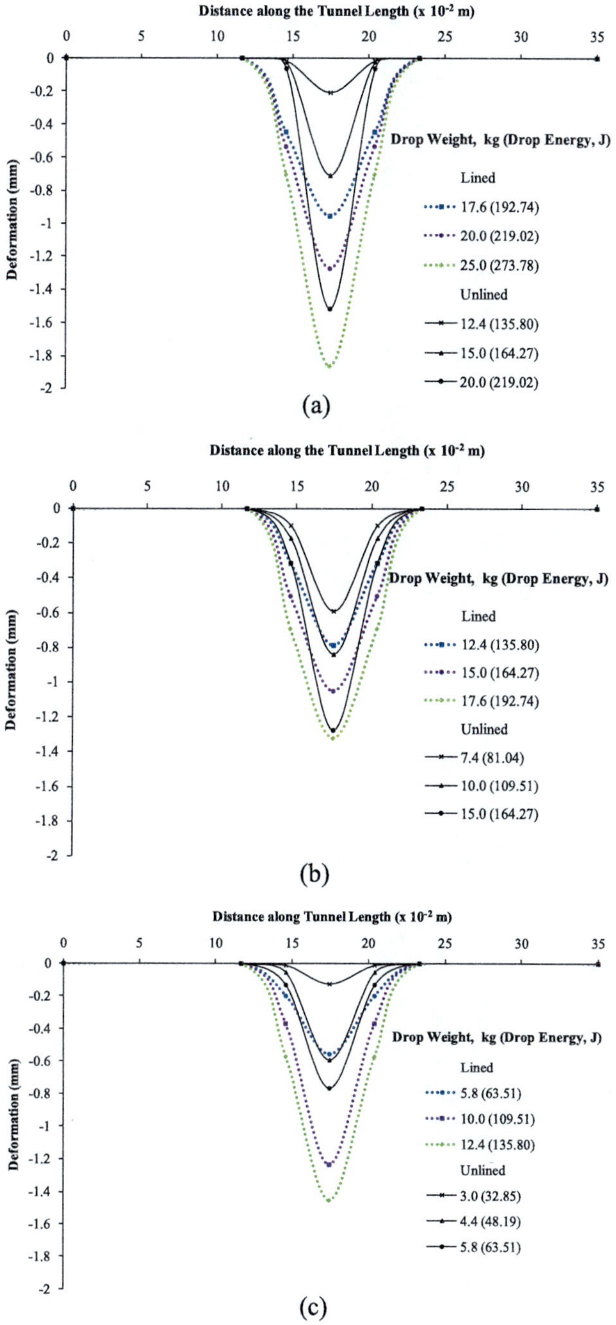

Fig. 7 Variation of deformations of tunnel crown in GM1 material under different impact loads for C/D ratios of (**a**) C/D = 1.0, (**b**) 0.7, and (**c**) C/D = 0.5

(a) *Before Testing* (b) *After Testing*

(c) *LVDT Position* (d) *Cracked Surface Inside Tunnel*

Fig. 8 Impact testing on unlined tunnel

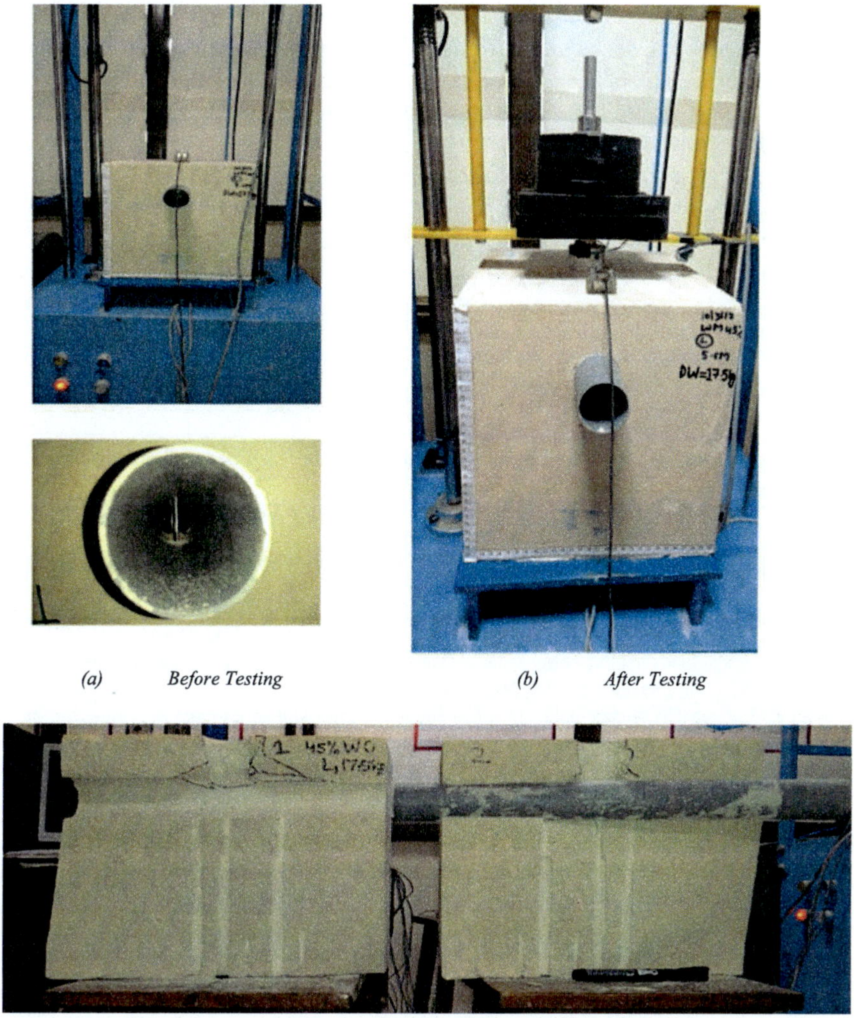

(a) *Before Testing* (b) *After Testing*

(c) *Crack Pattern in Failed Sample*

Fig. 9 Impact testing on lined tunnel (**a**) before testing (**b**) after testing (**c**) crack pattern in failed sample

rockmass properties and varying impact energies. In order to replicate the real in-situ response of tunnels and rockmass under blast and impact, it is important to formulate weak and soft rockmass in the laboratory subjected to similar loading conditions. The synthetic rockmass is prepared and named as GM1. The GM1 is made up of 50% PoP, 35% Badarpur sand, and 15% kaolinite clay with 0.45 water powder ratio, which resulted in the compressive strength of 3.52 MPa. It is observed that if any sudden load is impacted on the rock tunnel model, then, for lined tunnels, the impact energy is dissipated up to a larger extent in a rockmass with comparatively less displacement at the tunnel crown; therefore, the zone of influence is significant. The same impact energy is concentrated in the small area causing large deformations in unlined tunnels. This is because of the shielding provided by the lining against upcoming dynamic loads. The test result shows that the deformation curves exhibit punching behavior where rock tunnel models undergo tensile and shear failure. It is also observed that the peak load coming over the tunnel for C/D ratio of 1.0 is more than that of C/D ratio of 0.5 and 0.7. Similar behavior is noted for the surrounding rockmass with less dynamic strength. Further, the results presented can be used by the researchers to validate of numerical models.

References

1. Camacho GT, Ortiz M (1996) Computational modeling of impact damage in brittle materials. Int J Solids Struct 33:2899–2938
2. Xia K, Ahrens TJ (2001) Impact induced damage beneath craters. Geophy Res Lett 28:3525–3527
3. Grange S, Forquin P, Mencacci S, Hild F (2008) On the dynamic fragmentation of two limestones using edge-on impact tests. Int J Impact 35:977–991
4. Gao F, Hou H, Yang X (2010) Numerical analysis of dynamic mechanical properties for rock sample under strong impact loading. Int J Inform Eng Electron Bus 2:10–16
5. Cao P, Wang Y, Xia K (2011) Testing study on damage fracture responses of rock under dynamic impact loading. Adv Mater Res 255–260:1815–1819
6. Taylor LM, Chen EP, Kuszmaul JS (1986) Microcracking-induced damage accumulation in brittle rock under dynamic loading. Comput Methods Appl Mech Eng 55:301–320
7. Beus MJ, Iverson SR, Dreschler A, Scott VA (1999) Static and dynamic loads in ore and waste rock passes in underground mines. rock mechanics for industry. In: Proceedings of the 37th U.S. rock mechanics symposium, Vail, Colorado
8. Nazeri H, Mustoe GGW, Rozgonyi TG, Wienecke CJ (2002) Implementation of a discrete element methodology for the modeling of gravity flow of ore in ore passes. In: Proceedings of the 5th North American rock mechanics symposium and 17th tunneling association of Canada, Toronto, Canada.
9. Esmaieli K, Hadjigeorgiou J (2011) Selecting ore pass-finger raise configurations in underground mines. Rock Mech Rock Eng 44:291–303
10. Zhao HB, Long Y, Li XH, Lu L (2015) Experimental and numerical investigation of the effect of blast-induced vibration from adjacent tunnel on existing tunnel. KSCE J Civ Eng 1–9
11. Chen H, Zhou J, Fan H, Jin F, Xu Y, Qiu Y, Xie W (2014) Dynamic responses of buried arch structure subjected to subsurface localized impulsive loading: experimental study. Int J Impact Eng 65:89–101
12. Flathau WJ, Breckenridge RA, Wiehle CK (1959) Blast loading and response of underground concrete-arch protective structures. In: Army engineer waterways experiment station vicksburg Ms.
13. Grubaugh RE (1975) Full scale field tests of dome and arch structures. ITR-I425, Operation Plum bob Project 3.6. Albuquerque, N.M. Air Force Special Weapons Centre
14. Luo K, Yong W, Yuetang Z (2007) Numerical simulation of section subway tunnel under surface explosion. J PLA Univ Sci Technol 8(6):674
15. Green SJ, Perkins RD (1968) Uniaxial compression tests at varying strain rates on three geological materials. In: Proceedings of the 10th US symposium on rock mechanics (USRMS), American rock mechanics association
16. Lindholm US, Yeakley LM, Nagy A (1974) The dynamic strength and fracture properties of dresser basalt. Int J Rock Mechan Min Sci Geomech Abstr 11(5):181–191

17. Blanton TL (1981) Effect of strain rates from 10–2 to 10 sec-1 in triaxial compression tests on three rocks. Int J Rock Mechan Min Sci Geomechan Abstr 18(1):47–62
18. Li QM, Meng H (2003) About the dynamic strength enhancement of concrete-like materials in a split hopkinson pressure bar test. Int J Solids Struct 40(2):343–360
19. Chakraborty T, Mishra S, Loukus J, Halonen B, Bekkala B (2016) Characterization of three himalayan rocks using a split hopkinson pressure bar. Int J Rock Mech Min Sci 85:112–118
20. King MS (1966) Wave velocities in rocks as a function of changes in over burden pressure and pore fluid saturates. Geophysics 31(1):50–73
21. Tavares LM (2004) Optimum routes for particle breakage by impact. Powder Technol 142(2–3):81–91
22. Vanichkobchinda P (2004) Numerical simulation of the dynamic impact breakage testing of rock. Ph.D. Thesis, University of Nottingham
23. Mishra S (2019) Physical and numerical modeling of tunnels under blast and impact loads. Ph.D. Thesis. Department of Civil Engineering, Indian Institute of Technology Delhi
24. Mishra S, Rao KS, Gupta NK, Kumar A (2018) Damage to shallow tunnels in different geomaterials under static and dynamic loading. Thin-Walled Struct 126:138–49
25. Villaescusa E, Kusui A, Funatsu T (2016) Mechanical behaviour of scaled-down unsupported tunnel walls in hard rock under high stress. Tunnel Underground Space Technol 60:30–40

Biomimetic Design Approaches for Impact Applications: A Review

S. Dharani Kumar, S. Ravi Prasath, U. Magarajan, and S. Sibi Nandhan

Abstract The development of lightweight structures with exceptional toughness properties is critical for automotive, aerospace and defence applications. There is a need to reduce the weight and improve the mobility and fuel efficiency of these vehicles without compromising their impact properties. Biomimetic designs have been shown to offer advancement in impact properties over conventional engineering structures, as they are motivated by different biological structures seen in the natural world. This research article provides a board outline of recent developments in biomimetic designs and materials for impact applications. We have described the distinctive characteristics and impact resistance of different natural structures, which can be used to develop an efficient design. Next, the experimental and numerical aspects of present biomimetic structures with various configurations such as composite plates, sandwich panels and aluminium plates have been discussed. Additionally, we review the materials and manufacturing methods that are applied for processing biomimetic structures. We also focussed on insights into conventional manufacturing processes for biomimetic structures, as well as additive manufacturing. Moreover, we also discussed the different impact testing methods and computational techniques adopted for biomimetic design. Finally, we conclude with current defies and the upcoming outlooks of biomimetic designs for impact applications.

Keywords Biomimetic · Design · Impact resistance · Manufacturing · Applications

S. Dharani Kumar (✉) · S. Ravi Prasath · S. Sibi Nandhan
Department of Mechanical Engineering, KPR Institute of Engineering and Technology, Arasur, Coimbatore 641407, India
e-mail: sdharanikumarmech@gmail.com

U. Magarajan
Departement of Mechanical Engineering, Sri Venkateswara College of Engineering, Chennai 602117, India

R. Velmurugan et al. (eds.), *Dynamic Behavior of Soft and Hard Materials Volume 1*, Springer Proceedings in Materials 34, https://doi.org/10.1007/978-981-99-6030-9_20

1 Introduction

Recent years there is a need for structures which are light in weight and whose capacity in absorbing energy is high in few fields such as automobile, aerospace and defence. Lightweight and high-impact performance materials have great advantage for manufacturing of military tanks. However, mobility and fuel efficiency are two essential things that are required by the military tanks in the battle field. Other factors such as cost, weight and flexibility are also need to be considered while designing the tanks. Above following factors makes the problem even complicated. As a concern, few of energy-absorbing lightweight materials like aluminium alloy [1], composites [2], foams [3] and sandwich panels [4] have been studied recently. Most of the studies [1, 3] focussed on energy absorption capability of above-mentioned materials not on varying the structure [5]. For better improvement, the researchers trying to study and incorporate the nature inspired structures with armour applications. A lot of biological structures (plants and animals) have excellent mechanical properties [5, 6]. The synergistic effect between the mineral phases and biopolymer defines the mechanical properties of biological material [7]. In the common engineering material, toughness normally decreases suddenly with increases in stiffness. However, toughness is the important properties that are required for impact applications. Bio-inspired structured materials can resist the fracture even with increase in stiffness [6]. The unique structure that exists in the bio-inspired material was the reason for enhancement in the toughness properties [8]. Normally animal structures have excellent strength and energy absorption capability with low density [5]. However, it can be biomimetic to design and develop novel structures with significant impact properties. Figure 1 shows the different types of bio-inspired animal structures. Several animals possess protective against predatory attacks by the existing shell surrounding in the body and scales on the top layer of the skin.

Among, the scale protective structure is classified in two types as shown in Fig. 2. Both the scale structures have greater flexibility and sever protection [6]. The concepts in impact energy absorption are understood by scale animal bio-inspired structures.

For example, Naleway et al. [9] had analysed the mechanical properties and protective approaches of scale biological structure. Similarly, Tingyi et al. [1] developed and ballistic tested nacre scale biomimic AA7075-T651–based composite plate. The energy-absorbing mechanism of the composite was completely dominated by hierarchical structure. Fish scale also exhibited better impact resistance such as teleost fish scales [10] and elasmoid fish [11]. Similarly, the ballistic performance of box fish with hexagonal structures is reduced by less than 12% at the least impact region [6]. Few key understandings in high-impact absorption of energy is also studied using shell surrounded body structures. Sheep's horns are good example for shell surrounded body structure. On that time of collisions, mostly head-ons, the horns of grazers like sheep can resist the impact force of maximum magnitude 3.4 kN [12]. Another lesson can be studied from bovine hoofs, which shows the outer wall of bovine hoof has higher young's modulus value compared to inner wall [13]. Generally, shell and

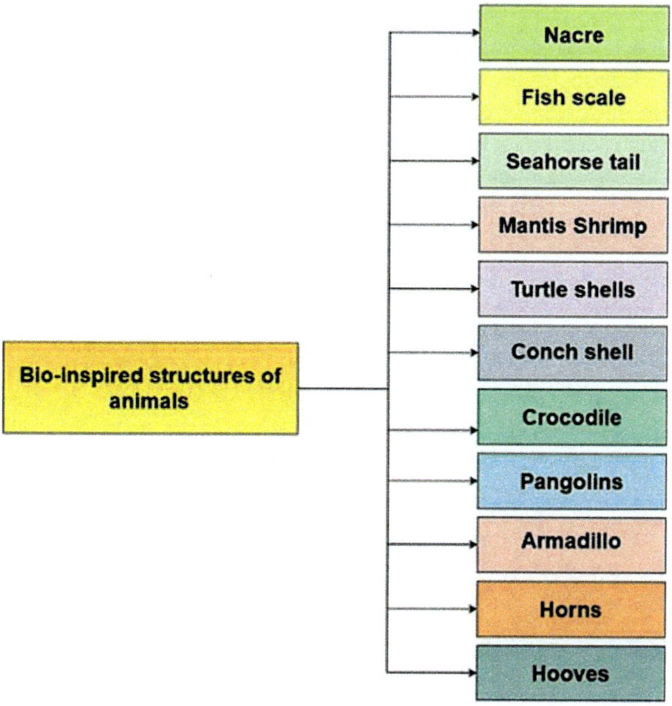

Fig. 1 Different bio-inspired structures

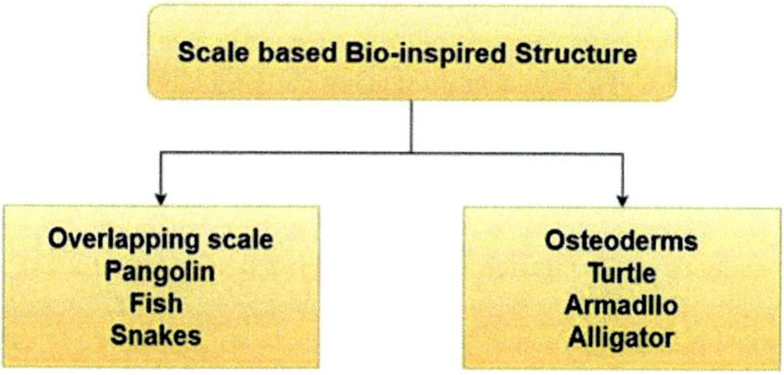

Fig. 2 Types of scale base bio-inspired structure

scale animal component and elemental structures showed strong ability of absorption of energy and gives motivation for the energy absorber design patterns. Bio-inspired structures from animal can be mimicked and used for impact applications for replacing conventional structures in future.

The main goal of the review is to furnish broad outline of recent improvements in ideals behind biomimicking terrestrial and marine animal's structure for impact applications. In particular, the scale and shell body bio-inspired structures of nacre, crocodile, pangolins, armadillo, conch shell, turtle shells, mantis shrimp, seahorse tail, fish scale, horns and hooves are reviewed and discussed. Furthermore, we also discussed in detail about materials and manufacturing and impact testing technique adopted for above-mentioned biomimic structure for impact applications. We also highlight the subtractive and additive manufacturing processes that are employed to build different biomimetic structures. Finally, the immediate challenges and prudential notions of the biomimetic structures are taken into the light of consideration.

2 Impact Studies of Selected Bio-Inspired Structures

In this section, the experimental methods employed for finding relationships between the structures and impact properties of selected biomimetic are reviewed and explored in detail.

2.1 Nacre

Nacre biomimetic structure is considered for developing armour structural materials which have outstanding impact properties [14]. Mollusk shell was found on the inner side of the nacre, which helps to protect the body from impact load [15, 16]. Normally, nacre structure was made up of two different materials such as one was soft (glues) and another one (aragonite tablets) was brittle material [17]. However, nacre structure exhibits higher toughness compared to aragonite [18]. Yin et al. [19] found that brick and mortar configured glass has higher impact resistance compared to tempered glass. Consequently, many researchers studied the impact resistance of nacre bio-inspired composite plates. Sarvestani et al. [20] concluded that nacre-like ceramics has lower stiffness and higher impact energy capability. Toughness of the nacre was completely dominated by the soft bio-polymeric matrix [21]. Grace et al. [22] revealed that mortar (soft layer) in the nacre structure played important in energy dissipation under dynamic condition. Hence, it is understood that impact properties of nacre structured composite material greatly depend upon soft material. The interfacial strength of the structure is important to enhance the impact resistance [5]. Therefore, Wu et al. [23] studied the effect of interfacial strength and impact properties of the 3D printed nacre sample made up of veromagenta (soft material) and tangblackplus (stiffer material). Their results concluded that optimum interfacial strength drops with the rise of impact velocity. Lee et al. [24] prepared the nacre composite plate by using epoxy resin and alumina particles by rolling method. Flores-Johnson et al. [25] investigate the ballistic performance of nacre-like -AA7075 /epoxy resin plates using Abaqus/Explicit software. The ballistic performance of nacre-like plate was better than flat and wavy layer plates. Sun et al. [26] investigated nacre-inspired structures and materials for impact applications in civil engineering, as well as the probable toughness designs of layered and staggered cement and clay-based composite materials.

2.2 *Crocodile*

Crocodile possess springy armoured skin in the arrangement of osteodrems. The finite hard plates surrounded in the softer tissues plays an important role in protection. The stretchy skin in the crocodile is puncture-resistant and it is repellent to projectile and arrows [26]. Chintapalli et al. [27] found that the puncture resistance of osteoderms glass–rubber system was increased 70% compared to continuous plate of ordinary glass–rubber of the same thickness. Yuan et al. [28] studied the impact behaviour of 3D printed crocodile structured stab-resistant plates. They concluded that laser sintered polyamide with carbon fibre of 6.5-mm plate thickness and 300 pyramid angles exhibited better impact resistance compared to conventional stab plates.

2.3 *Pangolins*

Pangolin also like a crocodile that whole body covered by sharp and plate scales. However, this scales help to protect themselves from predators in the wild. Teng Zhou et al. [29] studied the impact and damage responses of the pangolin scales. Their analysis results indicated that the absorbed energy, rebound velocity and absorbed rate showed linear relationship with immersion time. In old days, the armour coats were manufactured using pangolin scale structure [21, 30]. Suresh Kumar et al. [31] studied the effect of pangolin's scales and grooved mechanism on ballistic performance of aluminium alloy plates. The cantilever action offered by the pangolin scale pattern target increased the ballistic resistance as compared to curved targets.

2.4 *Conch Shell*

The three-tier hierarchical lamellar structure of the conch shell provides outstanding mechanical properties [32], which can prevent the attack from other animals. However, toughness of the shell mainly depends upon on damage controlling mechanism [32, 33]. Gu et al. [21] investigated the impact behaviour of the 3D printed biomimetic conch shell plate. They found that addition of cross-level lamellar enhanced the impact resistance compared to single-level structure. Williamson et al. [34] conch shell control the crack growth and surface area, which helps to prevent catastrophic failure.

2.5 Turtle Shells

Early in the development of synthetic polymers, turtle shell material was known a biological polymer [35]. The protein layer along with bony parts covers the top portion of the turtle shell. Numerous researchers have investigated the mechanical properties of the turtle shell. Yu et al. [36] studied the mechanical properties of turtle shell biomimicking SiCf-reinforced Ti-intermetallic multi-layers composite. Better interfacial connection between fibre (SiC) and matrix (Ti) bio-inspired could support higher load and withstand the fracture as compared the single layer composite plate. Rhee et al. [37] prepared and tested the turtle shell structured aluminium foams. Their results concluded that biological structured materials have higher mechanical properties compared to synthetic material. Wu et al. [4] compared the impact results of turtle shell and beetle forewings bio-inspired carbon fibre-reinforced plastics (CFRP) and aluminium honeycomb. They found that bare aluminium honeycomb was of inferior than energy absorption capability of CFRP sheets. Turtle shells could provide as a concept for future hard armour shells if mobility is not a problem [38].

2.6 Armadillo

Armadillo contains of overlapping protection layer in the outer skin. The external layer has dark brown and internal layer impart with hexagonal bony tiles. However, the hexagonal bony tiles are connected by collagen which is shown in Fig. 3.

The collagen fibre controls the flexibility of the armadillos [39]. Gonza´lez-Albuixech et al. [40] compared the ballistic performance of armadillo bio-inspired structure with the placoid and ganoid fish scale structure. The ballistic performance results were evaluated by using Abaqus/Explicit finite element software. Their results found that armadillo structured ceramic plate has higher ballistic resistance compared to fish scale. Similar numerical study was carried out by Miranda et al. [41]. They compared the ballistic performance of hexagonal structure of the armadillo and boxfish. They concluded that title size should keep always greater than the projectile size to minimize the penetration of the projectile.

2.7 Mantis Shrimp

Most of biological materials have capability to withstand the low velocity range. The dactyl club of the mantis shrimp, which strikes its target at a velocity of up to 23 m/s, is one of the most impressive impact-resistant biological structures [42]. Oriented chitin fibres generate a twisted Bouligand structure beneath the heavily mineralized surface of the dactyl club. Bouligand-type structures have outstanding

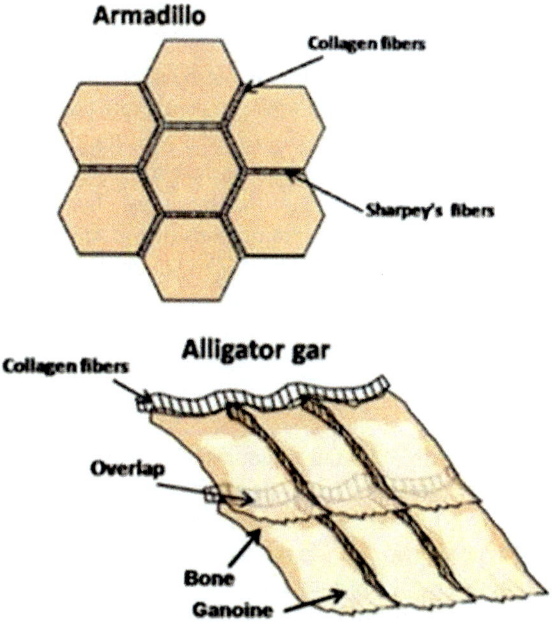

Fig. 3 Armadillo structure [39]

energy dissipation ability when subjected to impact loading [43]. Bouligand structure was made up of layers of fibres that are superimposed and rotated in relation to the layers above and below, resulting in a helical and stacked plywood configuration. Yang et al. [44] studied the effect of pitch angles on low velocity impact behaviour of Bouligand structured CFRP laminate by numerical approach. Their numerical results clearly indicated that greater fibre longitudinal modulus of elasticity and lesser pitch angles has higher impact resistance and enhanced energy absorption capability. In other research, Yang et al. [45] improved the impact and blast performance of Bouligand structure (helicoidal) by unique computational modelling method. Their results concluded that focussing on thickening of the Bouligand structures will not improve the impact and blast performance rather than configuration of the structure. Liu et al. [46] investigated the ballistic performance of Bouligand structured carbon, kelvar and polyethylene fibre-reinforced laminated composites. Combination of helicoidal and cross-ply-configured carbon/epoxy-laminated composite showed superior ballistic resistance compared to other laminated composites. The small inter-ply angle helped to improve the ballistic resistance of above-mentioned laminate composite.

2.8 Seahorse Tail

Seahorse tail is collected of subdermal bony plates arranged in form like ring segments. The tail structure has multifunctional benefits like structural support, protection and prehension [21]. The deformation mechanism of seahorse tail was quite different and it protect against hunters [5]. Porter et al. [47] found that unique hardness and better structural hierarchy of the plate offers better joint flexibility and it prevents seahorses from impact and crushing load. Biomimicking seahorse armour may inspire to develop flexible armour and fracture-resistant structures for defence applications. Seahorse bony plates are made of flexible and deformable materials [48]. Praet et al. [24] discuss that the bony plates in seahorses show an important role in axial bending and the prehensility of their tails. Overall it was noticed the seahorse's structures more flexible and better fracture-resistant structure.

2.9 *Fish Scale*

Fish have flexible dermal armour beside predatory attacks and protect their inner body. Hence, it can be motivated to design biomimetic protective equipment's with high ballistic resistance. Fish scale is one of the ultra-lightweight biological materials [49]. Signetti et al. [10] proposed creating ceramic composite panels for ballistic protection using a fish-inspired plate with a hard front and soft backing layers. Fish scale generally oriented as overlapping scale [21]. However, the scales are categorized into five types as bony plates, placoid scales, elasmoid scales, ganoid scales and carapace scutes [50, 51]. A novel composite architecture was proposed using the overlapping structure. Rudykh et al. [52] manufactured and tested the penetration resistance of 3D printed hybrid material using elasmoid fish bio-inspired structure. The penetration resistance of 3D printed samples increased by factor 40; however, the flexibility was decreased by less than 5 times. Onozato and Watabe [53] found that elasmoid scale has higher mechanical properties. Martini and Barthelat et al. [11] fabricated flexible fish inspired ceramic plate using laser engraving and stretch and release method.

3 Challenges and Future Directions

A huge number of research studies are inspired to use biomimetic-developed structures for impact applications compared with conventional structures. However, optimization and manufacturing of biomimetic structures are quite difficult. Table 1 shows the few outcomes of biomimetic processed parts and their testing. The significant challenges in the design, manufacturing and testing of biomimetic structures for impact applications are the following:

- It is difficult to select suitable design concepts from natural structures for impact energy absorption applications because they have many geometrical parameters.

Table 1 Outcome of different biomimetic structured materials

Biomimetic structure	Testing	Findings	Reference
Nacre	Freeze casting	Bidirectional freeze casting will certainly be extended to other materials systems for improving the mechanical properties The effect of adhesive and cohesive bonds revealed that shear strength and stiffness were the most important parameters	Gao et al. [14]
	Puncture tests	Laser engraving and lamination are low cost and simple to implement in protective structures	Yin et al. [19]
	Drop tower impact testing	The experiment and simulation results show that the nacre-like design is capable of preventing projectile perforation upon impact	Gu et al. [21]
	X-ray Diffraction, AFM, TEM	Understanding nacre's excellent combination of strength, stiffness and toughness, despite its high, for a biological material	Kakisawa et al. [15]
	Milling (CNC), uniaxial tension	This material has a highly sophisticated structure and mechanisms	Barthelat et al. [18]
Conch shell	Quasi-static and dynamic uniaxial compression tests	The dynamic fracture strength reached a startlingly high value of 600 MPa, representing a 67% improvement over the quasi-static fracture strength of 360 MPa	Li et al. [54]
	Drop tower testing	The conch shell architecture provides unrivalled solutions for the design of protective wear	Gu et al. [22]
	Three-point bending test	The findings suggest that an excellent composite material with higher ductility could be developed	Hou et al. [33]
	Dynamic compression and Three-point bend test	Conch shells' hierarchical structure significantly increases the toughness	Menig et al. [34]
Fish	Numerical simulation of scale-like protection system	The residual velocity results revealed that when, the increasing rate of residual velocity with an increase in impact velocity was significantly reduced	Liu et al. [35]

(continued)

Table 1 (continued)

Biomimetic structure	Testing	Findings	Reference
	Three-point bending tests	The ability of a given microstructure to provide a compliant matrix shearing deformation mechanism for flexibility, as well as a highly resistant plate bending mechanism to resist penetration	Rudykh et al. [53]
	Puncture test	Only segmented plates provide flexibility to the much softer skin and underlying tissues, as well as reducing flexural stresses and postponing fracture	Chintapalli et al. [28]
Armadillo	Numerical simulations	This study shows that segmentation reduces the size of the damaged area after an impact while maintaining ballistic protection in centred impacts	Miranda et al.[42]
	X-ray diffraction(XRD), X-ray fluorescence (XRF)	The tensile strength of dry mineralized tiles is greater than the strength of hydrated non-mineralized Sharpey's fibres	Chen et al. [61]
Crocodile	Fracture toughness test (numerical method)	PA/CF was the best material for creating new armour for the stab-resistant application	Yuan et al. [29]
Pangolin	Tensile and Bending test, Impact test	It was concluded tensile strength increased as immersion time increased, three-point bending experiment followed the same trend	Zhou et al. [31]
Turtle shell	Quasi-static compression test, High strain rate compression testing, Nano-indentation testing	High strain rate experiments revealed that some Al foams absorbed significantly more specific energy during impact than other foams Nano-indentation of the materials revealed comparable intrinsic material properties	Wang et al. [30]
Sea horse	Compression testing, Micro-hardness testing	It is concluded that after permanent deformation (> 50% deformation), the tail does not exhibit brittle fracture	Porter et al. [48]

(continued)

Table 1 (continued)

Biomimetic structure	Testing	Findings	Reference
Mantis scrimp	Numerical simulations:-low velocity impact	The fibre longitudinal elastic modulus and fibre volume fraction of the CFRPs have been found to have a significant influence on the peak load during the impact event The longitudinal elastic modulus of the fibre has the greatest influence on the energy dissipated	Yang et al. [45]
	Quasi-static load, impact resistance	According to the experimental and numerical results, multilayered architecture ceramics with lower stiffness absorb more energy during impact loads	Sarvestani et al. [20]

- The explicit impact analysis of biomimetic structures was very difficult, and it is greatly depended upon the selection of suitable computational and material model.
- A few possible research topics concern the usage of conventional materials for manufacturing the biomimetic structure. Recently, the functionally graded composite of nacre structure has shown better impact strength. Therefore, a lot of scope is there to study the impact behaviour of biomimetic structured functionally graded composites.
- In the future, machine learning techniques can be used to get optimized biomimetic structures for specific applications. However, it helps to minimize the material, cost and time.

4 Conclusions

This comprehensive review outlines biomimetic concepts that could have a capability for impact applications, especially defensive and protective structures, by replicating the animal biological structure. The protection mechanisms of different animal biomimetic structures have been studied in detail. The structural design and optimization techniques for scale base bio-inspired structures were also reviewed and discussed separately. The materials are used for biomimetic structures, including but not restricted to ceramics, polymers, non-ferrous metals and fibre-reinforced composites. This review also covered various numerical techniques for expanding the options for impact testing bio-inspired structures. However, it helps to reduce the experimental cost and time. We have emphasized additive manufacturing techniques to extend the scope of preparing and optimizing the bio-inspired structured components for real-time applications.

Acknowledgements I like to thank Chairman KPRIET Dr. K. P. Ramasamy, Principal KPRIET Dr. M. Akila, Head of the department Mechanical KPRIET Dr. S. Ramesh Babu for motivating and supporting me to write this article.

References

1. Miao T et al (2019) Ballistic performance of bioinspired nacre-like aluminium composite plates. Compos Part B: Eng 177:107382
2. Grujicic M, Snipes JS, Ramaswami S (2016) Ballistic impact behavior of nacre-like laminated composites consisting of B 4 C tablets and polyurea matrix. J Mater Eng Perform 25:977–994
3. Huang J, Durden H, Chowdhury M (2011) Bio-inspired armor protective material systems for ballistic shock mitigation. Mater Des 32(7):3702–3710
4. Wu Y et al (2017) Dynamic crash responses of bio-inspired aluminum honeycomb sandwich structures with CFRP panels. Compos Part B: Eng 121:122–133
5. San H, Ngoc L, Guoxing L (2020) A review of recent research on bio-inspired structures and materials for energy absorption applications. Compos Part B: Eng 181:107496
6. Islam MK et al Biomimetic armour design strategies for additive manufacturing: a review. Mater Design 205:109730
7. McKittrick J et al (2010) Energy absorbent natural materials and bioinspired design strategies: a review. Mater Sci Eng C 30(3):331–342
8. Naleway SE et al (2015) Structural design elements in biological materials: application to bioinspiration. Adv Mater 27.37:5455–5476
9. Naleway SE et al Structure and mechanical properties of selected protective systems in marine organisms. Mater Sci Eng C 59:1143–1167
10. Signetti S, Pugno NM (2014) Evidence of optimal interfaces in bio-inspired ceramic-composite panels for superior ballistic protection. J Eur Ceram Soc 34(11):2823–2831
11. Martini R, Balit Y, Barthelat F (2017) A comparative study of bio-inspired protective scales using 3D printing and mechanical testing. Acta Biomater 55:360–372
12. Kitchener A (1988) An analysis of the forces of fighting of the blackbuck (Antilope cervicapra) and the bighorn sheep (Ovis canadensis) and the mechanical design of the horn of bovids. J Zool 214(1):1–20
13. Huang W et al (2019) A natural energy absorbent polymer composite: the equine hoof wall. Acta Biomaterialia 90:267–277
14. Gao W, Wang M, Bai H (2020) A review of multifunctional nacre-mimetic materials based on bidirectional freeze casting. J Mech Behav Biomed Mater 109:103820
15. Kakisawa H, Taro S (2012) The toughening mechanism of nacre and structural materials inspired by nacre. Science Technol Adv Mater
16. Sun J, Bhushan B (2012) Hierarchical structure and mechanical properties of nacre: a review. RSC Adv 2(20):7617–7632
17. Cilento F, Martone A, Giordano M (2022) Insights on shear transfer efficiency in "brick-and-mortar" composites made of 2D carbon nanoparticles. Nanomaterials 12(8):1359
18. Barthelat F, Zhu D (2011) A novel biomimetic material duplicating the structure and mechanics of natural nacre. J Mater Res 26(10):1203–1215
19. Yin Z, Hannard F, Barthelat F (2019) Impact-resistant nacre-like transparent materials. Science 364(6447):1260–1263
20. Sarvestani HY et al (2019) Multilayered architectured ceramic panels with weak interfaces: energy absorption and multi-hit capabilities. Mater Design 167:107627
21. Gu Grace X et al (2016) Biomimetic additive manufactured polymer composites for improved impact resistance. Extreme Mechan Lett 9:317–323

22. Gu GX, Takaffoli M, Buehler MJ (2017) Hierarchically enhanced impact resistance of bioinspired composites. Adv Mater 29(28):1700060
23. Wu K et al (2019) Interfacial strength-controlled energy dissipation mechanism and optimization in impact-resistant nacreous structure. Mater Design 163:107532
24. Lee Y et al (2017) Nacre-inspired composite prepared by rolling method I: effect of particle orientation on deformation behavior. Comp Struct 182:549–554
25. Flores-Johnson EA et al (2014) Numerical investigation of the impact behaviour of bioinspired nacre-like aluminium composite plates. Compos Sci Technol 96:13–22
26. Chen PY, Joanna M, Marc André M (2012) Biological materials: functional adaptations and bioinspired designs. Progr Mater Sci 57.8:1492–1704
27. Chintapalli RK et al (2014) Fabrication, testing and modeling of a new flexible armor inspired from natural fish scales and osteoderms. Bioinspirat Biomimet 9.3:036005
28. Yuan MQ et al (2017) The application of PA/CF in stab resistance body armor. IOP Conferen Series Mater Sci Eng 213(1)
29. Wang B et al (2016) Pangolin armor: overlapping, structure, and mechanical properties of the keratinous scales. Acta biomaterialia 41:60–74
30. Zhou T et al (2020) Effect of hydration on mechanical characteristics of pangolin scales. J Mater Sci 55:4420–4436
31. Suresh Kumar S, Pranaav S (2022) Effect of bio-inspired surface pattern (Pangolin's scales) and grooved mechanisms on the high velocity ballistic performance of aluminum 6061-T6 targets. Mechan Adv Mater Struct 29.25:4151–4169
32. Hou DF, Zhou GS, Zheng M (2004) Conch shell structure and its effect on mechanical behaviors. Biomaterials 25(4):751–756
33. Menig R et al (2001) Quasi-static and dynamic mechanical response of Strombus gigas (conch) shells. Mater Sci Eng A 297(1–2):203–211
34. Williamson DM, Proud WG (2011) The conch shell as a model for tougher composites. Int J Mater Eng Innov 2(2):149–164
35. Chen IH et al (2011) Armadillo armor: mechanical testing and micro-structural evaluation. J Mech Behav Biomed Mater 4.5:713–722
36. Wenbo Y et al (2016) Bio-inspired design of SiCf-reinforced multi-layered Ti-intermetallic composite. Mater Design 101:102–108
37. Rhee H et al (2015) Structure-property responses of bio-inspired synthetic foams at low and high strain rates. Sci Eng Compos Mater 22.4:365–373
38. White ZW (2018) Fish-scales: the next step in soft body protection?. Diss. University of Colorado at Boulder
39. Yang W et al (2013) Natural flexible dermal armor. Adv Mater 25.1:31–48
40. González A, Vicente F et al (2019) Numerical analysis for design of bioinspired ceramic modular armors for ballistic protections. Int J Damage Mechan 28.6:815–837
41. Miranda P, Pajares A, Meyers MA (2019) Bioinspired composite segmented armour: numerical simulations. J Market Res 8(1):1274–1287
42. Trim MW et al (2011) The effects of water and microstructure on the mechanical properties of bighorn sheep (Ovis canadensis) horn keratin. Acta biomaterialia 7.3:1228–1240
43. Weaver JC et al (2012) The stomatopod dactyl club: a formidable damage-tolerant biological hammer. Science 336.6086:1275–1280
44. Yang F, Xie W, Meng S (2020) Global sensitivity analysis of low-velocity impact response of bio-inspired helicoidal laminates. Int J Mech Sci 187:106110
45. Yang F, Xie W, Meng S (2020) Impact and blast performance enhancement in bio-inspired helicoidal structures: a numerical study. J Mech Phys Solids 142:104025
46. Liu JL et al (2020) The response of bio-inspired helicoidal laminates to small projectile impact. Int J Impact Eng 142:103608
47. Porter MM et al (2013) Highly deformable bones: unusual deformation mechanisms of seahorse armor. Acta Biomaterialia 9.6:6763–6770
48. Porter MM et al (2015) Why the seahorse tail is square. Science 349.6243:aaa6683

49. Liu P et al (2016) Numerical simulation of ballistic impact behavior of bio-inspired scale-like protection system. Mater Design 99:201–210
50. Song J (2011) Multiscale materials design of natural exoskeletons: fish armor. Diss. Massachusetts Institute of Technology
51. Johnson A (2014) Establishing design characteristics for the development of stab resistant Laser Sintered body armour. Loughborough University, Diss
52. Rudykh S, Ortiz C, Boyce MC (2015) Flexibility and protection by design: imbricated hybrid microstructures of bio-inspired armor. Soft Matter 11(13):2547–2554
53. Onozato H, Watabe N (1979) Studies on fish scale formation and resorption: III. Fine structure and calcification of the fibrillary plates of the scales in Carassius auratus (Cypriniformes: Cyprinidae). Cell Tissue Res 201:409–422
54. Liu S et al (2020) Mechanical properties and failure deformation mechanisms of yak horn under quasi-static compression and dynamic impact. J Mech Behav Biomed Mater 107:103753

Ballistic Performance of Bio-mimicked Nacreous Protection System on Concrete Target: A Numerical Investigation

Abinas Nayak, Abinash Kumar Swain, and Vishwas A. Sawant

Abstract Inspired by the staggered pattern and global energy dissipation mechanism of nacre layers, an aluminum alloy (AA) composite layer structure was designed on the concrete surface to improve its ballistics performance in the numerical investigation. This low-weight 7075-T651 AA composite structure is of 12 mm thickness, having layers of sheets bonded together using stress criteria corresponding to epoxy resin. A numerical model was developed for reinforced concrete using an explicit dynamics solver and validated against experimental penetration depth and damage data. This model is extended to develop surface protection of AA plates. The total thickness of the AA composite plate is maintained at 12 mm. Parametric analysis is done for plain reinforced concrete, with an AA bulk plate and 4, 6 layers of AA nacre layers to study the ballistics performance. A significant residual velocity decreased for nacre layer structures compared to the plain bulk plate. Due to the layered and staggered arrangement, a larger area of delocalized damage and debonding between plates was observed in the nacre-like plates. Energy characteristics improved for the four-layer nacre structure compared to the bulk plate. This paper will provide insight into enhancing designs for better ballistics performances in future.

Keywords Impact · Concrete · 7075-T651 AA · Nacre · Ansys

1 Introduction

Concrete is used worldwide to construct surface and underground protective structures, such as nuclear, military containments, and other facilities, to survive the accidental or intentional impact loadings caused by blasts, projectiles, and other threats.

A. Nayak · A. K. Swain (✉)
Mechanical and Industrial Engineering Department, IIT Roorkee, Roorkee, India
e-mail: abinash.swain@me.iitr.ac.in

V. A. Sawant
Civil Engineering Department, IIT Roorkee, Roorkee, India

R. Velmurugan et al. (eds.), *Dynamic Behavior of Soft and Hard Materials Volume 1*, Springer Proceedings in Materials 34, https://doi.org/10.1007/978-981-99-6030-9_21

Though several efforts were made to enhance the dynamic performance of concrete structures by increasing compressive strength [1] and with reinforcements [2] under impact loadings, optimal designs of external reinforcements are still required to increase the concrete structure's impact strength. The confinement on the concrete's frontal and back-free surfaces helps protect the structure from impact. The frontal confining plates resist the projectile, while the rear liner effectively aids in restraining rear face damage and restricting the scabbing fragment [3, 4]. Although the ballistic performance of concrete slabs increases with liners, the overall weight increases with external reinforcements.

Structural natural materials like timber, bone, and shells [5] are a great source of inspiration for evolutionary designs and high-performance, lightweight composite frameworks. These natural materials deform through high-energy-absorbing methods, enhancing structural and mechanical characteristics like stiffness, strength, and toughness.

Nacre, also known as mother-of-pearl, is a naturally occurring substance with exceptional mechanical properties due to its hierarchical design that spans multiple scales. It is a brittle mineral composed of stacks of aragonite glued together in a brick-like pattern. Despite having a 95 percentage of aragonite composition, nacre has a hardness approximately 3000 times greater than aragonite [6]. This remarkable performance is credited to the structure's brick arrangement, the tablets' waviness, and the various tablet interfaces. Barthelat and Zhu [7] established the effect of tablet waviness on the toughness of a nacre-like composite. They also gave theoretical models for the best design of this bio-inspired material. Nacre performs better sustaining impact and blast loading than standard laminated composite or bulk plates. Knipprath [8] demonstrated that the impact response of B_4C ceramic could be improved by using a straightforward, nacre-like structural design that encourages fracture delocalization. Tran [9] also showed how a nacre-like structural design improved the blast performance of glass fiber/thermoset resin composites.

This research aims to understand better the surface protection system for reinforced concrete targets that employ nacre structures. Simulations of the projectile's impact against reinforced concrete targets were done in Ansys Autodyn for four conditions, i.e., reinforced concrete without plate, with plane aluminum alloy plate, and concrete with nacreous aluminum alloy plates with 4 and 6 layers. In all conditions, plates are placed on the frontal surface of the concrete.

2 Problem Description and Numerical Modeling

2.1 Problem Description

All impact phenomena are considered very complex, accompanied by non-linear pressure response, strain hardening, and strain rate hardening. So, to simulate the impact phenomena practically, the material models should be robust enough to accommodate all these responses. For all the simulations, commercial finite element package Autodyn has been used. Autodyn is an engineering software package that solves non-linear dynamic problems like impact and blast phenomena using Lagrange, Euler, ALE, and other mesh-free solvers like SPH. The finite element method is used to solve partial differential equations obtained from governing equations such as conservation of energy, mass, and momentum to determine the behavior of materials such as stress and strain. An experimental model by Hanchak [10] is selected to investigate the ballistic impact behavior of surface-protected reinforced concrete structures. Hanchak has conducted perforation experiments of 610 mm × 610 mm × 178 mm concrete specimen of compressive strength of 140 MPa and 48 MPa with a projectile weighing 0.5 kg, diameter 25.4 mm CRH of 3.0. Several projectile velocities starting from 301 to 1058 were used in experiments. The base model simulation uses a 48 MPa concrete slab with the above dimensions above which the nacreous layer will be mounted. Three layers of reinforcements with a gap of 76 mm in the crisscross dimension and the transverse direction are maintained. The diameter of the reinforcement bars is 5.69 mm. The schematic diagram of impact simulation is given in Fig. 1. The simulations used the quarter model to take advantage of the symmetry and reduce computation time. The projectile is considered rigid because our simulation focuses on the performance of the nacre structure.

The 3D full model's solid geometry was inspired by the structure of the nacre and consisted of a 600 mm × 600 mm plate built of alternative 7075-T651 AA layers. Simulated targets were 4, 6 layers, with a total thickness of 12 mm. These results are compared to solid bulk plates with a thickness of 12 mm. Each layer comprised thirty-six 50 mm × 50 mm square tablets which were moved concerning its upper or lower adjoining layer so that individual tablets covered 50 percent of the surface area (Fig. 2). This overlap is almost a third of the surface overlap observed in natural nacre [7]. The material property of 7075-T651 AA is given in Table 1 [11]. SolidWorks 2020 (Dassault Systems, SolidWorks Corp., France) was used to create the quarter solid geometry, subsequently imported into Ansys/Explicit (version 2021). An epoxy

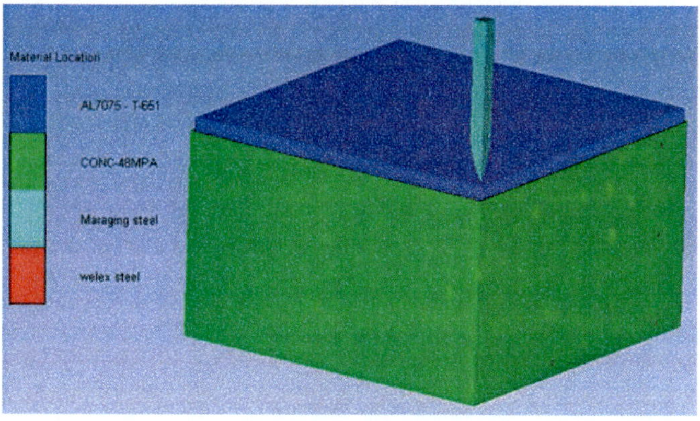

Fig. 1 Schematic model for impact simulation

resin adhesive was used to model the interface between AA layers and tablets [12]. The parameters in Table 2 are the stress criteria used for modeling the debonding of layers in case of failure.

Concrete and the plates have meshed in solid hexahedral elements, and projectile meshed using tetrahedral elements. Reinforcement is meshed using 1D beam elements. A mesh size of 2 mm is used for the concrete, reinforcements, and beam elements, while bulk plate and the nacre plates are meshed using 1 mm elements. The elements used in the simulation are 25 lakhs to 30 lakhs. Trajectory contact is used for the interaction between master and slave surfaces.

Fig. 2 A six-layer nacre-like AA plate

Table 1 Johnson–Cook parameters for 7075-T651 AA material [11]

Material properties	Unit	Value
Density, ρ	Kg/m^3	2700
Poisson's ratio		0.3
Young's modulus	GPa	70
Yield stress (A)	MPa	520
Hardening constant (B)	MPa	477
Hardening exponent (n)		0.52
Strain rate constant (c)		0.001
Thermal softening exponent (m)		1
Room temperature (T)	K	293
Melting temperature (T_m)	K	893
Damage parameters	Unit	Value
D_1		0.096
D_2		0.049
D_3		– 3.465
D_4		0.016
D_5		1.099
Ref. strain rate, $\mathring{\varepsilon}_0$	s^{-1}	5×10^{-4}

Table 2 Debonding criteria for epoxy material [12]

Material	Epoxy
Peak tensile stress (MPa)	27
Peak shear stress (MPa)	17

2.2 Material Models

The concrete simulation was based on the Riedel, Hiermaier, and Thoma (RHT) model [13]. The RHT model is a more advanced plasticity model for brittle materials considering concrete's pressure hardening, strain hardening, strain rate hardening, strain softening, and third invariant dependency. In stress space, the RHT model has three pressure-dependent surfaces. Elastic limit surface, failure surface, and residual surface are the three types of surfaces [14]. The residual surface expresses the strength of the completely crushed material as follows:

$$Y_{\text{residual}} = B \times \left(\frac{p}{f_c}\right)^m \tag{1}$$

B and m are two non-dimensional constant parameters established via curve fitting of experimental data. The residual strength of concrete is determined by *B* and *m*.

Tu and Lu [15] found that the residual strength surface in the typical RHT model has a circular deviatoric cross-sectional plane in the major stress space. After reaching the peak failure strength, the model hardens instead of softening, as is generally predicted. With various compressive strengths (26, 57, 77, and 92 MPa) of concrete specimens at a 300 m/s impact velocity, Abdel-Kader [16] investigated the projectile impacts on concrete targets with the default RHT model. Due to the high predicted residual strength value, the simulation did not predict radial cracks at both the impact and rear faces in all circumstances. Also, all specimens overestimated perforation limits due to the high assumed tensile strength value. So, the author calibrated the parameters concerning $f_{c,}$ in Eqs. (2–3) as follows:

$$\frac{f_t}{f_c} = 0.26 f_c^{\frac{-1}{3}} \tag{2}$$

$$B = 1.6(35/f_c)^{0.5} \tag{3}$$

The RHT model parameters are presented in Table 3 for 48 MPa concrete with modified B, M, (ft/fc) Eqs. (2–3). The projectile is made rigid with a density of 8020 kg/m^3. The reinforcements are made up of weldex steel [17], which uses the Johnson–Cook model for strength. No failure model is used for reinforcements because of no direct projectile interaction with reinforcements.

To simulate the constitutive response of the nacre layers, the Johnson–Cook material model was combined with the Johnson–Cook fracture criterion [18]. The Johnson–Cook (JC) constitutive model is an empirical model in which yield stress *Y* fluctuates as a function of strain, strain rate, and temperature in Eq. (4) as follows:

$$\sigma_{eq} = (A + B\varepsilon_{eq}^n)(1 + C \ln \dot{\varepsilon}_{eq})(1 - T^{*m}) \tag{4}$$

Table 3 RHT material model for concrete [16]

Material properties	Value	Material properties	Value
Reference density	2.7500 (g/cm^3)	Elastic strength/ft	0.7 (none)
Strength	RHT Concrete	Elastic strength/fc	0.53000 (none)
Shear modulus	1.67E + 07(kPa)	**Fractured strength const. B**	**1.27640(none)**
Compressive strength (fc)	4.80E + 04(kPa)	**Fractured strength exponent M**	**0.61 (none)**
Shear strength (fs/fc)	0.18 (none)	Compressive strain rate exp. (α)	0.03200 (none)
Tensile strength (ft/fc)	**0.0684 (none)**	Tensile strain rate exp. (δ)	0.03600 (none)
Intact failure surface expo. N	0.61 (none)	Use CAP on elastic surface?	Yes
Intact failure surface const. A	1.6 (none)		
Tens./comp. meridian ratio (Q)	0.68 (none)		
Brittle to ductile transition	0.01050 (none)	G (elas.)/(elas. -plas.)	2.00000 (none)

In the above, $A, B, C, n,$ and m are five material constants, and T_h is the homologous temperature defined as $T^{*m} = (T - T_{\text{room}})/(T_{\text{melt}} - T_{\text{room}})$, where T is absolute temperature, T_{melt} and T_{room} are the melting and room temperatures, respectively. ε_{eq}^n and $\dot{\varepsilon}_{eq}^n$ are effective plastic strain and normalized effective plastic strain rate, respectively. Here, $\dot{\varepsilon}_{eq}^n = \varepsilon_{eq}^n/\varepsilon_0$, where ε_0 is the user-defined strain rate.

The Johnson–Cook damage model is based on damage accumulation at an element integration point, and the Johnson–Cook (JC) fracture criterion is defined in Eqs. (5–6) as the ratio of the increment of the equivalent plastic strain during an integration cycle to a threshold equivalent fracture strain:

$$D_{JC} = \sum \frac{\Delta\varepsilon_{eq}}{\varepsilon_f^{JC}} \tag{5}$$

$$\varepsilon_f^{JC} = (D_1 + D_2 \exp(D_3\sigma^*)(1 + D_5 T^*) \tag{6}$$

Here, D_{JC} is the damage parameter and ε_f^{JC} the equivalent fracture strain in the stress triaxiality, where σ^* represents the hydrostatic stress. Materials $D_1 - D_5$ are damage parameters.

A mesh size of 2 mm is used to simulate the concrete and reinforcements. Simulations for different projectile velocities are shown to agree well with experimental results. A detailed mesh sensitivity analysis for concrete has been shown in Fig. 3. For mesh sensitivity, aluminum alloy (AA) plates have meshed with 0.5 mm, 1 mm, and 2 mm. The results have shown an appreciable change between 2-mm and 1-mm

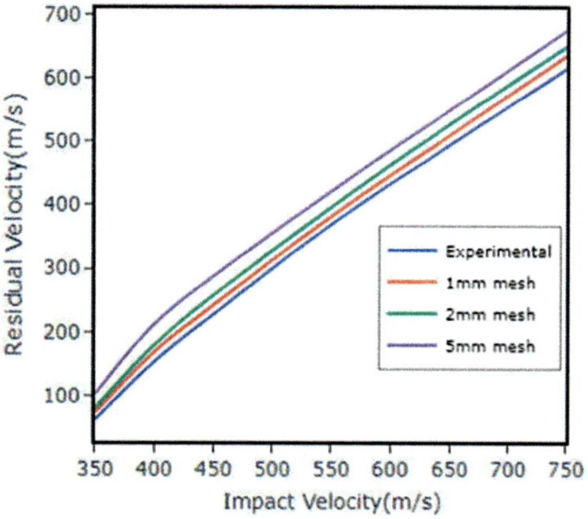

Fig. 3 Mesh sensitivity analysis of concrete

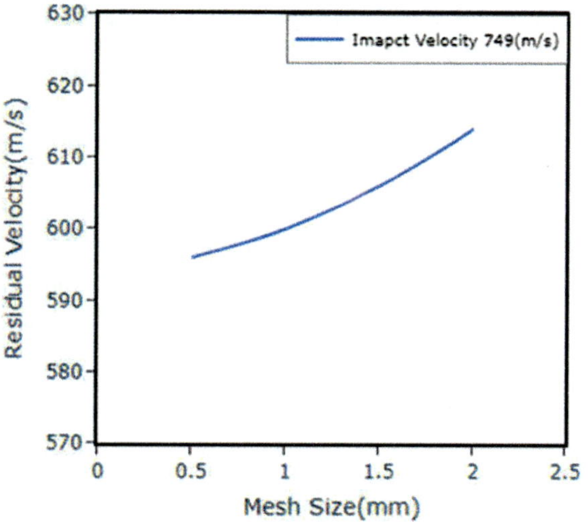

Fig. 4 Mesh sensitivity analysis of 7075-T651 AA

meshes in residual velocity but not between 0.5- and 1-mm mesh (Fig. 4). The AA plates have been meshed with 1 mm, utilizing the available computational resources.

3 Results

Numerical simulations were done using Ansys Autodyn to study the ballistic performance of the plain reinforced concrete, with a bulk plate of AA, with 4-layer and 6-layer AA nacre structured plates. A damage contour of a complete model with a six-layer nacre layer after the simulation is presented in Fig. 6 for understanding the whole phenomena. The residual velocity versus time of the projectile for all the four conditions is plotted in Fig. 5. It is evident from the graph that pure concrete has the highest residual velocity of 640 m/s, while bulk plate AA of weight of 2.916 kg could bring down the residual velocity to 616 m/s. In contrast, the residual velocity of 4-layer nacre AA plate is 603 m/s, and six-layer nacre reduces the velocity to 604 m/s. The bulk plate decreases the residual velocity, whereas the four-layer and six-layer nacre arrangements of the same thickness AA plate decrease the velocity further by 13 m/s and 12 m/s, respectively. This superior ballistics resistance of nacre is attributed to the more globalized plastic work and damage, as shown in Figs. 11, 12 and 13.

In the case of the bulk plate, we can observe localized damage near the penetration hole. The damage in nacre plates shows debonding between layers of nacre and between plates between layers. In Fig. 13, one of the top nacre plates in a 6-layer structure is completely deboned from other layers. It is not that much happened in 4-layer nacre, which shows that increasing the layers of the nacre while keeping the total thickness of the plate increases the damage in the form of debonding.

Figures 7, 8, 9 and 10 show different energy distribution graphs of the four types of structures. In the internal energy distribution of AA plate (Fig. 7), bulk plate absorbed maximum energy at the initial penetration stage while decreasing as time progressed below four-layer nacre. It is due to the initial ductile failure offered by bulk plates and the diverse energy-absorbing mechanisms activated in nacre at later stages of penetration. Six-layer nacre absorbed significantly less energy compared to both bulk and 4-layer nacre. This behavior is due to the excessive debonding and detachment of the nacre plates at the impact site, making the failure more brittle. This

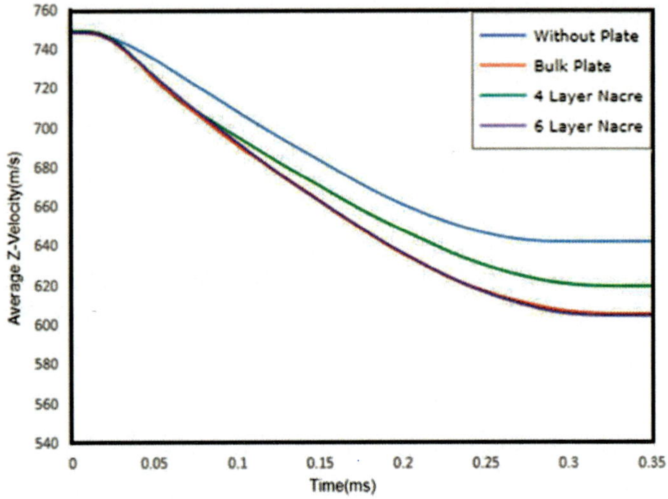

Fig. 5 Velocity profile of projectile for (i) without plate, (ii) with bulk plate, (iii) nacre 4 layers, (iv) nacre 6 layers

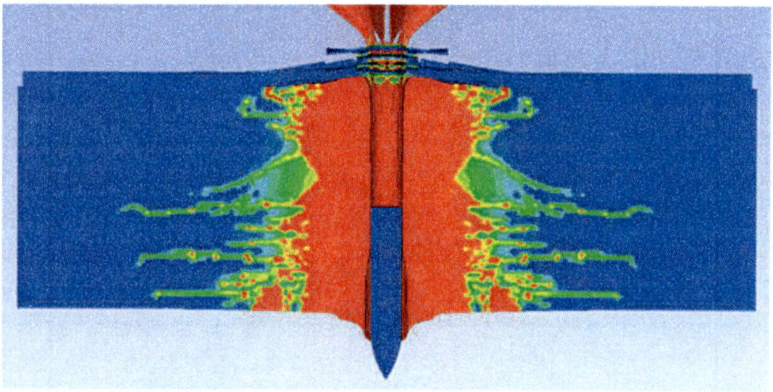

Fig. 6 Damage contour after impact simulation

behavior shows that the performance of the nacre depends heavily on the number and thickness of layers.

Similar behavior of AA plates is shown in the total energy graphs in Fig. 9, but more plastic and damage work are done in the case of 4-layer and six-layer nacre than in bulk plate, which is observed from the difference between total energy and internal energy absorbed.

Adding a front plate to concrete increased the energy absorption of the concrete (Figs. 9 and 10) in all cases by restricting the spalling of concrete at the front impact face. In the case of the nacre specimen, the concrete absorbed more internal energy than the bulk plate. It can be attributed to the more rate of decrease of velocity in

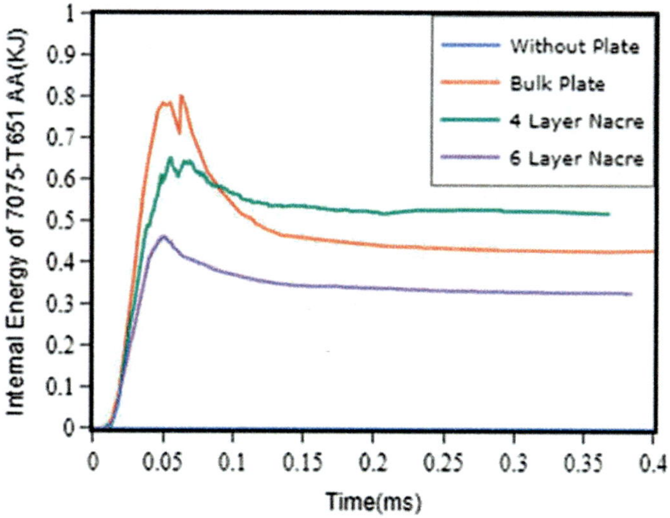

Fig. 7 Internal energy profile of 7075-T651 AA for (i) without plate, (ii) with bulk plate, (iii) nacre 4 layers, (iv) nacre 6 layers

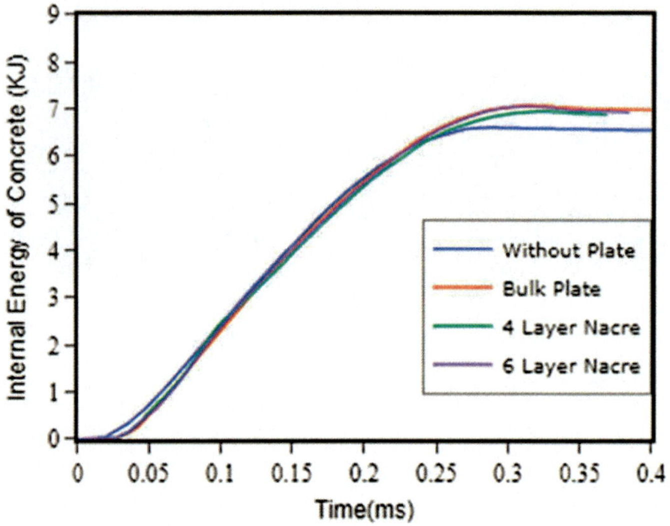

Fig. 8 Internal energy profile of concrete for (i) without plate, (ii) with bulk plate, (iii) nacre 4 layers, (iv) nacre 6 layers

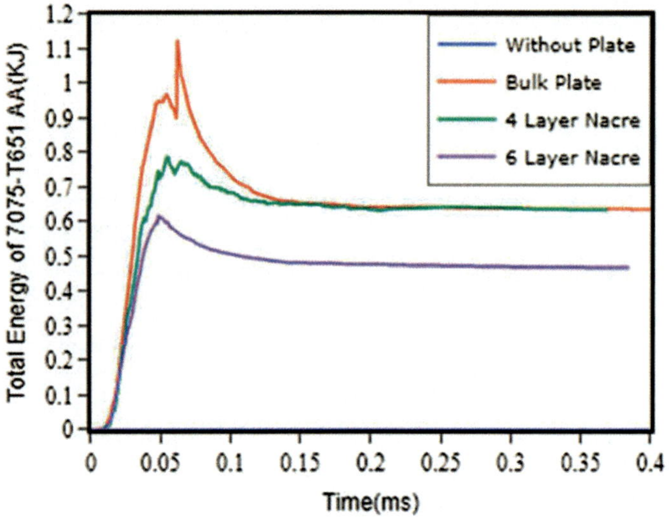

Fig. 9 Total energy profile of 7075-T651 AA for (i) without plate, (ii) with bulk plate, (iii) nacre four layers, (iv) nacre six layers

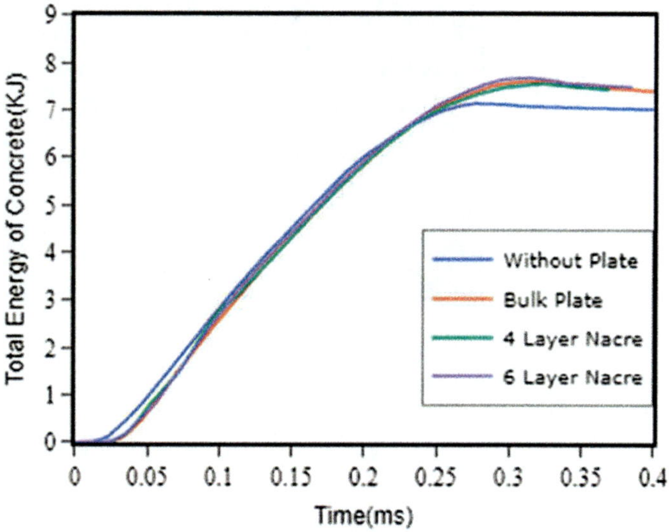

Fig. 10 Total Energy profile of concrete for (i) without plate, (ii) with bulk plate, (iii) nacre four layers, (iv) nacre six layers

Fig. 11 Damage contour of bulk plate

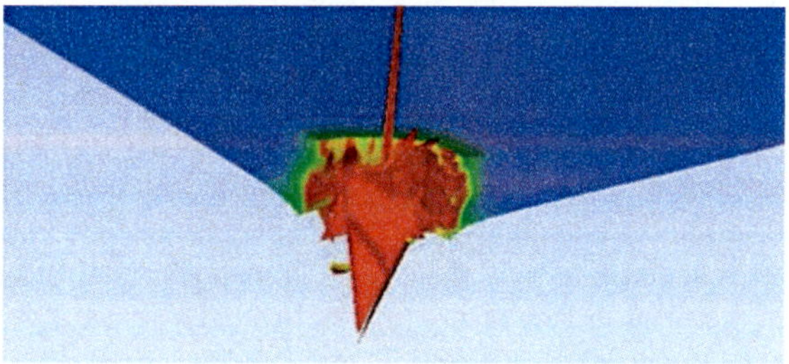

Fig. 12 Damage contour of 4-layer nacre plate

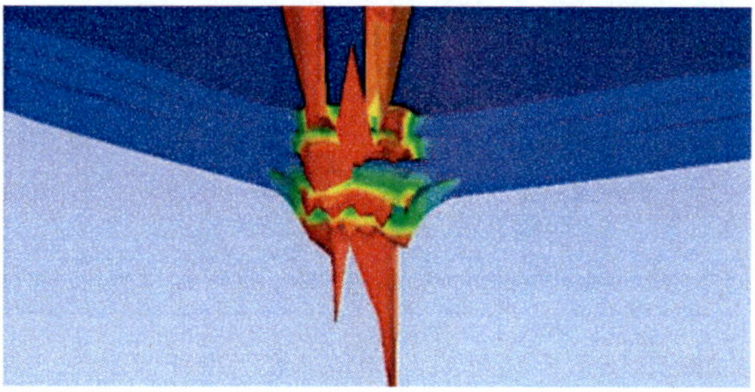

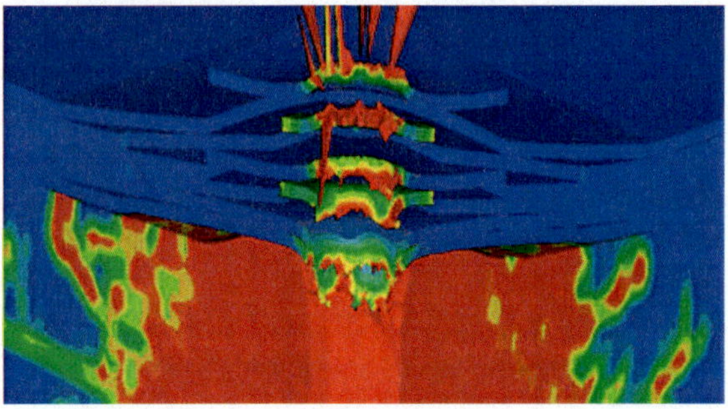

Fig. 13 Damage contour of 6-layer nacre plate

nacre specimens and less energy absorbed by the AA plate. The total energy absorbed by concrete in the case of 4 layers, six layers, and the bulk plate is more than the plain concrete by 2 kJ energy. Unlike bulk plates, nacre shows little effect on the total energy absorption characteristics.

4 Conclusion

Several simulations investigated the response of 12-mm AA bulk plate and nacre plates of 4 and 6 layers on the reinforced concrete specimen 176 mm thick. The addition of a bulk plate improved the ballistic performance of concrete significantly. Changing the structure of the surface plate from bulk to nacre structure improved the ballistic performance by an appreciable amount. This behavior shows that the nacre layer is superior to bulk plates. The four-layer nacre structure performed superior to the bulk plate in decreasing residual velocity and energy absorption characteristics. Due to structural disintegration, six layers performed superior in ballistics resistance but absorbed less energy than bulk plates. These results show that the number of layers, thickness, and bonding properties are vital in the energy absorption characteristic of nacre. This paper explains that nacre structures are better for ballistics performance than bulk plates. However, a more thorough study on the number of layers with varying plate thickness will provide more insight into improving nacre structures.

References

1. Zhang MH, Shim VPW, Lu G, Chew CW (2005) Resistance of high-strength concrete to projectile impact. Int J Impact Eng 31(7):825–841. https://doi.org/10.1016/j.ijimpeng.2004.04.009
2. Zhang X, Wu H, Zhang S, Huang FL (2022) Projectile penetration of reinforced concrete considering the effect of steel reinforcement: experimental study and theoretical analysis. Int J Impact Eng 144:103653. https://doi.org/10.1016/j.ijimpeng.2020.103653

3. Wu H, Fang Q, Peng Y, Gong ZM, Kong XZ (2015) Hard projectile perforation on the monolithic and segmented RC panels with a rear steel liner. Int J Impact Eng 76:232–250. https://doi.org/10.1016/J.IJIMPENG.2014.10.010
4. Remennikov A, Gan ECJ, Ngo T, Netherton MD (2019) The development and ballistic performance of protective steel-concrete composite barriers against hypervelocity impacts by explosively formed projectiles. Compos Struct 207(August 2018):625–644, 2019. https://doi.org/10.1016/j.compstruct.2018.09.060
5. Espinosa HD, Rim JE, Barthelat F, Buehler MJ (2009) Merger of structure and material in nacre and bone—perspectives on de novo biomimetic materials. Prog Mater Sci 54(8):1059–1100. https://doi.org/10.1016/j.pmatsci.2009.05.001
6. Sun J, Bhushan B (2012) Hierarchical structure and mechanical properties of nacre: a review. RSC Adv 2(20):7617–7632. https://doi.org/10.1039/c2ra20218b
7. Humburg H, Zhu D, Beznia S, Barthelat F (2012) Bio-inspired tapered fibers for composites with superior toughness. Compos Sci Technol 72(9):1012–1019. https://doi.org/10.1016/j.compscitech.2012.03.013
8. Knipprath C, Bond IP, Trask RS (2012) Biologically inspired crack delocalization in a high strain-rate environment. J R Soc Interface 9(69):665–676. https://doi.org/10.1098/rsif.2011.0442
9. Tran P, Ngo TD, Mendis P (2014) Bio-inspired composite structures subjected to underwater impulsive loading. Comput Mater Sci 82:134–139. https://doi.org/10.1016/j.commatsci.2013.09.033
10. Hanchak SJ, Forrestal MJ, Young ER, Ehrgott JQ (1992) Perforation of concrete slabs with 48 MPa (7 ksi) and 140 MPa (20 ksi) unconfined compressive strengths. Int J Impact Eng 12(1):1–7. https://doi.org/10.1016/0734-743X(92)90282-X
11. Flores-Johnson EA, Shen L, Guiamatsia I, Nguyen GD (2014) Numerical investigation of the impact behaviour of bioinspired nacre-like aluminium composite plates. Compos Sci Technol 96:13–22. https://doi.org/10.1016/j.compscitech.2014.03.001
12. Shen Y, Wang Y, Du S, Yang Z, Cheng H, Wang F (2021) Effects of the adhesive layer on the multi-hit ballistic performance of ceramic/metal composite armors. J Mater Res Technol 13:1496–1508. https://doi.org/10.1016/j.jmrt.2021.05.058
13. Riedel W et al (1999) Penetration of reinforced concrete by BETA-B-500 numerical analysis using a new macroscopic concrete model for hydrocodes. In: Proceedings of the 9th international symposium on the effects of munitions with structures, vol 315. Berlin-Strausberg Germany
14. Tu Z, Lu Y (2008) Evaluation of typical concrete material models used in hydrocodes for high dynamic response simulations. Int J Impact Eng 36(1):132–146. https://doi.org/10.1016/j.ijimpeng.2007.12.010
15. Tu Z, Lu Y (2010) Modifications of RHT material model for improved numerical simulation of dynamic response of concrete. Int J Impact Eng 37(10):1072–1082. https://doi.org/10.1016/j.ijimpeng.2010.04.004
16. Abdel-Kader M (2019) Modified settings of concrete parameters in RHT model for predicting the response of concrete panels to impact. Int J Impact Eng 132(June):103312. https://doi.org/10.1016/j.ijimpeng.2019.06.001
17. Holmen JK, Johnsen J, Jupp S, Hopperstad OS, Børvik T (2013) Effects of heat treatment on the ballistic properties of AA6070 aluminium alloy. Int J Impact Eng 57:119–133. https://doi.org/10.1016/j.ijimpeng.2013.02.002
18. Johnson GR, Cook WH (1985) Fracture characteristics of three metals subjected to various strains, strain rates, temperatures and pressures. Eng Fract Mech 21(1):31–48. https://doi.org/10.1016/0013-7944(85)90052-9

Characterization of Porcine Lung Parenchyma for Blunt Impact Loads

Yeswanth S. Pydi, Ved Vyas, Sudipto Mukherjee, Anoop Chawla, and N. V. Datla

Abstract Thoracic, specifically, lung injuries are common in blunt impact in an automotive crash. The current research focused on assessing the effect of blunt impact on lung parenchymal tissue. Tokens of an average size of $20 \times 20 \times 10$ mm^3 were prepared from the porcine lung parenchyma. It was ensured that all tokens were without bronchi to maintain uniformity and homogeneity. Unidirectional compression tests were carried out in both quasi-static and dynamic conditions. Quasi-static studies were carried out at a velocity of 1 mm/s, while the dynamic studies were conducted at velocities of 1.6, 3.0, and 5.0 m/s. Engineering stress versus strain curves was obtained for each impact velocity. A bilinear fit was generated for each stress–strain curve, and the toe and elastic modulus were evaluated. It was observed that an increase in the strain rate increases the elastic moduli of the lung tissue.

Keywords Blunt impact · Lung parenchyma · Dynamic compression · Bilinear stress–strain curve

1 Introduction

Blunt impact injuries are one of the most critical injuries, which involve the injuries to the head and thorax. Statistically, case studies on human subjected to blunt impact reveal that injuries to the thorax are dominant. Thorax has the largest surface area and contains the most critical organs of the human body. One of the critical organs is the lung, which occupies the significant volume in the entire thorax, and supplies oxygen to the body. Initial experimental studies were performed by Bayliss and Robertson [1] on the lung tissue to record its deformation by pumping air in a cyclic series of inhalation followed by exhalation. Based on load–deformation curves, the lung tissues are considered a viscoelastic material. Later, Fung et al. [2] performed the first experimental studies on human lung tissue in a confined biaxial state condition.

Y. S. Pydi (✉) · V. Vyas · S. Mukherjee · A. Chawla · N. V. Datla
Department of Mechanical Engineering, Indian Institute of Technology Delhi, Hauz Khas, New Delhi 110016, India
e-mail: pyeswantsai@gmail.com

R. Velmurugan et al. (eds.), *Dynamic Behavior of Soft and Hard Materials Volume 1*, Springer Proceedings in Materials 34, https://doi.org/10.1007/978-981-99-6030-9_22

Creep and relaxation tests were performed and observed that lung parenchyma is insensitive to the strain rate and formulated a pseudoelastic constitutive equation. Same Fung et al. [3] determined the stress–strain relationship and hysteresis of the rabbit lung when subjected to compression and re-expansion. The primary application of these studies was to analyze the pressure–volume (PV) relationship of lung tissue deformation with a trachea being closed and transform it into the stress–strain relationship. All the above studies are performed by considering the lung as a homogeneous tissue. With several material constituents like parenchyma, large bronchi, small bronchi, and pleural membrane, lung tissue is a heterogeneous tissue.

Studies were performed by Benjamin Weed et al. [4] to understand the isotropy of porcine lung parenchyma tissues. Lung parenchyma samples were dissected and aligned with one of the three anatomic planes, sagittal, frontal, and transverse, and then subjected to compressive loading of 0.1 s^{-1} strain rate. Statistical studies were carried out on experimental data and concluded that lung parenchyma is isotropic. Pinelopi Andrikakou et al. [5] performed compression and tensile test on freshly excised lung parenchyma obtained from both Sprague–Dawley rats and New Zealand white rabbits. These tests were conducted at three different strain rates of 0.25, 2.5, and 25 min^{-1}, and Van der Waals' hyper elastic material model was used to characterize the experimental data. Results exhibited asymmetric differences between the tension and compression test of both species. Although these studies were carried out at various strain rates, because of a narrow strain rate band, no conclusions were drawn regarding the rate sensitivity. Mona Eskandari et al. [6], for the first time, characterized the mechanical behavior of large and small bronchi of the porcine lung tissue. Uniaxial tensile tests were carried out on the trachea, large bronchi, and small bronchi in two different orientations (axial and circumferential). These studies were limited to 0.002, 0.01, and 0.05 s^{-1} static strain rates. From the above experimental data, they concluded that linear pseudo elastic modulus was significantly higher axially (30.5 $\pm$ 3.1 kPa) than circumferentially (8.4 $\pm$ 1.1 kPa). Among circumferential samples, small bronchi (12.5 $\pm$ 1.9 kPa) have a higher modulus than the trachea (6.0 $\pm$ 0.6 kPa) and large bronchi (6.6 $\pm$ 0.9 kPa).

Although studies were performed on both human and porcine lung tissues, but most of them were limited to the quasi-static regime. Moreover, with the increase in the utilization of computation models and finite element (FE) studies, it is feasible to understand tissue behavior under blunt and impact scenarios. These studies require the material parameters of the tissue at high strain rates. Roberts et al. [7] performed FE studies on a human torso model to provide a soft armor vest that prevents projectile penetration. Although similar trends and patterns were observed, the accelerations and peak pressures from the FEM did not match those from testing. One of the possible reasons was the lack of soft tissue material properties at high strain rates. Further studies by Perters et al.[8] and Brooks et al. [9] also identify the significant limitation of the high strain material properties of human tissues in performing the FE studies.

Specifically, the current research was focused to assess the effect of blunt impact on the lung parenchymal tissue. A series of experiments were performed on the

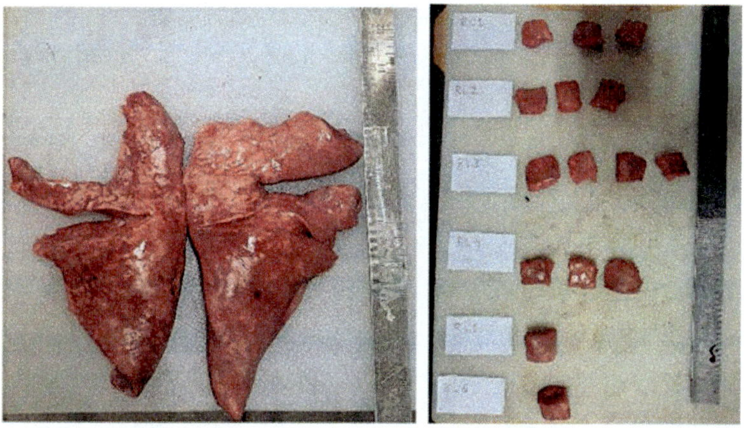

(a) (b)

Fig. 1 **a** Porcine lung segmented into right and left parts; **b** Token level specimens

porcine parenchymal tissue from quasi-static to low impact compression, and the corresponding material parameters were estimated.

2 Materials and Methods

2.1 Sample Preparation

Fresh porcine lung tissues (n = 4) were obtained from 9- to 15-month-old pigs from an abattoir. Lung tissues were collected approximately after 4 h of sacrifice, and experiments were performed within two days. Lung tissues were stored in the refrigerator at the temperature of −24 °C and thawed at room temperature in a buffered saline solution before the start of the experiment. Lung tissues were sectioned into a token level specimen using a surgical scalpel, as shown in the Fig. 1. Although it is difficult to achieve the exact dimensions, most samples were maintained with the same aspect ratio. Dimensions of the sample were measured at three different points using digital Vernier calipers. An average dimension for all the token level specimens is shown in the Table 1.

2.2 Dynamic Uniaxial Compression Setup

A spring-loaded compression setup is used to perform the dynamic compression on lung parenchyma. The test setup consists of a stepper motor that compresses the spring using the lead screw. Based on the desired velocity of impact, the spring

Table 1 Average dimensions of the sample and the corresponding impact velocities and strain rates

S. No.	Velocity of impact	Length	Width	Thickness	Strain rate (_/s)
1	0.001	19.6 ± 0.62	18.5 ± 0.25	10.7 ± 0.36	0.9
2	1.6 ± 0.01	24.9 ± 0.93	22.2 ± 1.67	11.9 ± 0.88	135
3	3.0 ± 0.04	24.6 ± 1.93	21.6 ± 1.82	11.9 ± 0.31	258
4	5.0 ± 0.11	22.8 ± 1.20	19.2 ± 1.33	10.1 ± 0.80	499

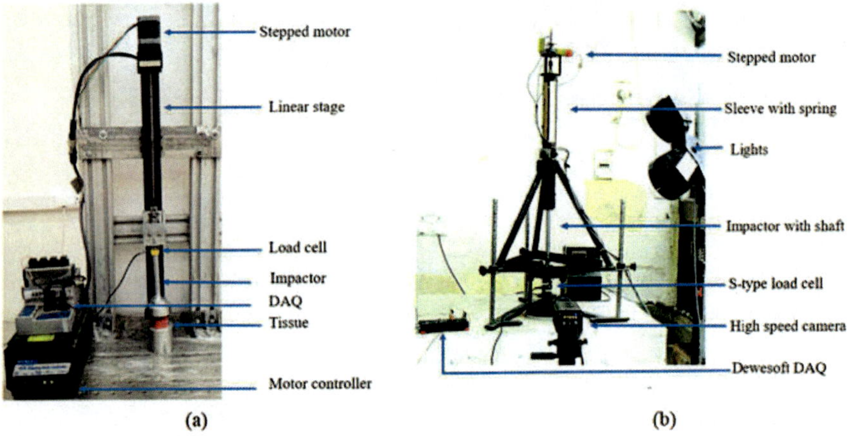

Fig. 2 **a** Quasi-static compression setup, **b** Spring-loaded compression setup

needs to be compressed accordingly. The other end of the spring is connected to the impactor, which will come in contact with the specimen. A high-speed camera was used to record the deformation of the tissue during the impact. A buffered switch triggered by the impact was used to trigger the high-speed camera and data acquisition system. It additionally synchronizes the recorded force and displacement data and thereby identifies the start of the event recorded using a data acquisition system (Dewesoft Corporation, Slovenia) as shown in the Fig. 2b. The sampling rates used to acquire load cell data and high-speed video samples are 20,000 frames/s.

2.3 *Quasi-Static Compression Setup*

Quasi-static compression setup consists of a displacement-controlled linear stepper motor (Velmex, Inc., Bloomfield, NY, USA) of 0.63 μm resolution connected to one end of the compression plate. A 111 N load cell (LSB200, Futek Inc., Irvine, CA, USA) with an accuracy of ± 3% in the range of 1–5 N was placed between the compression plate and the stepped motor as shown in the Fig. 2a. Being a displacement-controlled setup, the position and the velocity of the compression plate can be defined initially. A gradual downward movement of the compression plate until a visible change in the load cell readings is required to guarantee full contact between the compression plate and tissue. Additionally, a DSLR camera to examine the contact between the compression plate and the tissue and record the tissue's deformation pattern (Figs. 3 and 4).

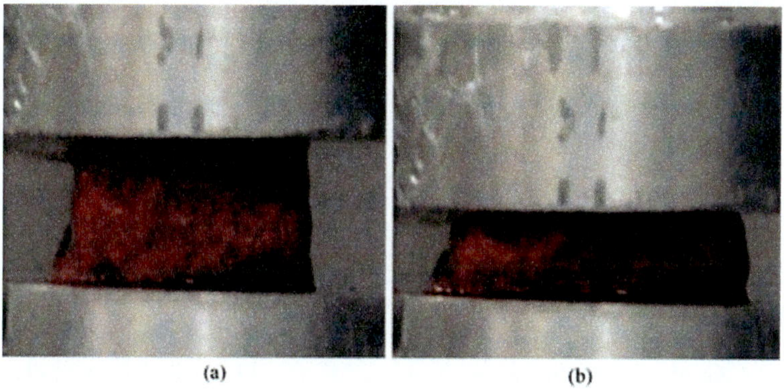

Fig. 3 Quasi-static compression of parenchyma **a** before, **b** after

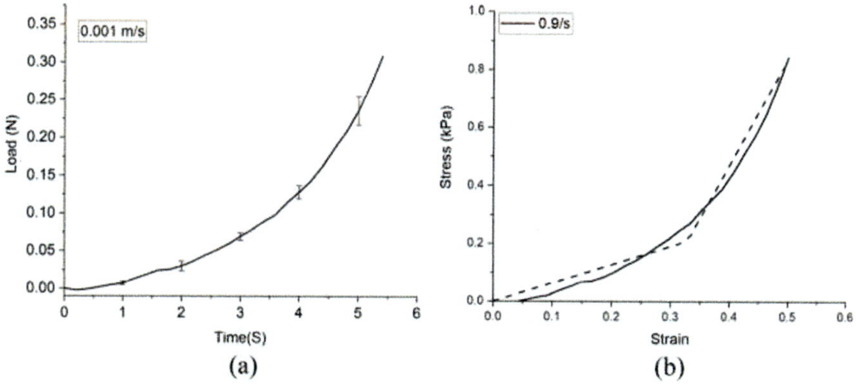

Fig. 4 Load time curve of parenchyma sample at 0.001 m/s; **b** corresponding engineering stress–strain curve with the bilinear curve

3 Results and Discussion

In total, 20 successful uniaxial compression tests were performed on porcine lung parenchyma, where ($n = 5$ samples) at a quasi-static regime (0.001 m/s) and ($n = 15$ samples) at a low range impact regime (1.6, 3, and 5 m/s). Data processing methodology is different for both the quasi-static and the spring-loaded compression setups. In quasi-static experiments, a constant velocity and the pre-calibrated displacement are provided to the impactor, which ensures controlled deformation of the specimen. Load displacement curves are obtained with the help of the load time data and the velocity of the impactor, whereas in the spring-loaded compression setup, analyzing the images from a high-speed camera yields the details about the start and velocity of the sample compression. Since the trigger switch allows both the camera and a load cell to record the data at the same time, we correlate the load values from the load cell data to displacement values obtained from image tracking software (Motion Studio, Integrated Design Tool Inc.) as shown in the Fig. 5.

Average load versus time plots was shown in the Fig. 6a. As expected, an increase in the impact velocity not only increases the maximum load but also decreases the time duration to reach the peak load. The following relations are used to understand and identify the parameters like stress, strain, and strain rate. Displacement divided by undeformed thickness gives the axial strain, whereas the load divided by the undeformed area provides the stress values. The ratio of impact velocity and the initial thickness of the specimen provides the strain rate. Engineering stress versus engineering strain plots was plotted for all the average load time curves with the corresponding average dimensions of the specimen as shown in the Fig. 6b; as the strain rate increases, the maximum stress values increase, which is in correspondence with the viscoelastic materials. A nonlinear stress–strain curve was observed due to the presence of fibrous substance elastin, collagen, and the ground substance proteoglycans [10].

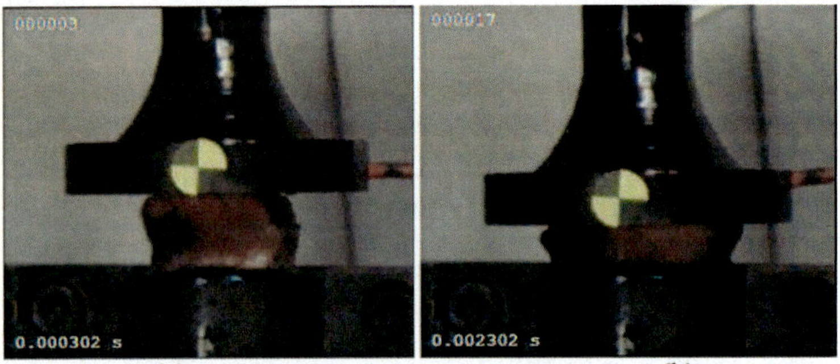

(a) (b)

Fig. 5 Dynamic compression of parenchyma **a** before; **b** after

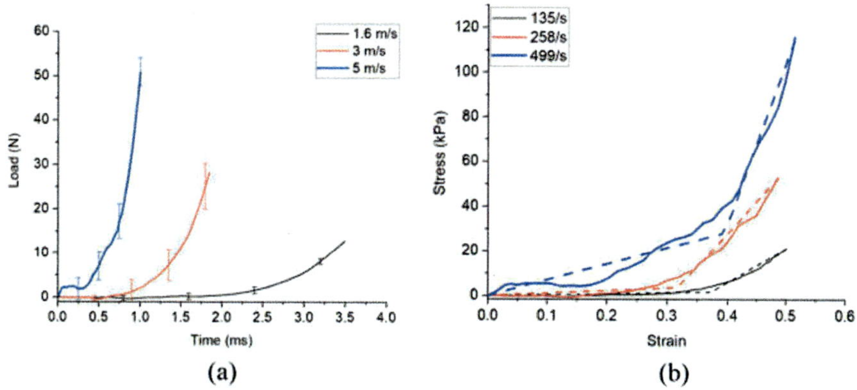

Fig. 6 **a** Load time curve of parenchyma samples; **b** corresponding engineering stress–strain curve with the bilinear curve

Many studies [11–13] have analyzed the stress–strain response using the bilinear curve, as shown in the Fig. 6b. The bilinear model consists of two regions, specifically the toe and elastic region. The toe region signifies the working of soft tissue in the physiologic region where the fibrous material will rearrange itself in the tissue, whereas the elastic region represents the deformation of tissue fibers. A bilinear curve was used to analyze the nonlinear stress–strain behavior of the lung parenchymal tissue. Using MATLAB (version R2021b, MathWorks, Inc.) minimum of unconstrained multivariable optimization, a bilinear fit was generated to minimize the root mean square of the error between the experimental and the bilinear models.

The bilinear material parameters obtained for both quasi-static and dynamic compression are tabulated in Table 2. It is observed that an increase in the strain rate from 0.9/s to 499/s increases both the toe modulus and elastic modulus from 0.63 to 73.05 and 3.17 to 712.85 kPa, respectively. An increase in the maximum stress from 0.84 kPa to 114 kPa is clear evidence of strain rate-dependent behavior of lung parenchymal tissue during compression. Moreover, the increase in the elastic modulus from 0.9 to 499/s is higher than the toe modulus. This reveals that the rate of fiber deformation increases with an increase in the strain rate. Although a clear trend is not being observed in the toe strain, the current data can be approximated as an increasing trend from 0.2 to 0.39 with the exception of 0.32 at the 258/s strain rate.

Table 2 Optimized bilinear material parameters for both quasi-static and dynamic compression

S. No.	Strain rate (_/s)	Toe modulus kPa	Elastic modulus kPa	Toe strain	Maximum stress kPa
1	0.9	0.63	3.71	0.2	0.84
2	135	4.42	150	0.37	21.07
3	258	11.46	325.16	0.32	52.79
4	499	73.05	712.85	0.39	114.3

4 Conclusion

The current study provides the material properties of the lung parenchymal tissues in quasi-static and low-velocity impact scenarios. Uniaxial compression test performed at 0.001, 1.3, 3, and 5 m/s. Nonlinear load time curves and corresponding stress–strain curves reveal that lung parenchymal tissue is a strain rate-dependent material. The obtained toe and elastic modulus data can be used as an input material parameter for the lung parenchymal material in the existing human body finite element models. These models can be helpful in performing FE simulations to evaluate the damage and injury criteria in blunt and automotive collisions.

Acknowledgements The authors acknowledge the financial support received from Joint Advanced Technology Centre, DRDO (DFTM/03/3203/M/01/JATC).

References

1. Bayliss LE, Robertson GW (1939) The visco-elastic properties of the lungs. Q J Exp Physiol Cogn Med Sci 29:27–47. https://doi.org/10.1113/expphysiol.1939.sp000792
2. Zeng YJ, Yager D, Fung YC (1987) Measurement of the mechanical properties of the human lung tissue. J Biomech Eng 109(2):169–174. https://doi.org/10.1115/1.3138661
3. Fung YC, Yen RT, Tao ZL, Liu SQ (1988) A hypothesis on the mechanism of trauma of lung tissue subjected to impact load. J Biomech Eng 110(1):50–56. https://doi.org/10.1115/1.3108405
4. Weed B et al (2015) Experimental evidence of mechanical isotropy in porcine lung parenchyma. Materials (Basel) 8(5):2454–2466. https://doi.org/10.3390/ma8052454
5. Andrikakou P, Vickraman K, Arora H (2016) On the behaviour of lung tissue under tension and compression. Sci Rep 6. https://doi.org/10.1038/srep36642
6. Eskandari M, Arvayo AL, Levenston ME (2018) Mechanical properties of the airway tree: Heterogeneous and anisotropic pseudoelastic and viscoelastic tissue responses. J Appl Physiol 125(3):878–888. https://doi.org/10.1152/japplphysiol.00090.2018
7. Roberts JC et al (2005) Modeling non penetrating ballistic impact on a human Torso
8. Peters AE, Akhtar R, Comerford EJ, Bates KT (2018) Tissue material properties and computational modelling of the human tibiofemoral joint: a critical review. PeerJ 2018(1):1–48. https://doi.org/10.7717/peerj.4298
9. Brooks T et al (2018) Finite element models and material data for analysis of infant head impacts. Heliyon 4(12):e01010. https://doi.org/10.1016/j.heliyon.2018.e01010

10. Suki B, Ito S, Stamenović D, Lutchen KR, Ingenito EP (2005) Biomechanics of the lung parenchyma: critical roles of collagen and mechanical forces. J Appl Physiol 98(5):1892–1899. https://doi.org/10.1152/japplphysiol.01087.2004
11. Gaur P et al (2016) Characterisation of human diaphragm at high strain rate loading. J Mech Behav Biomed Mater 60:603–616. https://doi.org/10.1016/j.jmbbm.2016.02.031
12. Lake SP, Miller KS, Elliott DM, Soslowsky LJ (2009) Effect of fiber distribution and realignment on the nonlinear and inhomogeneous mechanical properties of human supraspinatus tendon under longitudinal tensile loading. J Orthop Res 27(12):1596–1602. https://doi.org/10.1002/jor.20938
13. Mattucci SFE, Moulton JA, Chandrashekar N, Cronin DS (2012) Strain rate dependent properties of younger human cervical spine ligaments 10:216–226. https://doi.org/10.1016/j.jmbbm.2012.02.004

Understanding Impact Loading on Rocks and Its Implication: An Insight

K. Balakrishna, V. B. Maji, and C. Lakshmana Rao

Abstract Many of the applications in rock engineering require high-strain rate dynamic mechanical properties in the order of 100–1000/s, feasible to study using split-Hopkinson pressure bar (SHPB) apparatus. Unlike ductile materials like metals, $< 1\%$ is often the failure strain of brittle materials. Additionally, in many rock engineering applications, multi-axial loading is expected. Although SHPB is common equipment modified for material testing, certain fundamental concerns should be carefully considered when applied to rock specimens. Many literatures are available conducting dynamic impact tests on rocks using SHPB, performed using small diameter rods. Moreover, the multi-axial confining loading is highly relevant to underground rock engineering problems. In this presentation, preliminary work of high-strain dynamic testing on rocks is done using SHPB. A preliminary review of the impact of loading on rocks and related applications is also presented. Few preliminary test results with SHPB on rocks are also presented. A brief insight into the customization requirement, issues and challenges, and corresponding limitations with the SHPB is discussed. Experiments on artificial rock specimens (using gypsum plaster) under high-strain dynamic loading are carried out, and the corresponding results from preliminary experiments with rock-like material specimens are discussed.

Keywords Impact loading · Rock · SHPB · High-strain dynamic loading

1 Introduction

Dynamic loading frequency ranges vary and could be very low, from less than 10 Hz during earthquakes to the ultrasonic pulse (low strain) ranges in the order of 150 MHz (Fig. 1). There are many methods to replicate dynamic loading in the laboratory. The split-Hopkinson pressure bar (SHPB) is a pivotal piece of testing outfit for applying impact and dynamic loads to materials while examining their dynamic mechanical

K. Balakrishna (✉) · V. B. Maji · C. Lakshmana Rao
Indian Institute of Technology Madras, Chennai 600036, India
e-mail: balakrishnacivil112@gmail.com

R. Velmurugan et al. (eds.), *Dynamic Behavior of Soft and Hard Materials Volume 1*,
Springer Proceedings in Materials 34, https://doi.org/10.1007/978-981-99-6030-9_23

properties. To research the dynamic compressive reactions of engineering materials at rapid strain rates between 10^0 and 10^4 s^{-1}, the SHPB system by Kolsky [1, 2] has been effectively employed. Recently, this has also been frequently employed in investigating the dynamic mechanical responses of rocks and brittle materials that resemble rocks [3]. The SHPB compressive tests are based on two fundamental hypotheses: (1) the propagation of an elastic surge in one dimension (1-D) in the bars and (2) sample uniform distortion [1]. Using lengthy bars ensures the presumption of 1-D stress surge propagation, and by controlling the striker's striking velocity throughout the test bar's deformation elasticity is ensured. Inertial effects (namely, the axial and radial inertial effects) and the interfacial friction effect influence the uniformity of the sample deformation [4].

Loading in rock or civil engineering applications was found to be a little different than other mechanical engineering activities. In most cases, impact loads are applied to rocks, while tunneling, quarrying, etc., may encounter dynamic disturbances caused by impact loadings like rock bursts and blasting at various zones nearby, leading to instability. According to the distance to the hole, three zones may be distinguished within the rock mass around a subterranean aperture, as illustrated in Fig. 2. The confining stress states range from the mostly hydrostatic condition in the distant zone to the triaxial state in the middle zone to the condition of compression or tension in the near zone. Therefore, to fully assess the dynamic reactions of subterranean rocks, before dynamic loading, it is preferable to subject rock samples to all three of those stress regimes. Tunnel abutment zones usually undergo large dynamic stresses due to successive blast loadings during tunneling and mining. Rocks exhibit more complex dynamic mechanical properties when compared to rocks that are under static conditions. The response of tunnels and tunnel lining under impact loading is of great concern. An appropriate design of tunnel lining with optimum thickness is important to provide maximum blast resistance. The deformation in the lining and rock material is related to many factors that include rock mass and lining properties, its thickness, joint characteristics, the weight of the explosives, material type and angle of internal friction, the effect of the curvature, alignment of a tunnel, weight of explosives, and many other parameters. Deformations in rock are inversely proportional to the strength; therefore, higher strength of rock mass will be helpful to better blast resistance. The overburden depth also plays a significant role in the blast resistance; higher overburden depths are more resistant to the blast than lower ones [5]. Therefore, detailed attributes of rock strength parameters are essential in many rock

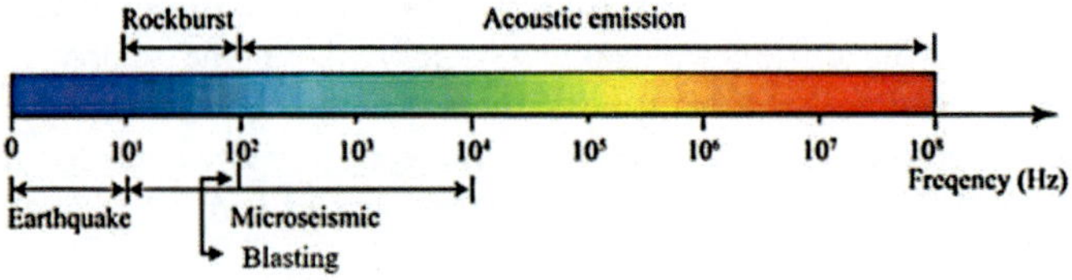

Fig. 1 Impact loads classification

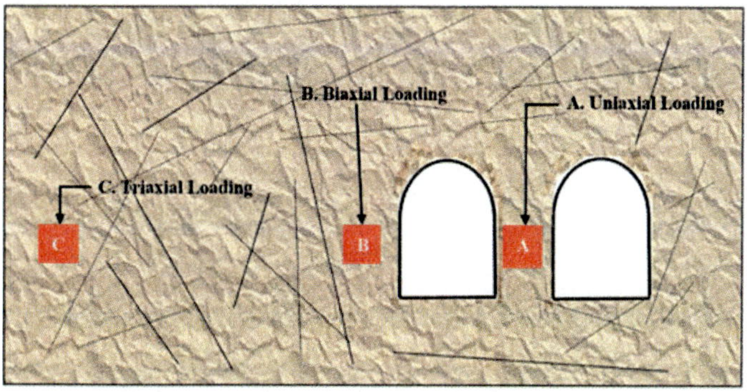

Fig. 2 Types of loading acting in tunnel surroundings

engineering applications in the presence of high rates of loading or dynamic circumstances. The dynamic elastic modulus, often referred to as the strength at high-strain rates in SHPB tests, is also examined in the current work.

Numerous in-depth investigations of the brittle materials' dynamic behaviors, like shale rock, concrete, and igneous rocks, have been carried out [4, 6–8]. The SHPB provided bar diameters fluctuate along with the dimensions of the specimen. In the International Society for Rock Mechanics (2014) suggested methods, an SHPB device of 50 mm is utilized. For small diameter specimens, ISRM (2014) suggests that specimens have an aspect ratio of 1.0 as opposed to a 50-mm-diameter bar, and specimens having a larger diameter equivalent to that size have an aspect ratio of 0.5 [9]. Currently, there is no comprehensive discussion of dynamic experimental methods for small diameter rock specimens produced artificially utilizing the SHPB setup. Consequently, the primary goal of the current study is to test artificial rock specimens (using gypsum plaster) using an SHPB device and give an insight into the test methods and results.

2 Methodology

2.1 About SHPB Setup

In 1914, Bertram Hopkinson published the first English description of a method for determining the form of a stress pulse of an impact load in a lengthy elastic bar. RM Davies and Herbert Kolsky built upon Hopkinson's invention in 1948 and 1949, who added displacement gauges and oscillographic recording methods to acquire impact stress pulses waveforms and full amplitudes in identical elastic bars. An arrangement of two bars with a short specimen that undergoes compression between the bars was employed by Kolsky. The test specimen, which is weaker than bars,

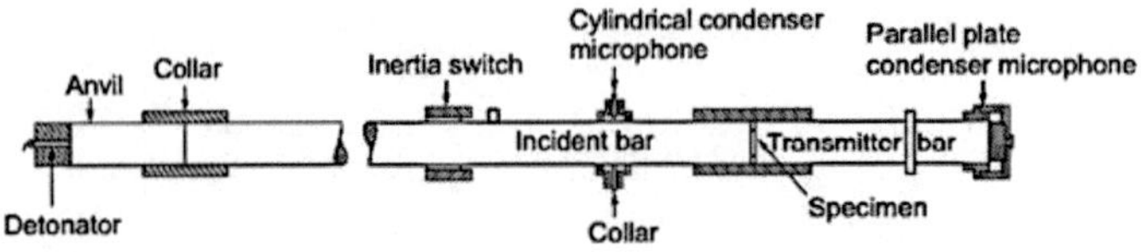

Fig. 3 Kolsky (1949) described a schematic for a SHPB device

deformed plastically. It is possible to derive strain as well as stress. The design eventually became the split-Hopkinson pressure bar, as illustrated in Fig. 3 [10].

2.2 *Advancements in Apparatus*

Kolsky's method was improved by Ulric Lindholm [5] of Southwest Research Institute, Texas, chiefly by changing the bars' lengths and the strain gauges' locations. He used electrical systems and modern strain gauge technologies to identify the transient stress pulse patterns on both bars that, with a single stroke, instantly produced the whole stress–strain curve. Additionally, he mechanically launched a striker bar rather than with explosive pellets. The impact stress pulse was produced when the incident bar and the striker bar collided. The striker bar's velocity and the intensity of the stress pulse are controlled by the impact speed and may be relatively precisely regulated by varying the compressed gas pressure. The striker bar's length affects how long the stress pulse lasts [10]. Figure 4 depicts an upgraded version of the SHPB device. To assess and describe metal behavior under various loading conditions, the SHPB approach has recently become a crucial component of materials research. Numerous different materials, such as geologic minerals, ceramics, polymers, and rubbers, are used with this technology [11].

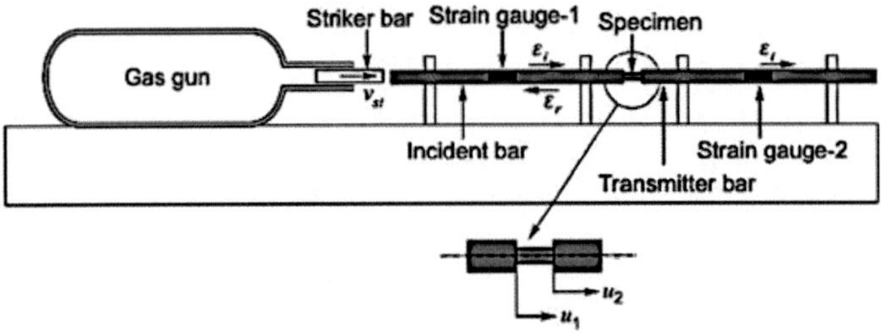

Fig. 4 A generic SHPB device's schematic

2.3 Principle and Theory of SHPB

The High-Rate Impact Measurements Lab, Department of Applied Mechanics, Indian Institute of Technology Madras, is where the SHPB apparatus used in this work is located. It mostly comprises three bars (striker, incident, and transmitted bar) and a gas pistol. Figure 5 shows the experimental configuration, and Table 1 lists the size of those bars. When the gas gun is charged and discharged, the striker bar is pushed by the needed energy, which collides with the incident bar's end. An incident wave results from a compressive impact stress pulse, traveling to the right and striking the bar before it reaches the specimen. As the wave approaches the interface between the bar and specimen, in two shorter waves, the wave gets separated. Transmitted wave is one of them, which enters the transmitter bar through the specimen and causes elastic deformation. The second wave, often called the reflected wave, rebounds away from the sample before returning to the incident bar. The strain gauges, which are labeled as incident strain (ε_i), reflected strain (ε_r), and transmitted strain (ε_t) and are situated on the input and output bars, respectively, measure the reflected and transmitted impact pulses; these are shown in Fig. 4. In order to analyze the data and ascertain the specimen's history of strain, recorded signals are employed. As illustrated in Fig. 6, the incident compressive pulse ε_i and its tensile reflection ε_r, passing through an incident bar, differ only in sign, making up the impact stress wave strain history. The following equations are then used to the observed strains to determine stress (σ_s), strain (ε_s), and strain rate ($\dot{\varepsilon}_s$), respectively.

$$\sigma = \frac{A}{A_s} = E\varepsilon_s\frac{A}{A_s} \tag{1}$$

$$\varepsilon_s = \frac{-2C_0}{L_s}\int_0^t \varepsilon_r dt \tag{2}$$

$$\dot{\varepsilon}_s = \frac{-2C_0}{L_s}\varepsilon_r \tag{3}$$

Here,

σ_s = stress inside the sample

ε_s = strain inside the sample

$\dot{\varepsilon}_s$ = rate of strain in the sample

C_0 = wave speed in bar

E = Young's modulus of elasticity of the bar

A = bar's cross-sectional area

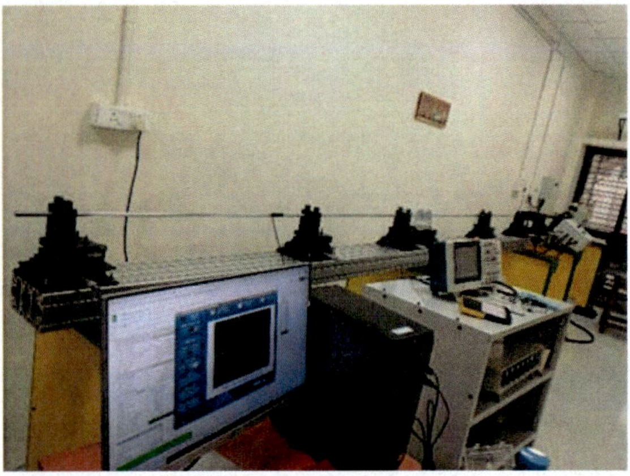

Fig. 5 Experimental setup available in IIT Madras

Table 1 SHPB bars' dimensions

S. No.	Bar	Symbol	Length (mm)
1	Striker bar	L_s	400
2	Incident bar	L_i	2000
3	Transmitted bar	L_t	2000
4	Diameter	D	12

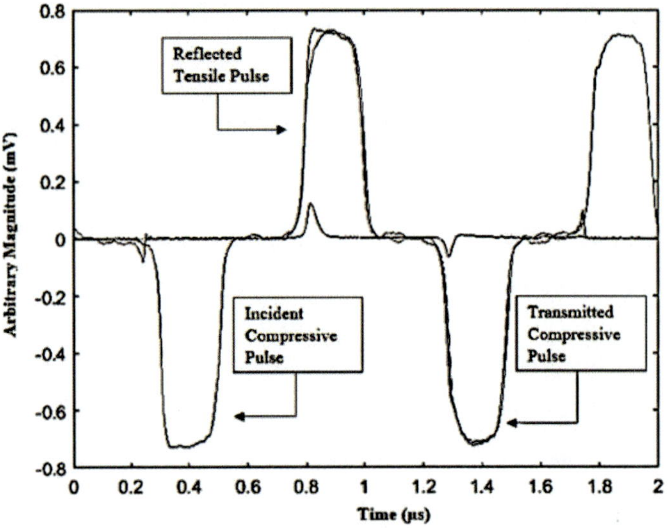

Fig. 6 Incident and transmitter bars' history of strain

A_s = area of the sample

L_s = length of the specimen

2.4 Pulse Shaping

In contrast to ductile materials, brittle materials have modest (1%) failure stresses. Hence, if the loading rate increases, the specimen may fail haphazardly during a standard SHPB test. The dynamic behavior of brittle materials, such as rocks, may be studied using the pulse-shaping approach [12]. It can be challenging to attain a stress equilibrium state under dynamic loading in those materials because the sample can rapidly break when the incident wave reaches its culmination and comes into contact with the incident bar. Since the specimen will experience quasi-static forces, the dynamic loading should be carried out slowly enough. A trapezoidal/ rectangular incident stress wave is amended to a ramping shape in the rock sample to achieve dynamic stress equilibrium using the appropriate pulse shaper, preventing any premature failure caused by the rapid increase of the incident pulse failure [4]. While in this study, dynamic elastic modulus and dynamic compressive strength of the rock specimens are also investigated in addition to the pulse shaper effects on the incident pulse shape. A grease of significant thickness is used as a pulse shaper to achieve dynamic stress equilibrium in the specimen.

2.5 Sample Preparation

Due to the simplicity of sample preparation utilizing an acrylic mold, the investigation used two kinds of gypsum powder. Gypsum specimens were made by mixing 25% plaster of Paris and 75% S-35 gypsum in a dry state [13]. The gypsum powders with the highest uniaxial compression strength (UCS) were then chosen, and their respective water contents (24% and 35%) were considered. Then, the corresponding amounts of water were added directly to the mixture of gypsum powders to create a slurry, which was subsequently transferred into the mold. The length to diameter was selected to have a slenderness ratio of 1. To lessen the dispersion effect of the pulse, the specimen has a diameter of 10 mm, which is a bit lower than the bars' diameter of 12 mm. After hardening, samples are removed from the mold, allowing for air-drying for one day, and oven-dried at 40 °C for 4 days. Prepared specimens are depicted in Fig. 7. As seen in Fig. 8, those samples are wedged between the incident and transmitted bars.

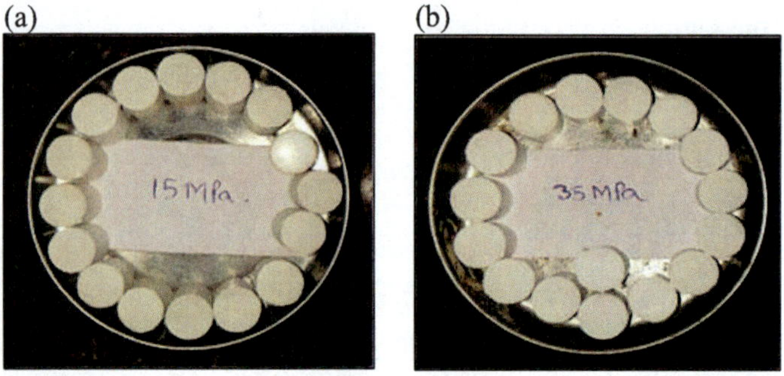

Fig. 7 **a** Set of samples having UCS of 15 MPa; **b** set of samples having UCS of 35 MPa

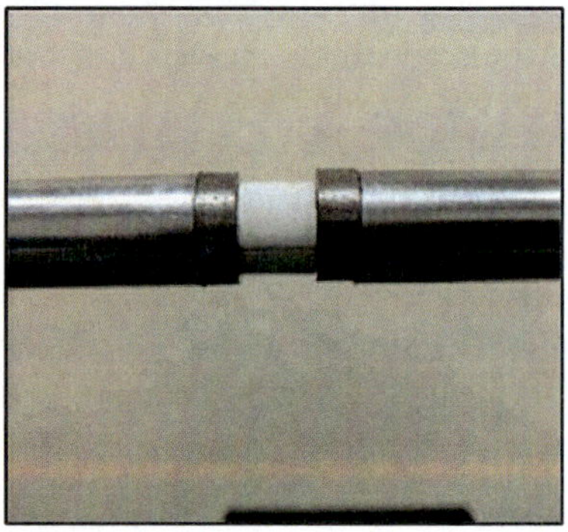

Fig. 8 Photograph of placing of samples

3 Experimental Results and Discussions

Changing the length of the striker bar, keeping the gas pressure constant, or changing the gas pressure while keeping the striker bar length constant might result in different strain rates. In the present study, by adjusting the pressure of the gas gun, different strain rates were attained, eventually changes the velocity of striker bar. Those strain rate conditions were applied to two groups of gypsum specimens during testing. Each experiment utilizes a striker bar that is the same length. The specimen's dimensions for each test are the same: L (length) = 10 mm and D (diameter) = 10 mm. For each group of samples, five different loading sets were used in the impact test. The range of samples' average strain rates, from low to high, was 117.206/s to 421.58/s for 15 MPa set of samples and 49.44/s to 260.1/s for 35 MPa set of samples, respectively. Corresponding plots are presented in the below-mentioned (Figs. 9 and 10). The peak strength of the 15 MPa set of samples was attained as 56.98 MPa when the strain rate is 193.61/s, and the corresponding peak strain is 0.0386 (mm/mm). The peak strength of the 35 MPa set of samples is 72.48 MPa when the strain rate is 145.482/s, and its peak strain is 0.0207/s. Variation of dynamic Young's modulus with regard to each strain rate is also studied, and in Table 2, those variations are

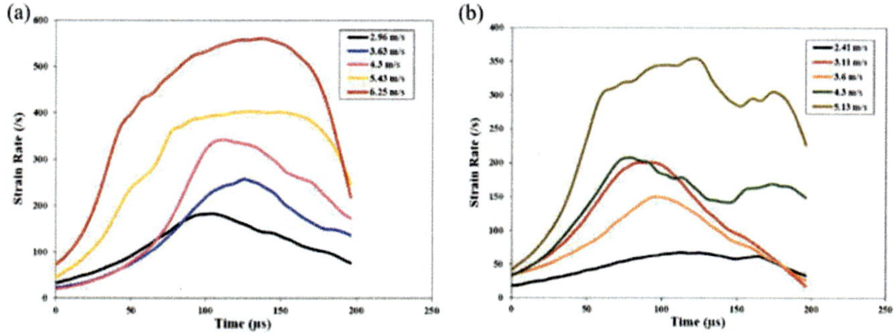

Fig. 9 Strain rate (/s) versus time (μs) plot of **a** 15 MPa, **b** 35 MPa

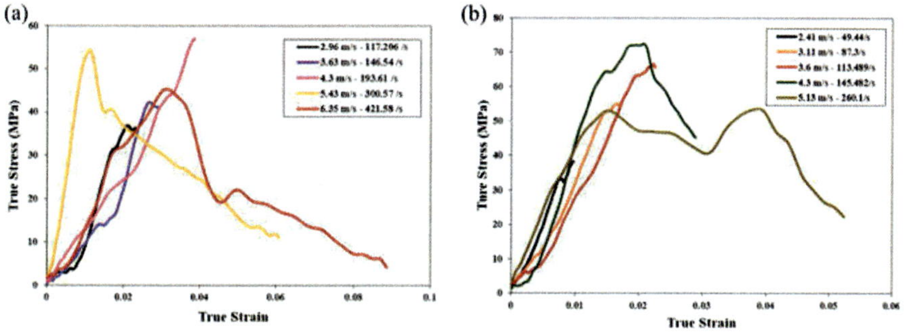

Fig. 10 True stress (MPa) versus true strain plot of **a** 15 MPa, **b** 35 MPa

listed. Following the rate of strain at which the corresponding samples attained their peak stress, the specimens are completely broken down into pieces with a further increase in strain rates. Figure 11 shows the peak strength with different strain rates, showing an optimum.

4 Conclusions

The behavior of brittle material like a rock under impact load is quite different from metals, and there needs some customization while using the SHPB test for rocks. In the present study, a preliminary study was conducted, and gypsum rock specimens were characterized under different loading rates. From the findings of the present investigation, a comparison was established with two sets of specimens. Both sets of rock samples showed an increase in compressive strength under uniaxial dynamic compressive tests, and obtained peak stress is about 3.7 times and 2.0 times for UCS of 15 MPa and 35 MPa, respectively. In the samples with higher UCS, dynamic peak strength is achieved at a lower strain rate than those with lower UCS. It depicts that

Table 2 Dynamic Young's modulus at 50% peak stress

UCS (MPa)	Static Young's modulus (GPa)	Strain rate (/s)	Dynamic Young's modulus at 50% peak stress (GPa)
15	1.037	117.206	2.934
		146.54	3.920
		193.61	4.917
		300.57	6.784
		421.58	3.323
35	1.927	49.44	3.856
		87.3	4.422
		113.489	5.507
		145.482	7.366
		260.1	4.758

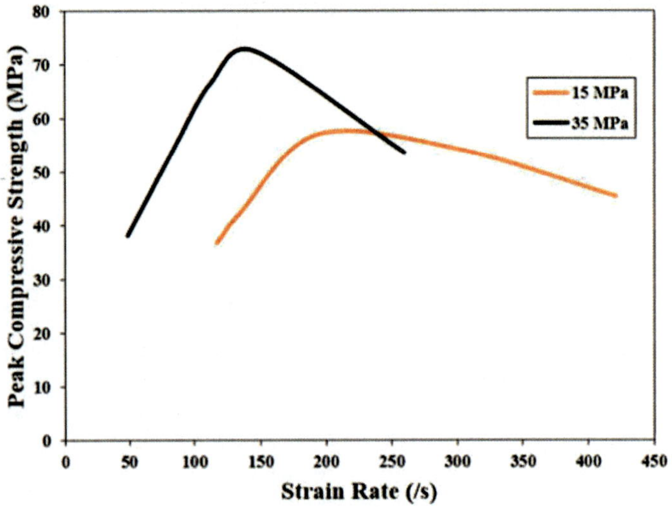

Fig. 11 The plot of peak compressive strength vs. strain rate

compressive strength increases as the strain rates increase compared to static loading, but up to a certain limit, it decreases with an increase in strain rate. Under impact load, one gets a higher modulus than the static modulus, and dynamic modulus increases with increasing strain rate.

Acknowledgements The authors acknowledge the assistance provided during the SHPB test by Ashish Pandey and V B Bramhadathan, Ph.D. scholars in the Department of Applied Mechanics, IIT Madras.

References

1. Kolsky H (1949) An investigation of the mechanical properties of materials at very high rates of loading. Proc R Soc A Math Phys B62:676–700
2. Kolsky H (1953) Stress waves in solids. Clarendon Press, Oxford
3. Sun B, Chen R, Ping Y, Zhu Z, Wu N, He Y (2022) Dynamic response of rock-like materials based on SHPB pulse waveform characteristics. Materials 15:210. https://doi.org/10.3390/ma15010210
4. Dai F, Huang S, Xia K, Tan Z (2010) Some fundamental issues in dynamic compression and tension tests of rocks using split Hopkinson pressure bar. Rock Mech Rock Eng 43:657–666. https://doi.org/10.1007/s00603-010-0091-8
5. Aghayan S, Bieler S, Weinberg K (2022) Determination of the high-strain rate elastic modulus of printing resins using two different split Hopkinson pressure bars. Mech Time-Depend Mater. https://doi.org/10.1007/s11043-021-09511-2
6. Xia K, Yao W (2015) Dynamic rock tests using split Hopkinson (Kolsky) bar system—a review. J Rock Mech Geotech Eng 7:27–59.https://doi.org/10.1016/j.jrmge.2014.07.008
7. Tahir A (2020) Investigation of compressive strength of shale rock under impact loading 19:162–167. http://ijpsat.ijsht-journals.org/
8. Mishra S, Chakraborty T, Seshagiri Rao K (2021) Dynamic response of two extrusive igneous rocks using split Hopkinson pressure bar test. J Mater Civ Eng 33(6): 04021133(1–14). https://doi.org/10.1061/(ASCE)MT.1943-5533.0003725
9. Ganorkar K, Goel MD, Chakraborty T (2021) Specimen size effect and dynamic increase factor for basalt fiber–reinforced concrete using split Hopkinson pressure bar. J Mater Civ Eng 33(12):04021364 (1–15). https://doi.org/10.1061/(ASCE)MT.1943-5533.0003992
10. Zhou YX, Xia K, Li XB, Li HB, Ma GW, Zhao J, Zhou ZL, Dai F (2015) Suggested methods for determining the dynamic strength parameters and mode-I fracture toughness of rock materials. Int J Rock Mech Mining Sci 49:105–112
11. Lakshmana Rao C, Narayanamurthy V, Simha KRY (2016) Applied impact mechanics. Wiley. ISBN: 978-11-1924-180-5
12. Song B, Chen W (2003) One-dimensional dynamic compressive behavior of EPDM rubber. J Eng Math Tech 125:294–301
13. Sivakumar G, Maji VB (2021) Crack growth in rocks with pre-existing narrow flaws under uniaxial compression. Int J Geomech 21(4):04021032 (1–18). https://doi.org/10.1061/(ASCE)GM.1943-5622.0001960
14. Frew DJ, Forrestal MJ, Chen W (2001) A split Hopkinson pressure bar technique to determine compressive stress-strain data for rock materials. Exp Mech 41(1):40–46
15. Ma Q, Gao C (2018) Effect of basalt fiber on the dynamic mechanical properties of cement-soil in SHPB test. J Mater Civ Eng 30(8):04018185 (1–10). https://doi.org/10.1061/(ASCE)MT.1943-5533.0002386
16. Li X, Hong L, Yin T, Zhou Z, Ye Z (2008) Relationship between diameter of split Hopkinson pressure bar and minimum loading rate under rock failure. J Cent South Univ Technol 15:218–223. 10.1007/s11771ί008ί0042ί7
17. Yu Y, Ding JW, Xu Q, Zhu BT, Xu CJ, Tong LH (2020) Damage of sandstone induced by repetitive impact loading. Int J Geomech 20(7):04020090 (1–10), https://doi.org/10.1061/(ASCE)GM.1943-5622.0001736
18. Chen X, Ge L, Zhou J, Wu S (2016) Experimental study on split Hopkinson pressure bar pulse-shaping techniques for concrete. J Mater Civ Eng 28(5):04015196 (1–10). https://doi.org/10.1061/(ASCE)MT.1943-5533.0001494
19. Zou C, Wong LNY, Cheng Y (2012) The strength and crack behavior of the rock-like gypsum under high strain rate. https://www.researchgate.net/publication/261673418

20. Liu Yi, Dai F (2021) A review of experimental and theoretical research on the deformation and failure behavior of rocks subjected to cyclic loading. J Rock Mech Geotech Eng 13(5):1203–1230. https://doi.org/10.1016/j.jrmge.2021.03.012
21. Zaid M, Sadique MR (2020) The response of rock tunnel when subjected to blast loading: finite element analysis. Eng Rep. https://doi.org/10.1002/eng2.12293

Optimization of Multi-layered Composite Structures Against Impact Loading

Shashwat Kapoor and Sonalisa Ray

Abstract Protection of military bunkers against projectile impact caused by arms or blast-induced fragments in the hilly region is of serious concern. Protective structures made up of homogeneous concrete or fiber-reinforced concrete might not be sufficient to resist repeated projectile impacts. Therefore, there is a need to develop and analyze functionally graded layered composite structures with varying material properties. Additionally, conducting experimental studies on such full-scale layered composite structures is extremely time-consuming and expensive. Analysis of such structures by developing a computational framework will provide realistic solutions. Therefore, the present study focuses on performing numerical simulations for providing feasible solutions for optimizing the configuration of the layered composite structure. In this research, the composite structure involves silty–sand soil, RCC panel, mild steel plate, and EPDM elastomer. The target has been subjected to the projectile impact of 1.62 kg ogive-nosed hard steel cylindrical projectile having a 52 mm diameter and length of 234 mm. Holmquist–Johnson–Cook model for concrete, elastoplastic model for reinforcement bars and mild steel, Mohr–Coulomb model for soil, Ogden model for EPDM elastomer, and Johnson–Cook plasticity model for projectile have been used in the numerical simulation. The effect of the strength of RCC and thickness of various layers of the target have been varied and compared to the penetration capability of the projectile. Based on the proposed study, an optimized configuration of the multilayered composite structure has been derived in terms of various output parameters such as deformation (penetration), stresses, and damage parameters.

Keywords High-velocity projectile impact · ABAQUS/Explicit · Numerical simulations · Crater · Spalling · Scabbing

---

S. Kapoor (✉) · S. Ray
Civil Engineering Department, Indian Institute of Technology, Roorkee, India
e-mail: shashwat_k@ce.iitr.ac.in

R. Velmurugan et al. (eds.), *Dynamic Behavior of Soft and Hard Materials Volume 1*,
Springer Proceedings in Materials 34, https://doi.org/10.1007/978-981-99-6030-9_24

1 Introduction

Hard and profoundly covered structures are generally utilized in the defense establishment of numerous nations. These underground designs are intended to hide the individual, gear, order control, weapons of mass annihilation, as well as crucial components inside the organizations that help or are upheld by such offices. Safeguarding military bunkers against projectile impact in the hilly region is of serious concern. These bunkers are developed utilizing locally accessible materials. With improvements in innovation and materials, new types of security can be created paired with existing dugouts. Reinforced concrete (RC) can be utilized as development material for bunkers because of its high compressive strength, and durability as well as low development and support costs. If it is adequately thick and satisfactorily built-up with steel support bars, RC is generally equipped for forestalling perforation and catastrophic failure of the structure. Regardless, concrete being a quasi-brittle material has an inborn shortcoming in resisting tensile stresses which are appeared through the fragmentations, spalling, and scabbing at the concrete surfaces. The subsequent pieces can go at high speed during the effect occasion and comprise a danger to the security of the tactical staff inside the area of the bunker. Upgrading the tensile capacity and ductility can extensively work on the exhibition of RC defensive designs. The most recent twenty years have seen the advancement of cement with upgraded mechanical properties like higher tensile strength, fracture energy, and ductility, opening new aspects for working on the exhibition of RC defensive designs. Many examinations have been led to explore the effect obstruction of cement/cementitious materials, going from ordinary strength cement to high-strength concrete, fiber-reinforced concrete (FRC), and functionally graded cementitious composite (FGCC). It has been seen by analysts that the exhibition of functionally graded cementitious composite (FGCC) under influence stacking is greatly improved when contrasted with high-strength cement and fiber-reinforced concrete (FRC). Thus, layered composite is carried out in this review to explore its impact in involving it as a development material for bunkers in the hilly region subjected to impact loading. These are multilayered structures principally comprised of soil, reinforced concrete slabs, steel plates, etc. Understanding the projectile impact for penetration and perforation against these multilayered targets is of extraordinary importance as far as overcoming the objective. The ongoing information on the way of behaving of such typical multilayered structures is negligible. The greater part of the research papers published on the ballistic impact forces on the penetration behavior of reinforced concrete members. As various layers are added to the defensive designs, a different mode of failure shows up and material properties change with the depth of penetration. Furthermore, directing an experimental study on such full-scale layered composite designs is very tedious and costly. Hence, numerical simulations would give conceivable plausible arrangements for advancing the layered composite design to give obstruction against high-velocity projectile impact. By and large, concrete being a profoundly heterogeneous composite material shows complex conduct under the action of loading. The issue gets intensified under the activity of high-velocity impact, in this way requiring the reception of strong numerical models to foresee the way of behaving of the experimental tests. Consequently, in this research, an endeavor will be had to examine the impact resistance of concrete composite materials through numerical simulations. Consequently, there is a need to design and develop high-velocity kinetic energy projectiles to defeat these hardened targets. Concrete is a composite combination of concrete, sand, and totals (rock). Concrete is a composite mixture of cement, sand, and aggregates (gravels). Depending upon the concrete mix proportions, concrete is classified as lean, standard, high-strength, and ultra-high performance concrete.

Recently, ultra-high-performance steel fiber-reinforced concrete provides excellent mechanical properties such as high compressive, tensile, and flexural strength, improved workability, and durability properties. It is considered to be a promising material under impact, shock, and explosive loadings. Often experimental investigations are performed on concrete specimens to achieve information on the ballistic limit, residual velocity, depth of penetration, and load-time history. These results can be suitably used to develop empirical/semi-empirical/analytical formulations and to validate theoretical models and numerical simulations. Moreover, these formulations can be used for similar experimental conditions without repeating the experiment. Based on such studies, a large number of empirical/semi-empirical/analytical equations and engineering models have been proposed over the years. Further, it is well known that experimental studies on the impact of projectiles on different targets are expensive in terms of cost and time. In particular, experimentation is more challenging and expensive when the target is in double layers, such as soil and concrete. Predicting the behavior of the projectile at the soil-structure interface through experimentation may not be possible.

The above-stated limitations on experimental investigations and empirical/ analytical formulations have motivated the researcher to focus more on numerical simulations. It is a very challenging job to incorporate different failure modes at various stages of the simulation. Nonetheless, the majority of analytical expressions are centered around obtaining scabbing thickness and penetration depth expecting that the impact zone to be exceptionally confined and target to be semi-infinite [1–4]. Concrete is a material that offers resistance even after the complete failure of the material called residual strength. In the finite element analysis after the failure, failed elements are deactivated from the analysis; thus, they cannot offer residual strength. Using the same modeling strategies and analysis settings for thin concrete targets does not yield accurate results. This is because the material properties like residual strength, analysis settings like failure criteria, and contact properties need to be updated according to the thickness of the concrete. The velocity of the projectile also plays an important role in deciding failure modes. At low velocities, cracks appear on distal and proximal faces without penetration of the projectile. At medium velocities, spalling occurs where some material fails at the proximal face without penetration. At high velocities, scabbing happens where some material from both faces fails, and the projectile penetrates slightly into the material. For very high-velocity ballistic impact, cone cracking appears in which the crater's diameter at the distal face is larger than the proximal face making a cone-shaped hole in the material.

Concrete is not the only material used for constructing the protective structure. There are many advantages of using soil or sand layer over the concrete structure as the soil makes camouflage and helps the protective structure to blend with the surroundings and offer a layer of protection. Boulder mixed soil acted as an anti-penetration layer, the RCC panel acted as spalling resistance layer, and the rest acted as a crack resistance layer. The thick layer of soil and sand reduces the velocity of ballistic projectiles/missiles before hitting the concrete. Thus, there were an increase in the strength of the structure and a reduction in damage from the attack. There are very few publications available on the dynamic analysis of soil layers. Most of them are limited to dynamic responses on structure in the event of earthquakes. In recent years, numerical modeling has made significant advances in computational resources by modeling complex phenomena and interactions during rapid penetration of projectiles in soil and concrete double-layered structures. Recently, a theoretical model was proposed by Xiao et al., which described penetration into soil and concrete double-layered targets [5]. It has been observed that transmitted and reflected waves reduced penetration velocity when passed through soil/concrete interface, which provides a significant role in penetration depth.

2 Materials and Methods

Numerical simulations were performed in ABAQUS/Explicit solver in three-dimensional space on 400 × 400 mm quarter target with monolayer and multilayered configurations subjected to the projectile of 1.62 kg hard steel cylindrical projectile having 52 mm diameter and 234 mm length. Targets (monolayer and multilayered configurations) were modeled using 3-D continuum brick elements with reduced integration (C3D8R), and the projectile was modeled using 4-node 3-D bilinear rigid quadrilateral elements (R3D4). For reducing computational time, the quarter model was analyzed because of the symmetry about the center axis of the projectile and target. The mesh size of the target was increased with an increase in distance from the impact zone. Four simulations were performed to compare monolayer and multilayer targets with different configurations subjected to projectile impact. In the first variant, a reinforced concrete monolayer with a thickness of 700 mm having a compressive strength of 25 MPa and reinforcement bars of diameter 16 mm spaced at 100 mm in both directions with a cover of 20 mm were modeled, while in other variants, multilayer with three different configurations having thickness 735 mm, 755 mm, and 755 mm, respectively, consisting of different layers such as sandy soil, MS plate, RCC, EPDM elastomer layer was modeled to compare the response of target layers subjected to projectile impact (Fig. 1).

Holmquist–Johnson–Cook model for concrete, classical metal plasticity model for reinforcement bars and MS plate, Mohr–Coulomb model for sandy soil, Ogden model for EPDM elastomer, and VUSDFLD subroutine for incorporating damage parameter in EPDM elastomer have been used in the numerical simulation. A mesh convergence study was conducted to find out the size of the element that shall be used

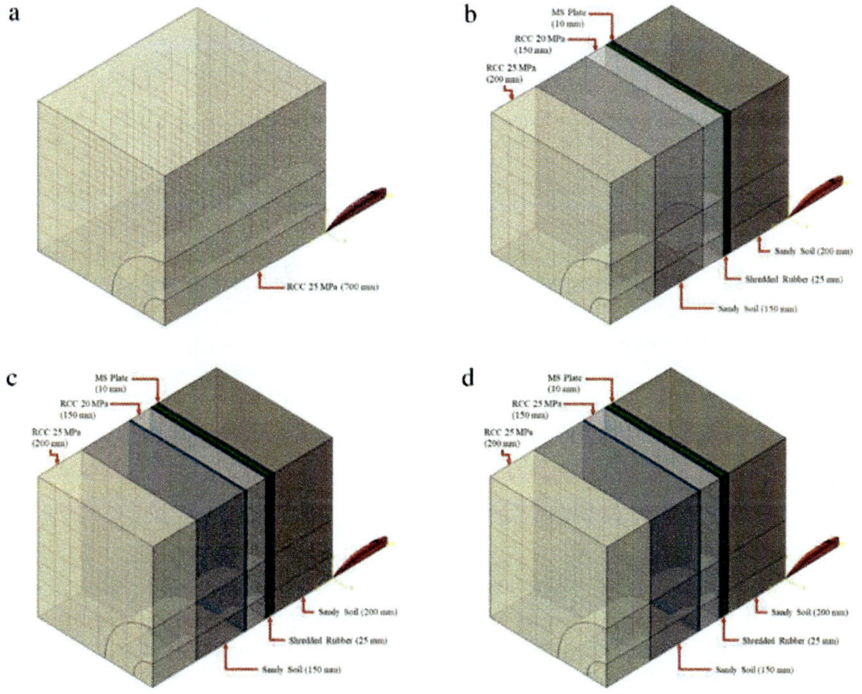

Fig. 1 Assembly of the **a** monolayer; **b** multilayer configuration-1; **c** multilayer configuration-2; **d** multilayer configuration-3; targets and projectile

in the meshing of the target while optimizing analysis run time. As seen in Fig. 2a, the residual velocity of the projectile started converging for the finest mesh sizes of 20, 15, 10, 5 and 2 mm, and based on these observations, it was determined that finer element size of 2 mm up to 50 mm radius of the impact zone, element size of 5 mm up to 150 mm radius of the impact zone, and rest element size of 10 mm provided accurate results optimizing analysis run time (Fig. 2b).

3 Material Models

The background of the material models of different layers used in this work are briefly discussed next.

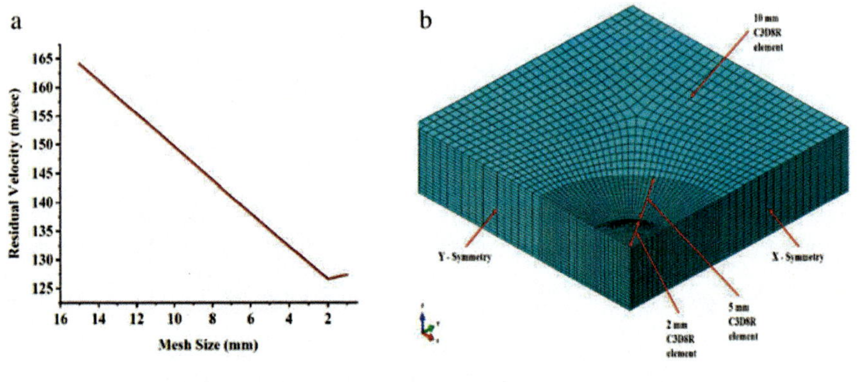

Fig. 2 **a** Mesh convergence study; **b** finite element model of each layer of target

3.1 Holmquist–Johnson–Cook Model

The HJC model for concrete [6] assumes that the normalized strength of the material can be expressed as functions of pressure and strain rate as follows;

$$\boldsymbol{\sigma}^{*} = \left(\boldsymbol{A}(1 - \boldsymbol{D}) + \boldsymbol{B}\boldsymbol{P}^{*\boldsymbol{N}}\right)\left(1 + \boldsymbol{C}\ln\dot{\epsilon}^{*}\right) \leq \boldsymbol{\sigma}^{\boldsymbol{max}} \quad (1)$$

here, A, B, N, and C are material constants, D is the damage variable, and $\dot{\varepsilon}^{*} = \dot{\overline{\epsilon}}^{pl}/\dot{\epsilon}_0$ is dimensionless strain rate, where $\dot{\overline{\epsilon}}^{pl}$ is equivalent plastic strain rate and $\dot{\epsilon}_0$ is reference strain rate. The normalized strength of the material is defined as $\boldsymbol{\sigma}^{*} = \boldsymbol{\sigma}/\boldsymbol{f}'_{c}$, where $\boldsymbol{\sigma}$ is von Mises equivalent stress and $\boldsymbol{f}'_{c}$ is uniaxial compressive strength.

The damage variable accumulates both with equivalent plastic strain and volumetric plastic (or compaction) strain according to

$$D = \sum \frac{\Delta\overline{\epsilon}^{pl} + \Delta\mu^{pl}}{\overline{\epsilon}_f^{pl}(P^{*})}$$

With

$$\overline{\epsilon}_f^{pl} = D_1\left(P^{*} + T^{*}\right)^{D_2}, \quad \overline{\epsilon}_{f,\min}^{pl} \leq \overline{\epsilon}_f^{pl} \leq \overline{\epsilon}_{f,\max}^{pl} \quad (2)$$

here, $\Delta\overline{\epsilon}^{pl}$ is the increment in equivalent plastic strain, $\Delta\mu^{pl}$ is the increment in volumetric compaction strain, and $\overline{\epsilon}_f^{pl}(P^{*})$ is the equivalent plastic strain to fracture under constant pressure. $T^{*} = T/f_{c'}$ is normalized maximum tensile hydrostatic stress; and D_1 and D_2 are material constants.

Pressure is given as a function of the volumetric strain measure as

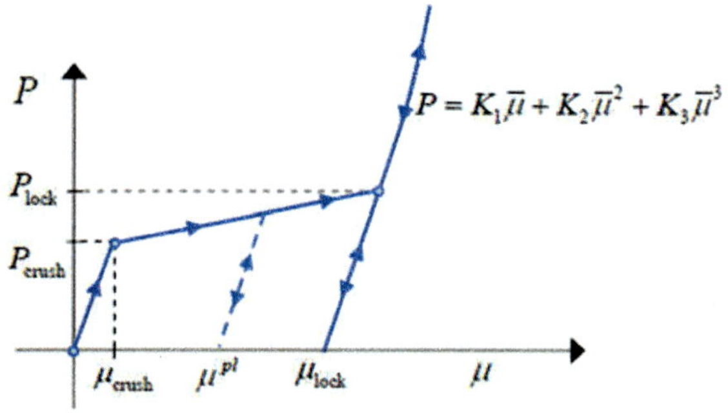

Fig. 3 Pressure–volume response for the HJC concrete model

$$\mu = \frac{\rho}{\rho_0} - 1 \tag{3}$$

here, ρ is the current density, and ρ_0 is reference density.

In the first region, the pressure increases linearly with μ from zero to P_{crush} corresponding to the onset of crushing (compaction) when $\mu = \mu_{\text{ crush}}$. The second region is associated with the crushing behavior of concrete. In this region, the material accumulates plastic volumetric strain, μ^{pl}, , until air voids are fully compressed, corresponding to $(\mu_{\text{lock}}, P_{\text{lock}})$. Beyond this point, in the third region, the material is assumed to be fully dense (all air voids have been removed), and pressure is given by the nonlinear elastic relations (Fig. 3).

$$P = K_1\overline{\mu} + K_2\overline{\mu}^2 + K_3\overline{\mu}^3,$$

with

$$\overline{\mu} = \frac{\mu - \mu_{\text{lock}}}{1 + \mu_{\text{lock}}} \tag{4}$$

here, K_1, K_2, and K_3 are material constants.

3.2 Classical Metal Plasticity Model

The plastic behavior is described by its yield point and post-yield hardening. Since measured data are nominal stress–strain data, conversion from true stress to logarithmic strain is done as follows;

$$\sigma_{\text{true}} = \sigma_{\text{nom}}(1 + \varepsilon_{\text{nom}}) \tag{5}$$

$$\varepsilon_{ln}^{pl} = \ln(1 + \varepsilon_{\text{nom}}) - \sigma_{\text{true}}/E \tag{6}$$

here, σ_{true} is true stress; ε_{ln}^{pl} is log strain; σ_{nom} and ε_{nom} are nominal stress and nominal strain, respectively.

3.3 Ogden Model

The constitutive behavior of hyperelastic (EPDM elastomer) material is derived from strain energy potential (SEP) function [7]. The strain energy potential function is the summation of deviatoric and volumetric strains.

$$U = U_{\text{dev}} + U_{\text{vol}} \tag{7}$$

here, U is strain energy potential, and U_{dev} and U_{vol} are deviatoric and volumetric strain energy potential, respectively.

Volumetric strain energy potential is given as follows;

$$U_{\text{vol}} = \sum_{i=1}^{N} \frac{1}{D_i} (J_{el} - 1)^{2i} \tag{8}$$

here, N, D_i, and J_{el}, represent strain energy potential order, compressibility index, and elastic volumetric strain of material, respectively.

Volumetric strain energy potential is given as follows;

$$U_{\text{dev}} = \sum_{i=1}^{N} \frac{2\mu_i}{\alpha_i^2} \left(\overline{\lambda}_1^{\alpha i} + \overline{\lambda}_2^{\alpha i} + \overline{\lambda}_3^{\alpha i} - 3 \right) \tag{9}$$

here, $\bar{\lambda}_i$ is the principal stretch ratio. μ_i and α_i are material constants, which describe its shear behavior.

3.4 Mohr–Coulomb Plasticity Model

The yielding function is in the form of [8];

$$F = R_{mc} q - p \tan \varphi' - c' = 0 \tag{10}$$

here,

$$R_{mc}(\Theta, \varphi') = \frac{1}{\sqrt{3} \cos \varphi'} \sin\left(\Theta + \frac{\pi}{3}\right) + \frac{1}{3} \cos\left(\Theta + \frac{\pi}{3}\right) \tan \varphi' \tag{11}$$

φ' is friction angle (slope of the yield surface in p-$R_{mc}q$ stress planes).
c' is the cohesion of the material.
Θ is the deviatoric polar angle defined as follows:

$$\cos(3\Theta) = \left(\frac{r}{q}\right)^3 \quad (12)$$

$$p = -1/3(\sigma_{11} + \sigma_{22} + \sigma_{33}) \quad (13)$$

$$q = \sqrt{\frac{3}{2} S_{ij} S_{ij}} \quad (14)$$

here, S_{ij} is deviatoric stress tensor and is defined as follows:

$$S_{ij} = S\sigma_{ij} + p\delta_{ij} \quad (15)$$

here, δ_{ij} is Kronecker delta.
And,

$$r = \left(\frac{9}{2} S \cdot S : S\right)^{1/3} = \left(\frac{9}{2} S_{ij} \cdot S_{jk} : S_{ki}\right)^{1/3} \quad (16)$$

The strain response follows a non-associative flow rule. A flow function is defined with a smooth elliptic function in the deviatoric stress plane as follows;

$$P = \sqrt{(\varepsilon c' \tan \Psi)^2 + (R_{mw} q)^2} - p \tan \Psi \quad (17)$$

here,

$$R_{mw}(\Theta, e) = \frac{4\left(1 - e^2\right) \cos^2 \Theta + (2e - 1)^2}{2\left(1 - e^2\right) \cos \Theta + (2e - 1)\sqrt{4\left(1 - e^2\right) \cos \Theta + 5e^2 - 4e}}$$

$$R_{mc}\left(\pi/3, \varphi'\right)$$

$$R_{mc}\left(\pi/3, \varphi'\right) = \frac{3 - \sin \varphi'}{6 + \cos \varphi'} \quad (18)$$

here, φ' and Ψ are friction angle and dilation angle measured in p-Rmwq plane, respectively. c' is initial cohesion yield stress; Θ remains as defined in yield criterion and e is a deviatoric eccentricity that describes the 'out-of-roundedness' of the deviatoric section in terms of the ratio between shear stress along the extension meridian ($\Theta = 0$) and shear stress along the compression meridian ($\Theta = \pi/3$).

4 Validation of Numerical Simulation Model

Heckötter et al. conducted three impact tests in which a cylindrical steel shell filled with concrete rigid missiles with an impact velocity of 135 m/s impacted RC slabs [9]. RCC slab having a width of 2100 mm and thickness of 150 mm having a compressive strength of 67 MPa and reinforcement bars of the diameter of 10 mm spaced at 90 mm in both directions with a cover of 20 mm and rigid missile having a length of 640 mm and diameter 168 mm were used in impact tests. In their work, they determined the residual velocity of the missile, mass of ejected concrete, scabbing, and spalling areas. In the present work, ABAQUS FE software was used for simulating the experiment. The constitutive modeling for simulating the concrete having a compressive strength of 70 MPa was carried out using concrete damage plasticity (CDP) model because this model represents both tensile cracking and compressive crushing failure. The material behavior of reinforcement bars was incorporated using classical metal plasticity model. Both concrete slab and reinforcement bars were meshed using 8-noded linear 3-D brick elements (C3D8R) having element sizes of 15 mm and 10 mm, respectively. Material properties of the slab and missile used in the FE model are shown in Table 1. Embedded region constraint was used to constrain reinforcement bars in concrete, and a friction coefficient of 0.3 was used to define interactions between the missile and RCC slab (Fig. 4).

It has been seen that the FE model determined close values corresponding to the residual velocity of the missile and mass of ejected concrete from experiments. The residual velocity of the projectile was found to be 41.91 m/sec (Fig. 5b) with an error of 8.51% when compared to the experimental results as listed in Table 2, validating the numerical simulation models. Additionally, the FE model represented spalling area, scabbing area, and damage pattern of the damaged concrete slab.

Table 1 Material properties of slab and reinforcement used in FE model

Parameters	Concrete slab	Reinforcement
Density (kg/ m^3)	2098.32	7131.4
Young's modulus (MPa)	27,850	200,000
Poisson's ratio	0.15	0.3
Dilation angle	54	–
Eccentricity	0.1	–
Fb0/Fc0	1.16	–
Kc	0.64	–
Tensile strength (MPa)	4.04	605.3
Compressive strength (MPa)	70	–
Yield stress in tension (MPa)	–	540

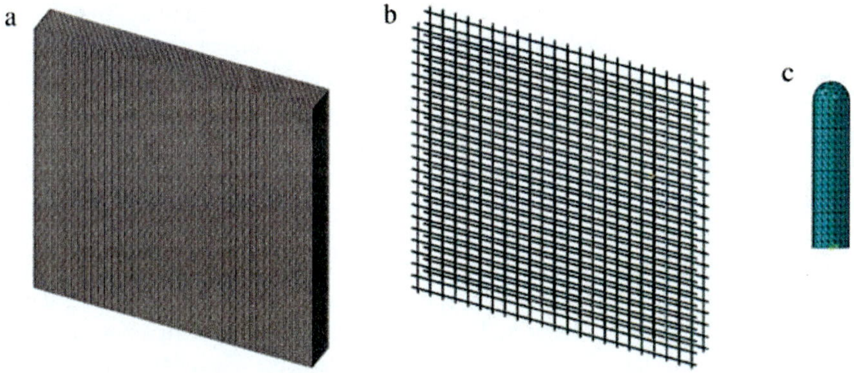

Fig. 4 **a** Concrete slab; **b** reinforcement; **c** missile FE model of IRIS experiment

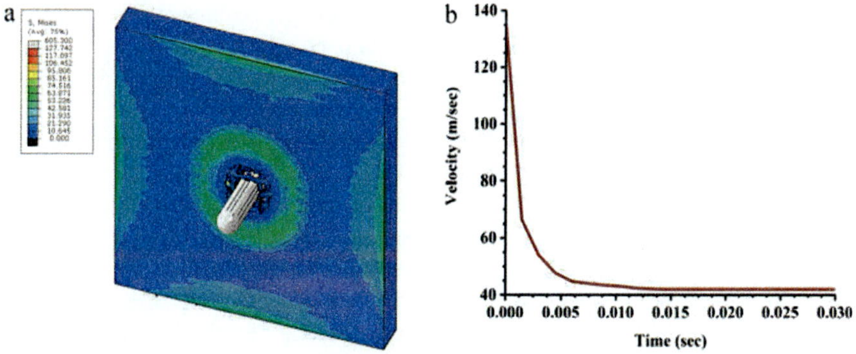

Fig. 5 **a** Von Mises stress in back face of RCC slab; **b** missile velocity; during impact

Table 2 Comparison between experimental and finite element method results

Parameters	Experimental	Finite element method	Error (%)
Residual velocity (m/sec)	45.80	41.91	8.51
Scabbing area (m^2)	1.00	0.915	8.50
Spalling area (m^2)	0.103	0.092	10.68

5 Numerical Simulation and Results

Numerical simulations were performed to compare monolayer and multilayer with different configurations in ABAQUS/Explicit solver. Four simulations were reproduced to obtain the response of target layers subjected to projectile impact. In the first variant, a reinforced concrete monolayer was modeled, while in other variants, multilayers with different configurations were modeled in order to compare the response of target layers subjected to projectile impact. Targets (monolayer and multilayered configurations) were modeled using 3-D continuum brick elements with reduced integration (C3D8R), and the projectile was modeled as a rigid body using 4-node 3-D bilinear rigid quadrilateral elements (R3D4). To reduce the computational time of the study, the quarter model was analyzed because of symmetry about the center axis of the projectile and target. A mesh convergence study was conducted to find out the size of the element to be used in the target while optimizing analysis run time, and it was found that finer element sizes of 2 mm up to 50 mm radius of the impact zone, element size of 5 mm up to 150 mm radius of the impact zone, and rest element size of 10 mm provided accurate results optimizing analysis run time. The angular velocity of the projectile was neglected because elements that are lying on planes of symmetry denied translational and rotational displacements. Boundary conditions were set so that the mathematical model ought to reflect actual elements of target layers. Energy dissemination is also one of the significant boundaries which ensnare its resistance when exposed to projectile impact. The external energy includes viscous and frictional energy dissipation in the material. The velocity profile of the projectile, stress profile, strain profile, crater diameter, and damage of different layered targets were determined to compare the results of monolayer and multilayer with different configurations.

The constitutive modeling for simulating the concrete layer having a compressive strength of 25 MPa was carried out using the HJC model, and material parameters of the concrete layer have been taken from Hafezolghorani et al. [10]. The material behavior of reinforcement bars was incorporated using classical metal plasticity model, and material parameters have been taken from Paul et al. [11]. The material behavior of the sandy soil layer was incorporated using the Mohr–Coulomb model, and material parameters have been taken from Mosadegh et al. [12]. The material behavior of EPDM elastomer was incorporated using Ogden Model, material parameters have been taken from Feichter et al. [13], and the damage parameter was incorporated using VUSDFLD subroutine. The material behavior of MS plate was incorporated using classical metal plasticity model, and material parameters have been taken from Xiang et al. [14] and shown in Table 3. A friction coefficient of 0 was used to define interactions between the projectile and the target.

The damage induced in monolayer and multilayer with different configuration targets was evaluated by estimating the volume of material eliminated from the front and back target surface. The equivalent radius of monolayer and multilayer with different configurations has been determined by taking the mean of the radius of the affected area after perforation in both front and back faces and listed in Table 4. It could be observed that the crater radius of the front face was found to be almost the same for all configurations of multilayer targets, whereas the crater radius of the back face in the multilayer configuration-3 target (84.53 mm) was found to be lower than other configurations. Damage dissipation energy of multilayer targets of configuration-1, configuration-2, and configuration-3 was found to be 86.48, 82.79, and 84.92, respectively, times higher than a reinforced concrete monolayer. Similarly, the internal energy of multilayer targets of configuration-1, configuration-2, and configuration-3 was found to be 31.25, 47.14, and 47.50, respectively, times higher

Table 3 Material properties of concrete, reinforcement bars, sandy soil, MS plate, and EPDM elastomer used in FE model

Parameters	Concrete	Reinforcement bars	Sandy soil	MS plate	EPDM elastomer
Density (kg/ m^3)	2400	7800	1920	7800	589
Young's modulus (MPa)	25,000	2.1E + 05	182	200,000	–
Poisson's ratio	0.2	0.3	0.3	0.3	–
Angle of friction	–	–	46	–	–
Dilation angle	–	–	4	–	–
Maximum tensile strength (MPa)	–	367	–	580	–
Yield stress in tension (MPa)	–	500	–	330	–
μ_1	–	–	–	–	0.99492924
μ_2	–	–	–	–	0.060157293
μ_3	–	–	–	–	0.077564639
α_1	–	–	–	–	3.36255675
α_2	–	–	–	–	9.76206145
α_3	–	–	–	–	−4.61426912

than a reinforced concrete monolayer. The residual velocity of the projectile after impact on monolayer is plotted in Fig. 6a, multilayer with different configuration targets is plotted in Fig. 6b–d, and comparisons between different configurations are listed in Table 5. It was observed that the projectile completely perforated all the targets after 0.005 s, and the residual velocity of the projectile in multilayer configuration-1 (173.56 m/sec) was found to be the maximum; residual velocity of the projectile in multilayer configuration-3 (133.01 m/sec) was found to be minimum than other multilayer configurations. Hence, multilayer configuration-3 is the best configuration to replace the RC monolayer.

Table 4 Equivalent crater radius of reinforced concrete monolayer and multilayers with different configurations

Target	Surface	R1 (mm)	R2 (mm)	R3(mm)	Equivalent radius (mm)
Reinforced concrete monolayer	Front face	85.71	88.92	80.00	84.88
	Back face	85.84	88.96	80.00	84.93
Multilayer configuration-1	Front face	102.38	109.67	110.00	107.35
	Back face	85.71	95.47	90.00	90.39
Multilayer configuration-2	Front face	102.38	109.67	110.00	107.35
	Back face	85.71	88.94	80.00	84.88
Multilayer configuration-3	Front face	102.38	109.67	110.00	107.35
	Back face	85.23	88.35	80.00	84.53

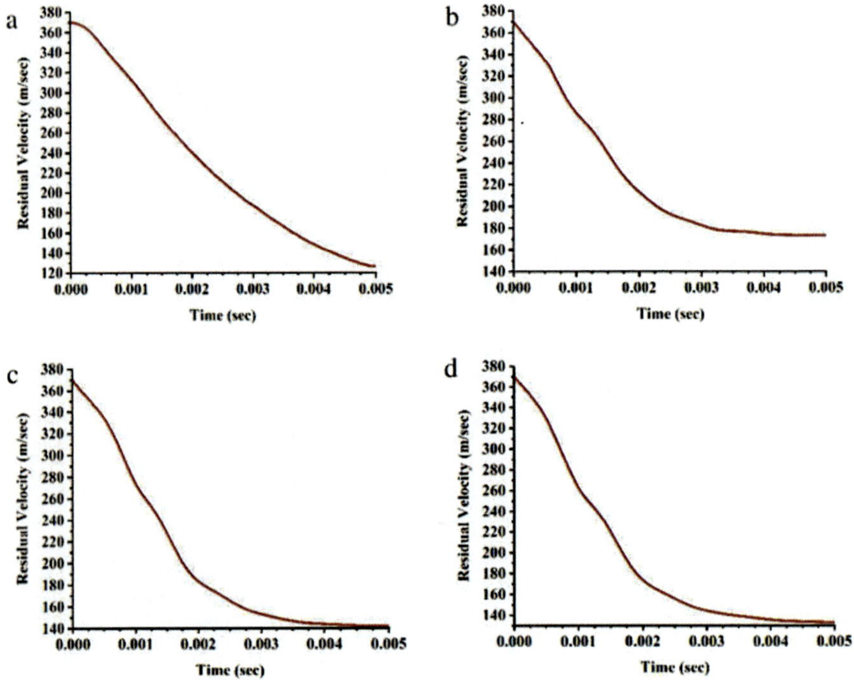

Fig. 6 Velocity profile of projectile in **a** RCC monolayer; **b** multilayer configuration-1; **c** multilayer configuration-2; **d** multilayer configuration-3; w.r.t. time

Table 5 Comparison of the residual velocity of the projectile after impact between monolayer and multilayer with different configurations

Target	Thickness (mm)	Impact velocity (m/sec)	Residual velocity (m/sec)
Reinforced concrete monolayer	700	370	126.57
Multilayer configuration-1	735	370	173.56
Multilayer configuration-2	755	370	142.04
Multilayer configuration-3	755	370	133.01

6 Conclusions

The present study investigated the projectile impact on reinforced concrete taking 25 MPa compressive strength monolayer target of quarter dimensions 400 × 400 mm with thickness 700 mm and multilayer with three configurations having same quarter dimensions 400 × 400 mm and thickness 735 mm, 755 mm, and 755 mm, respectively. From the simulation's results, it was concluded that:

- Damage dissipation energy of multilayer targets of configuration-1, configuration-2, and configuration-3 was found to be 86.48, 82.79, and 84.92, respectively, times higher than a reinforced concrete monolayer. Similarly, the internal energy of multilayer targets of configuration-1, configuration-2, and configuration-3 was found to be 31.25, 47.14, and 47.50, respectively, times higher than a reinforced concrete monolayer. Also, the residual velocity of the projectile in multilayer configuration-3 was found to be lower than in other configurations.
- The crater radius of the front face was found to be almost the same for all configurations of multilayer targets, whereas the crater radius of the back face in the multilayer configuration-3 target (84.53 mm) was found to be lower than in other configurations.

Hence, the multilayer configuration-3 target was found to optimum multilayer composite structure to replace the reinforced concrete monolayer as it pursues higher projectile penetration with minimum spalling and scrabbling in comparison to other multilayer configuration targets to replace RC monolayer.

Acknowledgements Acknowledgements and Reference heading should be left justified, bold, with the first letter capitalized but have no numbers. Text below continues as normal.

References

1. Hughes G (1984) Hard missile impact on reinforced concrete. Nucl Eng Des 77:23–35
2. Li QM, Ye ZQ, Ma GW, Reid SR (2007) Influence of overall structural response on perforation of concrete targets. Int J Impact Eng 34:926–941
3. Corbett GG, Reid SR, Johnson W (1986) Impact loading of plates and shells by free-flying projectiles: a review. Int J Impact Eng 18:141–230
4. Li QM, Reid SR, Wen HM, Telford AR (2005) Local impact effects of hard missiles on concrete targets. Int J Impact Eng 32:224–284
5. Xiao Q, Huang Z, Jia X, Zu X, Zhu Q (2017) Shaped charge penetrator into soil–concrete double-layered target. Int J Impact Eng 109:302–310. https://doi.org/10.1016/j.ijimpeng.2017.07.003
6. Holmquist TJ, Johnson GR, Cook WH (1993) A computational constitutive model for concrete subjected to large strains, high strain rates, and high pressures. In: The 14th international symposium on Ballistics. Feichter
7. Ogden RW, Roxburgh DG (1999) A pseudo–elastic model for the Mullins effect in filled rubber. In: Proceedings of the royal society of London; Series A: mathematical, physical and engineering sciences; 1998, pp 2861–2877. https://doi.org/10.1098/rspa.1999.0431
8. Robert DJ (2017) A modified Mohr-Coulomb model to simulate the behavior of pipelines in unsaturated soils. Comput Geotech 91:146–160
9. Heckötter C, Vepsä A (2015) Experimental investigation and numerical analyses of reinforced concrete structures subjected to external missile impact. Prog Nucl Energy 84:56–67. https://doi.org/10.1016/j.pnucene.2015.02.007

10. Hafezolghorani M, Hejazi F, Vaghei R, Jaafar MSB, Karimzade K (2017) Simplified damage plasticity model for concrete. Struct Eng Int 27(1):68–78. https://doi.org/10.2749/101686616x1081
11. Paul SK, Majumdar S, Kundu S (2014) Low cycle fatigue behavior of thermo-mechanically treated rebar. Mater Des 58:402–411. https://doi.org/10.1016/j.matdes.2014.01.079
12. Mosadegh A, Nikraz H (2015) Bearing capacity evaluation of footing on a layered-soil using ABAQUS. J Earth Sci Climatic Change 06(03). https://doi.org/10.4172/2157-7617.1000264
13. Feichter C, Major Z, Lang RW (2006) Deformation analysis of notched rubber specimens. Strain 42(4):299–304. https://doi.org/10.1111/j.1475-1305.2006.00286.x
14. Yun X, Gardner L (2017) Stress-strain curves for hot-rolled steels. J Constr Steel Res 133:36–46. https://doi.org/10.1016/j.jcsr.2017.01.024

Overall Response of Fluid-Filled Elastomers Subjected to High Loading Rates

Vivek Singh and Vikranth Racherla

Abstract For a long time, porous materials have been investigated for application in impact protection equipment as packaging, helmets, armors, etc. With repeated loading, the elastomeric cell walls lose their elasticity and cannot return to initial state. In the present work deformation behavior of porous elastomer filled with a non-Newtonian fluid on the overall response of the composite is obtained for high loading rates. Elastomer is modeled as an incompressible, neo-Hookean material, whereas the fluid is modeled as an incompressible, power-law fluid. When the load is applied, deformation of the elastomer causes fluid to flow out of the pore and upon unloading the fluid moves back into the pore. The analytical results are compared with numerical simulation results. Numerical simulations have been carried out in COMSOL Multiphysics software. A good match is obtained between the analytical and numerical results.

Keywords Effective behavior · Fluid–structure interaction · Newtonian fluid · Porous elastomer

1 Introduction

Porous elastomers have wide range of applications, and therefore, they have gained a lot of attention from researchers. They find application in packaging [1, 2], automobile [3, 4], noise attenuation [5] and protection equipment [6–8]. The elastomers are capable of storing energy when external load is applied and releasing the energy when external load is removed. In most of the applications, these porous elastomers undergo repeated loading and unloading cycles. While undergoing multiple cycles of loading and unloading, the cell wall may get damaged and the porous elastomer will not be able to sustain further loading. To solve this problem, we propose to fill the pores with a non-Newtonian (power-law) fluid. The addition of fluid provides viscous

V. Singh (✉) · V. Racherla
Department of Mechanical Engineering, Indian Institute of Technology Kharagpur, Kharagpur 721302, India
e-mail: viveksingh@iitkgp.ac.in

R. Velmurugan et al. (eds.), *Dynamic Behavior of Soft and Hard Materials Volume 1*, Springer Proceedings in Materials 34, https://doi.org/10.1007/978-981-99-6030-9_25

dissipation in addition to storage of energy and also it helps reduce the damage to elastomer. In the present work, we investigate the effect of adding power-law fluid to porous elastomer on the overall response of the composite when subjected to high loading rates.

Shear thickening fluid (STF)-filled elastomers have attracted attention for a long time for application in impact protection equipment because of their improved capabilities as compared to conventional materials. STF is like a dense suspension that, when subjected to high shear rates, undergoes a rapid viscosity increase and behaves like a solid. STF returns to its initial liquid state upon removal of the applied load. STF-filled fabrics, foam or sandwich panels have been studied in the past for ballistic protection [9–13]. Addition of STF keeps the composite material to be flexible and also improves the ballistic protection of the composite material, as the fraction of incident energy that is absorbed by STF-impregnated composite is much higher than that without STF. Gürgen [14] studied the performance of STF-impregnated composite under stabbing condition. STF-impregnated composite performed better than the neat fabrics. Sheikhi and Gürgen [15] studied the impact behavior of multi-layer composite consisting of cork, warp knitted spacer fabrics and foam filled with STF. Addition of STF improved the energy absorption and reduced the impact force transmitted.

In the present work, we obtain theoretical estimates for the overall response of STF-filled elastomer under high loading rates. The elastomer has a square arrangement of cylindrical pores. Theoretical estimates obtained for macroscopic stored energy density, macroscopic viscous dissipation density and overall response are compared with the results obtained from numerical simulations. COMSOL Multiphysics has been used to execute numerical simulations.

2 Problem Statement

2.1 Geometry

Figure 1[1] shows the porous elastomer having periodic square arrangement of cylindrical pores. The porous elastomer is kept in an outer container filled with a power-law fluid such that the surrounding fluid and the fluid saturating the pores are identical. The bracket helps restrict the longitudinal deformation of elastomer so that the elastomer experiences plane strain loading condition. The load is applied through the piston. As the load is applied, the fluid in the pores moves out, and when the load is removed, the fluid moves back into the pores and the pores regain their original shape. A unit cell having a cuboidal elastomeric block with a cylindrical void in the center has been considered for analytical results and for numerical simulation.

[1] Reprinted from semi-analytical estimates for overall response of porous elastomer filled with power-law fluid, 148, Vivek Singh, Vikranth Racherla, with permission from Elsevier.

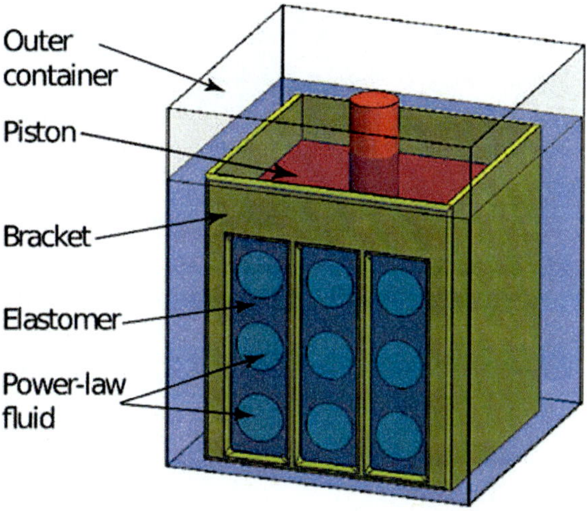

Fig. 1 Schematic of a power-law fluid-filled elastomer subjected to plane strain compression

2.2 *Material Behavior*

- Solid Phase

The elastomer is taken to be incompressible and hyperelastic, whose strain energy density is given by neo-Hookean model as follows:

$$W^{(El)} = \frac{G}{2}(I_1 - 3)$$

Here, G is the shear modulus of elastomer in ground state and I_1 is the first principal invariant of the right Cauchy–Green deformation tensor $\mathbf{C} = \mathbf{F}^{\mathrm{T}}\mathbf{F}$. $\mathbf{F}$ is the deformation gradient tensor. The elastomer satisfies the equilibrium equation as follows:

$$\nabla.\boldsymbol{\sigma} = 0$$

Here, the inertial effects and body forces have been neglected, and $\boldsymbol{\sigma}$ represents Cauchy stress tensor.

- Fluid Phase

The fluid considered in the present work is taken to be incompressible and non-Newtonian. The fluid flow in the pores is governed by continuity equation as follows:

$$\nabla.\mathbf{v} = 0$$

and the Navier–Stokes equation

$$\rho_f\left(\frac{\partial \mathbf{v}}{\partial t}+\mathbf{v}.\nabla\mathbf{v}\right)=-\nabla p+\nabla.(\mu\dot{\boldsymbol{\gamma}})$$

Here, $\mathbf{v}$ is fluid velocity vector, p is fluid pressure, ρ_f is fluid density, and μ is the apparent fluid dynamic viscosity given as follows:

$$\mu(\dot{\gamma})=\mu_o|\dot{\gamma}|^{n-1}$$

Here, μ_o is the consistency factor, n is power-law index, and $\dot{\gamma}=\sqrt{\frac{1}{2}\dot{\boldsymbol{\gamma}}:\dot{\boldsymbol{\gamma}}}$ and $\dot{\boldsymbol{\gamma}}=\left(\nabla\mathbf{v}+(\nabla\mathbf{v})^{T}\right)$.

2.3 Boundary Conditions

The outer boundaries of the porous elastomer are subjected to affine displacement boundary conditions of the form:

$$\overline{\mathbf{u}}=\left(\overline{\mathbf{F}}-\mathbf{I}\right)\mathbf{X}$$

The pore ends are subjected to constant pressure condition. The no-slip condition is applied at fluid–solid interface.

2.4 Numerical Simulation Setup

Coupled fluid–structure interaction simulations have been executed in COMSOL Multiphysics software. The solid domain was discretized using quadratic serendipity elements. Fluid domain is discretized using $P2 + P1$ method, where second-order elements are used for velocity and first-order elements for pressure. Time stepping was carried out using backward differentiation formula (BDF), and a direct solver was used. Arbitrary Lagrangian–Eulerian method was used for fluid–solid interaction. Velocity and pressure along pore axis were considered for convergence criteria. Details of the mesh used for numerical simulation are given below (Table 1).

2.5 Theoretical Estimates

Figure 1 shows a schematic representation of fluid-filled elastomer composite. When the load is applied, piston compresses the elastomer and fluid squeezes out of the pore, and when load is released, the fluid gets sucked back into the pore. The analytical estimates for response of the composite have been derived in a previous work [16].

Table 1 Mesh parameters of the mesh used for numerical simulation in COMSOL [16][2]

Parameters	Mesh
No. of domain elements	184,883
No. of boundary elements	22,738
No. of edge elements	972
Min. solid element size (mm)	0.18
Max. solid element size (mm)	1.00
Min fluid element size (mm)	0.0436
Max. fluid element size (mm)	0.135
Boundary layers	2

Here, we present the results for the response of composite under high loading rates which have not been presented before.

3 Results and Discussion

In this section, a comparison of theoretical estimates and numerical simulation results is presented. The results have been plotted for initial porosity for values 0.4 and 0.6, elastomer shear modulus G is 369.4 kPa, fluid consistency index μ_o is 1 Pa.s, and power-law index n is 1.3.

Figure 2a, b show the macroscopic response of the power-law fluid-filled elastomeric composite, when it is subjected to an impact load having loading rate 1 s^{-1} and 10 s^{-1}. In both the cases, two initial porosities 0.4 and 0.6 have been considered. As seen from the figure, there is a very good match for 1 s^{-1} loading rate, but for 10 s^{-1}, there is difference at start but later on both the results match well. The contribution of elastomer to overall response being independent of rate term is not affected by the loading rate considered. Therefore, the difference in response can be attributed to fluid flow. For 1 s^{-1} loading rate, the response varies linearly with the applied deformation for the entire deformation, and the numerical results match very well with the theoretical results (Fig. 2a). For 10 s^{-1} loading rate, as observed from Fig. 2b, theoretical results overpredict the response. The numerical simulations show that the fluid undergoes a transient phase at the beginning, and after some time, the dissipation must be similar to that predicted using the theoretical estimates; therefore, the results match well after a short duration. A deviation in overall response is observed for higher loading rates above 1 s^{-1}.

[2] Reprinted from semi-analytical estimates for overall response of porous elastomer filled with power-law fluid, 148, Vivek Singh, Vikranth Racherla, with permission from Elsevier.

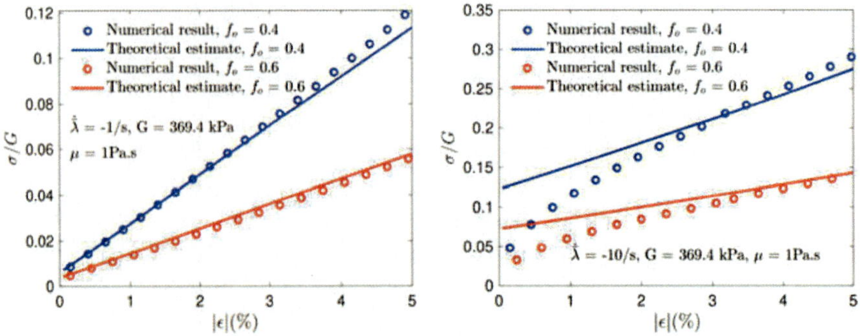

a b

Fig. 2 Stress–strain curve for power-law fluid-filled elastomer subjected to loading rate 1 s^{-1} and 10 s^{-1}

4 Conclusions

The present study of power-law fluid-filled elastomer subjected to high loading rates has been carried out for small deformation. The overall macroscopic response has been estimated for the composite as a function of initial porosity and external loading rate. For loading rate 1 s^{-1}, the response is linear, but with increase in loading rate, i.e., for 10 s^{-1}, the response starts to get non-linear. The obtained results are promising and show that the response of such composite under higher loading rates can also be estimated using these theoretical results.

References

1. Shuttleworth R, Shestopal V, Goss P (1985) Open-cell flexible polyurethane foams: comparison of static and dynamic compression properties. J Appl Polym Sci 30(1):333–343
2. Sims G, Bennett J (1998) Cushioning performance of flexible polyurethane foams. Polym Eng Sci 38(1):134–142
3. Deng R, Davies P, Bajaj A (2003) Flexible polyurethane foam modelling and identification of viscoelastic parameters for automotive seating applications. J Sound Vib 262(3):391–417
4. Zhang L, Dupuis R (2011) Measurement and identification of dynamic properties of flexible polyurethane foam. J Vib Control 17(4):517–526
5. Imai Y, Asano T (1982) Studies of acoustical absorption of flexible polyurethane foam. J Appl Polym Sci 27(1):183–195
6. Bazhenov S (1997) Dissipation of energy by bulletproof aramid fabric. J Mater Sci 32(15):167–4173
7. Carr´e M, James D, Haake S (2004) Impact of a non-homogeneous sphere on a rigid surface. Proc Inst Mech Eng Part C: J Mech Eng Sci 218(3):273–281
8. Ramirez B, Gupta V (2018) Evaluation of novel temperature-stable viscoelastic polyurea foams as helmet liner materials. Mater Des 137:298–304
9. Lee Y, Wetzel E, Wagner N (2003) The ballistic impact characteristics of kevlar® woven fabrics impregnated with a colloidal shear thickening fluid. J Mater Sci 38(13):2825–2833

10. Decker M, Halbach C, Nam C, Wagner N, Wetzel E (2007) Stab resistance of shear thickening fluid (stf)-treated fabrics. Compos Sci Technol 67(3–4):565–578
11. Majumdar A, Butola BS, Srivastava A (2013) Optimal designing of soft body armour materials using shear thickening fluid. Mater Des 46:191–198
12. Hasanzadeh M, Mottaghitalab V (2014) The role of shear-thickening fluids (STFs) in ballistic and stab-resistance improvement of flexible armor. J Mater Eng Perform 23(4):1182–1196
13. Chatterjee VA, Verma SK, Bhattacharjee D, Biswas I, Neogi S (2019) Enhancement of energy absorption by incorporation of shear thickening fluids in 3d-mat sandwich composite panels upon ballistic impact. Compos Struct 225:111148
14. G¨urgen S (2019) An investigation on composite laminates including shear thickening fluid under stab condition. J Compos Mater 53(8):1111–1122
15. Sheikhi MR, G¨urgen S (2022) Anti-impact design of multi-layer composites enhanced by shear thickening fluid. Compos Struct 279:114797
16. Singh V, Racherla V (2023) Semi-analytical estimates for overall response of porous elastomer filled with power-law fluid. Int J Non-Linear Mech 148:104253

Behaviour of Reinforced and Prestressed Concrete Slabs Under Multiple Impacts

Vimal Kumar

Abstract In the present study, the performance of reinforced and prestressed concrete slabs was investigated against multiple drop impacts. The slab specimens were cast using two different concretes having 28 days strength of 48 and 68 N/mm^2. The span length of the slabs and thickness were kept the same for all the slabs as 0.80 m and 100 mm, respectively. The level of prestressed in the slabs varied between 0 to 30%. A known mass of 242.8 kg was lifted to 1.0 and 0.5 m height and dropped freely to hit the target slabs multiple times. The deflection, impact force, reaction, progress of damage, energy, acceleration and strain were measured through different sensors and compared. Under the first impact, the magnitude of impact force and reaction increased in the prestressed concrete by 20 and 36%, respectively, compared to reinforced concrete. On the other hand, the deflection was reduced by 28% in prestressed concrete. Under multiple impacts, the impact force and reaction were noticed to have reduced and increased, respectively. The deflection increased under multiple impacts due to the accumulation of damage. A nominal influence on prestressed concrete was witnessed in the energy absorbed by the slabs. The energy was reduced by 10% with an increase in the strength of concrete. The reinforced concrete witnessed flexure–shear deformation, while prestressed concrete witnessed splitting–shear deformation under multiple impacts.

Keywords Reinforced concrete · Prestressed concrete · Slab · Multiple impact · Damage · Dynamic response

1 Introduction

Most of the structures and infrastructures for civilian or defense purposes are performed using reinforced concrete or prestressed concrete. These structures are exposed to extreme loading during their service life resulting from seismic events,

V. Kumar (✉)
Department of Civil Engineering, National Institute of Technology Hamirpur, Hamirpur 177005, India
e-mail: panchariya.vimal@gmail.com; vimalkumar@nith.ac.in

R. Velmurugan et al. (eds.), *Dynamic Behavior of Soft and Hard Materials Volume 1*, Springer Proceedings in Materials 34, https://doi.org/10.1007/978-981-99-6030-9_26

explosions, impact and rockfall which can led to a much greater load than their the design load [1]. In the past couple of decades, several techniques were evolved to explore the performance of concrete and concrete structures which include the Charpy impact test, split-Hopkinson pressure bar test and drop-weight impact setup [2]. Although studies are available in the literature, most of those were focused on a single impact on the structural members [3–6]. Some multiple impact studies are available in the literature that was focused on the prestressed concrete sleeper [7–9]. Thus, the response of these structures is well understood against static loading but still not known completely against multiple impacts. In this study, the impact behavior of reinforced and prestressed concrete slabs was investigated under multiple drop impacts. The response of the slabs such as impact force, reaction, deflection, energy absorption, damage and acceleration was measured, studied and compared.

2 Experimental Program

2.1 Material and Test Specimens

The impacts were performed on slab specimens cast using concrete of strength 48 and 68 N/mm^2. The concrete was prepared in a mechanical mixture using OPC, river sand, 10 mm basalt aggregate, water, chemical and mineral admixture. The slabs were having a span length of 0.8 m along length and breadth and a uniform thickness of 0.1 m. The slabs were reinforced using 8-mm-diameter Fe 500 grade steel provided along both the spans. In prestressed concrete, the same number of tendons was used which stressed to the same magnitude of tensile force to induce prestress. The provided tendons induced 20 and 30% prestress of concrete strength 68 and 48 N/mm^2, respectively.

2.2 Impact Experiments

The impact experiments were performed on the casted slabs under a drop impact testing facility. The casted slabs were held in the testing setup, and different sensors were attached to record forces, deflections and accelerations. A known steel mass of 242.85 kg was then lifted to 1.0 m height and dropped on those slabs at the center to develop impact load. The generated response of the slabs was recorded in a data logger at a very high sampling rate of 10 k samples per second per sensor. Later, the same steel mass was raised to 0.5 m height and dropped multiple times, and the responses were measured multiple times to study the residual performance of that slab. For discussion on the response in this manuscript, the reinforced concrete slabs are RC-48 and RC-68, while the prestressed concrete slabs are named PC-48 and

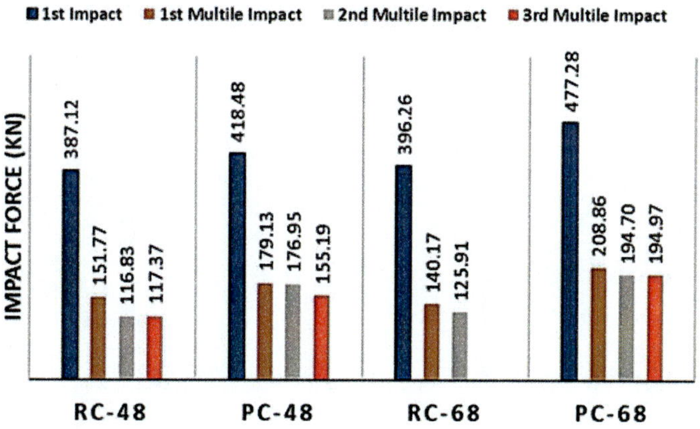

Fig. 1 Impact force for reinforced and prestressed concrete against multiple impacts

PC-68. The term 48 and 68 in their names is showing the strength of the concrete at 28 days.

3 Results and Discussion

3.1 Impact Force

The impact force measured during the impact is shown in Fig. 1. The magnitude of the impact force was noticed to have increased in prestressed concrete compared to reinforced concrete irrespective of the strength of concrete. The increase was also noticed under multiple impacts. The maximum increase in the impact force was about 20% under the very first impact. The increase in the peak impact force of the prestressed concrete could be due to the improved strength/stiffness of prestressed concrete. On the other hand, the magnitude of the impact force was reduced for a given concrete and its strength under multiple impacts. This reduction was due to the accumulation of damage under successive impacts.

3.2 Reaction

The reaction measured during the impact is shown in Fig. 2. The magnitude of the reaction was noticed to have increased in prestressed concrete compared to reinforced concrete irrespective of the strength of concrete. The increase was also noticed under multiple impacts. The maximum increase in the reaction was about 14.5 and 36.5%

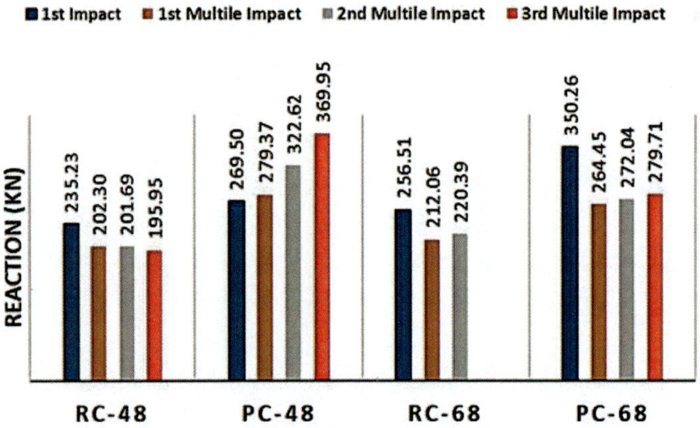

Fig. 2 Reaction for reinforced and prestressed concrete against multiple impacts

under the very first impact for slabs with concrete strength 48 and 68 N/mm^2, respectively. The increase in the reaction of the prestressed concrete under the first impact could be due to the improved strength/stiffness of prestressed concrete. In general, the magnitude of the reaction increased under multiple impacts from 0.5 m drop height could be due to the higher bending deformation due to the accumulated damage.

3.3 *Deflection*

The deflection measured during the impact is shown in Fig. 3. The magnitude of the deflection was noticed to have reduced in prestressed concrete compared to reinforced concrete irrespective of the strength of concrete. Under the first impact, the deflection in prestressed concrete was reduced by 28 and 26%, respectively, in slabs with 48 and 68 N/mm^2 concrete. Under multiple impacts, the deflection in the slab was noticed to have increased due to the accumulated damage in the slabs. In general, the PC-68 slab witnessed the smallest, while slab RC-48 witnessed the highest deflection under single as well as multiple impacts.

3.4 *Energy Absorption*

The energy absorbed by the slabs under single and multiple impacts has been presented in Fig. 4. For a given concrete, only a nominal difference (3–4%) was noticed in the energy absorbed by reinforced and prestressed concrete. On the other hand, the energy was reduced with the increase in the strength of concrete. The energy reduction was 3.5 and 10%, respectively, in reinforced and prestressed concrete. Further, the energy was noticed to have increased under multiple impacts. In general,

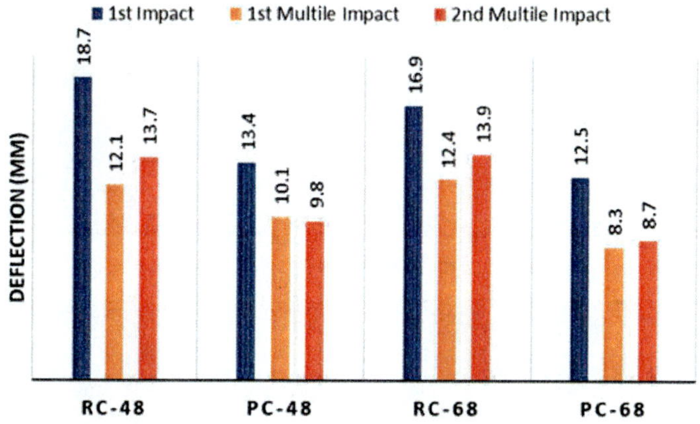

Fig. 3 Deflection for reinforced and prestressed concrete against multiple impacts

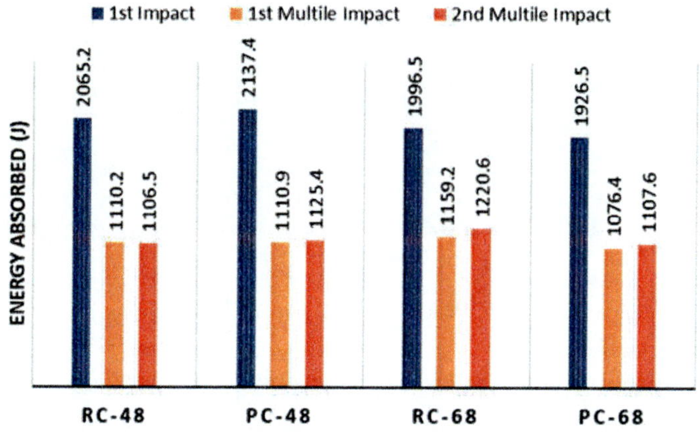

Fig. 4 Energy absorbed by reinforced and prestressed concrete against multiple impacts

the energy absorbed under the second drop from 0.5 m was noticed higher by 15–60 J compared to the energy absorbed under the first drop from 0.5 m.

3.5 Damage Under Multiple Impacts

The damage in the reinforced and prestressed concrete slabs has been shown in Fig. 5. Under the first drop of the hammer, the reinforced concrete slab RC-48 witnessed flexural deformation showing a relatively uniform distribution of the cracks and a greater number of cracks. On the other hand, the splitting damage was more pronounced in the prestressed concrete (slab PC-48 and PC-68). The relatively wider splitting damage in PC-48 compared to PC-68 under the first drop might led to more

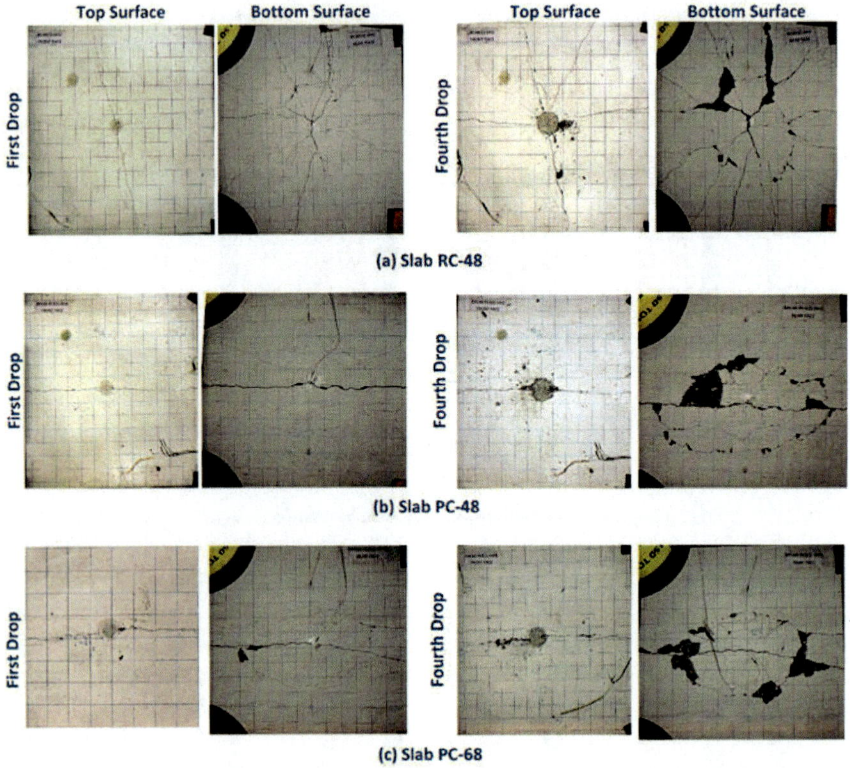

Fig. 5 Damage in slab **a** RC-48, **b** PC-48 and **c** PC-68 under multiple impacts

rotation of the slab at the support under multiple impacts. This could be the possible reason for the increased reaction in PC-48 compared to PC-68 (Fig. 2). On the other hand, if the punching damage inside the slab is more dominant, there is a possibility of reduced reaction force. However, the inner damage and rotation of the slab were not measured during the multiple impact experimentations. With an increase in the number of drops, the damage was increased in both the concretes, such that reinforced concrete witnessed flexure–shear deformation, while prestressed concrete witnessed splitting–shear deformation. The damage indicates that slab PC-48 suffered more damage compared to slab PC-68.

4 Conclusions

The reinforced and prestressed concrete slabs were tested against single and multiple drop impacts under an instrumented drop impact setup. The slabs were having identical geometrical configurations however differ in the strength of concrete and level of prestressed. The slabs were first impacted from 1.0 m height and successively impacted from 0.5 m height to study the residual performance. Under the first impact, the magnitude of impact force and reaction increased in the prestressed concrete by 20 and 36%, respectively, compared to reinforced concrete. On the other hand, the deflection was reduced by 28% in prestressed concrete. Under multiple impacts, the impact force and reaction were noticed to have reduced and increased, respectively. The deflection increased under multiple impacts due to the accumulation of damage. A nominal influence on prestressed concrete was witnessed in the energy absorbed by the slabs. The energy was reduced by 10% with an increase in the strength of concrete. The reinforced concrete witnessed flexure–shear deformation, while prestressed concrete witnessed splitting–shear deformation under multiple impacts.

Acknowledgements The author sincerely thank CBRI Roorkee India and IIT Roorkee India for extending their laboratory support for this study.

References

1. Kezmane A, Chiaia B, Kumpyak O, Maksimov V, Placidi L (2017) 3D modelling of reinforced concrete slab with yielding supports subject to impact load. Eur J Environ Civ Eng 21:988–1025
2. Zhang XX, Ruiz G, Yu RC (2010) A new drop-weight impact machine for studying fracture processes in structural concrete. Strain 46:252–257
3. Ruiz G, Zhang XX, Yu RC, Porras R, Poveda E, del Viso JR (2011) Effect of loading rate on fracture energy of high-strength concrete. Strain 47:518–524
4. Saatci S, Vecchio F (2009) Effects of shear mechanisms on impact behavior of reinforced concrete beams. ACI Struct J 106:78–86
5. Nuruddin MF, Azmee NM, Chang KY (2014) Effect of MIRHA and fly ash in ductile self-compacting concrete on abrasion and impact performance. Appl Mech Mater 567:393–398
6. Hossain MM, Karim MR, Islam MA, Zain MFM (2014) Crack chronology of reinforced concrete beam under impact loading, middle-east. J Sci Res 21:1663–1669
7. Kaewunruen S, Remennikov AM (2010) Dynamic crack propagations in prestressed concrete sleepers in railway track systems subjected to severe impact loads. ASCE J Struct Eng 136:749–754
8. Kaewunruen S, Remennikov AM (2009) Dynamic flexural influence on a railway concrete sleeper in track system due to a single wheel impact. Eng Fail Anal 16:705–712
9. Kaewunruen S, Remennikov AM (2011) Experiments into impact behaviour of railway prestressed concrete sleepers. Eng Fail Anal 18:2305–2315

Impact Mitigation in a Conico-Cylindrical Projectile During Sub-ordnance Velocity Impact

Ankur Trigunayak, Rohit Sankrityayan, Anoop Chawla, and Devendra K. Dubey

Abstract This work demonstrates effective impact load mitigation methods which can be used to provide protection of subsystem inside a conico-cylindrical projectiles during impact at sub-ordnance velocity ranges (25–500 m/s). The projectiles under a short duration impact with hard soil or concrete structure lead to large deformation due to high 'g' deceleration inside projectile subsystem. Typically, these projectiles contain sophisticated electronics packages and high energetic materials inside at different locations. Evidently, high 'g' deceleration loads can severely affect their intended function due to uncontrolled deformation mode. In view of this, present study investigates the energy mitigation effect in terms of pressure–time history of energetic material, deceleration of the projectiles, and critical parameters for detonation like energy fluence at various places inside projectile using finite element simulation in ANSYS/AUTODYN® using explicit dynamics approach. The respective methodology for energy absorption/dissipation by different materials deformation, use of collapsible rods and bimetallic rods inside cones were simulated and compared to enhance the crashworthiness of the projectile. Plastic deformation of collapsible and bi-metallic rods induces weak points by buckling and bending of rods, leading to higher energy dissipation, hence providing better and safer impact mitigation to protect projectile from damage due to accidental impact.

Keywords High 'g' deceleration · Sub-ordnance velocity · Explicit analysis · Energy absorption · ANSYS/Autodyn

A. Trigunayak
Terminal Ballistics Research Laboratory (TBRL) DRDO, Chandigarh, India

R. Sankrityayan · A. Chawla · D. K. Dubey (✉)
Department of Mechanical Engineering, Indian Institute of Technology, New Delhi, India
e-mail: dkdubey@mech.iitd.ac.in

R. Velmurugan et al. (eds.), *Dynamic Behavior of Soft and Hard Materials Volume 1*, Springer Proceedings in Materials 34, https://doi.org/10.1007/978-981-99-6030-9_27

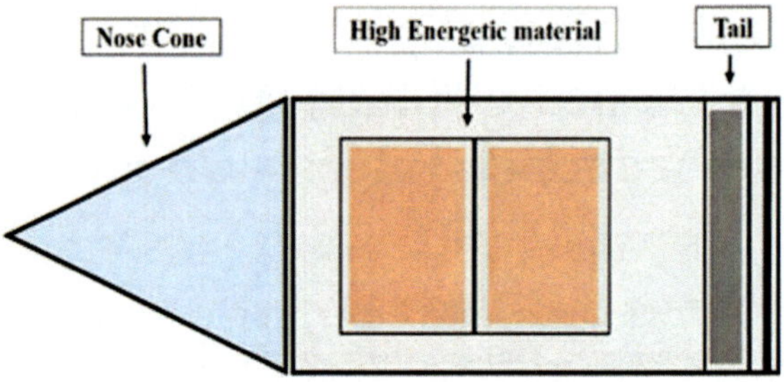

Fig. 1 Conico-cylindrical projectile

1 Introduction

The penetration of projectile into concrete target is current area of interest for both defence scientists and researchers to provide better and safe kill mechanisms as well as for civilians to save structure from impact. The penetration process of a projectile subjected to high-velocity impact is highly complex, and most of the research in this area is experimental due to its interdisciplinary aspect. Due to the high cost of experimentation, many investigators and researchers have been compelled to rely on simple analytical and engineering models that can be iterated through impact simulations in explicit solvers of hydrocodes like AUTODYN. Many strategies can be used to categorize the phenomena of projectile impact, such as the angle of incidence of the projectile, the material characteristics of the projectile, as well as the target and range of initial velocities. Velocities lower than 500 m/s can be categorized as sub-ordnance velocities (25–500 m/s).

[1]In this study, a projectile as shown in Fig. 1 is analysed at a terminal velocity of 300 m/s just before impact. To serve the intended function, this conico-cylindrical projectile contains various electronics packages and high energetic material assemblies. Electronics packages are primarily integrated into the tail section or aft end of projectiles, with high energetic material occupying the mid-section. Shock mitigations are primarily necessary in these two locations (projectile tail portion and mid-section), so that the functionality of these packages is not affected during impact. Electronic circuitry serves as the brain of modern ammunition.

In impact circumstances, some of these subsystems or components are susceptible to the initial high-shock levels. During impact, electronic components in weapons armaments are subjected to shock loads of up to 15,000 g. Solder joints in electronic components suffer from voids and cracks as a result of such severe loading. After multiple investigations, it has been reported that if no energy dissipation measures are used, high-deceleration or acceleration shock pulses impair the functionality of electronic subsystems [1].

Such projectiles are sometimes meant to impact concrete surfaces in order to destroy the target, although they may hit hard surfaces by accident, obtaining a terminal velocity of 300 m per second. Due to accidental impact, shock waves are

[1] Mr. Ankur Trigunayak carried this work as part of his master's thesis at IIT Delhi.

generated as projectile carries high energetic material inside, which is shock/impact sensitive material. Shock can cause detonation of high energetic material, resulting in unintentional damage at an undesirable location. Deceleration of up to 50,000 g is experienced as a result of the projectile's impact/penetration in the target [2]. The electronic package and its intended function may be compromised as a result of the high deceleration. Moxley et al. [3] had modelled the impact of a scaled bullet on a concrete target analytically. However, computer simulations using FE simulations are necessary to obtain thorough understanding of deformations in a 3D projectile including high-energy materials inside. Jones (1980) [4] studied impact of regular geometrical structures. These theories have been used as a preliminary step for a variety of impact studies in software packages such as AUTODYN, ABAQUS EXPLICIT, and LS-DYNA, amongst others.

For calculation of deceleration at 300 m/s velocity impact case, impact simulations were performed in AUTODYN explicit solver for 125 mm diameter projectile impact with 35 MPa concrete, a 16,000 g shock peak was observed for 6 ms, shown in Fig. 2. This level of deceleration pulse, which occurs for short periods of time, transfers a very high level of shock to the mid-section, which contains a large amount of high-energy material of the projectile. It is necessary to reduce the level of this shock to provide safe impact mitigation mechanism. In presented work, a number of different mitigating strategies have been suggested. These projectiles are put into motion according to a set of commands, resulting in a terminal velocity of 300 m per second.

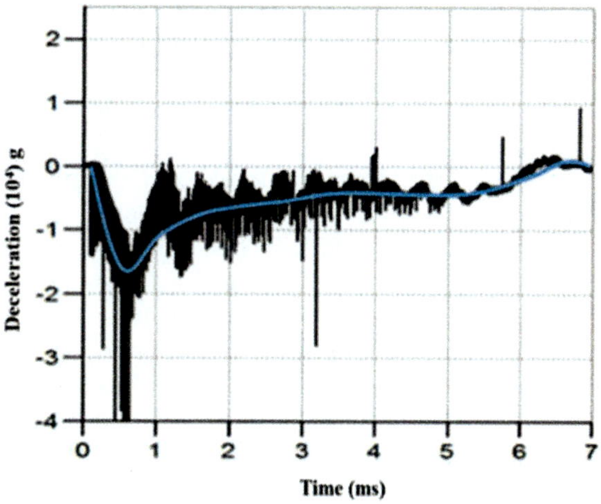

Fig. 2 Computed deceleration vs time profile (Ansys/Autodyn) of projectile impact

2 Literature Review

Several energy-absorbing materials are being used to convert the kinetic energy of the projectile impact to other forms of energy, resulting in a safe deceleration of the structure and its occupants. In current work, projectile is considered as a vehicle, with high-energy materials and electronics serving as its occupants, despite the fact that velocity of impact is considerably higher.

In literature, four techniques are suggested to mitigate mechanical shock in the Printed Circuit Board (PCBs) of Electronic packages [5]. Structural support may be provided to electronic components by encapsulating with potting material which dampens the shock levels. During shock trials, bilayer potting was found to be the most effective method, in which an inner layer of hard materials such as epoxies and polyester keep all electronic components intact with the PCB whilst the outer layers absorb the amplitude of the shock pulse and are composed of comparatively soft materials such as neoprene and silicones [6]. Underfills encapsulation enables load sharing with the solder and protection from shock. This careful design of the load path from the external shell of an electronic product to the internal electronic packages mitigates shock level in which, active or passive isolators may be used. Components should be oriented in such a way that the PCB supports the component-PCB joint during shock load.

Various mechanical devices and elements, made of metals, polymers, and composite materials, are designed to absorb impact energy under different loading conditions. An important requirement is that, in the event of collision, these structural members must be able to dissipate large amounts of energy by controlled/progressive collapse. Dissipated energy basically depends on deformation pattern of structural components. The energy-absorbing capacity differs from one part or component to others in a manner that depends on the mode of deformation involved and the material used [7].

Lu et al. [8] demonstrate in detail about energy absorption in structures and in materials. They have explored the methodology for analysing energy absorbing capacities of materials under dynamic loading, resulting in large deformation in their work. They demonstrated ring system, laterally constrained tubes, and thin-walled members under transverse loading. It can be concluded from their book that axial crushing, local deformation, axial splitting, and tube inversion are the various modes where energy can be absorbed by structures for higher velocity impact cases. Olabi et al. [9] give a description of metallic tube type energy absorbers, in which he mostly used steel and aluminium tubes for energy absorption and illustrated modes of deformation such as axial and lateral compression using frusta. These metallic tubes can be used to mitigate effect of impact mitigation. As roles of cylindrical tubes were found very important in crashworthiness, Wang et al. [10] have conducted several experiments and FE simulations to predict the mechanism of deformation as progressive collapse or dynamic plastic buckling. They have demonstrated higher velocity impact cases, in which shell wall thickening or mushrooming phenomenon are observed which may lead to dynamic tensile fracture. Alghamdi et al. [11] have reviewed the common shapes of collapsible energy absorbers and the different modes of deformation of the most common ones. Common shapes include circular tubes, square tubes, frusta, struts, honeycombs, and sandwich plates. Common modes of deformation for circular tubes include axial crushing, lateral indentation, lateral flattening, inversion, and splitting. Alghamdi et al. [11] observed that load deflection curve of the conical frusta is more uniform than cylindrical and thick frusta and these can be deformed in axisymmetric manner.

The need for impact mitigation is crucial in society due to the possibility of significant loss of life and property damage in the event of an explosion, whether caused by natural disasters, industrial accidents, or malicious acts such as terrorism. The potential for widespread destruction and loss of life from such incidents emphasizes the importance of implementing measures to mitigate their impact. There are only a few literature resources available that discuss effective mitigation strategies for sub-velocity impact cases. Additionally, there is a lack of literature that addresses the impact mitigation techniques for accidental blasts from high-energy projectiles within the sub-velocity range. The current study is driven by the need to overcome the limitations of existing mitigation strategies by enhancing the crashworthiness of the projectile and implementing better, safer impact mitigation mechanisms that can protect structures from accidental impacts.

3 Finite Element Study

In this study, Finite Element Analysis (FEA) is conducted using ANSYS/AUTODYN. The equations of motion are integrated with time explicitly using central differences. For stable solutions, the method requires very small timesteps. As a result, it is particularly well suited for impact and crash simulations, and the governing equations do not require global matrix assembly or inversion. Using explicit analysis in the ANSYS/Autodyn hydrocode, a 3-Dimensional computational model of the projectile and target is generated and analysed. In explicit analysis, role of time step is very crucial as if the time step is very small (especially in severely distorted areas), as a result, the computational time increases significantly. The numerical erosion technique is employed to deal with these significantly deformed zones. Deformation and stress propagation pattern are studied for the impact of the projectile with various targets like rigid surface, sand, concrete, and hard soil. To define the material behaviour for large deformation impacts, such as shock situations, precise selection of Equation of State (EOS), strength, and failure models should be made, taking into account a variety of available models. To simulate this millisecond impact phenomenon, Mie-Gruenisen Shock-EOS is selected, strength model for the plastic deformation in ductile materials of projectile is defined by Johnson–Cook Model and failure criteria is defined for failure strain by J-C damage model in ANSYS/AUTODYN. To define target as semi-infinite hard soil, compaction model is chosen. The detailed methodology to simulate sub-ordnance velocity impact of conico-cylindrical projectile with semi-infinite target has been demonstrated elsewhere by

Trigunayak et al. [12]. Each of these simulations was carried out using ANSYS/ AUTODYN® commercial license (version 14.0).

3.1 Simulation Setup

The solid geometry of the structure was created using Solidworks®. Once the assembly was complete, the geometry was saved in a portable format like .iges or .step. To ensure a smooth import process, the number of edges, faces, and holes were minimized before importing any free-form geometry. After importing the geometry into ANSYS with design modeller, the whole geometry was reviewed and cleaned up. The current impact analysis employed Ansys/Autodyn hydrocodes, which are explicit finite element codes utilizing a combined Lagrangian and Eulerian formulation.

3.2 Material and Material Model

The objective of constitutive model is to simulate the strength behaviour of materials subjected to large strains, high-strain rates, and high temperatures. Such behaviour may occur in situations involving extreme impulsive loading caused by high-velocity impact and explosive detonation, as explained by Johnson and Cook, 1985, 1983 [13]. The yield stress Y is defined by Eq. (1)

$$Y = \left[A + B\varepsilon_p^n\right]\left[1 + c\log\varepsilon_p^*\right]\left[1 - T_H^m\right] \tag{1}$$

where

ε_p = effective plastic strain.

$\varepsilon_p{}^*$ = normalized effective plastic strain rate.

T_H = homologous temperature = $(T - T_{\text{room}})/(T_{\text{melt}} - T_{\text{room}})$.

The five material constants are A, B, C, n and m.

The expression in the first set of brackets gives the stress as a function of strain when $\varepsilon_p{}^* = 1.0\ \text{s}^{-1}$ and $T_H = 0$ (i.e. for laboratory experiments at room temperature). The constant A represents the yield stress at low stresses, whilst B and n reflect strain hardening. The formulae within the second and third sets of brackets represent, respectively, the effects of strain rate and temperature. Specifically, the latter relationship represents the thermal softening so that the yield stress becomes zero at the melting temperature T_{melt}.

In addition to stress triaxiality, Johnson and Cook proposed the failure criterion by adding the effects of strain route, strain rate, and temperature into the fracture strain expression. The fracture criteria are based on the damage evolution, in which it is considered that material damage occurs when the damage parameter in Eq. (2), $\omega \geq$;

$$\omega = \sum\left(\frac{\Delta\varepsilon^{pl}}{\varepsilon_f^{pl}}\right) \quad (2)$$

where $\Delta\varepsilon^{pl}$ is an increment of the equivalent plastic strain, ε_f^{pl} is the strain at failure, and the summation is performed over all the increments throughout the analysis. The strain at failure ε_f^{pl} is assumed to be dependent on a non-dimensional plastic strain rate,$\dot{\varepsilon}_f^{pl}$, a dimensionless pressure-deviatoric stress ratio, $\frac{\sigma_m}{\sigma}$ (where σ_m is the mean stress and σ is the equivalent von Mises stress), and the non-dimensional temperature, T, defined earlier in the Johnson Cook hardening model. Failure criteria proposed can be shown as Eq. (3);

$$\varepsilon_f^{pl}\left(\frac{\sigma_m}{\sigma}, \dot{\varepsilon}_f^{pl}, T\right) = \left[D_1 + D_2 exp\left(D_3\frac{\sigma_m}{\sigma}\right)\right]\left[1 + D_4 \ln\left(\frac{\dot{\varepsilon}^{pl}}{\dot{\varepsilon}_0}\right)\right][1 + D_5 T] \quad (3)$$

Here D_1, D_2, D_3, D_4 are the damage coefficients.

For the explosive, the pressure as a function of relative volume, V, and internal energy per unit volume, E can be defined by using the Jones-Wilkins-Lee (JWL) [14] equation of state given by;

$$\boldsymbol{P} = \boldsymbol{A}\left(1 - \frac{\boldsymbol{\omega}}{\boldsymbol{R}_1\boldsymbol{V}}\right)\boldsymbol{e}^{-\boldsymbol{R}_1\boldsymbol{V}} + \boldsymbol{B}\left(1 - \frac{\boldsymbol{\omega}}{\boldsymbol{R}_2\boldsymbol{V}}\right)\boldsymbol{e}^{-\boldsymbol{R}_2\boldsymbol{V}} + \frac{\boldsymbol{\omega}}{\boldsymbol{V}}\boldsymbol{E}$$

where A, B, $R1$, and $R2$ are constants that depend upon the explosive characteristics used.

The model for the target block included sand, concrete, and hard soil. Laine [15] considered the compaction model effective for predicting pressure wave propagation and attenuation since it employs the density dependency on wave velocity. Plastic compaction curve defines compaction EOS. Material and Material model for different part of the assembly are mentioned in Table 1.

Table 1 Material and material model

Component	Material	Material model
Nose section (conical)	Steel 1006	Johnson–Cook [13]
Mid-section (Cylindrical casting)	Steel 1006	Johnson–Cook [13]
Casted explosive	COMP-B	JWL [14]
Explosive casting	AL 2024-T351	Johnson–Cook [13]
Tail section	Steel 4340	Johnson–Cook [13]
Target	Sand	Compaction [15]

3.3 Initial and Boundary Conditions

In Autodyn software, the projectile was provided an initial velocity input. The part was then filled with the previously defined material card and velocity initial conditions. Because the target was semi-infinite, the initial velocity was set to be zero, and the individual faces were assumed to be fixed [12].

3.4 Contact and Interactions

In this simulation, the Lagrangian domain is defined as the domain of particles or objects that move with the flow, whilst the Eulerian domain is defined as the domain of fluid or gas that is modelled as a continuous medium. The interaction between Lagrangian domains is modelled using a bonded contact algorithm, which applies a penalty-based approach to enforce contact constraints between particles or objects.

On the other hand, the interaction between Eulerian and Lagrangian domains is modelled using a fluid–structure interaction coupling algorithm, which allows for the exchange of momentum and energy between the fluid and the particles or objects [16]. Implementing Euler–Lagrange and Lagrange-Lagrange interactions in FEM generally entails using a Lagrangian or Eulerian formulation to derive the equations of motion for each particle or system of particles, and discretizing these equations using FEM to obtain a set of equations that describe the behaviour of each finite element.

4 Results and Discussion

4.1 Redesigning of Nose Section of Projectile

Three iterative changes were tried in nose cone configuration and then simulation results were compared to the initial configuration. In initial configuration, the cone shell thickness was 2 mm and cylinder thickness was 10 mm. The first trial simulation was performed by increasing the shell thickness of cone from 2 to 3 mm and cylinder thickness from 10 to 12.5 mm.

Figure 3 shows comparison of change in energy fluence through high energetic material amongst initial projectile configuration at 200 and 300 m/s velocity impact cases and after changed thickness in projectile with 300 m/s velocity. In initial configuration case, when impact happens at 300 m/s, 150 kJ of energy was available at the high energetic material front, which was found to be sufficient crucial for this large surface area of high energetic material. Critical energy fluence values were found to be 201.8 J/cm^2 for high energetic materials like COMP-B [17] and 0.06 MJ/m^2 for PETN [18].

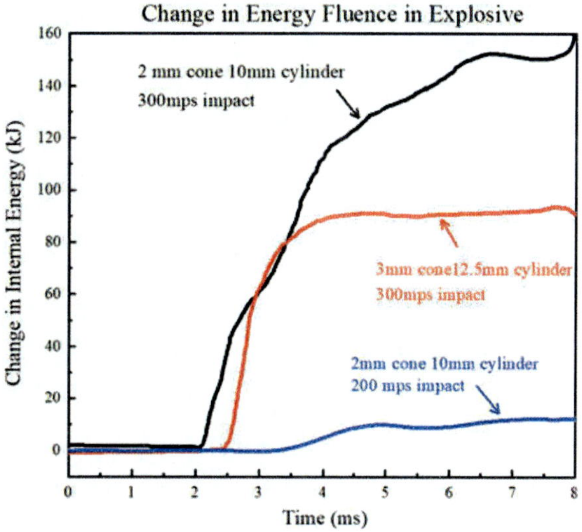

Fig. 3 Comparison in energy fluence in high energetic material by varying shell thickness

4.2 Effect of Collapsible Rods as Energy Absorber

As energy absorbers, circular metallic rods can be employed [11]. As a result, in this iteration, the energy absorbing rods are directly attached to the bulkhead of the nose-cylinder assembly. These circular hollow rods are attached to the middle nose plates to provide the nose cone more structural strength. A total of 12 circular rods are attached and evenly distributed. Each round rod has a 20 mm outside diameter and a 10 mm inner diameter. Under compression, these rods buckle and bend, enabling them to collapse like struts.

Figure 4a shows solid model of modified configuration of cone with 12 rods and Fig. 4b shows cone without collapsible rod. FE model was prepared for cone with collapsible rod and impact simulation was performed with similar approach as earlier in ANSYS/ AUTODYN. Figure 5a shows deformation pattern of modified nose cone (with collapsible rods) at different time events.

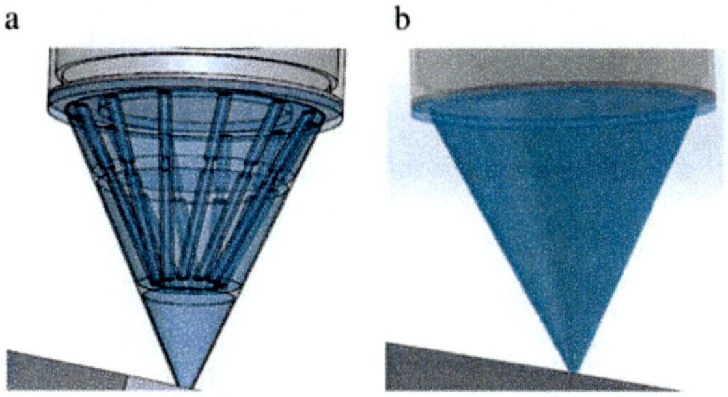

Fig. 4 **a** Cone with collapsible rod **b** Cone without collapsible rod

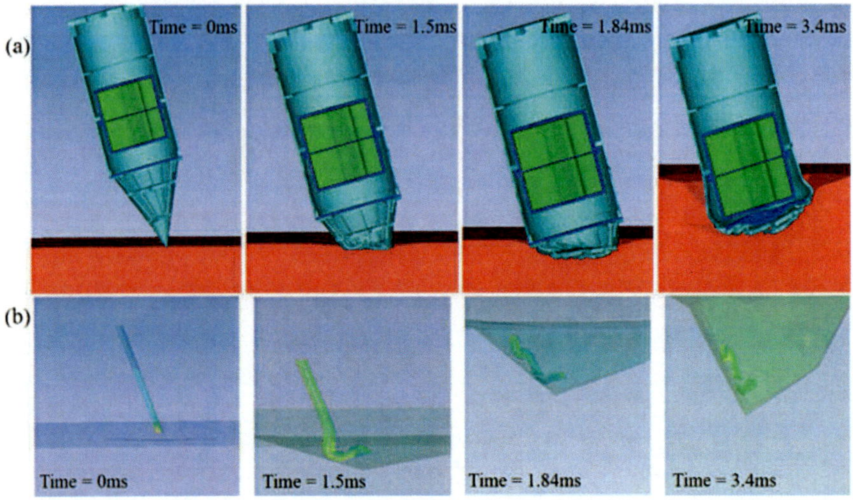

Fig. 5 **a** Events of deformation of projectile with collapsible rods in cone **b** Events of deformation of single collapsible rod

It has been observed that when a compressive stress wave occurs, the hollow rod strut first buckles, and then some of the rod come to rest with soil following penetration. Deformation of a single collapsible rod with time is shown in Fig. 5b. Although plastic deformation occurred until the fracture in the rod, the moving portion of the rod deformed again by buckling in the remaining rod. So, it starts to collapse progressively in 'shell mode', and then it begins to buckle in global bending mode, which may be considered as 'Rod Mode'. This transition in axial compression is described in [19]. In post-processing, the energy absorbed by the modified nose cone is computed using plastic work done and plotted as internal energy. Figure 6 illustrates the absorbed energy in the plastic deformation of a modified cone with 12 collapsible rods in contrast to the initial design (i.e. without collapsible rods).

As previously stated, the cone in initial design absorbs approximately 174 kJ energy in 2 ms in a 300 m/s velocity impact event. The combined plastic deformation of collapsible rods and cone absorbs 656 kJ energy in the modified cone arrangement, see Fig. 6. Each rod absorbs about 40 kJ of additional energy through plastic deformation. Total KE at initial point of impact is found to be 13,176 kJ, indicating that this modified nose cone case absorbs 5.97% of total K.E., compared to only 1.47% in the initial configuration, which is a cone without collapsible rod. With a collapsible cone configuration, the energy absorbed in the cone is nearly four times higher. Energy flow through high energetic material was computed by post-processing history of COMP-B (high energetic material). After combined deformation in rods and cones, the stress wave reached high energetic material assembly, and strains contributed to an increase in the material's internal energy. Figure 7 depicts the pressure–time history of high energetic material inside the projectile computed in ANSYS/AUTODYN for both with and without collapsible rod structure. When

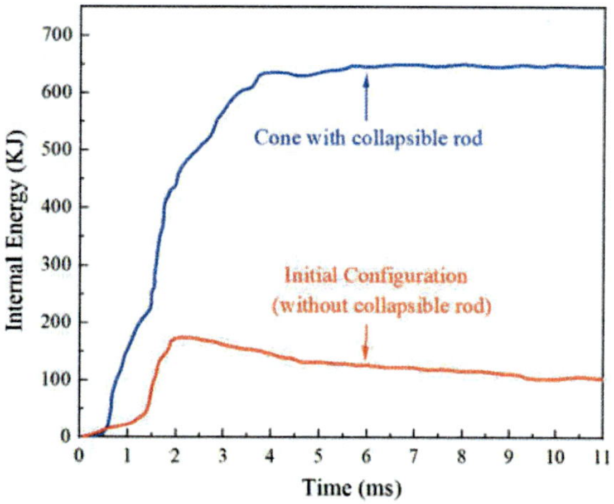

Fig. 6 Energy absorbed in nose cone

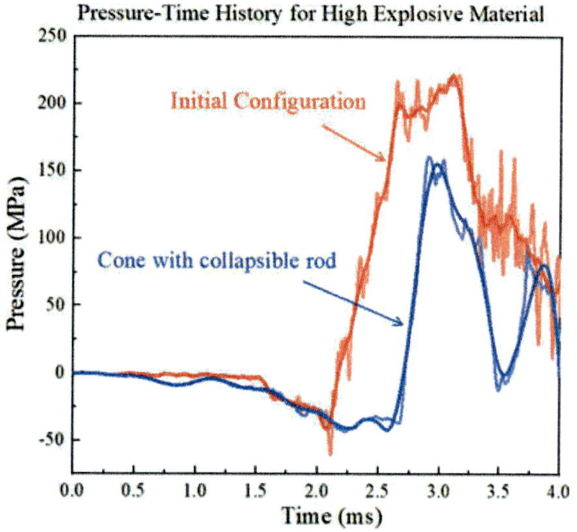

Fig. 7 Pressure versus time history for high energetic material

compared to a cone without a collapsible rod, a cone with a collapsible rod transmits 25% less pressure for a shorter period of time to the energetic material.

4.3 *Effect of Bimetallic Rod*

As in the previous case, hollow metallic rods have been used as energy absorbers in the nose cone. Another hypothesis is the influence of two materials that can exhibit strength and hydrodynamic behaviour at the same time. Hsu and Jones [20] investigated response of three materials, stainless steel, mild steel, and aluminum alloy subjected to dynamic axial loading. According to their findings, stainless steel tubes absorb highest energy per unit volume, whereas aluminum has the highest energy-absorbing efficiency. Bimetallic rods of the same outer dimension of 20 mm diameter were modelled for the following trial, and inside Al was filled with a small opening towards the rear end to flow out aluminum material in case of large deformation and temperature increases over melting point (see Fig. 8). So, it was postulated that if the interior material changes from solid to liquid, the energy-absorbing efficiency would increase. The Al material inside rods was grouped for Euler–Lagrange interactions with other materials whilst defining the algorithm for impact simulation, whereas all other components were defined as Lagrange-Lagrange interactions. A 300 m/s velocity impact with a hard solid was simulated. In comparison to hollow collapsible rods, which dissipated 656 kJ, the modified cone with bimetallic rods dissipated 850 kJ in the form of plastic deformation (as shown in Fig. 9).

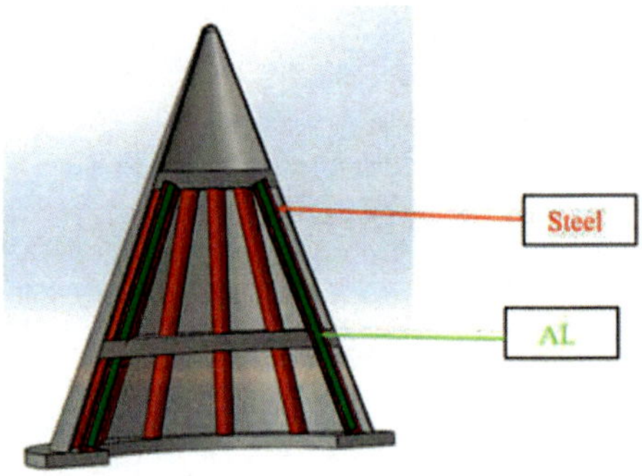

Fig. 8 Bimetallic rods in nose cone

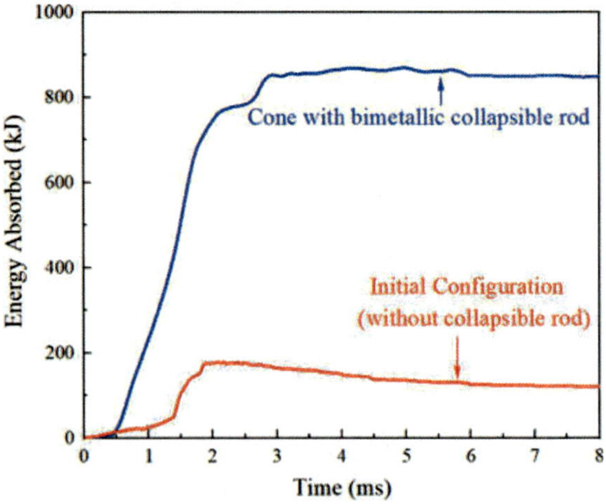

Fig. 9 Energy absorbed in plastic deformation of modified cone

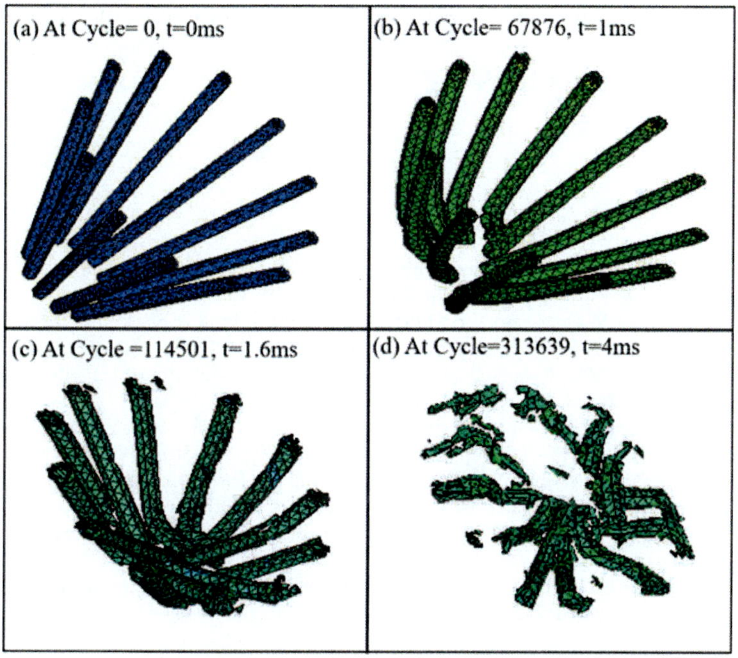

Fig. 10 Deformation pattern of bimetallic rods

The deformation pattern of bimetallic rods during the impact event is shown in Fig. 10. The events commence with the buckling of the Bi-metallic rod and progress with material flow until the Bi-metallic rod completely collapses.

5 Conclusion

In the presented work, three different structural configurations inside projectile nose cone were analysed using finite element simulations for energy absorption and impact mitigation. In the first scheme, thicknesses of nose cone and of the cylinder were increased to 3 mm and 12.5 mm, respectively. The second and third schemes proposed twelve evenly spaced collapsible circular rods with and without metallic infill, respectively. These three configurations were compared to the baseline model (initial configuration) in terms of energy dissipation and deformation mode. When appropriately integrated with the structure, the addition of hollow cylindrical rods to the nose cone absorbs 482 kJ more energy as compared to the initial configuration. Bimetallic rods were found to absorb 30% more energy compared to hollow collapsible rod configuration. These impact mitigation measures have been found to be more suitable for impact energy dissipation. Effect of mass of these rods on projectile's aerodynamics will require a separate study to be done in future. Plastic buckling of collapsible rods or bi-metallic rods absorbs or dissipates energy, making the projectile to be more crashworthy and offering better and safer impact mitigation mechanisms to protect the structure from accidental impact.

References

1. Singh A et al (2021) Strain hardening and shock mitigation response of polyurethane under high strain rates. AIP Adv 11(11):115306. https://doi.org/10.1063/5.0068403
2. Zhang W, Zu J, Li Y, Qin L, Song S (1995) The intelligent missile black-box. IEEE Trans Instrum Meas 44(3):824–827. https://doi.org/10.1109/19.387342
3. Moxley RE, Adley MD, Rohani B (1995) Impact of thin-walled projectiles with concrete targets. Shock Vib 2(5):355–364. https://doi.org/10.1155/1995/861749
4. Jones N (2003) Several phenomena in structural impact and structural crashworthiness. Europ J Mech A-solids 22(5):693–707. https://doi.org/10.1016/s0997-7538(03)00077-9
5. Berman MS (2006) Electronic components for high-g hardened packaging. Weapons and Materials Research Directorate, ARL, Army Research Laboratory ARL-TR-3705, Jan 2006. [Online]. Available: https://citeseerx.ist.psu.edu/viewdoc/download?doi=10.1.1.101.9472&rep=rep1&type=pdf
6. Meguid SA, Zhuo C, Yang F (2014) Effective mitigation of shock loads in embedded electronic packaging using bilayered potting materials. J Electron Packag 136(4):041010. https://doi.org/10.1115/1.4026542
7. Mamalis AG, Robinson M, Manolakos DE, Demosthenous GA, Ioannidis MB, Carruthers J (1997) Crashworthy capability of composite material structures. Compos Struct 37(2):109–134. https://doi.org/10.1016/S0263-8223(97)80005-0
8. Lu G, Yu TX (2003) Energy absorption of structures and materials. Boca Raton, Cambridge, England: CRC Press , Woodhead Pub., 2003. Accessed 06 May 2022. [Online]. Available: http://www.crcnetbase.com/isbn/9780203484128
9. Olabi AG, Morris E, Hashmi MSJ (2007) Metallic tube type energy absorbers: a synopsis. Thin-Walled Struct 45(7–8):706–726. https://doi.org/10.1016/j.tws.2007.05.003
10. Wang B, Lu G (2002) Mushrooming of circular tubes under dynamic axial loading. Thin-Walled Struct 40(2):167–182. https://doi.org/10.1016/S0263-8231(01)00057-X
11. Alghamdi AAA (2001) Collapsible impact energy absorbers: an overview. Thin-Walled Struct 39(2):189–213. https://doi.org/10.1016/S0263-8231(00)00048-3
12. Trigunayak A, Dubey DK, Chawla A (2017) Computational modelling of sub-ordnance velocity impact of conico-cylindrical projectile with semi-infinite target. Proc Eng 173:339–346. https://doi.org/10.1016/j.proeng.2016.12.027
13. Johnson GR, Cook W (1983) A constitutive model and data for materials subjected to large strains, high strain rates, and high temperatures, pp 541–547
14. Dobratz BM (1985) LLNL explosives handbook, properties of chemical explosives and explosive simulants
15. Laine L, Sandvik A (2001) Derivation of mechanical properties for sand. In: Proceedings of the 4th Asia-Pacific conference on shock and impact loads on structures, CI-Premier PTE LTD, Singapore, ANSYS Inc.., Jan. 2001. vol 361. pp 368
16. Birnbaum NK, Francis NJ, Gerber BI (1999) Coupled techniques for the simulation of fluid-structure and impact problems. Comput Assisted Mech Eng Sci 295–311
17. Yadav HS (2006) Initiation of detonation in explosives by impact of projectiles. Defence Sci J 56(2):169–177
18. Cooper PW (2022) Explosives engineering. 2018. Accessed 06 May 2022. [Online]. Available http://rbdigital.rbdigital.com

19. Karagiozova D, Alves M (2004) Transition from progressive buckling to global bending of circular shells under axial impact—Part II: theoretical analysis. Int J Solids Struct 41(5–6):1581–1604. https://doi.org/10.1016/j.ijsolstr.2003.10.006
20. Hsu SS, Jones N (2004) Quasi-static and dynamic axial crushing of thin-walled circular stainless steel, mild steel and aluminium alloy tubes. Int J Crashworthiness 9(2):195–217. https://doi.org/10.1533/ijcr.2004.0282

Impact Analysis of Uncontained Engine Rotor Debris on Rotorcraft Structure

Ramkumar Kaushik, Kalinga Gulbarga, M. Ramesh Babu, and Dineshkumar Harursampath

Abstract Engine failure hazard effect analysis has to be demonstrated by all rotorcraft manufacturers. To meet the safety objective, the hazardous effects of uncontained rotor burst debris on the aircraft structure and occupants are evaluated. A qualitative assessment for each critical system/component is studied for uncontained rotor burst fragments including translational and spread risk angles using the geometrical method as per AC 20-128A. In this paper, the critical power turbine uncontained disc fragment is considered for hazard assessment as per FAR/CS 29.903(d). The LS-Dyna nonlinear transient code is used as a part of risk analysis to show airworthiness compliance. A simulation model using LS-Dyna is developed for structural elements like the engine deck, instrument panel, and floorboard with detailed FE idealization and contact definitions to accurately predict debris penetration through structural elements. The energy absorption in each structural member is calculated.

Keywords Rotorcraft · Hazardous · Rotor debris · LS-Dyna · FAR 29 · AC 20-128A · Energy absorption

1 Introduction

Uncontained gas turbine engine disk failures have been a direct hazard to the aircraft, crew, and passengers because high-energy disk fragments can penetrate and cause damage to the cabin, fuel tanks, flight control surfaces, and hydraulic lines. Hence, design precautions ensure that uncontained fragments are in the safe trajectory envelope, not because catastrophic failure of aircraft and sufficient structural/system redundancy is maintained. Engine rotor failures occur due to high speeds/jamming of gear, high resonant vibrations, material defects, misalignments, over temperatures,

R. Kaushik (✉) · K. Gulbarga · M. Ramesh Babu
RWRDC, HAL, Bangalore 560017, India
e-mail: ramkaushik3@gmail.com

R. Kaushik · D. Harursampath
Department of Aerospace Engineering, Indian Institute of Science, Bangalore 560012, India

R. Velmurugan et al. (eds.), *Dynamic Behavior of Soft and Hard Materials Volume 1*, Springer Proceedings in Materials 34, https://doi.org/10.1007/978-981-99-6030-9_28

fretting fatigue, and foreign object damage. Federal Aviation Regulation 29.903(d) with AC 20-128A states design precautions must be taken to minimize the hazards to the rotorcraft in the event of an engine rotor failure [1, 2]. The typical rotorcraft engine consists of an axial compressor, centrifugal compressor, gas generator turbine, power turbine, and cooling fan. In the majority of cases, the power turbine disc failure causes catastrophic hazards [3]. Fengchan et al. [4] studied the cause of uncontained turbine engine rotor failure, and the hazard validation method, and presented the airworthiness procedure. Carney et al. [5] worked on the risk of fragment penetration outside the engine casing studied by testing and nonlinear FE analysis by using alternate geometry casing configurations, correlating tests, and FE results. Senthil et al. [6, 7] performed the nonlinear explicit dynamics software ABAQUS to simulate perforation, impact forces, and correlation of analysis and test results. The accurate FE modeling of honeycomb core with face sheets and face sheet-to-core debonding phenomena were also investigated using the LS-Dyna code [8–10].

In this paper, the uncontained power turbine disc fragment is modeled in LS-Dyna software with a 1/3rd disc fragment and 1/3rd bladed disc mass with a spread angle of $\pm$ 3° as per AC 20-128A. The critical structural parts like the engine deck, composite panels, and floorboard are accurately idealized with appropriate material properties assigned and contact definitions between parts. The impact simulation has been carried out and energy absorption is calculated for each structural part.

2 Uncontained Power Turbine Fragment Model

The following methodology was adopted for uncontained power turbine disc fragment analysis as schematically shown in Fig. 1. The risk envelope zone was arrived at based on engine integration and its turbine fragment projectile path. A projectile disc fragment model was made as per standard. Residual energy assessment and design mitigation plan were carried out for each component.

2.1 Engine Power Turbine Fragment Hazard Zone

The schematic diagram of the engine power turbine fragment hazard zone is shown in Fig. 2. The critical structural parts impacted by the power turbine fragment trajectory path and spread angle with $\pm$ 3° include the engine deck, instrument panel, and floorboard. The fragment spread angle is the angle measured fore and aft, from the center of the plane of rotation at the engine shaft centerline. It is assumed that failure, which produced blade fragments, occurred when the engines were operating at their design rotation per minute. The highest fragment energy condition is reached during the take-off thrust of the rotorcraft. Most of the engine containment literature concludes that fragments, especially the heavy fragments, exit the engine in the plane of disc rotation and have a rotation about the fragment center of gravity. Since the

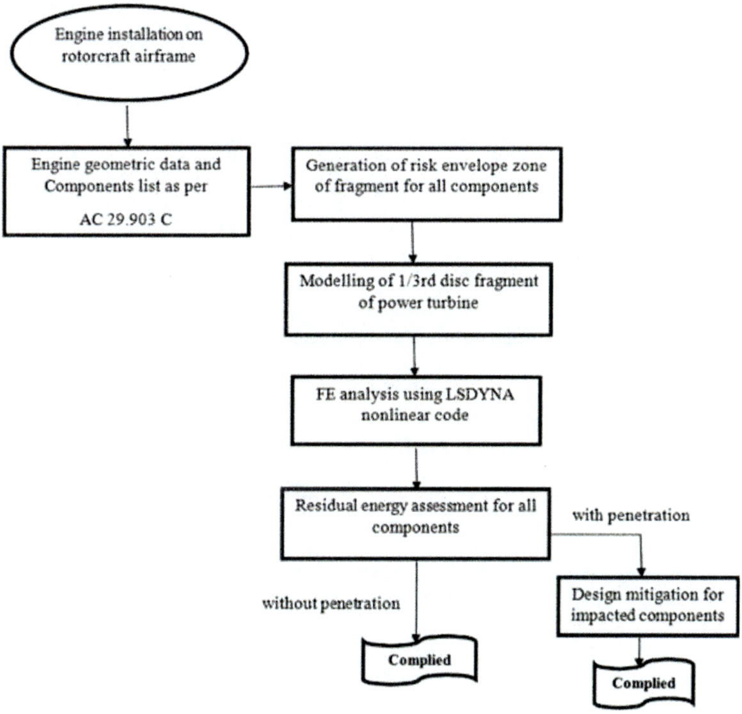

Fig. 1 Flow chart for uncontained disc fragment analysis

rotational energy of the fragment is low when compared to the translational energy, in the present analysis, it is assumed that all the fragment energy is considered only translation.

2.2 *Uncontained Power Turbine Fragment Model*

In the present analysis, it is assumed that the one-third disc fragment has the maximum dimension corresponding to one-third of the disc with one-third blade height and a fragment spread angle of ± 3° is considered. The detailed disc fragment and model description as per AC 20-128A are shown in Fig. 3a, b, respectively. The details of the fragment's mass, spread angle, and dimension are given as shown in Table 1.

The Power turbine is one-third of the disc fragment model shown in Fig. 4. The shape of the disc fragment is a piece of a circular disc. Having a rigid body mass of 1.323 kg, a spread angle of ± 3° and an initial fragment kinetic energy of 29.175 kJ are considered for the analysis. The fragment velocity at the time of failure is calculated from the below-mentioned equation.

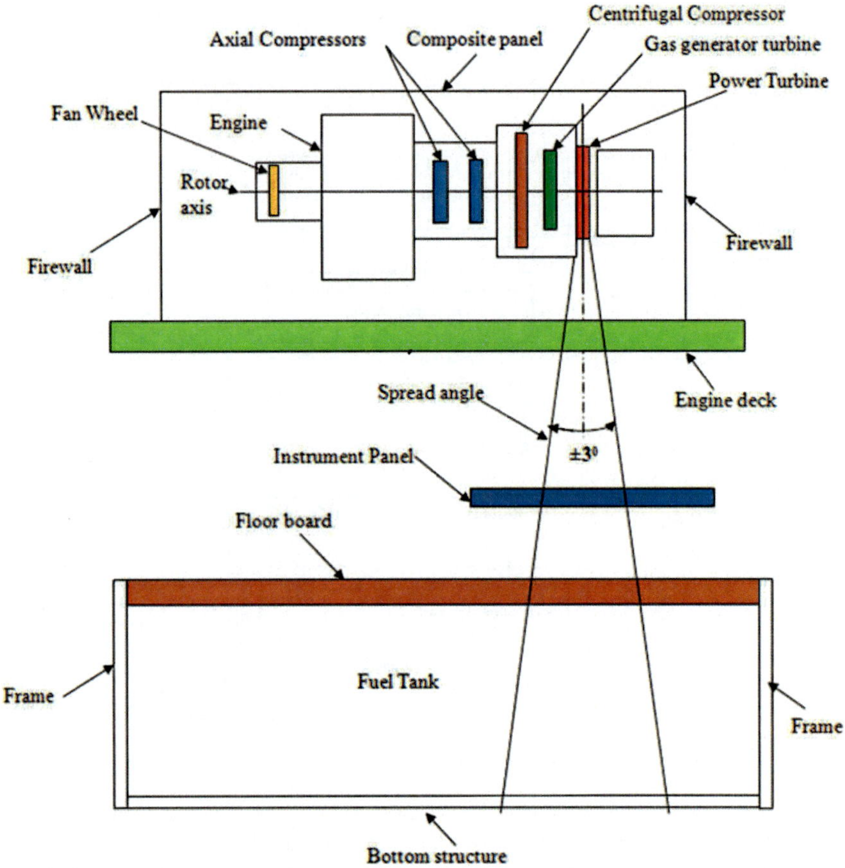

Fig. 2 Engine power turbine fragment impact zone

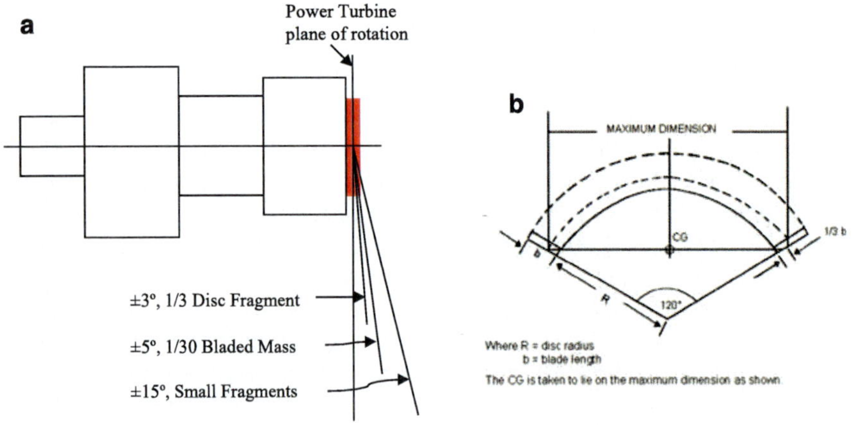

Fig. 3 **a** The detailed fragment description. **b** The disc fragment model description

Table 1 Details of uncontained turbine disc blade

Type	Maximum dimension	Mass	Spread angle
1/3 disc fragment	1/3*b + R	1/3 bladed disc mass	± 3°
Intermediate fragment	1/3*(b + R)	1/30 bladed disc mass	± 5°
Small fragments	1/3*b	1/3 blade tip part	± 15°

Note b = blade length, R = disc radius

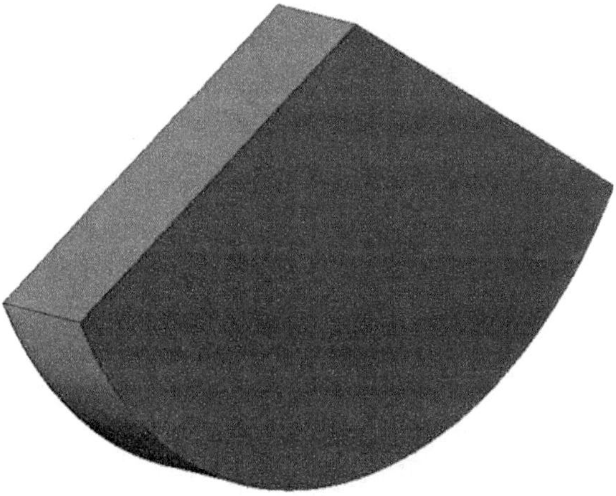

Fig. 4 Power turbine one-third of disc fragment model

$$E = \frac{1}{2}mv^2 = \frac{W}{2g}(2\pi rw)^2$$

where

v velocity at the fragment center of gravity considering the blades and disc serrations as a single mass, in m/s.
m Mass of the selected fragment, in kg.
g gravitational constant, in m/s^2
w speed in revolutions/second.
r radius from the center of rotation to the center of gravity of the fragment mass, in mm.
W weight of the selected fragment, in N.
E Fragment energy, in J.

3 Impact Analysis Using LS-Dyna Nonlinear Transient FE Code

LS-Dyna is a commercial nonlinear transient FE code widely used in the analysis of engine rotor burst containment applications. The rotor burst impact involves complex FE idealization, material modeling of metallic and composite airframe parts, and capabilities of LS-Dyna to model such impact events are used in the present analysis.

3.1 CAD Modeling

The uncontained power turbine disc fragment envelop zone was studied. The trajectory path of the fragment impacts first to the engine deck, instrument panel (composite sandwich), and floorboard (metal sandwich) consecutively. The detailed CAD model of structural parts has been made sequentially based on the location in the aircraft as shown in Fig. 5 and corresponding FE models were generated as shown in Fig. 6.

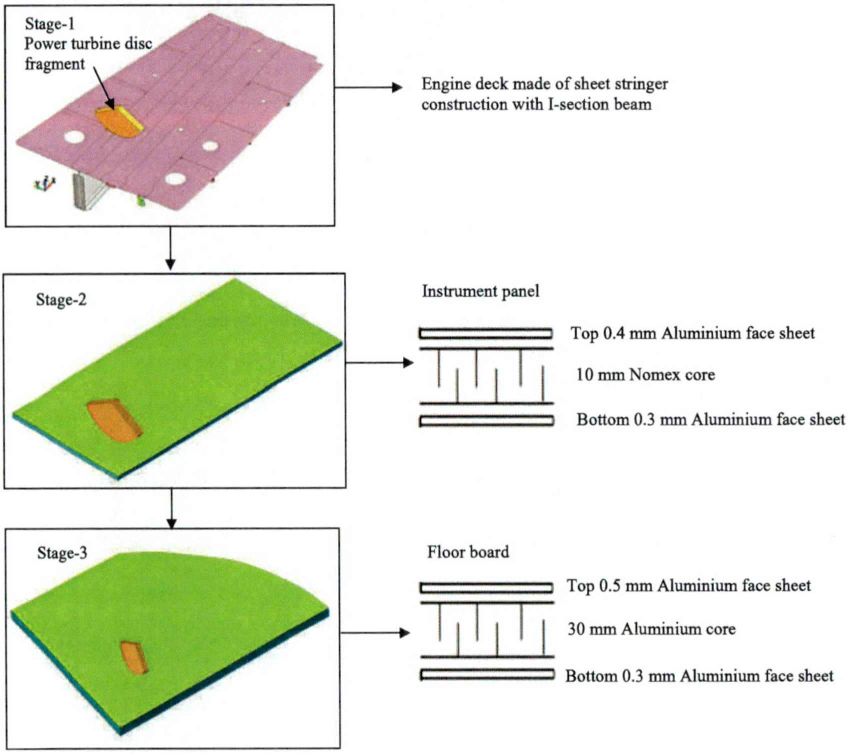

Fig. 5 Sequential fragment impact CAD model

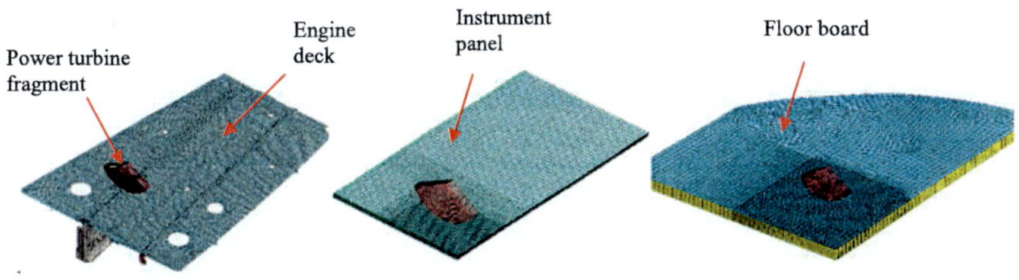

Fig. 6 FE models of structural components

3.1.1 Structure FEM Idealization

The structure parts namely engine deck sheet, sheet stringers, I-section, and sandwich face sheets are modeled with 2D shell elements using SECTION_SHELL card and sandwich cores are modeled with solid elements using SECTION_SOLID card. The rotor fragment is considered as a rigid body and meshed with 8 noded brick elements considering 1/3rd disc shape and is oriented to impact the structural target at $\pm 3°$ impact obliquity as per the spread angle defined.

3.1.2 Material Modeling

The engine deck sheet, sheet stringers, I-section beam, sandwich face sheets idealized as 2D Shell elements are assigned with the Johnson–Cook material model as a thermo-visco-plastic material coupled with a nonlinear equation of state and an accumulated damage evaluation algorithm for the numerical simulation. The rotor fragment modeled as 3D solid element is assigned using the 'MAT_RIGID' material card. The sandwich Nomex and aluminum core are modeled with MAT_ COMPOSITE_ FAILURE_SOLID_MODEL and properties assigned with reference to the material coordinate system.

3.1.3 Contact Definitions

The contact *CONTACT_AUTOMATIC_SINGLE_SURFACE is used between rotor fragment and structural parts where, in this contact, all the components are included in a single set for contact checking and sandwich panels a *CONTACT_ TIEBREAK_SURFACE_TO_SURFACE is defined between face sheet and core.

3.1.4 Boundary Conditions

The maximum ejection velocity of the power turbine disc fragment 210 m/s is given as the initial condition for rotor fragment in trajectory direction *INITIAL_VELOCITY card. The rotational velocity and temperature effects are not considered in this analysis. The boundaries of the target structures are constrained by translations and rotations.

3.1.5 FE Model Units

The standard LS-Dyna model units in N, mm, milli sec, kg/mm^3, and GPa are followed for defining the dimensions and properties.

4 Simulation Results

4.1 Engine Deck Assembly

The initial impact of the engine power turbine disc fragment of rigid body mass 1.323 kg, while impacting the engine deck with an initial velocity of 210 m/s is shown in Fig. 7. When the fragment just touches the top panel at the moment of 0.1 ms, the top panel along with the top flange of the I-section beam is subjected to shear failure under the impact of the fragment and there is plugging in the top panel along with beam flange around the fragment. At the same time, the fragment begins to shear the web of the I-section beam till 0.51 ms. The fragment plugs the bottom of flange beams at 0.86 ms and completely penetrates out of the structure. The fragment's initial and final kinetic energy after perforation through the engine deck assembly is 29.175 kJ and 25.265 kJ shown in Fig. 7. The energy absorption in the engine deck is 3.91 kJ. The energy plot of the fragment is shown in Fig. 8.

4.2 Instrument Panel

The exit velocity of the engine power turbine fragment of rigid body mass 1.323 kg perforating the engine deck assembly, impacting to the instrument panel with an initial velocity of 195.43 m/s is shown in Fig. 9. At the moment of 0.08 ms, the top aluminium face sheet is subjected to shear failure under the impact of the fragment as shown in Fig. 9. At the same time, the fragment begins to compress the nomex core. At the moment of 0.15 ms, the fragment continuously compresses the nomex core, and the bulge begins to appear on the backing surface of the nomex core. When the moment is 0.23 ms, the nomex core is completely penetrated. When the moment is 0.30 ms, the bottom aluminium face sheet is completely damaged with petals and the sandwich structure has been completely penetrated. The initial and then final kinetic

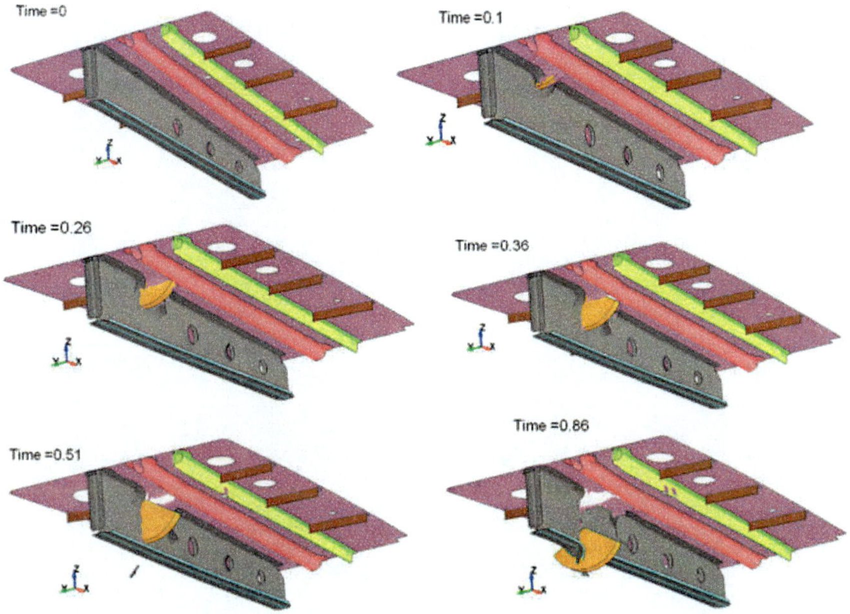

Fig. 7 Penetration of power turbine disc fragment through engine deck assembly

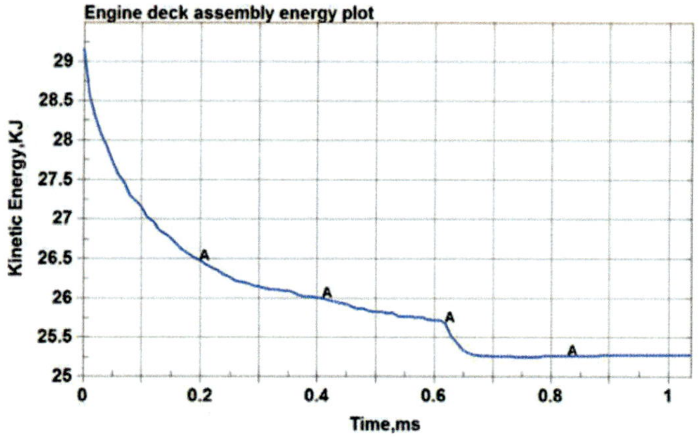

Fig. 8 Engine deck assembly energy plot

energy of the fragment after perforation through the instrument panel is 25.265 and 24.76 kJ. The energy absorption in the instrument panel is 0.505 kJ. The energy plot of the fragment is shown in Fig. 10.

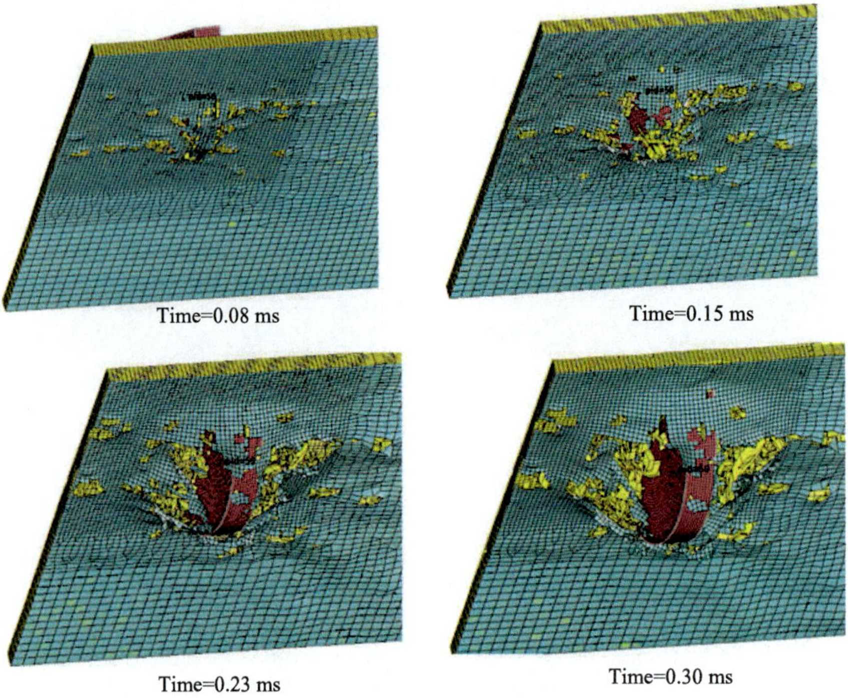

Fig. 9 Penetration of power turbine fragment through the instrument panel

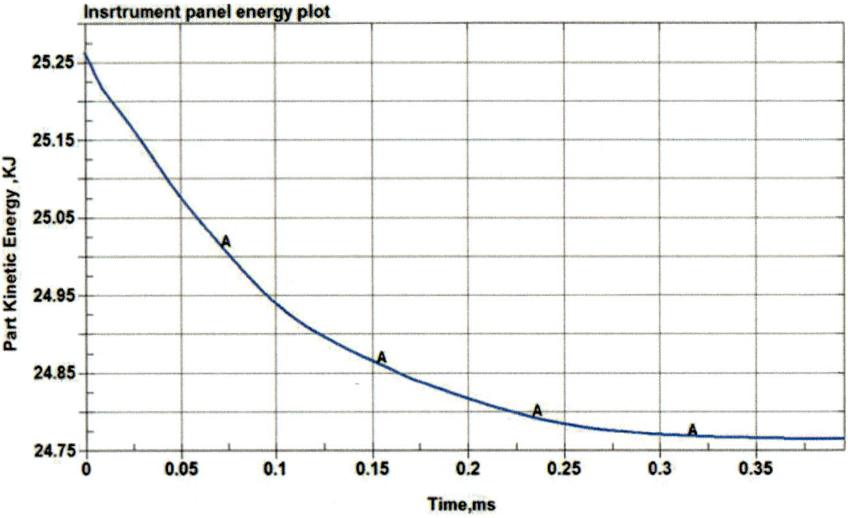

Fig. 10 Instrument panel energy plot

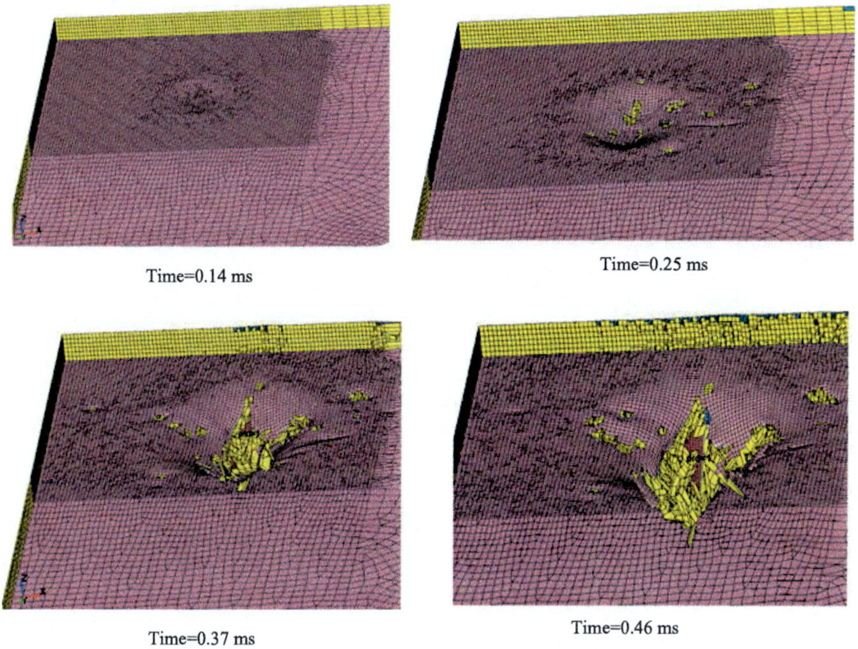

Fig. 11 Penetration of power turbine fragment through the floor board

4.3 Floor Board

The exit velocity of the engine power turbine disc fragment of rigid body mass 1.323 kg perforating from the instrument panel impacting to the floorboard with an initial velocity of 193.46 m/s is shown in Fig. 11. The damage pattern of the floorboard is similar to the instrument panel, the energy absorption of the floorboard board is higher than the instrument panel because the floorboard has a high-density metal core with higher thickness. The initial and final kinetic energy of the fragment perforating from the floorboard is 24.76 and 23.64 kJ. The energy absorption in the floorboard is 1.12 kJ. The energy plot of the fragment is shown in Fig. 12.

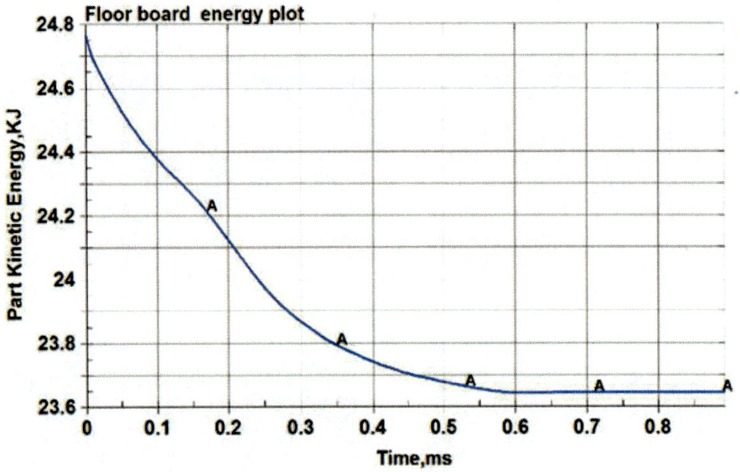

Fig. 12 Floor board energy plot

5 Conclusion

The uncontained rotor fragment analysis summarizes the methodology followed for the 1/3rd disk fragment definition and the risk targets as per AC 20-128A. The evaluation of structural damage is performed as per the criteria in rotor burst requirements. LS-Dyna nonlinear finite element software is successfully used in this paper to model high-velocity impact on engine deck assembly, instrument panel, and floor board.

The analysis following impacting of the power turbine disc fragment having initial energy of 29.175 kJ to the airframe structure is followed sequentially as per the design layout in rotorcraft. The metallic engine deck assembly energy absorption is 3.91 kJ, the energy absorption of a sandwich instrument panel made of a Nomex core with an aluminium face sheet is 0.505 kJ, and the energy absorption of a sandwich floor board made of the metallic core with aluminium face sheet is 1.12 kJ, the total energy absorption is 5.535 kJ.

Mostly, the aircraft systems falling in the power turbine fragment envelope like hydraulics, flight controls, fire protection, fuel feed lines, and electrical power are studied and found to be redundant. The structure maintained a minimum strength level when damaged under power turbine fragments (70% limit flight maneuver loads) and met the airworthiness standards. The LS-Dyna analysis predicted the uncontained power turbine fragment failure analysis and met the safety standards as per AC 20-128A and reduced the design cost of complex testing and time.

6 Scope of the Future Work

The possible way of improving energy absorption can be further achieved using advanced composite materials and structural modifications at the impact zone. The validation of analysis can be part of the future scope of work.

References

1. Federal Aviation Regulation, Part 29-Airworthiness Standards: Transport Category Rotorcraft
2. AMC 20-128A, Design Considerations for Minimizing Hazards Caused by Uncontained Turbine Engine and Auxiliary Power Unit Rotor Failure
3. Juan-juan LI, Hai-jun XUAN, Lian-fang LIAO, Wei-rong HONG, Rong-ren WU (2009) Penetration of disk fragments following impact on thin plate. J Zhejiang University, Hangzhou, China 10(5):677–684
4. Fengchan WANG (2011) B, SUN Youchaoa, ZENG Haijuna, Study of uncontained turbine engine rotor failure airworthiness compliance verification method, ISAA 2011. Proced Eng 17:531–541
5. Carney KS, Pereira JM, Revilock DM, Matheny P (2009) Jet engine fan blade containment using an alternate Geometry. Int J Impact Eng 720–728
6. Senthil K, Arindam B, Mittal R, Iqbal MA, Gupta NK (2017) Numerical investigations on the impact of hemi spherically tipped projectiles on thin plates, implast 2016. Proc Eng 173:1926–1931
7. Senthil K, Arindam B, Mittal R, Iqbal MA, Gupta NK, Ballistic response of, (2024) Aluminium plates against blunt nose projectiles, Implast 2016. Proc Eng 173(2017):1926–1931
8. Heimbs S, Middendorf P, Maier M (2006) Honeycomb sandwich material modeling for dynamic simulations of aircraft interior components, 9th International LS-DYNA users conference
9. LS-DYNA Keyword User's Manual, Volume I, Version 971, Livermore Software Technology Corporation, California, February 2012
10. LS-DYNA Keyword User's Manual, Volume II, Version 971, Livermore Software Technology Corporation, California, February 2012

Influence of Reinforcement Bar on the Performance of Reinforced Concrete Slab Under Impact Loading

K. Senthil, Rachit Sharma, S. Rupali, M. A. Iqbal, and N. K. Gupta

Abstract The study is focused to investigate the behaviour of reinforced concrete slabs under low velocity impact loads. The experiment was performed on conventional reinforced concrete slab of size 1200 × 1200 mm with a thickness of 50 mm with reinforcement ratio of 0.23%. The slab was impacted under an impacted energy of 741.02 J through the hemispherical nose impactor with a mass of 60 kg. The measured impact force was found to be 57.37 kN. The numerical study was conducted using finite element software ABAQUS/ EXPLICIT. The material model for concrete as Concrete Damage Plasticity (CDP) model based on fracture energy criterion was used whereas the Johnson–Cook material model for steel reinforcement was adopted. The numerical simulation was performed to validate the peak impact force with the experimental results and found in good agreement with a difference of 3.73%. In order to evaluate the suitable sensitive reinforcement, the different reinforcement ratio (0.47 and 0.63%), radial reinforcement spacing (250, 150 and 100 mm c/c) and diameter of reinforcement (8Ø and 10Ø) were adopted in the finite element simulations. The predicted results were compared to the experimental results and the most effective solution for reinforced concrete slab strengthening was postulated.

Keywords Reinforced concrete slab · Reinforcement ratio · Radial reinforcement bar · Diameter of reinforcement bar · Impact loading

Nomenclature

CDM Continuum Damage Mechanics
CDP Concrete Damage Plasticity

K. Senthil (✉) · R. Sharma · S. Rupali
Department of Civil Engineering, NIT Jalandhar, Jalandhar, Punjab 144011, India
e-mail: kasilingams@nitj.ac.in

M. A. Iqbal
Department of Civil Engineering, IIT Roorkee, Roorkee, Uttarakhand 247667, India

N. K. Gupta
Department of Applied Mechanics, Indian Institute of Technology, Delhi, New Delhi 110016, India

R. Velmurugan et al. (eds.), *Dynamic Behavior of Soft and Hard Materials Volume 1*, Springer Proceedings in Materials 34, https://doi.org/10.1007/978-981-99-6030-9_29

f_c'	Compressive strength
f_t	Peak tensile strength
σ_t	Tensile strength
w	Crack opening
w_c	Critical crack opening
G_F	Fracture energy
$\tilde{\varepsilon}_t^{cr}$	Cracking strain in tension
w	Crack opening
L_c	Characteristic length of element
$\tilde{\varepsilon}_c^{pl}$	Plastic strain in compression
$\tilde{\varepsilon}_t^{pl}$	Plastic strain in tension
D_c	Damage indices in compression
D_t	Damage indices in tension
E_c	Initial modulus of elasticity
ρ	Density of concrete
υ	Poisson's ratio.
f_b'/f_c'	Second stress invariant
B	Dilation angle
K_c	Shape parameter
m	Eccentricity

1 Introduction

The use of thin reinforced concrete slabs for high rise structures has a few major benefits over beam supported slabs or slabs with column capitals including provisions for more space for a given story height and drop in dead weight of the building [1]. Most often, these reinforced concrete structures are under static loads even so the likelihood of reinforced concrete (RC) structures subjected to impact loads such as accidental drop weights during construction of high rise building on floor system cannot be omitted. These structures may experience extreme time dependent loadings such as earthquake, blast, impact, rock fall and explosive loading which generate stresses much higher than their design limit [2]. Under impact loading, the response of concrete structures is determined by the nature of the event and the properties of the constituent materials.

Earlier studies focused on dynamic effects of reinforced concrete structures were on nuclear containment walls and internal concrete barriers to evaluate their response to soft and hard missile impact at higher velocities [3, 4]. The main objective of these studies was to estimate the required slab thickness for resisting the perforation phenomenon for military applications. However, the penetration and perforation phenomenon observed in finite targets has a complex mechanism and no convincing findings were offered in the existing literature. The damage mechanisms occurring in the penetration of finite thin slab elements include compaction, pressure dependent behaviour under compression and tensile cracking [5]. Hummeltenberg et al. [6]

compared the behaviour of RC slabs (1000 × 1000 × 150) mm to different types of concrete. The configuration of reinforcement was changed by using C-shaped stirrups to improve shear resistance as well as additional fibre mesh reinforcement. Results indicate that the addition of stirrups had negligible effect on the slab resistance to scabbing; however, the use of fibre reinforcement prevented the perforation of concrete slabs. Further, it was observed that the behaviour of reinforced concrete sequentially changed from local to global when fibre mesh reinforcement was used in the slab. Yilmaz et al. [7] found that increasing rebar ratio in two-way slab increases bending strength, stiffness and toughness characteristics of RC slab. Further, higher localized damage with increase in the impact energy of the impactor at the location of impact was observed in the RC slab. Under large-scale testing of free fall impact on reinforced concrete slab, the specially designed steel supports may act as expendable fuse to absorb impact energy on designing the large-scale testing of RC structures under impact loading from rocks or boulders [8].

The continuum damage mechanics (CDM) is based on continuum mechanics which allows us to characterize and model the behaviour of structures under effects of damage initiation, evolution and propagation. Integrating the damage variable in force equilibrium can accurately represent damage in material by means of an equivalent damage force. Many finite element softwares are commercially available based on CDM approach such as ABAQUS, LS-Dyna, ADINA etc. However, parametric calibration based on characteristics equations is required to accurately predict their usages under impact analysis, whilst a few studies were found which described the in-depth numerical analysis of RC structures under low velocity impact. Delhomme et al. [9] tested the structural response of infrastructure built along steep slopes and used MATLAB-based mass spring model and numerical method of smeared element to predict its response. It was concluded that simultaneous use of both the approaches can make it possible to finalize the design process of these protective structures. Kezmane et al. [10] found that mesh density in finite element analysis plays an important role in local and global responses for concrete. The CDP model and steel reinforcement have the capability to replicate the behaviour of material. Further, some sliding was observed in concrete before cracking between steel bars and it was proposed that the effective contact law between them may be considered.

Based on the detailed literature survey, it was observed that there were limited studies on influence of reinforcement bar on the impact resistance of thin reinforced concrete slab. The experimental and numerical study on thin reinforced concrete slab with the effect of reinforcement is still restricted. Therefore, an attempt was made to perform experimental and numerical study on the thin reinforced concrete slab under low velocity impact loading and to perform numerical study on the effect of reinforcement on impact resistance. Section 2 highlights the experimental investigation and test setup. Section 3 highlights the material modelling using finite element software ABAQUS to validate the behaviour of the reinforced concrete slab under low velocity impact. Section 4 highlights the parametric study for the effect of reinforcement on the impact resistance of reinforced concrete slab.

2 Experimental Investigation

The reinforced concrete slab was designed for M40 grade of concrete as per BIS: 10,262: 2019 [11]. The mix ratio of 1: 1.423: 1.927 (cement: sand: coarse aggregate) and cement quantity of 450 kg/m^3 with water-cement ratio of 0.35 were adopted for making concrete. To ease the flowability of concrete onto the mould, the superplasticizer was employed for 0.6% of weight of binder. The sample was water cured to a target unconfined compressive strength of $f_c' = 48$ MPa for 28 days. The tension reinforcement used is Fe500 bar (Ø = 6 mm) with a clear cover of 20 mm as shown in Fig. 1. The overall dimensions of RC slab were (1200 × 1200) mm with a thickness of 50 mm and the spacing of reinforcement mesh for main and distribution steel was 250 mm centre to centre.

The experiment was performed using impact testing frame as shown in Fig. 2a. The test specimen was simply supported on one side and kept free on the other side, see Fig. 2b. The slab was impacted with hemispherical nose shape impactor with a mass of 60 kg. The mass was released using steel wire which was further connected to a winch. The winch was designed for manual operation and used to release impactor at desired angle from zero to 90°. The impact event was captured using slow motion video record at 240 Hz and further processing the distance travelled by the impactor in each time frame to calculate velocity using open-source software "Tracker". In order to measure the impact force, dynamic load cell with a capacity of 250 kN was attached to the impactor.

Fig. 1 Dimension and reinforcement detail of specimen

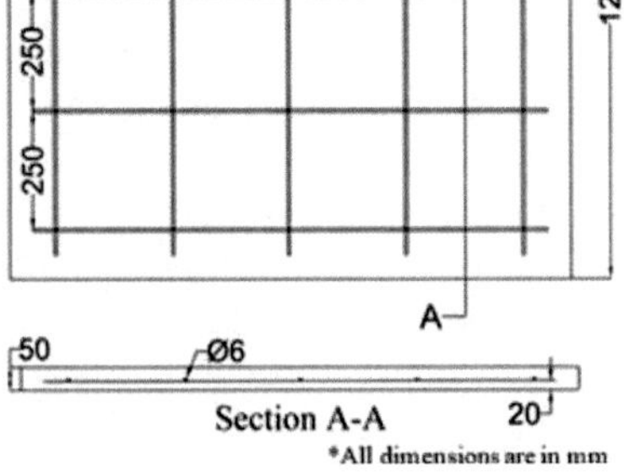

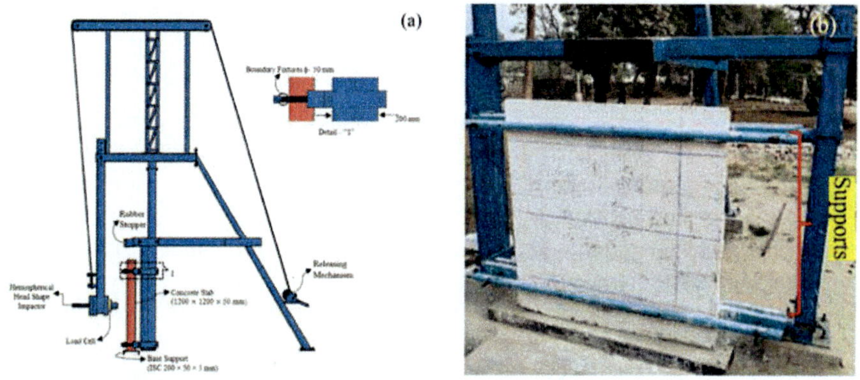

Fig. 2 **a** Test setup and **b** slab specimen

3 Constitutive Behaviour and Numerical Validation

The concrete damage plasticity model (CDP) was used to define the behaviour of concrete. The model was developed by Lubliner et al. [12] and further modified by Lee and Fenves [13] for cyclic and dynamic loading. The model was able to predict the elasto-plastic characteristics of concrete, the damage-dependent loading and unloading, stiffness degradation in tension and compression as well as strain rate effects. The behaviour under uniaxial compression was based on the stress–strain relationship proposed by Aslani and Jowkarmeimandi [14]. The strain rate effects in compression were based on the study by Malwar and Crawford [15]. The uniaxial compressive behaviour for concrete material is shown in Fig. 3a.

The uniaxial tensile behaviour proposed by Reinhardt et al. [16] based on static tensile test results is of the form:

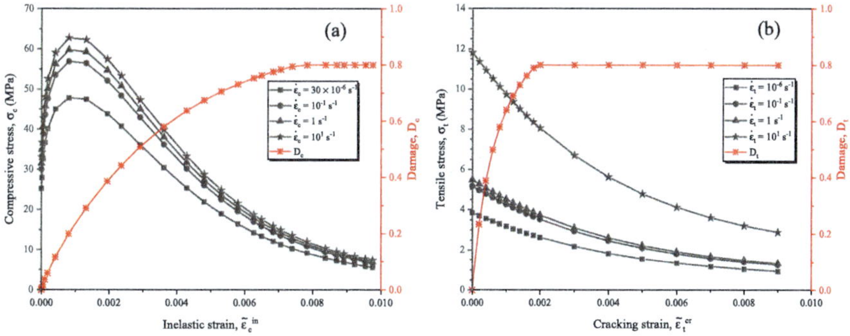

Fig. 3 The uniaxial behaviour in **a** compression and **b** tension defined in CDP model

$$\frac{\sigma_t}{f_t} = \left(1 + \left(c_1 \frac{w}{w_c}\right)^3\right) \exp\left(-c_2 \frac{w}{w_c}\right) - \frac{w}{w_c}\left(1 + c_1^3\right) \exp(-c_2) \tag{1}$$

where $c_1 = 3$; $c_2 = 6.93$, f_t as peak tensile strength and w_c the critical crack opening displacement with near zero stress is determined by $w_c = 5.136G_F/f_t$, in which $G_F \approx 2.5G_f$ is the fracture energy that equals the area under entire stress-crack opening displacement curve. The value of area under the initial tangent of the softening stress-separation curve of cohesive crack model, which governs the maximum loads of structure, G_f was based on the statistical study done by Bažant and Becq-Giraudon [17] as 54 N/m. To solve partially the problems and to keep mesh objectivity, mesh regularization based on fracture energy criterion was chosen. The dissipated energy at tensile cracking is then kept constant with element characteristic length (L_c) upon mesh refinement [10]. The cracking strain $\tilde{\varepsilon}_t^{\text{cr}}$ is then regularized as $\tilde{\varepsilon}_t^{\text{cr}} = w/L_c$. The strain rate effects under uniaxial tensile strength presented by Malvar and Ross [18] were used. The uniaxial tensile strength behaviour incorporated in CDP model is shown in Fig. 3b.

The compressive and tensile damage indices $D_c\left(\tilde{\varepsilon}_c^{\text{pl}}\right)$ and $D_t\left(\tilde{\varepsilon}_t^{\text{pl}}\right)$ that characterize the degraded elastic compressive and tensile stiffness are defined as $(1 - D_c)E_c$ and $(1 - D_t)E_c$, respectively. Li et al. [19] suggested the tensile and compressive damage index calibrated using $\lambda E_c\tilde{\varepsilon}_t^{\text{cr}}/\left(\lambda E_c\tilde{\varepsilon}_t^{\text{cr}} + \sigma_t\right)$ considering the $\lambda = 0.2$, whereas, the compressive damage index considering the $\lambda = 0.5$, used for fitting the equation". The damage indices used for the CDP model based on the study [19] as:

$$D_c\left(\tilde{\varepsilon}_c^{\text{in}}\right) = \frac{\tilde{\varepsilon}_c^{\text{in}}}{\tilde{\varepsilon}_c^{\text{in}} + \frac{2\sigma_c}{E_c}} < 0 \cdot 8 \tag{2}$$

$$D_t\left(\tilde{\varepsilon}_t^{\text{cr}}\right) = \frac{\tilde{\varepsilon}_t^{\text{cr}}}{\tilde{\varepsilon}_t^{\text{in}} + \frac{5\sigma_t}{E_c}} < 0 \cdot 8 \tag{3}$$

where $\tilde{\varepsilon}_c^{\text{in}} = \varepsilon_c - \sigma_c/E_c$ is the inelastic compressive strain; $\tilde{\varepsilon}_t^{\text{cr}} = \varepsilon_t - \sigma_t/E_c$ the cracking strain; and E_c the initial modulus of elasticity. The plastic properties of CDP model were taken from the study by Senthil et al. [20] and shown in Table 1.

Table 1 CDP properties of concrete

Elastic parameters				
Density (ρ)	Elastic modulus (E_c)		Poisson's ratio (υ)	
2400 kg/m^3	31.461 GPa		0.166	
CDP parameters				
Second stress invariant, f_b'/f_c'	Dilation angle, β	Eccentricity, m	Shape parameter, K_c	Viscosity Parameter
1.108	37.91	0.1	0.644	0

The model for steel reinforcement was elasto-plastic with ductile damage. The properties of steel reinforcement were taken from the study of Senthil et al. [20], shown in Table 2.

The finite element simulation was based on ABAQUS/ EXPLICIT scheme. The concrete part and impactor are modelled as three-dimensional deformable body and the boundary conditions were modelled as discrete rigid bodies. The boundary conditions were modelled as discrete rigid bodies because no deformation was observed in boundaries, and the discrete rigid body boundary was considered to save the computational cost. The mesh sensitivity analysis was shown in Fig. 4 and based on the mesh sensitivity analysis, a mesh size of 7.7 mm or lower shows convergence in terms of peak impact force. The schematic representations of the assembly and the type of element used for mesh generation are shown in Fig. 5. The mesh is divided into two zones based on the mesh convergence study as shown in Fig. 5a. The inner zone has a mesh size of 5 mm and outer zone has mesh size of 10 mm whereas the mesh size along the thickness direction is kept constant, 5 mm. The details of the impactor used in the study are shown in Fig. 5b.

The concrete slab is modelled as C3D8R element considering eight node linear brick element with reduced integration and viscous hourglass control. The steel reinforcement is modelled as B31 element, a 2-node linear beam in space with an element size of 10 mm. "EMBEDDED" constraint was used to define the behaviour of steel reinforcement in concrete slab. The impactor has a mass of 8.2 kg and added mass of 51.8 kg was applied to it to keep the mass at 60 kg similar to the experiment. General contact was given between impactor and RC slab. Tangential behaviour with friction coefficients of 0.3 between impactor and RC slab and 0.5 for other contacts has been used. "Hard" contact was used to define the normal behaviour between the surfaces [21].

The results in terms of impact force were predicted and compared, see Fig. 6a. The erosion parameters for validation purposes were used as $\tilde{\varepsilon}_t^{pl}$ and $\tilde{\varepsilon}_c^{pl}$ as 0.05 and 0.3 [22]. The failure strain for compressive strain was kept higher to avoid the excessive punching of the impactor onto the RC slab. The peak force for experimental results was found out to be 57.37 kN whereas the model predicts the peak impact force

Table 2 Material properties of reinforcement using elasto-plastic model, Senthil et al. [20]

Elastic properties		
P	*E*	*υ*
7850 kg/m^3	200 GPa	0.3

Average stress and plastic strain relationship of reinforcement

Plastic strain	Average stress (MPa)			
	Strain rate 0/s	Strain rate 1/s	Strain rate 5/s	Strain rate 15/s
0	296	429	455	473
0.1	605	647	655	661
0.2	605	647	655	661

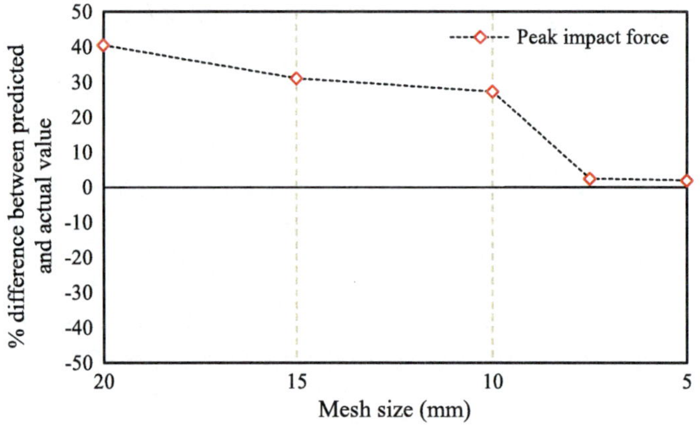

Fig. 4 Mesh size sensitivity analysis

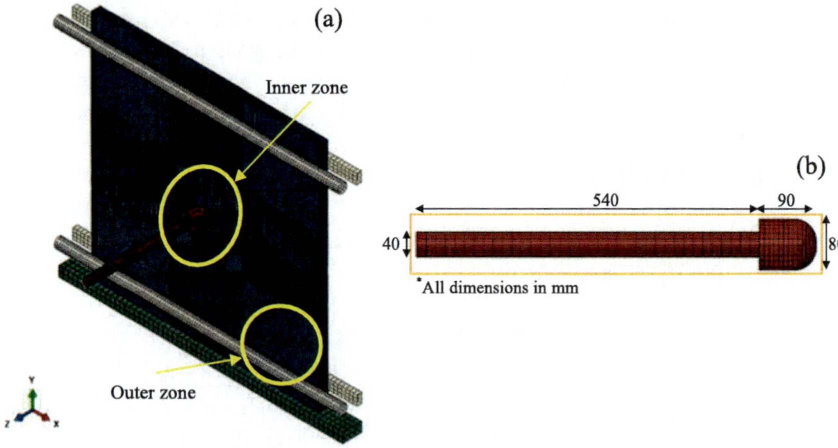

Fig. 5 Modelling of **a** impact testing assembly and **b** impactor specifications

to be 55.23 kN. The model has a satisfactory prediction for the peak force during impact analysis with a difference of 3.73%. The numerical midpoint displacement for the RC slab is shown in Fig. 6b. The residual displacement observed from the experiment and simulation was 17 and 19 mm, respectively. The damage to the front face of the slab experiences spalling at the impacted zone along with compression damage, see Fig. 7a-i. Similar behaviour was observed for numerical results where punching damage and horizontal cracks were predicted by the model, see Fig. 7a-ii. The damage to the rear face of the slab is characterized by scabbing of concrete with several flexural cracks, see Fig. 7b-i. The damage to the rear face was also seen in numerical results where the horizontal cracks were accurately predicted by the

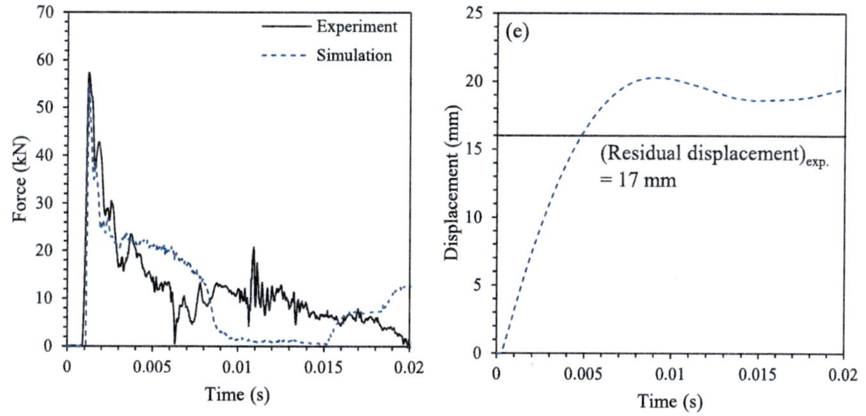

Fig. 6 Validation of simulated impact force with experimental results

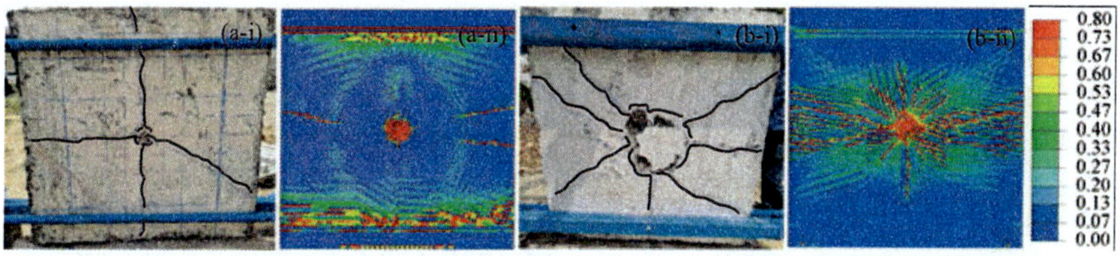

Fig. 7 Damage to the **a** front and **b** back of reinforced concrete slab for (i) experimental and (ii) simulation results

model. The scabbed area was under predicted by the model, which may be due to the smaller mesh size leading to higher fracture energy in the elements.

4 Parametric Study on Reinforcement Effect on Impact Resistance

Parametric study on effect of reinforcement on impact resistance of reinforced concrete slab was performed. Three aspects which are covered are: (1) effect of reinforcement percentage i.e., 0.47 and 0.63, (2) effect of reinforcement diameter i.e., 8Ø and 10Ø mm and use of radial reinforcement for strengthening at 250, 150 and 100 mm centre-to-centre with innermost reinforcement equal to diameter of impactor. The radial reinforcement detailing and arrangement are shown in Fig. 8.

The result for the parametric study is shown in Table 3. It was observed that the resistance of reinforced concrete slab was decreased on increasing the reinforcement ratio for both cases. The decrease in resistance was due to absence of reinforcement at the impact location. The resistance offered by these slabs was mainly because of

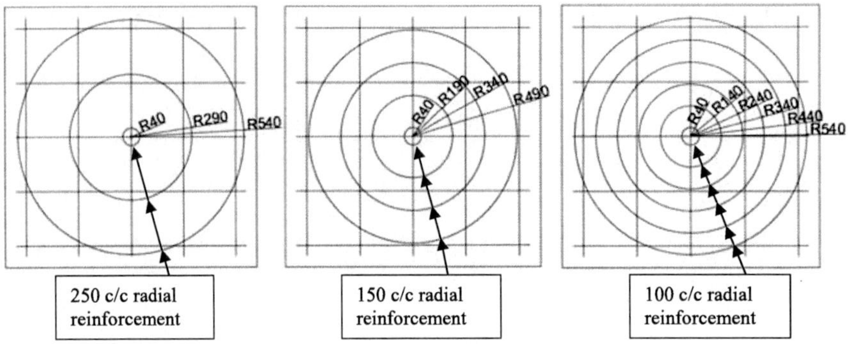

Fig. 8 Arrangement of radial reinforcement

concrete. On increasing the diameter of the reinforcement, the enhancement in impact resistance was increased to 2.77% and 7.39% for 8Ø and 10Ø mm, respectively. The diameter of reinforcement is a viable option to increase the impact resistance of slab. The radial reinforcement was found to have marginal enhancement on the impact resistance of slab. It was suggested that the use of radial reinforcement did not result in any significant improvement in impact resistance of the slab.

The damage to the rear face of the specimen did not change even after using different diameter of reinforcement as shown in Fig. 9. Therefore, it was concluded that the diameter of reinforcement has no significant change on the crack pattern of reinforced concrete slab under impact loading. Similarly, the use of radial reinforcement found no significant change in crack pattern as the damage phenomenon is localized to the impact zone. Therefore, it is suggested to keep the strengthening of reinforced concrete slab at the impact location for future studies using fibre polymer sheets.

Table 3 Parametric study on reinforcement effect on impact resistance

Parameter	Impact force (kN)	Rebar displacement (mm)	Enhancement (%)
0.47% reinforcement	52.10	–	−5.67
0.63% reinforcement	52.27	–	−5.34
8Ø reinforcement	56.76	25.70	2.77
10Ø reinforcement	59.31	24.81	7.39
250 mm c–c radial reinforcement	55.49	30.46	0.47
150 mm c–c radial reinforcement	55.68	30.33	0.81
100 mm c–c radial reinforcement	55.46	29.94	0.42

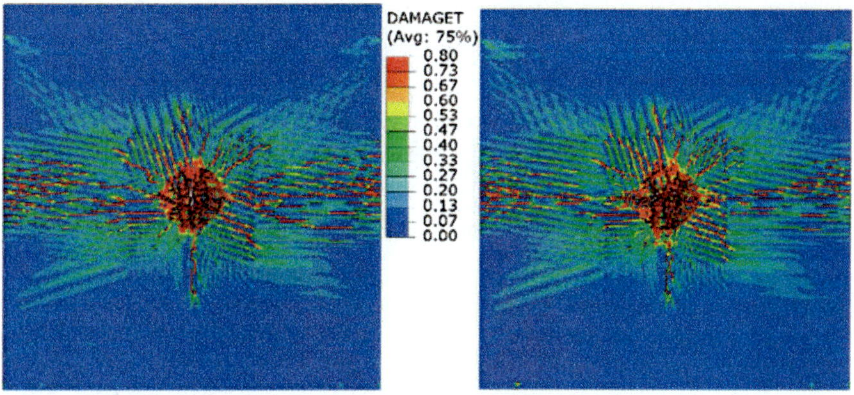

Fig. 9 Damage to the rear face of slab for **a** 8Ø and **b** 10Ø reinforcement

5 Conclusions

Based on the study, the following conclusion were drawn:

- The dynamic effects of low velocity impact loading on reinforced concrete slab were accurately predicted using CDP model with a difference of 3.73%.
- The radial reinforcement found to have no significant improvement in improving the impact resistance of reinforced concrete slab.
- On increasing the diameter of reinforcement bar, the enhancement in impact resistance was found to be 7.33%.

References

1. Torabian A, Isufi B, Mostofinejad D, Ramos AP (2019) Behavior of thin lightly reinforced flat slabs under concentric loading. Eng Struct 196:109327
2. Kennedy RP (1976) A review of procedures for the analysis and design of concrete structures to resist missile impact effects. Nucl Eng Des 37:183–203
3. Williams MS (1994) Modeling of local impact effects on plain and reinforced concrete. ACI Struct J 91:178–187
4. Corbett GG, Reid SR, Johnson W (1996) Impact loading of plates and shells by free-flying projectiles: a review. Int J Impact Eng 18:141–230
5. Polanco-Loria M, Hopperstad OS, Børvik T, Berstad T (2008) Numerical predictions of ballistic limits for concrete slabs using a modified version of the HJC concrete model. Int J Impact Eng 35:290–303
6. Hummeltenberg A, Beckmann B, Weber T, Curbach M (2011) Investigation of concrete slabs under impact load. Appl Mech Mater 82:398–403
7. Yılmaz T, Kıraç N, Anil Ö, Erdem RT, Kaçaran G (2020) Experimental Investigation of impact behaviour of RC slab with different reinforcement ratios. KSCE J Civ Eng 24:241–254
8. Mougin J-P, Perrotin P, Mommessin M, Tonnelo J, Agbossou A (2005) Rock fall impact on reinforced concrete slab: an experimental approach. Int J Impact Eng 31:169–183

9. Delhomme F, Mommessin M, Mougin JP, Perrotin P (2007) Simulation of a block impacting a reinforced concrete slab with a finite element model and a mass-spring system. Eng Struct 29:2844–2852
10. Kezmane A, Chiaia B, Kumpyak O, Maksimov V, Placidi L (2017) 3D modelling of reinforced concrete slab with yielding supports subject to impact load. Eur J Environ Civ Eng 21:988–1025
11. BIS: 10262. Concrete Mix Proportioning—Guidelines. Bur Indian Stand Delhi.
12. Lubliner J, Oliver J, Oller S, Oñate E (1989) A plastic-damage model for concrete. Int J Solids Struct 25:299–326
13. Lee J, Fenves GL (1998) Plastic-damage model for cyclic loading of concrete structures. J Eng Mech 124:892–900
14. Aslani F, Jowkarmeimandi R (2012) Stress–strain model for concrete under cyclic loading. Mag Concr Res 64:673–685
15. Malvar LJ, Crawford JE (1998) Dynamic increase factor for concrete. In: 28th department of defense explosives safety seminar, Orlando, FL
16. Reinhardt HW, Cornelissen HAW, Hordijk DA (1986) Tensile tests and failure analysis of concrete. J Struct Eng 112:2462–2477
17. Bažant ZP, Becq-Giraudon E (2002) Statistical prediction of fracture parameters of concrete and implications for choice of testing standard. Cem Concr Res 32:529–556
18. Malvar LJ, Ross CA (1998) Review of strain rate effects for concrete in tension. ACI Mater J 95:739–735
19. Li X-X (Lambert), Wang C, Sato J (2020) Framework for dynamic analysis of radioactive material transport packages under accident drop conditions. Nucl Eng Des 360:110480
20. Senthil K, Kubba Z, Sharma R, Thakur A (2021) Experimental and numerical investigation on reinforced concrete slab under low velocity impact loading. IOP Conf Ser Mater Sci Eng 1090:012090
21. Senthil K, Thakur A, Singh AP, Iqbal MA, Gupta NK (2023) Transient dynamic response of brick masonry walls under low velocity repeated impact load. Int J Impact Eng 174:104521
22. Chung C, Lee J, Jung R (2015) Numerical simulations of missile impacts on reinforced concrete plates: IRIS-2010/2012 benchmark project. *Nucl Eng Des*. Epub ahead of print 2015. DOI: https://doi.org/10.1016/j.nucengdes.2015.04.031

Low-velocity Impact Response of Jute/Banana Fiber in Natural Rubber-Based Hybrid Composites: FE Approach

Kartik Kumbhare, Vishwas Mahesh, and Sharnappa Joladarashi

Abstract Green composites are proposed as environmentally friendly, easily recycled, and reusable advanced composite materials. The present study aims at studying the damage done by low-velocity impact (LVI) of jute and banana fiber-based green composites using commercial finite element (FE) software. The LVI response is evaluated for flat, hemispherical, and conical impactors at three velocities of 5, 10, and 15 m/s. Hybrid composites are modeled in two stacking sequences: jute-rubber-banana-rubber-jute (JRBRJ) and banana-rubber-jute-rubber-banana (BRJRB). These hybrid green composites are compared to their pure fiber counterpart composites, i.e., jute-rubber-jute-rubber-jute (JRJRJ) and banana-rubber-banana-rubber-banana (BRBRB). The ABAQUS Finite Element Modeling software is used to model, and the explicit dynamic solver is used to simulate these proposed composites. The absorbed energy at 5 m/s for flat impactor for JRJRJ and BRBRB is 3.5 J and 0.52 J, respectively, whereas for JRBRJ and BRJRB is 2.3 J and 1.4 J, respectively. Similar results are obtained for 10 and 15 m/s. The energy absorbed follows a sequence JRJRJ > (JRBRJ, BRJRB) > BRBRB. The flat impactor has more damage due to its larger contact area and high energy absorption at higher velocities. Impact due to conical impactor shows local penetration and lower energy absorption. Results show that the proposed composites exhibit better energy absorption due to a flexible matrix and more resistance to damage due to the involvement of a hybrid structure which makes the composite stiffer.

Keywords Banana-Jute hybrid · Low-velocity impact · Natural rubber · Damage study · Green composites

---

K. Kumbhare · S. Joladarashi (✉)
Department of Mechanical Engineering, National Institute of Technology Karnataka, Surathkal, Karnataka 575025, India
e-mail: sharnappaj@nitk.edu.in

V. Mahesh
Department of Industrial Engineering and Management, Siddaganga Institute of Technology, Tumakuru, Karnataka 572103, India

R. Velmurugan et al. (eds.), *Dynamic Behavior of Soft and Hard Materials Volume 1*, Springer Proceedings in Materials 34, https://doi.org/10.1007/978-981-99-6030-9_30

1 Introduction

A composite material is formed with a combination of two or more distinct materials to create a new material with enhanced properties. Composites consist of two or more compatible materials that may differ in constitution and attributes embedded in a continuous phase. The embedded materials are usually more rigid and more robust than the continuous phase, called the reinforcing material, whereas the continuous phase is termed the matrix. Although lightweight, composites are prone to impact damage, which severely degrades the structural stiffness, firmness, and load-carrying capacity. In prevailing times, composite structures subjected to impact loads have been widely examined through experiments and mathematical models [1]. As we expand our use of composite structures in engineering applications, their relatively poor resistance to impact load is an area of research that has been and is still being pursued. The composite structures consisting of two or more discontinuous phases are defined as hybrid composites or more than one type of fiber (in the case of fiber-reinforced composites). Different methods combine these fibers, which involve stacking layers of fibers, coalescing fibers, selective placements of fiber in a layer or in the matrix itself, and in a specified orientation [2]. In recent years, rigid composites have substantially improved their physical and chemical characteristics, but they are inflexible, which negatively influences their impact behavior. Rigid composites are generally made with thermosetting materials, which cannot be restructured as per our needs. As a result, to improve the impact characteristics, flexible matrices such as rubber or a combination of other materials are being explored. Natural rubber (N.R.), which is inherently flexible, might be a possible substitute for thermosetting materials. Natural rubber is a polymeric substance that possesses visco-elastic properties. Mahesh et al. 2019c successfully validated that N.R. could be effectively used as a core material to improve the impact response of jute-epoxy and jute-epoxy-rubber sandwich composite [3]. Natural and synthetic rubbers are used in a wide range of applications and are not limited to shock absorbers and impact-resistant panels [4–8]. The "flexible composites" are based upon elastomeric polymers of which the operating range of deformation is very high than those of the conventional thermoset or thermoplastic composites [9].

Various researchers have carried out studies on natural and synthetic fiber reinforced with epoxy and poly-lactic acid [10]. Mahesh et al. 2019a [11] studied the impact behavior of jute epoxy (J.E.) composite and experimented to understand the influence of various parameters such as the thickness of the composite plate, impactor velocity, and shape of the impactor. The J.E. composite laminate with thicknesses ranging from 6 to 10 mm, subjected to low-velocity impact (LVI) at 2 to 8 m/s velocities using conical shaped, hemispherical, and flat impactors that the thickness of the composite laminate might affect the energy absorption and impact resistance. Thicker composites showed a high impact resistance to the impact. Another study [12] looked at the LVI response of a flexible green composite made of sisal and N.R., as well as the influence of impactor profile. The same author also conducted a parametric FE investigation on the impact behavior of sisal and cenosphere-reinforced

N.R.-based hybrid composites [13]. They came to the conclusion that the contact area has a direct impact on the energy absorption potential and the level of damage to the composite laminate when it comes to LVI. Few investigations on jute and banana hybrids using N.R. fiber composites have been conducted, particularly for impact studies. Jute and banana fibers have better mechanical properties than other natural fibers and can be used to replace synthetic fibers. They have good specific mechanical characteristics, notably in terms of stiffness and their low density. Epoxy resin is used as the matrix material in almost all polymer matrix composites. Despite its high energy absorption capacity and excellent puncture and tear resistance, there are limited studies on natural rubber-based matrix material composites, especially for impact applications.

A thorough literature survey on topics such as LVI, impact studies, and FE modeling of impact was done, and it is found that not much work has been reported on the FE analysis of impact behavior of the flexible composites with jute and banana fiber hybrid and natural rubber subjected to LVI. There also have been reports on enhancing the bond between the fibers and the matrix by various methods but not specifically for impact studies. The authors tried to explore only naturally available materials for the green composites, thereby taking care not to harm nature.

2 Modeling

The simulation was created using ASTM D7136/7136 M-07 standards, which is a standard test procedure for determining a fiber-reinforced polymer matrix composite's damage resistance to a drop-weight impact event. Several factors influence the damage resistance properties generated by this test method, including geometry, layup, impact energy, impactor shape, and boundary conditions. As a result, results are rarely transferrable to different configurations and are typically limited to the geometric and physical characteristics studied [14]. We modeled the composite laminates to be deformable bodies with dimensions of 125 mm $\times$ 75 mm. The deformable or rigid characteristic of a body is defined at the start of the modeling. The impactors (conical, flat, and hemispherical) are modeled as rigid bodies with dimensions according to [15]. The impactor is placed just 1 mm above the laminate. The composite laminates are assigned an encastered boundary condition at the edges such that they are constrained with all degrees of freedom (DOF). Material properties of both the impactor and the laminate are defined in the "property" section of ABAQUS. Firstly, each material is created with the defining properties. The laminate has a 5-layer stack, as seen in Fig. 1c. Secondly, each layer of the stack has been specified as a material accordingly. The impactors travel only in one direction, which is in the normal direction of the laminate plane. All composite laminates are modeled with the thickness of fibers as 1 mm, and Natural rubber as 2 mm. So, the volume fraction of the fibers (V_f) in the proposed composite is 43%, and the volume fraction of Natural rubber (V_m) is 57%. The tensile strength and tensile modulus are directly dependent on the fiber volume fraction. Thus, it is kept constant for each composite so as to compare the composites effectively. Necessary constraints are defined for the laminate and impactors. Step size is taken as 10–20 ms. Step size is intentionally kept small to reduce the simulation time. The front surface of the laminate and the outer surface of the impactors are defined by contact interaction properties. A contact interaction property is something that defines tangential behavior, i.e., friction, and normal behavior, i.e., hard or soft contact. A friction-tangential behavior with a friction coefficient of 0.3 and a hard contact normal behavior is defined for the fiber-matrix interaction. Surface-to-surface contact is considered here that describes contact between two deformable bodies or between a deformable surface and a rigid surface. All the rotational and translational DOFs have been constrained except for directions normal to the plane of the laminate for the impactor. While modeling, a reference point (RP) is set on the impactor, which will be constrained in every direction except in the direction to assign velocity. A point mass of 1 kg is also given to the RP. After all of the boundary conditions have been established, the impactor's reference point is given a preset field of velocity.

The medium-mesh composite plate has SC8R elements, which illustrates an 8-node hexahedron, general-purpose, finite membrane strain conventional shell. Meshing is done independently on the plate and impactor and is defined as is while modeling. The impactor also has a medium-mesh with R3D4 elements, which illustrates a 4-node 3D bilinear rigid quadrilateral. A medium-size mesh is used for both the impactor and the laminate as it reduces the simulation time. The (a) assembly and the boundary conditions (BCs), (b) the meshing of the laminate, and (c) a common ply-stack plot of the proposed composites are shown in Fig. 1a–c, respectively. The material properties used in the FE analysis are taken from the [8, 16, 17] and are

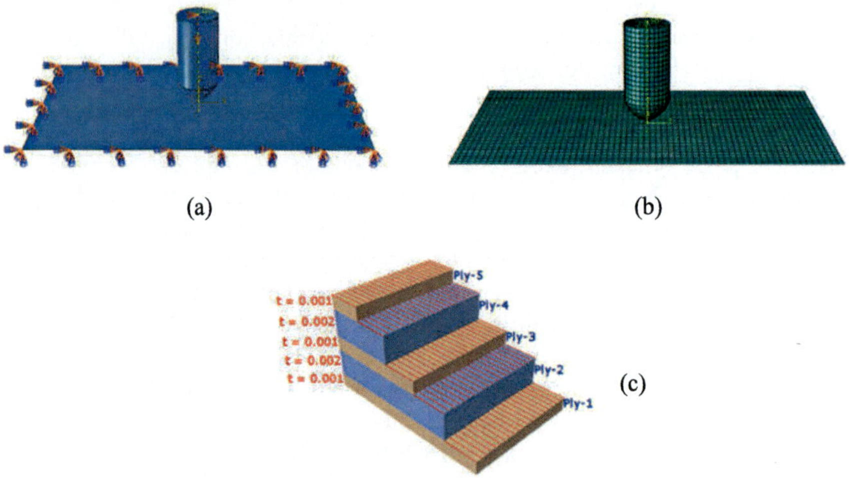

Fig. 1 **a** Assembly with BCs, **b** Mesh of the assembly, **c** A common ply-stack of the proposed composites

Table 1 Material properties [8, 16, 17]

Properties	Banana	Jute	Natural rubber
Density (kg/m^3)	1300	1400–1500	987.18
Youngs modulus (GPa)	28–29	10–30	4.5 × 10–4
Elongation at break (%)	2–5	1.8	11.1
Tensile strength (MPa)	200–800	300–700	0.05

Table 2 Nomenclature of the proposed composites

Proposed composites	Nomenclature
Jute + N.R	JRJRJ
Banana + N.R	BRBRB
Jute/banana hybrid _1	JRBRJ
Jute/banana hybrid _2	BRJRB

tabulated in Table 1. The terminologies of the proposed composites are provided in Table 2.

3 Analysis Results

The results of the energy absorbed by the JRJRJ, BRBRB, JRBRJ, and BRJRB composites at $V = 5$, 10, and 15 m/s when impacted by the three impactors (conical, hemispherical, and flat) are shown graphically in Figs. 2, 3, and 4, respectively.

3.1 Effect of Hybridization

The JRJRJ laminate absorbed the maximum energy for all velocities, as we can see from the peaks of Figs. 2, 3, and 4, while the BRBRB laminate absorbed the least. The hybrid laminates absorption capacity was that between the pure fiber counterpart laminates. The JRJRJ composite absorbed on an average 12% more than the hybrid laminates and 20% more than the BRBRB laminate. At the same time, the hybrid laminates absorbed on an average 11% more than the BRBRB laminate. This suggests that pure jute laminate is superior in absorbing energy to pure banana laminate and hybrid laminates. These results indicate that the hybridization of jute and banana fiber decreases the absorption capacity while increasing the stiffness. This might be because jute fibers have better mechanical properties than banana fibers.

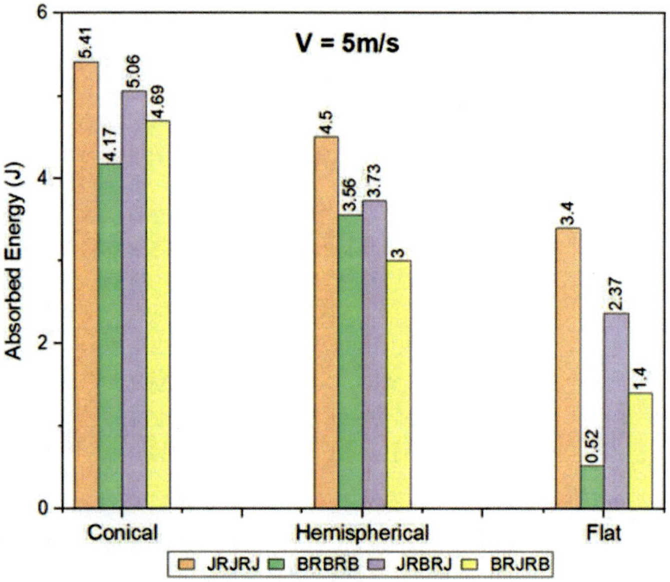

Fig. 2 Absorbed energy at $V = 5$ m/s

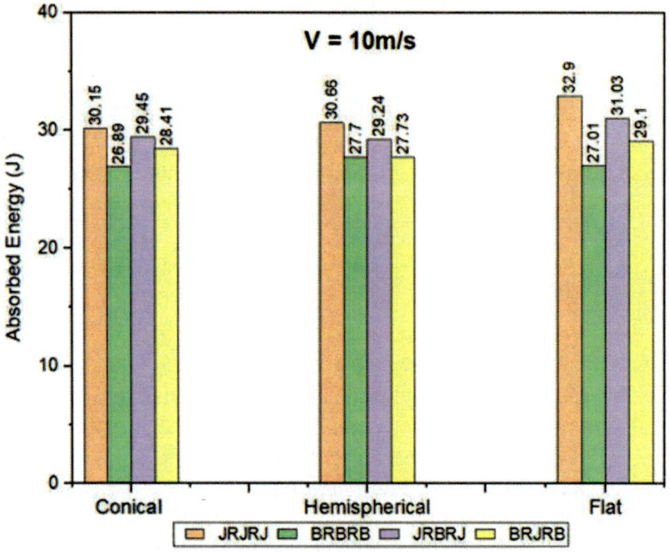

Fig. 3 Absorbed energy at $V = 10$ m/s

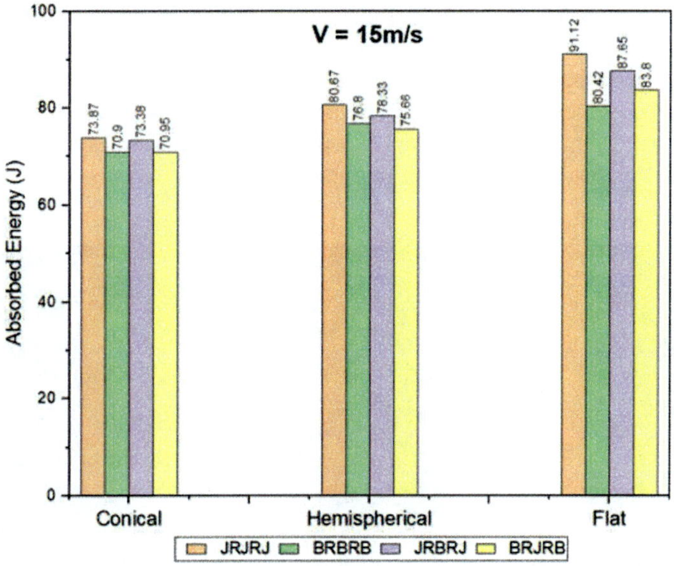

Fig. 4 Absorbed energy at $V = 15$ m/s

3.2 *Effect of Impactor Shape*

We found that the energy absorption capacity of the proposed composites is affected by the impactor shape and is shown graphically in Figs. 2, 3, and 4. The energy absorbed is maximum for conical impactor for $V = 5$ m/s (Fig. 2), but the energy absorbed is maximum for flat impactor at increased velocities, i.e., $V = 10$, 15 m/s (Figs. 3 and 4). This suggests that the impactor's energy absorption capacity and the level of damage to the laminate may be affected by the impactor's contact area with the laminate [13]. At 5 m/s, the conical impactor absorbed 1.2–8 times the energy absorbed by flat and hemispherical impactors. At 10 m/s, the flat impactor absorbed 1.1–1.25 times the energy absorbed by conical and hemispherical impactors.

3.3 *Damage Studies*

Figures 5, 6, and 7 present the damage induced by the three different impactors (conical, hemispherical, and flat) in some of the proposed laminates. Three different impactor velocities, i.e., 5, 10, and 15 m/s with a mass of 1 kg, are assumed for the damage study. The results reveal that the penetration by the conical impactor is much more than H.S. and flat impactors (Fig. 5). This could be because of the less area at the point of contact of the conical impactor. Flat impactor has the largest area of impact when compared to H.S. and conical impactors, which can be clearly seen in Fig. 6.

For the conical impactor, the damaged area is the least owing to local penetration. It can be summed up as follows: the more the contact area between impactor and laminate, the more the damage and vice versa.

As the velocity increases, the damaged area also increases. But there must be a limit to which the laminate can deform and will start to break apart as a result of high velocity. This can be a future scope of this study.

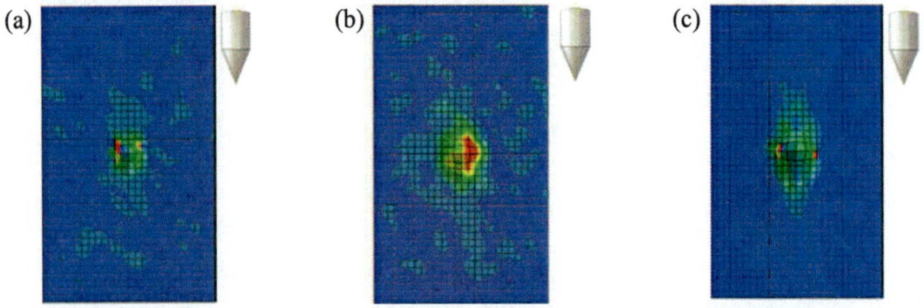

Fig. 5 Damage in laminate by conical impactor for velocities **a** 5 m/s **b** 10 m/s **c** 15 m/s

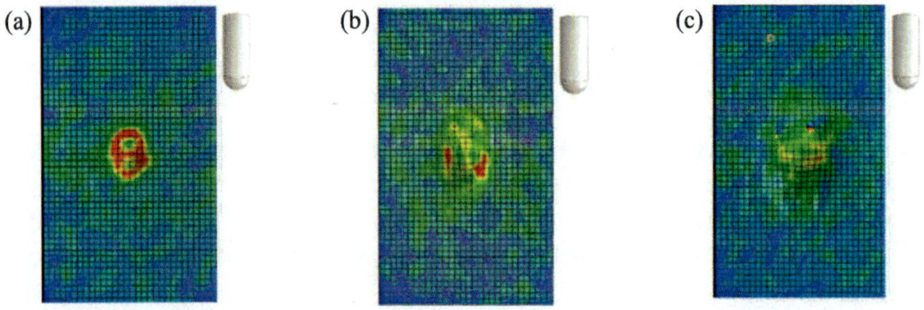

Fig. 6 Damage in laminate by hemispherical impactor for velocities **a** 5 m/s **b** 10 m/s **c** 15 m/s

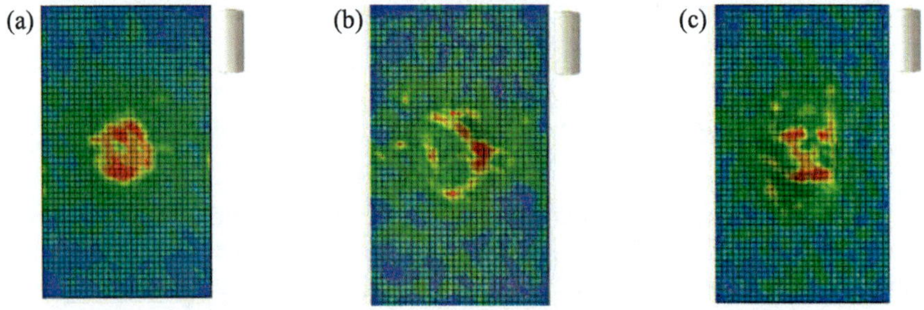

Fig. 7 Damage in laminate by flat impactor for velocities **a** 5 m/s **b** 10 m/s **c** 15 m/s

3.4 Coefficient of Restitution (CoR)

The coefficient of restitution is described as the ratio of the residual velocity of the object after the impact and the initial velocity of the object [18]. CoR can be expressed as

$$\text{CoR} = Vr/Vi \tag{1}$$

where Vr and Vi are the residual and initial velocities, respectively.

The values of the coefficient of restitution for velocities 5, 10, and 15 m/s are shown graphically in Fig. 8.

It is found that CoR is reduced when the impact velocity increases. The trend for all impactors in Fig. 8 shows just this. This indicates that the composite goes under plastic deformation and absorbs more energy. Also, for all composites and for all velocities, the conical impactor has the least CoR, which suggests that the energy absorbed is high for the conical impactor, around 25% more than the HS impactor and 40% more than the flat impactor. For the hemispherical impactor, the CoR follows Hybrid laminate > Pure fiber laminate, 30% more, which indicates that the hybrid composites show more resistance to the impactor with minor damage. Whereas, for conical and flat impactors, the CoR is greatest for BRBRB laminate.

The energy absorbed can be found by knowing the impactors' initial velocity, residual velocity, and mass. Energy absorbed can then be expressed as

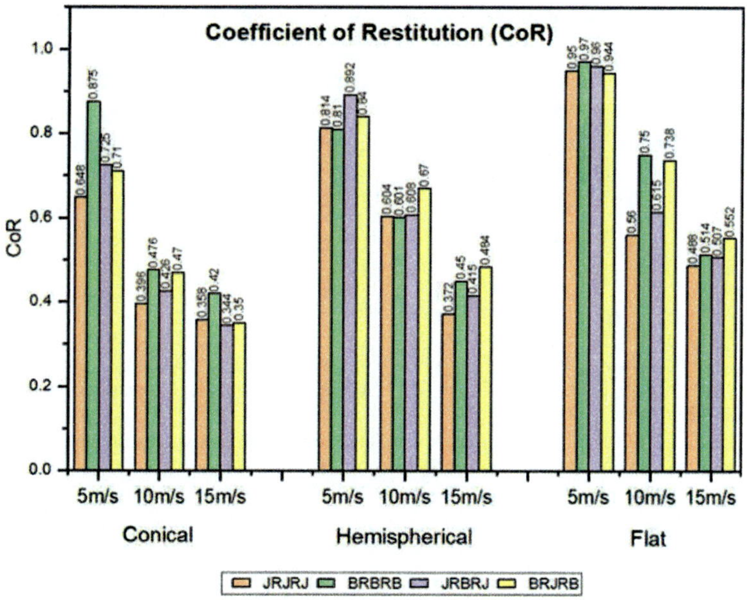

Fig. 8 Coefficient of restitution (CoR) for all impactors

$$\text{Energy Absorbed} = 0.5 * m * \left(Vr^2 - Vi^2\right) \quad (2)$$

where Vr and Vi are the residual and initial velocities, respectively, and m is the mass of the impactor

4 Conclusions

This study used FE analysis to investigate the low-velocity impact response of jute, banana, and their hybrid composite with natural rubber as a matrix, using three different impactors: flat, conical, and hemispherical.

When the laminates are impacted at higher velocities (10, 15 m/s) by a flat impactor, it shows 1.1–1.25 times energy absorption. Thus, the energy absorption capability depends on the area of contact. The JRJRJ composite absorbed, on average, 12% more than the hybrid laminates and 20% more than the BRBRB laminate. The hybrid laminates absorbed an average of 11% more than the BRBRB laminate, which shows that the hybridization of jute and banana fiber decreases the absorption capacity while increasing the stiffness. The CoR of the conical impactor is 25% more than the HS impactor and 40% more than the flat impactor, which suggests that the energy absorbed is more for the conical impactor. The CoR is up to 30% more for the hybrid composites than the pure fiber composites, suggesting that hybrid composites resist more damage.

As a result, the contact area between the impactor and the composite laminate is shown to have a substantial influence on the energy absorption capacity and the damage extent for low-velocity impacts. And the hybridization of the jute and banana fibers makes the composite stiffer and can be used for sacrificial structural applications. At the same time, there are other ways to make the composite stiffer such as using a stronger fiber material and making sandwich panels, which can be explored ahead.

References

1. Célino A, Fréour S, Jacquemin F, Casari P (2014) The hygroscopic behavior of plant fibers: a review. Front Chem. 1:43. PMID: 24790971; PMCID: PMC3982556. https://doi.org/10.3389/fchem.2013.00043
2. Chai GB, Zhu S (2011) A review of the low-velocity impact on sandwich structures. Proc Inst Mech Eng Part L: J Mater Des Appl 225(4):207–230. https://doi.org/10.1177/1464420711409985
3. Vishwas M, Joladarashi S, Kulkarni SM (2019c) Investigation on effect of using rubber as core material in sandwich composite plate subjected to low velocity normal and oblique impact loading. Scientia Iranica: Trans Mech Eng (B) 26(2):897–907 https://doi.org/10.24200/sci.2018.5538.1331

4. Rajak DK, Pagar DD, Menezes PL, Linul E (2019) Fiber-reinforced polymer composites: manufacturing, properties, and applications. Polymers 11(10):1667. https://doi.org/10.3390/polym11101667
5. Khodadadi A, Liaghat G, Ahmadi H, Reza A, Yavar B, Razmkhah O, Asemeni S (2019a) Numerical and experimental study of impact on hyperelastic rubber panels. Iranian Polymer J 28(2):113–22. Springer Berlin Heidelberg. https://doi.org/10.1007/s13726-018-0682-x
6. Mahesh V, Harausampath D, Mahesh V (2021) An experimental study on ballistic impact response of jute reinforced polyethylene glycol and nano silica based shear thickening fluid composite. In Defence Technology, Elsevier Ltd. 2214–9147. https://doi.org/10.1016/j.dt.2021.03.013
7. Mahesh V, Joladarashi S, Kulkarni SM (2020) Influence of thickness and projectile shape on penetration resistance of the compliant composite. Defence Tech 17(2):245–256. https://doi.org/10.1016/j.dt.2020.03.006
8. Mahesh V, Joladarashi S, Kulkarni SM (2020) Tribo-mechanical characterization and optimization of green flexible composites. Emerg Mater Res 9(3):887–896. https://doi.org/10.1680/jemmr.19.00145
9. Mahesh V, Joladarashi S, Kulkarni SM (2020) A comprehensive review on material selection for polymer matrix composites subjected to impact load. Def Tech 17(1):257–277. https://doi.org/10.1016/j.dt.2020.04.002
10. Chou T-W (1989) Flexible composites. J Mater Sci 24(3):761–783. https://doi.org/10.1007/BF01148757
11. Fadzullah SSM, Ramli SNN, Mustafa Z, Razali AS, Sivakumar D, Ismail I (2020) Low-velocity impact behaviour of pineapple leaf fiber reinforced polylactic acid biocomposites. J Adv Manuf Technol 14:5854. https://jamt.utem.edu.my/jamt/article/view/5854/3889
12. Mahesh V, Joladarashi S, Kulkarni SM (2020) Influence of laminate thickness and impactor shape on low-velocity impact response of jute-epoxy composite: F.E. study, Mater Today: Proc 28, Part 2:545–550. ISSN 2214-7853. https://doi.org/10.1016/j.matpr.2019.12.216
13. Mahesh Vishwas JS, Kulkarni SM (2019) Physio-mechanical and wear properties of novel jute reinforced natural rubber based flexible composite, Mater Res Express https://doi.org/10.1088/2053-1591/ab0164
14. Moustafa K, Aldajah S, Hayek S, Alomari A, Haik Y (2014) Role of nanofillers in low-speed impact enhancement of composites. J Compos Mater 48(14):1735–1744. https://doi.org/10.1177/0021998313490538
15. Rajkumar D, Mahesh V, Joladarashi S, Kulkarni SM (2022) A novel flexible green composite with sisal and natural rubber: investigation under low-velocity impact. J Nat Fib. https://doi.org/10.1080/15440478.2022.2036292
16. Mahesh V, Joladarashi Kulkarni SM (2020) Influence of laminate thickness and impactor shape on low-velocity impact response of jute-epoxy composite: F.E. study. Mater. Today Proc. [Internet]. 2020; Available from https://doi.org/10.1016/j.matpr.2019.12.216
17. Bismarck A, Mishra S, Lampke T (2005) Plant fibers as reinforcement for green composites. Nat Fib Biopoly Biocompos. https://doi.org/10.1201/9780203508206.ch2
18. Mahesh V, Joladarashi S, Kulkarni SM (2019) An experimental investigation on low-velocity impact response of novel jute/rubber flexible bio-composite. Compos Struct 225(111190):1–12. https://doi.org/10.1016/j.compstruct.2019.111190

Visualization of Flow and Impact Process of Oils Used in Shirodhara Treatment on a Flat Plate Under the Influence of Gravity

Rishabh, B. S. V. Patnaik, and Chebolu Lakshmana Rao

Abstract Shirodhara is an Ayurvedic process for treating various psychological ailments like depression, anxiety, mental stress, insomnia, headache, etc. It requires a narrow laminar low-speed stream of a medicinal fluid, in most of the cases, Mahanarayana or Ksheerabala Oil to be impacted on to the human forehead under the influence of gravity for a prescribed period of time, under controlled temperature (Pavana and Manoj in K. Rajagopalan ayurveda series 3:18, 2010 [1]; Vagbhaṭa, Ashtanga hridaya-with commentary of aruna dutta and hemadri. Reprint ed. Varanasi: Chowkhambha Sanskrit Series, 1946 [2]). Shirodhara is a dynamic process and despite being a very popular treatment method, just like other Ayurvedic treatments, it also relies on empirical rules. Hence, there is a need for a systematic scientific study of this treatment process. To understand the complete functioning of Shirodhara process, there is a necessity to visualize the fluid flow process during the treatment. We perform simple visualization of the fluid flow process under the influence of gravity coming out of a hanging Shirodhara pot (usually made up of brass). The two most commonly used oils in this treatment process are Ksheerabala and Mahanarayana oils and hence, those two have been visualized. Since the flow is axisymmetric, we have considered only one half of the flow from the images. The height differential between the outlet of Shirodhara pot and the impact plate is not very large and the velocities are also low, therefore, the images are captured using the low-speed camera as the disturbances are of negligible order. The images are processed using a MATLAB script to draw the contours of the boundary of the flow and then they are compared to the existing solutions.

Keywords Shirodhara · Ayurvedic · Ksheerabala · Mahanarayana · Flow visualization · Fluid Impact

Rishabh (✉) · B. S. V. Patnaik · C. Lakshmana Rao
Applied Mechanics Department, Indian Institute of Technology Madras, Chennai 600036, India
e-mail: rishabh.022298@gmail.com

C. Lakshmana Rao
e-mail: lakshman@iitm.ac.in

R. Velmurugan et al. (eds.), *Dynamic Behavior of Soft and Hard Materials Volume 1*, Springer Proceedings in Materials 34, https://doi.org/10.1007/978-981-99-6030-9_31

1 Introduction

Shirodhara is an ayurvedic treatment process, in which a medicinal fluid is allowed to impact on to the human forehead, see Fig. 1. It is considered to be a relaxation therapy which is commonly employed to cure various mental-related ailments like migraine, headaches, insomnia, etc. [1–4]. It is a very popular treatment procedure and runs completely upon the various empirical rules prescribed thousands of years ago [1, 2], few of those rules are:

- Fluid should be poured from the height of about 4 anguls i.e., approximately 3–4 inches.
- Vessel should be suspended with ropes made of jute.
- Material should be made up of either superior metals or earthenware.
- Hole in the vessel should be roughly of the same circumference as the little finger.
- Wick should be used and should be of the similar size.
- Temperatures are also vaguely defined for various conditions.
- Stream should be continuous.

Various studies [1, 3–10] have taken place to study the effectiveness of the Shirodhara. Few of those assessed the EEG and ECG of the patients along with few other parameters like skin temperature, blood sample, urinary sample, etc. [3, 5–9]. But most of the research regarding the effectiveness of Shirodhara is based on the responses of the patients to a pre-designed questionnaire [4, 7, 10]. Those questionnaires were used to evaluate the various psychological parameters associated with anxiety, depression, insomnia, etc. Researchers have also attempted to measure the impact forces due to the impact of medicinal oils on a flat region [11] and characterize the electromechanical response due to the fluid impact during the treatment on forehead tissue [12].

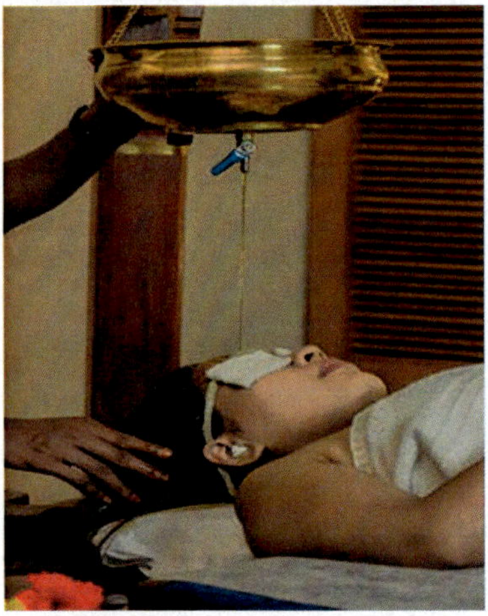

Fig. 1 A patient undergoing Shirodhara treatment

Despite that, there is hardly any scientific explanation available about why the abovementioned rules are required to be followed as it is without changing any parameters. That is why there is a necessity to explore the mechanics of the treatment process scientifically. The first step to study the mechanics of Shirodhara is to identify and characterize the type of flow for the prescribed conditions and fluids.

The flow in itself is very simple and its impact has already been studied for various tapping conditions [13], it depicts the flow falling through a nozzle under the influence of gravity from a container. These kinds of flow have been studied extensively in the past and are readily available in the textbooks. Few of the research articles related to such flow have been referenced [14–17]. Similarly, there are many resources available which deal with the visualization of fluid flow and flow pattern development like [18–22].

This paper deals with the visualization of the flow and the impact process of the Mahanarayana and Ksheerabala Oil when they come out of a statically hanging tank (Shirodhara pot) through a nozzle and impact a flat plate, see Fig. 2. We have developed a scheme which tracks and maps the boundary of the flow. Using the data associated with the boundary of the flow, we have calculated the velocities, cross-sectional area, and the volumetric flow rate near the tip of the developing flow. These parameters are used to characterize the type of the flow that actually takes place. The region near the impact zone has also been analyzed to understand the spread of the oils due to impact. This will be helpful in studying the Fluid–Solid Interaction process of these oils on human forehead.

The flow has been visualized for 2 oils, namely Mahanarayana Oil and Ksheerabala Oil. Their properties are as follows, see Table 1:

It has to be noted that the density and viscosity of water are 1000 kg/m^3 and 0.001 kg/m-s, respectively, so using water as a reference is not a very good approach.

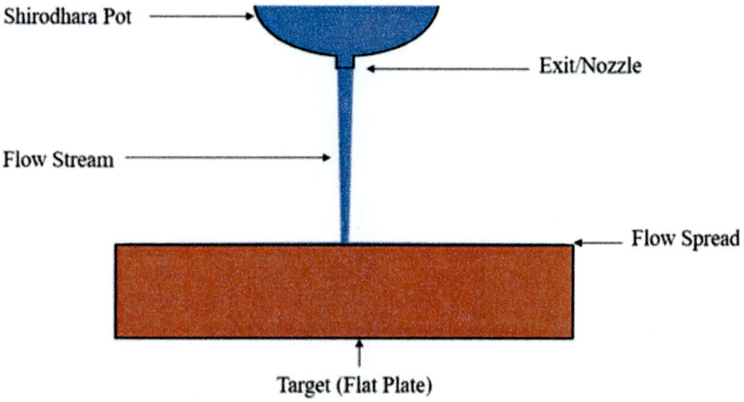

Fig. 2 Schematic diagram of Shirodhara treatment

Table 1 Properties of the oils used in Shirodhara treatment

Oil	Density (kg/m^3)	Viscosity (kg/m-s)
Mahanarayana	800	0.056
Ksheerabala	810	0.051

2 Experimentation and Analysis

Experimental Setup consists of a Shirodhara Pot with a nozzle attached to it at the bottom, as shown in Fig. 1. Instead of human forehead, a flat plate is used to study the spread of the impact of the fluid, see Fig. 2. The Oils are allowed to drop under the influence of gravity from the nozzle opening. Since the cross-sectional area of the pot is very large compared to the nozzle opening, it is assumed that for the shorter duration of time (less than 0.2 s), the fluid level in the Shirodhara pot is constant, hence giving the constant velocity field. The experiment takes place at the room temperature. The schematic of the experimental is shown below, see Fig. 3. Whole system is axisymmetric.

The development of flow has been recorded and analyzed. The frames are analyzed at the intervals of 1/30 s each from the initiation of the flow to the complete development of the flow. The developing flow for the Mahanarayana Oil has been shown, see Fig. 4.

Once the frames have been obtained, then they are processed using an Algorithm (see Appendix) to draw the boundary of the developing flow. Since it is an axisymmetric case, only one half of the images has been considered for the processing and drawing the boundaries. The images have been pre-processed using Image Region Analyzer and Image Viewer applications available in MATLAB R2021b Software.

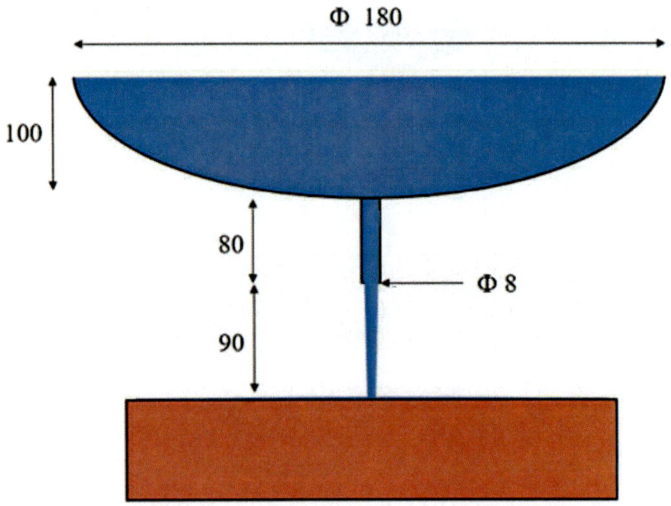

Fig. 3 Schematic diagram of Experimental Setup (All dimensions are in mm)

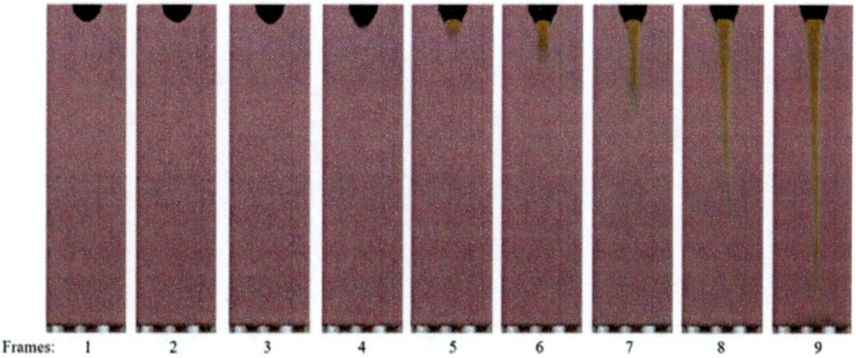

Fig. 4 Development of flow for Mahanarayana Oil (Each frame has a time difference of 1/30 s from previous ones)

Fig. 5 Methodology used to visualize fluid flow

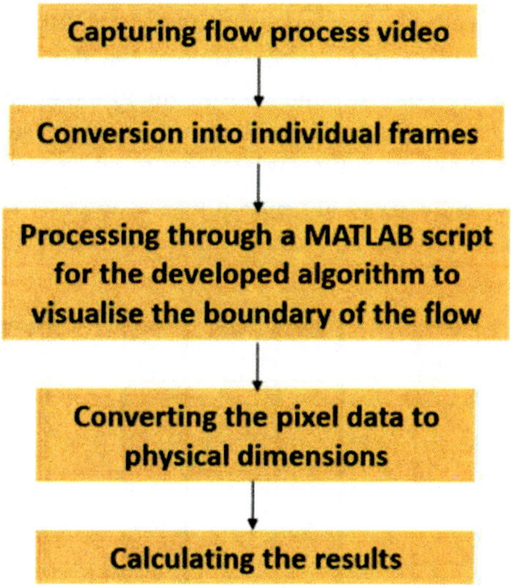

Same software has been used to write the script to implement the algorithm. The Methodology has been described in Fig. 5.

3 Development of Flow and Its Boundary

The boundary of the flow has been mapped as described in Sect. 2 and Appendix. For Mahanarayana Oil, flow and the boundary develop as shown in Fig. 6. The time difference between each frame is of 1/30 s.

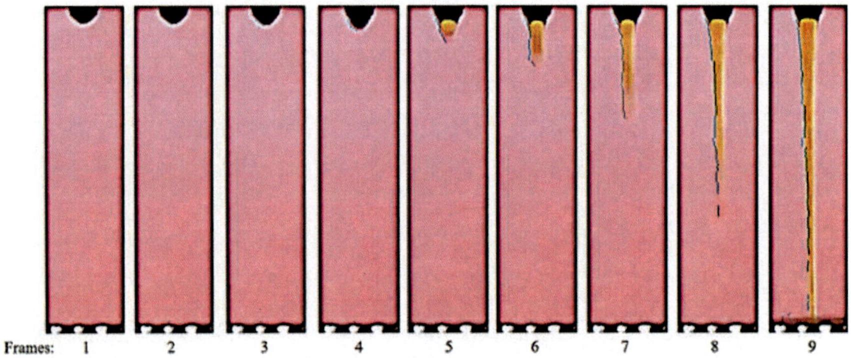

Fig. 6 Development of flow and its boundary for Mahanarayana Oil

Similarly, flow boundary has been mapped for Ksheerabala Oil, see Fig. 7. The information collected using the algorithm (see Sect. 2 and Appendix) is used to calculate velocity of the front of the flow during its development, the area of the front, and the volumetric flow rate of the fluid (see Sect. 4). Using those parameters, the flow has been characterized for each case.

Furthermore, the region closer to the plate has also been observed, see Fig. 8 for what happens to the fluid after the impact. It is evident from observing the region that there is no sign of any major hydraulic jump or separation.

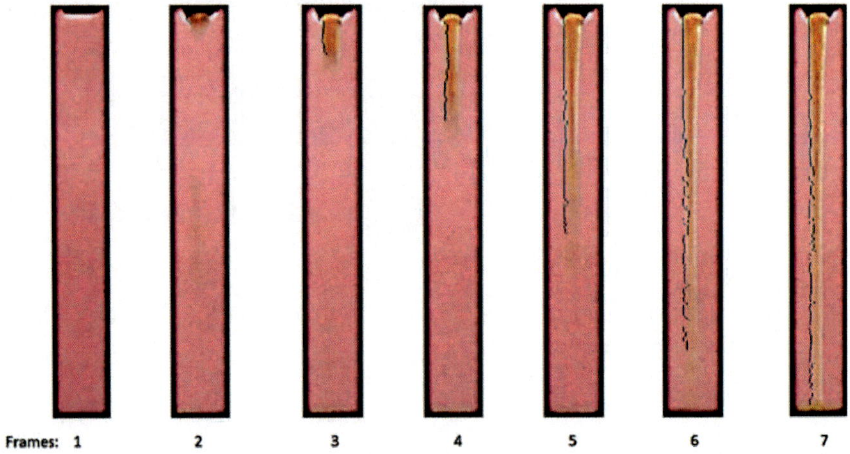

Fig. 7 Development of flow and its boundary for Ksheerabala Oil

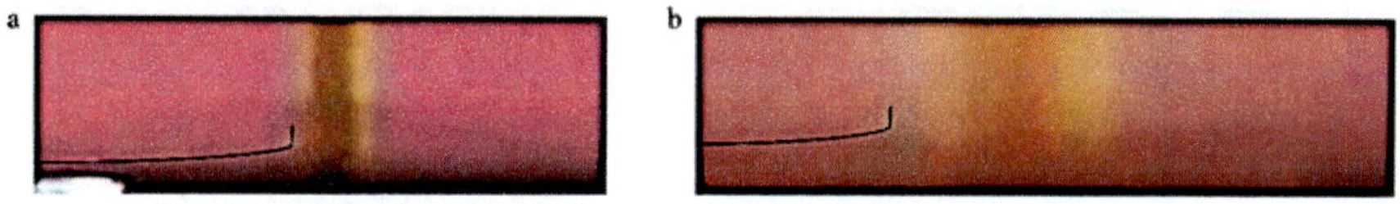

Fig. 8 Closer look at the impact region for **a** Mahanarayana Oil; **b** Ksheerabala Oil

4 Results and Discussion

Using the algorithm and the boundary data, velocity of the front has been calculated as the difference between the vertical value of the pixel location of the tip of the flow from the previous frame and the current frame divided by the time step of 1/30 s. The velocities for the Mahanarayana Oil and the Ksheerabala Oil have been plotted, see Fig. 9. Similarly, area has been calculated by taking the pixel data at the beginning of the meniscus at the tip of the flow. The difference between the horizontal location of the previously mentioned pixel and the center of the image corresponds to the radius. With the help of the radius, the area versus time has been plotted, see Fig. 10. Furthermore, volumetric flow rate has been calculated by multiplying the area and velocity to plot the volumetric flow rate versus time, see Fig. 11.

Velocity versus time plot shows that the velocity of the tip is somewhat increasing linearly with time, which is to be expected under the influence of constant force, i.e., gravity.

The areas are decreasing hyperbolically in both the cases, which is to be expected since the velocity is increasing linearly.

There are hardly any variations in volumetric flow rates and hence, we can consider it to be constant for the abovementioned experimental conditions.

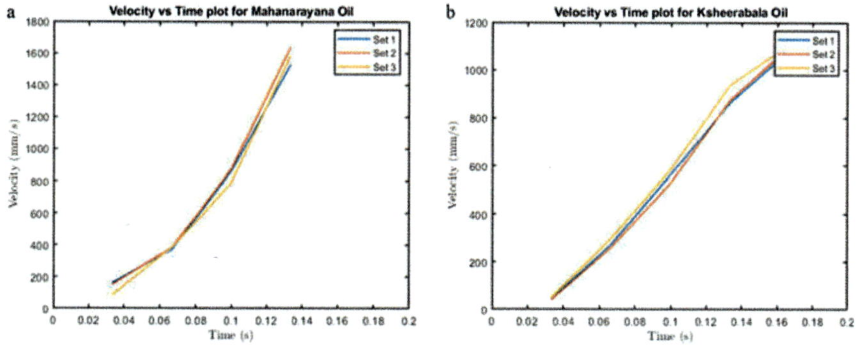

Fig. 9 Velocity versus Time plot at the tip of the developing flow for **a** Mahanarayana Oil; **b** Ksheerabala Oil

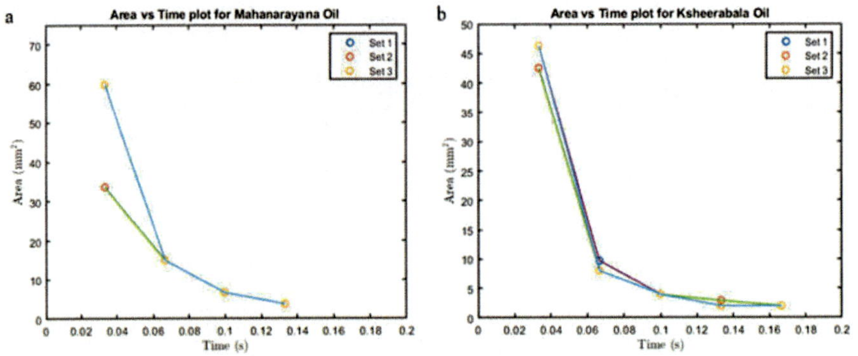

Fig. 10 Area versus Time at the tip of the developing flow plot for **a** Mahanarayana Oil; **b** Ksheerabala Oil

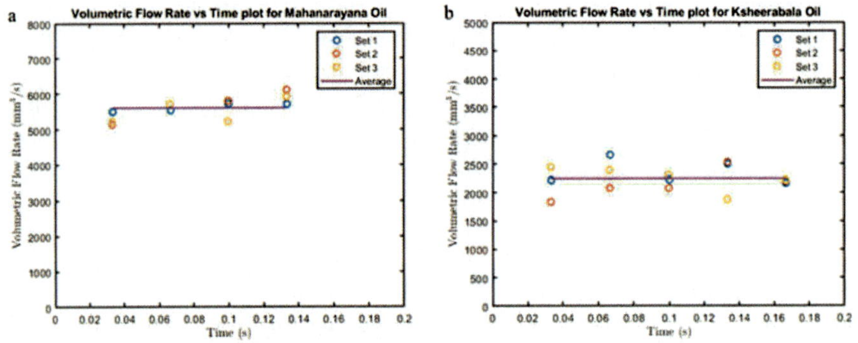

Fig. 11 Volumetric flow rate versus Time at the tip of the developing flow plot for **a** Mahanarayana Oil; **b** Ksheerabala Oil

5 Conclusion

In this paper, a method has been suggested to visualize and characterize flows with lower velocities. Flow and impact processes have been visualized for the predefined configuration of Shirodhara treatment. It is observed that the velocities are increasing linearly and there is no indication of separation of flow, forming of bubbles and vortices. The area is also decreasing hyperbolically as we move away from the nozzle outlet and the volumetric flow rate is found to be constant. All of this indicates that the rules which were mentioned in the Sect. 1 are there to ensure that the flow is laminar and the treatment process is smooth.

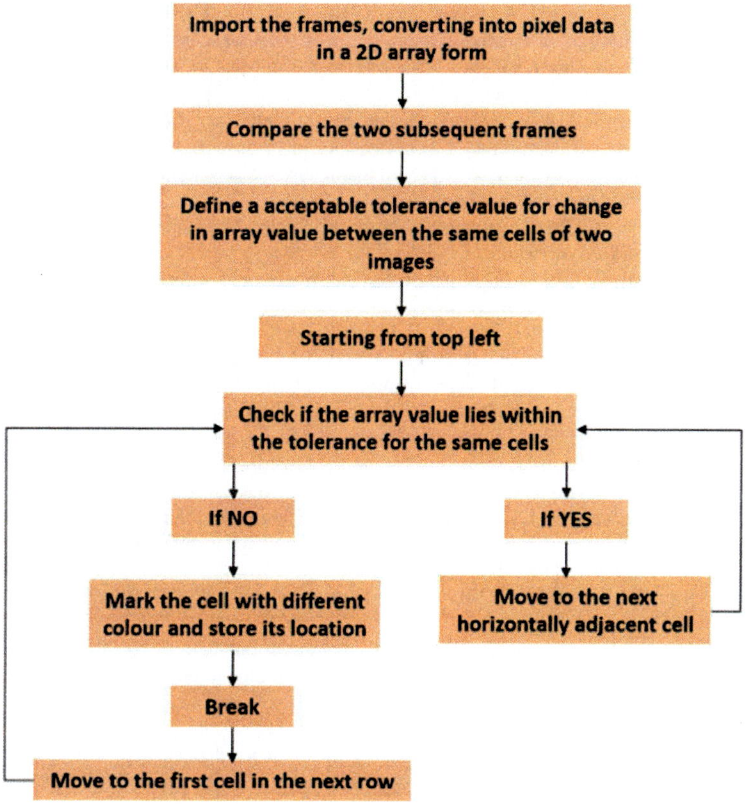

Fig. 12 Algorithm for mapping and storing the information about the boundary of the flow

Appendix: Algorithm for Mapping the Boundary of the Flow

The schematic for the algorithm implemented for mapping and storing the information of the boundary of the flow has been shown in the Fig. 12. Its implementation finds the boundary of the flow, marks it, and stores the location of the pixel associated with it.

References

1. Pavana J, Manoj S (2010) Keraliya Chikitsa paddhati: Padmasri DR. K. Rajagopalan ayurveda series 3, 18
2. Vagbhaṭa (1946) Ashtanga hridaya-with commentary of aruna dutta and hemadri. Reprint ed. Varanasi: Chowkhambha Sanskrit Series

3. Uebaba K, Xu FH, Tagawa M, Asakura R, Itou T, Tatsuse T et al (2005) Using a healing robot for the scientific study of shirodhara, Altered states of consciousness and decreased anxiety through Indian dripping oil treatments. IEEE Eng Med Biol Mag 24(2):69–78
4. Tokinobu A, Takashi Y, Toshihide T, Hiroyuki D (2016) Effects of ayurvedic oil-dripping treatment with sesame oil vs. with warm water on sleep: a randomized single-blinded crossover pilot study. J Alternative Compl Med 22(1):52–58
5. Dhuri KD, Bodhe PV, Vaidya AB (2013) Shirodhara: a psychophysiological profile in healthy volunteers. J Ayurveda Integrat Med 4(1):40
6. Pokharel S, Sharma AK (2010) Evaluation of Insomrid tablet and Shirodhara in the management of Anidra (Insomnia). Ayurveda 31(1):40
7. Thatte UM, Gogtay NJ (2018) Traditional medicine: a review of work in India (2012–2017). Proc Indian Natl Sci Acad 84(1):267–279
8. Xu FH, Uebaba K, Ogawa H, Tatsuse T, Wang BH, Hisajima T, Venkatraman S (2008) Pharmaco-physio-psychologic effect of Ayurvedic oil-dripping treatment using an essential oil from Lavendula angustifolia. J Alternat Complement Med 14(8):947–956
9. Uebaba K, Xu FH, Ogawa H, Tatsuse T, Wang BH, Hisajima T, Venkatraman S (2008) Psychoneuroimmunologic effects of Ayurvedic oil- dripping treatment. J Alternat Complement Med 14(10):1189–1198
10. Tubaki BR, Verma A, Sasidharan A, Sulekha S, Sathyaprabha TN, Sudhakar D, Chandrashekar CR, Lavekar GS, Kutty BM (2016) Manasamitra Vataka and Shirodhara treatments preserve slow wave sleep and promote sleep continuity in patients with generalized anxiety disorder and co-morbid generalized social phobia. Current Sci 111(2)
11. Meenraj S, Rao CL, Venkatesh B (2019) Impact force measurement in Shirodhara. J Ayurveda Integrat Med 10(3):192–197
12. Meenraj S, Rao CL, Balasubramanian V (2018) Characterisation of electromechanical response of forehead tissues due to fluid impact during shirodhara treatment. Int J Mater Struct Integrity 12(4):353–366
13. Meenraj S, Rao CL, Venkatesh B (2018) Fluid impact under various tapping conditions for biomedical application (Shirodhara). In: ASME International mechanical engineering congress and exposition, vol 52026. American Society of Mechanical Engineers, p V003T04A095
14. Dias F, Vanden-Broeck JM (1990) Flows emerging from a nozzle and falling under gravity. J Fluid Mech 213:465–477
15. Clarke NS (1968) Two-dimensional flow under gravity in a jet of viscous liquid. J Fluid Mech 31(3):481–500
16. Vanden-Broeck JM, Keller JB (1982) Jets rising and falling under gravity. J Fluid Mech 124:335–345
17. Dias F, Christodoulides P (1991) Ideal jets falling under gravity. Phys Fluids A Fluid Dyn 3(7):1711–1717
18. Merzkirch W (2012) Flow visualization. Elsevier
19. Ravigururajan TS, Bergles AE (1994) Visualization of flow phenomena near enhanced surfaces asmedigitalcolllection.asme.org, pp 54–57
20. Merzkirch W (1987) Techniques of flow visualization. Advisory Group for Aerospace Research And Development Neuilly-Sur-Seine (France)
21. Smits AJ (2012) Flow visualization: techniques and examples. World Scientific
22. Thurow BS, Lynch KP (2009) Development of a high-speed three-dimensional flow visualization technique. AIAA J 47(12):2857–2865

Oblique Impact Simulation Study of Ti-6Al-4 V Alloy Plates for Analysis of Blade-Off Event in Turbofan Aero-Engine

Saurabh Mangal, Sri Krishna Sudhamsu Kambhammettu, and C. Lakshmana Rao

Abstract Aero-engines employ heat-resistant, high-strength alloys such as the titanium-based superalloy Ti-6Al-4 V, in the construction of engine fan blades. Due to several sustained, fatigue and impact loads, these blades are prone to fracture during their operation. In such an event, the kinetically energized fan blade may fly away to impact with the casing or the containment ring of the engine. This could lead to a major catastrophe, if these flying parts manage to penetrate through the engine casing and impact other structural components of the aircraft. In this paper, a simplified analysis of such blade-off event is performed to understand the role of impact velocity, the rotational velocity of the fan blade, blade orientation and the angle of oblique impact of the blade on the engine casing or containment ring. For this, the fan blade and the containment ring are modeled as rectangular plates of appropriate sizes. The problem is analyzed using explicit finite element analysis of the impact of a moving blade on a fixed stationary containment ring at different orientations and velocities of the blade. Standard Johnson–Cook material parameters and fracture criteria have been used to characterize the titanium alloy behavior and failure in impact loading. The results provide useful insights into the impact problem and will help toward improving the design for better safety of the aircraft.

Keywords Titanium-based superalloy · Bade-off event · Impact

Nomenclature

v	Linear velocity for impact (m/s) = {250, 300, 350, 400}
v_n	Component of linear velocity perpendicular to target = v cos Ø
v_t	Component of linear velocity parallel to target = v sin Ø
v_{res}	Residual velocity after impact

S. Mangal (✉) · S. K. S. Kambhammettu · C. L. Rao
Indian Institute of Technology Madras, Chennai 600036, India
e-mail: smangal1308@gmail.com

R. Velmurugan et al. (eds.), *Dynamic Behavior of Soft and Hard Materials Volume 1*, Springer Proceedings in Materials 34, https://doi.org/10.1007/978-981-99-6030-9_32

v_0 Linear resultant velocity of a fan blade which takes care linear velocity and rotational velocity both
ω Initial rotational velocity (rad/s) = {0, 50, 100, 150, 200, 250, 300}
α Blade orientation angle = {0°, 15°, 30°, 45°, 60°, 75°}
Ø Angle of oblique impact = {0°, 15°, 30°, 45°, 60°, 75°}
a Perforation size = Area in mm^2

1 Introduction

Blade-off event is an undesirable event that signifies the impact of a broken "flying-off" blade with the casing or the containment ring of the engine. It is important to ensure that such an accident doesn't exacerbate into a catastrophe. One such concern is when the broken blade, after impact, perforates through the casing hitting other structural components of the aircraft. Hence, it is important to ensure that the broken blade is contained inside the engine casing in case of a blade-off event. Recent advances in gas turbine engines employ a containment ring that captures the kinetic energy and rotational energy of a released fan blade and stops it from perforating the casing. The containment ring is placed peripherally inside the casing of the turbine. Any blade of a gas turbine engine may prone to fracture but the fan blade is the most critical blade owing to its high weight leading to higher kinetic energy when compared to blades of compressor and turbine.

The main objective of this paper is to simulate the impact of a blade on the containment ring during the blade-off event at different blade linear velocities, rotational velocities, blade orientations and oblique angles. The insights obtained from this study will help in improving the blade design for better safety during a blade-off event.

The impact due to blade-off event (blade shedding) of the gas turbofan engine is a rotational oblique impact event as shown in Fig. 1. In this study, the perforation of the containment ring is simulated using the finite element dynamic-explicit tool. In the literature, analytical and experimental work [1] has been done, in spite of the predictive capability of finite element analysis, no one has used numerical methods significantly. In the literature, numerical studies are very less and that too consider only 45° of oblique impact [1] for the simulation of the blade-shedding event. Studying numerical simulations with a range of angles can give more insights and scope of study that's what's focused on in this paper.

Figure 1 shows a typical impact event of a blade with the containment ring which happens at different values of α and Ø. When Ø = 0°, the impact is called a normal impact. Variation of angle α that governs the blade orientation against the containment ring has not been studied in the literature so far. However, it is important to study angle α, hence in this paper, we are going to study its effect corresponding to different angular velocities and linear velocities.

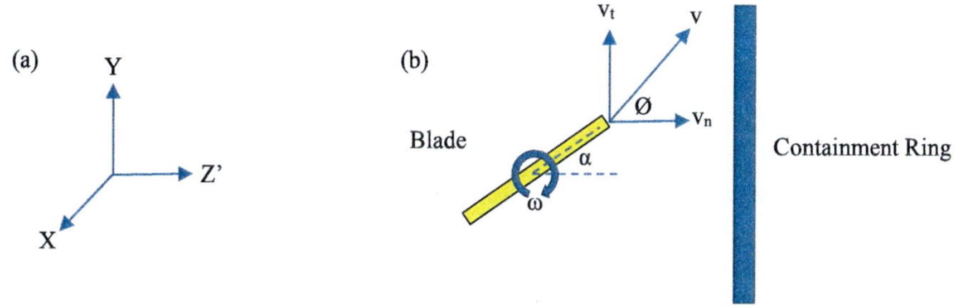

Fig. 1 **a** Cartesian coordinate system followed in Abaqus 6.19; **b** Geometrical representation of the impact problem addressed

2 Components of Aero-Engine

2.1 Material, Its Composition and Applications

Aero-engine fan blade and fan casing is manufactured using Ti-6Al-4 V, ASTM Grade-5 (American standard Ti64 and Chinese standard TC4) which is a titanium-based alloy [2]. High strength to weight ratio, corrosion resistance and creep resistance are the primary characteristics that make this material employable for this application.

Protective containment ring is manufactured with energy-absorbing materials i.e. Ti-6Al-4 V, aluminum foam 10%, kevlar, alumina or any optimized combination [3].

Fan blades, low-pressure compressor and low-pressure turbine are connected with a low-speed shaft (2500–3000 rpm), while the high-pressure compressor and high-pressure turbine are connected with a high-speed shaft (10,000–15,000 rpm). Jet propulsion and power production are the applications that require high-strength material at elevated temperatures and Ti-6Al-4 V is a suitable candidate to perform in the odd conditions.

2.2 Modes of Failure of the Blade

Bird ingestion, hailstones, rain-drops, runway debris, volcanic ash and broken fuselage parts are the foreign objects which can damage the fan blade. Material defects, creep loading, fatigue loading, corrosion and resonance are the possible reasons that reduce the structural integrity. Non-idealized testing methods, poor blade retaining capability, improper testing, poor scheduling and poor standards of maintenance are the reasons which increase the chances of failure of components.

Table 1 Physical constants of material Ti-64

E (GPa)	ν	ρ (kg/m^3)	T_m (K)	T_r (K)	C_p (J kg^{-1} K^{-1})
113	0.33	4430	1878	293	580

Table 2 Johnson–Cook constitutive relation constants for material Ti-64

$\acute{\varepsilon}_o$ (s^{-1})	A (MPa)	B (MPa)	n	C	m
1	1098	1092	0.93	0.014	1.1

Table 3 Johnson–Cook fracture criterion constants for material Ti-64

D_1	D_2	D_3	D_4	D_5
-0.090	0.270	−0.480	0.014	3.870

3 Numerical Model Incorporated and Tools Used

3.1 Material Model and Fracture Criterion

In the literature [1], the circular containment ring with a general blade profile is modeled as a planar containment ring (1500 mm × 1500 mm × 4.51 mm) with a planar blade (300 mm × 10 mm × 500 mm) profile for simulations.

The accuracy of any numerical simulation is reliant upon the accuracy of the employed material model. Ti-6Al-4 V is a strain rate sensitive material [4], Johnson–Cook constitutive relation can describe the behaviour of material subjected to larger strain, high strain rates and high temperature. Material Ti-6Al-4 V is modeled using the Johnson–Cook material model and for damage, the Johnson–Cook fracture criterion is used [5], which agrees the operating temperature for the fan −50 °C and for the turbine 1700 °C. Tables 1, 2 and 3 show the properties and material constants for Ti-6Al-4 V [5].

3.2 Finite Element Analysis Solver and Image-Processing Software

For the simulation, commercially available dynamic-explicit analysis code Abaqus 6.19 is used. Displacement at failure in the damage-evolution sub option (type-displacement) is taken as 0.1. In assembly, the clearance kept between the fan blade and the thin target is 0.01 mm. Step time is taken as 3 ms for all the simulations. General contact is chosen for fan blade and thin target interaction. Contact property is defined by tangential and normal behavior with 0.15 penalty. The containment ring has fixed boundary conditions and free to fail across the impact surface. For loading, the rotational angular velocity is given to the axis that passes through the centroid of the cuboidal fan blade and is parallel to 300 mm side. The element size is taken 5 mm for the entire fan blade. The containment ring is finely meshed near the area of impact and the element size is kept coarse, away from the centroid of the ring. 5, 10 and 20 mm elements are employed for the target to study impact simulations. Element shape Hex with structured technique is selected. C3D8R- 8-node linear brick, reduced integration with hourglass control is chosen for the element type. Element deletion is on for all the simulations.

For the image processing, Adobe Photoshop CS3 Extended Version 10.0 and ImageJ 1.53 k software are used.

4 Parametric Study

The interaction of the blade and casing or containment ring shows the bulge formation and perforation in simulations which has been observed in literature [1] too (Fig. 2).

4.1 *Variation of a versus v|α = Specific, ω = 200 rad/s, Ø = 0° (Normal Impact)*

In this normal impact study, the perforation size of the containment ring is studied with respect to the sub-ordnance impact linear velocity of the fan blade while keeping the rotational velocity of the fan blade constant i.e. 200 rad/s. The fan blade is oriented at different angles with respect to the containment ring. The following results are observed in Table 4, Figs. 3 and 4:

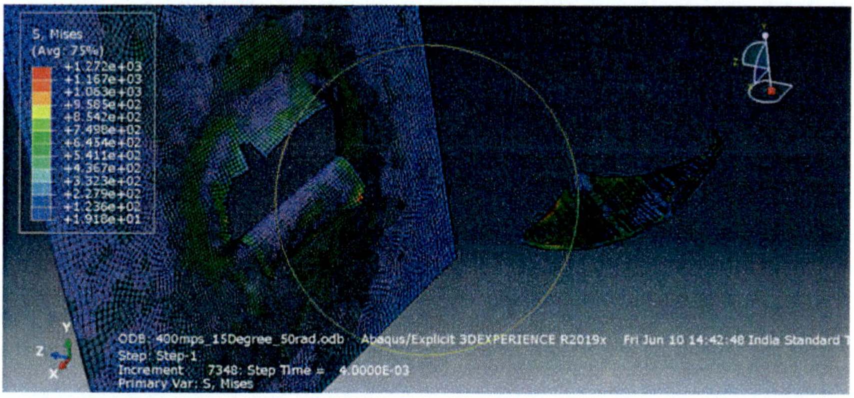

Fig. 2 Simulation shows the bulge formation and perforation of the ring

Table 4 Variation of a (mm^2) with respect to v (m/s) | α = Specific, ω = 200 rad/s, Ø = 0° (Normal Impact)

v	a | α = 0°	a | α = 15°	a | α = 30°	a | α = 45°	a | α = 60°	a | α = 75°
250	68,705.652	0	0	0	0	0
300	29,279.783	206,653.245	0	0	0	0
350	57,377.597	253,708.322	0	0	0	889.658
400	54,335.232	246,741.806	0	0	280,996.704	0

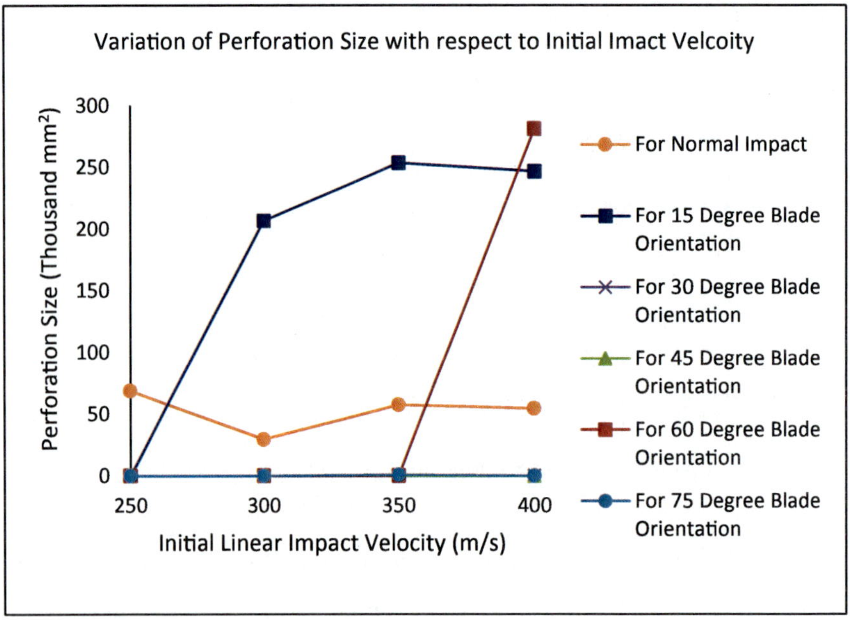

Fig. 3 Graph shows the variation of a versus v | α = Specific, ω = 200 rad/s, Ø = 0° (Normal Impact)

- 15-degree blade orientation gives the highest probability of perforation and the highest damage to the containment ring.
- Blade orientation in the range of 30 to 45 degrees gives the least probability of perforation of the containment ring.
- Only at a significantly higher linear impact velocity i.e. 400 m/s, the containment ring can be perforated by a 60-degree-oriented blade.
- At high angles of blade orientation (>30 degrees), a significant amount of kinetic energy gets distributed over a more exposed impact area and is not concentrated. Hence, perforation of the containment ring is not achievable in the velocities range (250–400 m/s) simulated.
- At lower velocity, only the normally oriented blade can perforate the containment ring. All the other orientations can rebound the blade and deform the ring plastically without any significant damage.

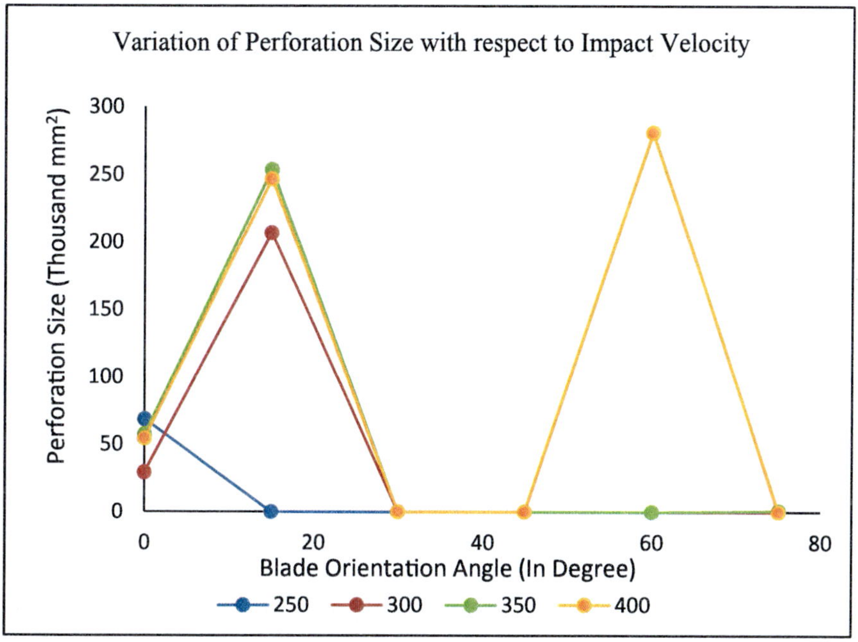

Fig. 4 Graph shows the variation of a versus α | v = Specific, ω = 200 rad/s, Ø = 0° (Normal Impact)

4.2 *Variation of a versus ω|α = Specific, v = 400 m/s, Ø = 0° (Normal Impact)*

In this normal impact study, the perforation size of the containment ring is studied with respect to the rotational velocity of the fan blade while keeping the impact linear velocity of the fan blade constant i.e. 400 m/s. The fan blade is oriented at different angles with respect to the containment ring. The following results are observed in Table 5 and Fig. 5:

Table 5 Variation of a (mm^2) versus ω (rad/s) | α = Specific, v = 400 m/s, Ø = 0° (Normal Impact)

ω	a | $\alpha = 0°$	a | $\alpha = 15°$
0	36,979.741	102,246.707
50	35,100	141,884.911
100	40,538.863	136,204.863
150	44,759.795	162,729
200	54,335.232	246,741.806
250	66,174.566	247,737
300	90,443	384,460.56

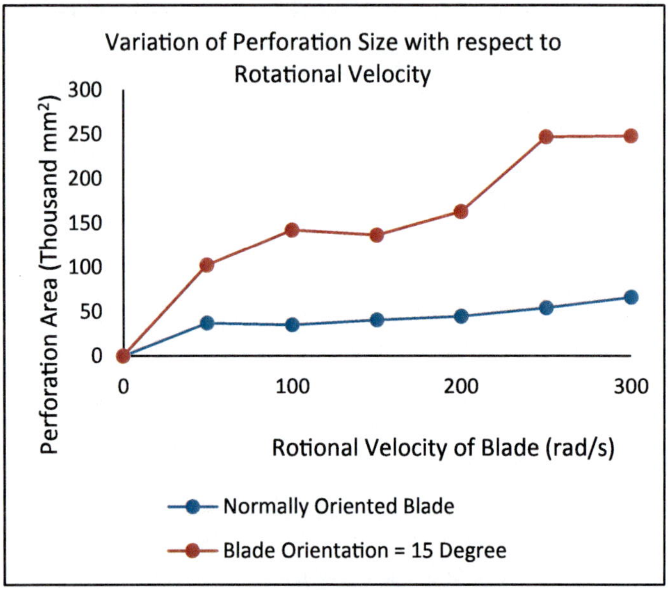

Fig. 5 Graph shows the variation of a versus ω | α = Specific, v = 400 m/s, Ø = 0° (Normal Impact)

- The size of the perforation of the containment ring depends on the rotational velocity of the fan blade. With the increase in the rotational velocity, the size of the perforated area increases. The angular rotational velocity of the blade tends to impart more surface area contact of the blade to the containment ring leading to an increment in the perforation size.
- The effect of the rotational velocity is significantly higher for the high-angle-oriented blades.

4.3 *Variation of a versus v|Ø = Specific, ω = 0 rad/s, α = 0° (Normally Oriented Fan Blade)*

In this oblique impact study, the perforation size of the containment ring is studied with respect to the impact linear velocity of the fan blade while giving no initial rotational velocity to the fan blade. The linear velocity is given to fan blade at different oblique angles to the normal. The following results are observed in Table 6 and Fig. 6:

- For a normally oriented released fan blade without initial rotational velocity, an increment in the impact velocity increases the size of the perforated area of the containment ring.
- Oblique impact angles of 15 degrees and 30 degrees show a higher dependency on the impact velocity as there are significant changes in the perforated area.

Table 6 Variation of a (mm^2) versus v (m/s) | Ø = Specific, $\omega = 0$ rad/s, $\alpha = 0°$ (Normally Oriented Fan Blade)

v	a | Ø = 0°	a | Ø = 15°	a | Ø = 30°	a | Ø = 45°	a | Ø = 60°	a | Ø = 75°
250	13,351.315	101,535.842	0	0	0	0
300	14,488.465	110,684	190,813.623	0	7217.794	0
350	16,510.644	89,766.574	224,549.659	0	8441.854	0
400	33,852.401	103,730.623	230,368.712	2583.367	7492.163	0

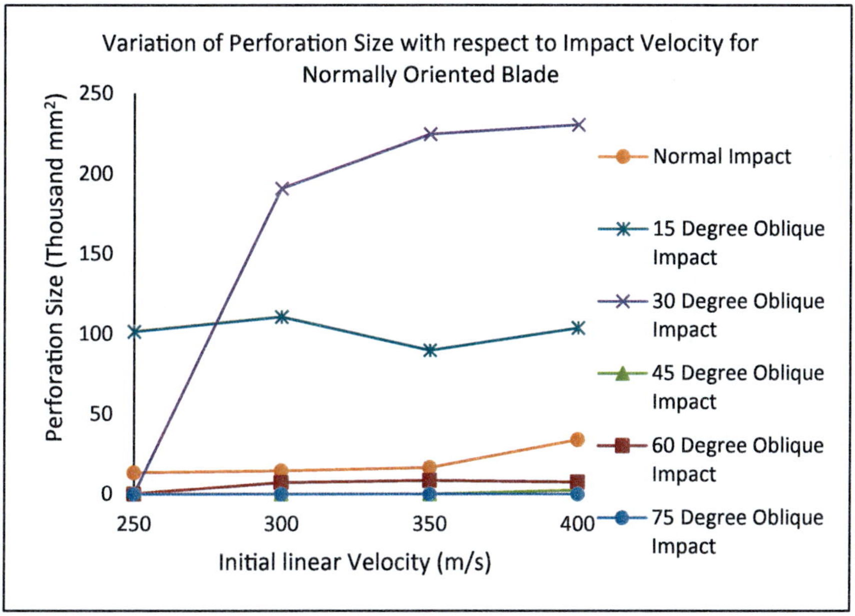

Fig. 6 Graph shows the variation of a versus v | Ø = Specific, $\omega = 0$ rad/s, $\alpha = 0°$ (Normally Oriented Fan Blade)

- At lower velocities, only lower angle oblique impact i.e. normal impact and 15 degrees are capable of perforating the containment ring.
- 45-degrees and 75-degrees oblique impact angles contribute very less to the normal component of the impact velocity. Hence, the perforation area is very small and sometimes, it leads to no perforation.
- For significant perforation, the impact velocity should be high enough for oblique impact of a 60-degree angle and more than that.

4.4 Variation of a Versus ω|Ø = Specific, v = 400 m/s, α = 0° (Normally Oriented Fan Blade)

In this oblique impact study, the perforation size of the containment ring is studied with respect to the initial angular velocity of the fan blade moving with 400 m/s initial linear velocity. The initial rotational velocity is given to fan blade at different oblique angles to the normal. The following results are observed in Table 7 and Fig. 7:

Table 7 Variation of a (mm^2) versus ω (rad/s) | Ø = Specific, v = 400 m/s, α = 0° (Normally Oriented Fan Blade)

ω	a | Ø = 0°	a | Ø = 15°	a | Ø = 30°	a | Ø = 45°	a | Ø = 60°	a | Ø = 75°
0	33,852.401	103,730.623	230,368.710	2583.367	7492.163	0
50	38,133.376	72,400.914	219,765.740	2538.837	6426.825	0
100	30,467.590	73,333.000	48,801.429	0	6921.594	0
150	45,319.427	57,642.144	128,609.040	0	5962.963	0
200	52,187.836	56,483.417	148,484.12	0	5883.498	0
250	61,524.383	69,193.304	0	0	6269.058	0
300	72,530.588	75,418.214	0	0	6091.682	0

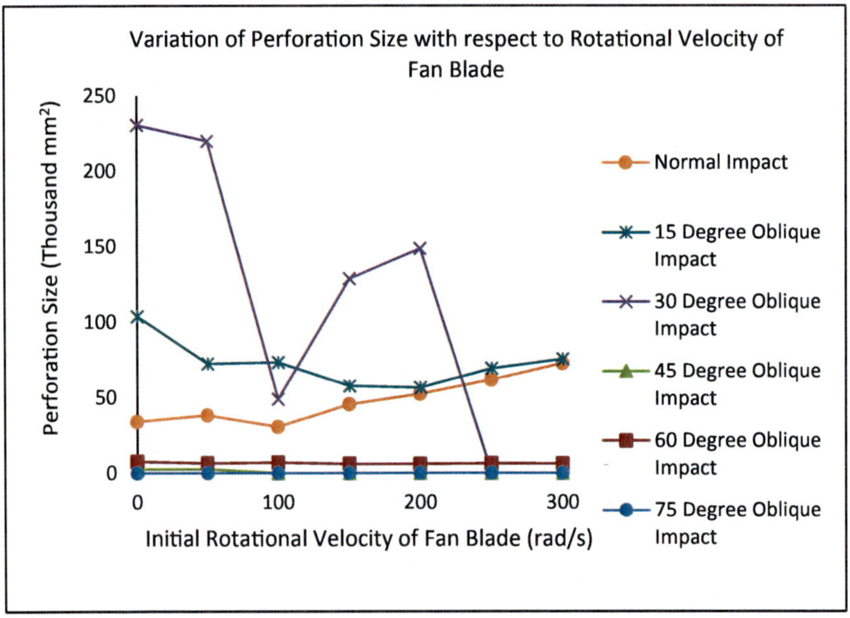

Fig. 7 Graph shows the variation of a versus ω | Ø = Specific, v = 400 m/s, α = 0° (Normally Oriented Fan Blade)

- For all the initial rotational velocities, the perforation size is significant for the oblique impact of 30-degree angle and less than that.
- Normal impact, 15-degree oblique impact and 30-degree oblique impact are highly dependent on the rotational velocity of the fan blade.
- 45-degree, 60-degree and 75-degree oblique impact angles are not greatly responsive to the change in the initial rotational velocity of the fan blade.
- For the normal impact, an increment in the rotational velocity of the fan blade increases the size of the perforation area of the containment ring.
- For 400 m/s linear impact velocity, normal impact and 15-degree oblique impact give the almost same perforation area at higher rotational velocities.
- At a lower rotational velocity of the blade, the angle of oblique impact influences the size of the perforation significantly.

4.5 Calculation of % Reduction in Velocity and Variation of v_{res} Versus Ø for a Corner Node of Fan Blade | $\omega =$ Specific, v = 400 m/s, $\alpha = 0°$ (Normally Oriented Fan Blade)

In this study, the blade is oriented normally to the containment ring. The residual velocity is simulated for the corner node of the fan blade with respect to different oblique impact angles for six values of initial rotational velocities of the fan blade which is moving with 400 m/s initial linear velocity. Following this, the reduction in the velocity is also calculated for the same node of the fan blade. The initial resultant velocity is calculated which is a resultant of rotational and linear velocity and acts as a reference to calculate the percentage reduction. The following results are observed in Tables 8, 9, 10 and Figs. 8, 9:

- For the 45-degree oblique impact angle, the residual velocity is minimum or the percentage reduction is maximum and almost same corresponding to all the rotational velocities.

Table 8 Variation of v_0 versus Ø for the corner node of the fan blade | $\omega =$ Specific, $v = 400$ m/s, $\alpha = 0°$ (Normally Oriented Fan Blade)

Ø	$v_0 \mid \omega = 0$	$v_0 \mid \omega = 50$	$v_0 \mid \omega = 100$	$v_0 \mid \omega = 150$	$v_0 \mid \omega = 200$	$v_0 \mid \omega = 250$	$v_0 \mid \omega = 300$
0	400	400.445	401.280	402.501	404.105	406.088	408.445
15	400	397.191	394.192	392.710	391.051	389.785	388.918
30	400	394.119	388.551	383.310	378.409	373.863	369.684
45	400	391.442	383.100	374.990	367.127	359.527	352.206
60	400	440.437	378.820	368.410	358.134	348.004	338.032
75	400	388.007	376.046	364.122	352.238	340.398	328.607

Table 9 Variation of v_{res} versus Ø for the corner node of the fan blade | ω = Specific, v = 400 m/s, α = 0° (Normally Oriented Fan Blade)

Ø	$v_0 \mid \omega = 0$	$v_0 \mid \omega = 50$	$v_0 \mid \omega = 100$	$v_0 \mid \omega = 150$	$v_0 \mid \omega = 200$	$v_0 \mid \omega = 250$	$v_0 \mid \omega = 300$
0	372.424	259.124	332.567	340.515	365.286	362.271	359.465
15	332.639	335.906	339.192	365.634	345.793	305.961	121.550
30	319.584	134.233	186.056	313.024	326.193	159.577	59.675
45	161.487	118.007	134.233	169.161	161.522	123.094	192.805
60	463.021	389.354	375.104	313.234	355.630	282.304	234.488
75	530.308	500.293	475.700	439.185	407.314	381.998	361.555

Table 10 Variation of % reduction in linear velocity versus Ø for the same node | ω = Specific, v = 400 m/s, α = 0° (Normally Oriented Fan Blade)

Ø	$v_0 \mid \omega = 0$	$v_0 \mid \omega = 50$	$v_0 \mid \omega = 100$	$v_0 \mid \omega = 150$	$v_0 \mid \omega = 200$	$v_0 \mid \omega = 250$	$v_0 \mid \omega = 300$
0	6.894	35.291	17.123	15.400	9.606	10.790	11.992
15	16.840	15.430	13.953	6.895	11.573	21.505	68.747
30	20.104	65.941	52.115	18.337	13.800	57.317	83.858
45	59.628	69.853	64.961	54.889	56.004	65.762	45.258
60	−15.755	11.598	0.981	14.977	0.699	18.879	30.631
75	−32.577	−28.939	−26.501	−20.615	−15.636	−12.221	−10.027

- Zero rotational velocity gives higher v_{res} compared to the rotating blade for all the oblique impact angles.
- Residual velocity decreases and then increases with the increment in the oblique impact angle.
- There is a less reduction in the kinetic energy of the blade with a high oblique impact angle compared to a low oblique impact angle.
- For the high oblique impact angle, the residual velocity decreases with the increase in the rotational velocity.
- All the residual velocities at all the oblique angles seem to be distributed in the bilinear band.

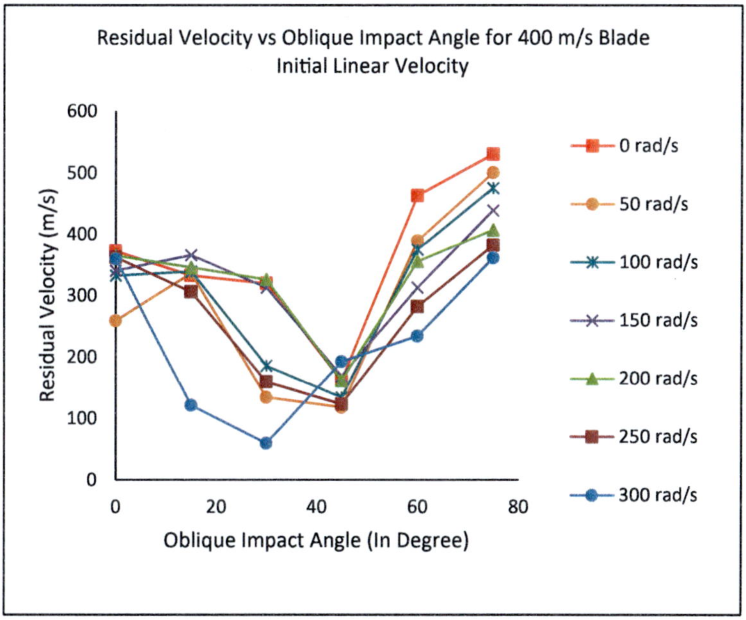

Fig. 8 Graph shows the variation of v_{res} versus Ø for the corner node of the fan blade | ω = Specific, v = 400 m/s, $\alpha = 0°$

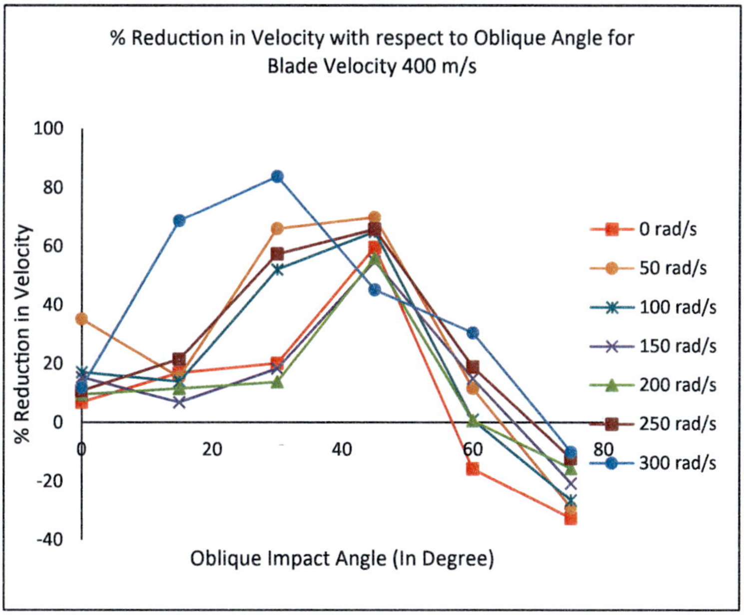

Fig. 9 Graph shows the variation of % reduction in velocity versus Ø for the corner node of the fan blade | ω = Specific, v = 400 m/s, $\alpha = 0°$

5 Conclusion

The simulation study is a compilation of the observations from the interaction of the flying-off blade and containment ring.

5.1 Importance of Blade Orientation

Blade orientation plays a vital role in perforating the containment ring. 15-degree blade orientation gives the highest probability of perforation and the highest damage to the containment ring while the range 30–45 degrees gives the least probability of perforation of the containment ring.

5.2 Importance of Oblique Impact Angle

For all the initial rotational velocities, the perforation size is significant for the oblique impact of 30-degree angle and less than that. At a lower rotational velocity of the blade, the angle of oblique impact influences the size of the perforation significantly. For the 45-degree oblique impact angle, the residual velocity is minimum or the percentage reduction is maximum and almost same corresponding to all the rotational velocities. 45-degrees and 75-degrees oblique impact angles contribute very less to the normal component of the impact velocity. Hence, the perforation area is very small and sometimes, it leads to no perforation. Residual velocity decreases and then increases with the increment in the oblique impact angle.

5.3 Importance of Initial Linear Velocity of the Blade

For a normally oriented released fan blade without initial rotational velocity, an increment in the impact velocity increases the size of the perforated area of the containment ring. Oblique impact angles of 15 degrees and 30 degrees show a higher dependency on the impact velocity as there are significant changes in the perforated area.

5.4 *Importance of Initial Blade Rotational Velocity of the Blade*

The size of the perforation of the containment ring depends on the rotational velocity of the fan blade. With the increase in the rotational velocity, the size of the perforated area increases. The effect of the rotational velocity is significantly higher for the high-angle-oriented blades. For the normal impact, an increment in the rotational velocity of the fan blade increases the size of the perforation area of the containment ring. Zero rotational velocity gives higher residual velocity compared to the rotating blade for all the oblique impact angles. Normal impact, 15-degree oblique impact and 30-degree oblique impact are highly dependent on the rotational velocity of the fan blade.

References

1. Ridley B (2018) Impact studies of containment ring during blade shed in turbofan engines
2. "Materials for Aircraft Engines TAKEHIRO OKURA Contents"
3. Roy PA, Meguid SA (2020) Modeling and characterization of bilayer containment ring in gas turbine engine. Int J Comput Methods Eng Sci Mech 21(2):96–108
4. He Q, Xie Z, Xuan H, Hong W (2016) Ballistic testing and theoretical analysis for perforation mechanism of the fan casing and fragmentation of the released blade. Int J Impact Eng 91:80–93
5. He Q, Xuan H, Liu L, Hong W, Wu R (2013) Perforation of aero-engine fan casing by a single rotating blade. Aerosp Sci Technol 25(1):234–241

Numerical Investigation of Behind the Armour Ballistic Trauma of Ceramic-Composite Armour System

Ashish Mishra, Vagish D. Mishra, and Luv Verma

Abstract A lightweight ceramic-based armour system must arrest bullet penetration and dissipate a large amount of impact energy. If the momentum transferred to the human body, also known as behind armour ballistic/blunt trauma (BABT), exceeds the threshold value, a non-penetrated target may cause serious injury or even death of the wearer. In this study, the impact response of body armour comprised of silicon carbide (SiC) plate backed with unidirectional ultrahigh molecular weight polyethylene (UD-UHMWPE) fibre composite struck at an impact velocity of 500 m/s is numerically investigated. A gelatin block is used as a surrogate model to simulate the human torso. The commercial finite element software ASNSYS/AUTODYN® is used to numerically analyse three-dimensional non-linear deformations of the system, utilizing the full integration rule to determine the elemental matrices. The Johnson–Cook, Johnson-Holmquist, and orthotropic material behaviour were utilized for the projectile, ceramic, and composite materials, respectively. Gelatin is modelled as a viscoelastic material and follows Mie-Grüneisen equation of state and Johnson–Cook failure model. Values of the material parameters were taken from the available literature. The deformation mechanism of the ceramic-composite target was evaluated. It is found that the maximum depth of cavity formed in the gelatin block is 11.65 mm. The impact pressure associated with momentum transferred to the human tissue is also calculated. The peak pressure of 14.2 MPa was observed near the cavity region, which gradually decreases at the rear face of the gelatin block. The maximum pressure and depth of the temporary cavity generated in the gelatin computed from the present analysis can be used to determine the degree of the body injury induced by impact.

A. Mishra (✉)
Department of Aerospace Engineering, UPES, Dehradun, India
e-mail: ashish.v2110@gmail.com

V. D. Mishra
Department of Aerospace Engineering, IIT Madras, Chennai, India
e-mail: vagishdm@gmail.com

L. Verma
Department of Applied Mechanics, IIT Madras, Chennai, India

R. Velmurugan et al. (eds.), *Dynamic Behavior of Soft and Hard Materials Volume 1*,
Springer Proceedings in Materials 34, https://doi.org/10.1007/978-981-99-6030-9_33

Keywords Behind armour blunt trauma (BABT) · Gelatin · SiC · UD-UHMWPE · Numerical simulation

1 Introduction

The ballistic armours are designed to stop the bullet impact and dissipate the energy and momentum without injuries to the wearer. A ballistic projectile hitting armour results in the dynamic transverse displacement and deformation of the target. Even in case of no perforation, the impact may result in a blow to the wearer, which is generally termed as behind armour blunt trauma (BABT) [1, 2]. Conducting an experiment to study the BABT is ethically not possible; therefore, it can be assed using numerical simulation [3, 4]. Additionally, the numerical models provide the details of the deformation, which is otherwise not possible through experiments.

Analysis of BABT involves modelling the complex dynamic response of multi-material systems that include protective layers, projectile, and backing materials. A proper material model is required to capture the damage and deformation of the projectile and protective layers. However, the modelling of the backing material is more challenging as it requires material representing human tissue to reasonably predict the pressure and momentum transfer to the wearer. Ballistic gelatine is frequently considered a relevant material to model the human tissue [5]. Additionally, targets backed with gelatin are used to predict the human injuries when projectile perforates the targets [6]. To model an object's impact into gelatin, a high strain rate and temperature-dependent material properties are required, however, they are not available in the open literature for an event involving relatively high impact speeds.

The modern hard armour consists of a front ceramic plate; generally, silicon carbide (SiC), boron carbide (B_4C), and alumina ($Al_2\ O_3$), which is backed with fibre reinforced composites (FRCs). The commonly employed composites are made of high-performance fibre such as ultrahigh molecular weight polyethylene (UHMWPE), Kevlar and glass etc.. UHMWPE fibre has high specific strength and modulus and is hence preferred over other composites [7, 8]. Further, due to the absence of undulation in UD fabrics, their impact resistance is superior to woven composites [9].

Experimental and numerical studies are performed to study the damage caused to humans due to ballistic impact [10–12]. The injuries caused to human tissues are assessed by using gelatin as a substitute material in the study. However, very limited studies are available on the BABT of the ceramic-composite armour. The front layer of the ceramic-composite amours blunts the tip of the projectile and reduces the velocity, and backing layer traps the fragmented ceramic and absorbs the residual energy of the projectile. Wen et al. [11] tested alumina backed UHMWPE composite at high speed and assessed the BABT by determining the temporary cavity and pressure value in the gelatin block. Experimental findings were compared to the numerical results, and quantitatively, a good correlation was found. More recently, Batra et al. [13] have studied the effect of adding a thin polymer layer covering on ceramic-composite armour at low velocity on BABT. A reduction in the pressure transferred to the gelatin block was reported with the use of polyether-ether-ketone (PEEK) layer in the front of the ceramic.

The present study focuses on to evaluate the deformation mechanism of ceramic-composite armour backed with a gelatin block, and quantify the momentum transfer to the wearer impacted against 4340 steel projectile using numerical simulation. The ceramic-composite target consists of SiC backed with UD-UHMWPE fibre composites. A ballistic gelatin layer was used as a surrogate model to replicate the human body. The failure mechanism of the ceramic-composite target is analysed at an impact velocity of 500 m/s. The maximum depth of the cavity in the gelatin block and the amount of pressure transferred are also computed.

2 Numerical Model

The whole model of ceramic-composite target backed with gelatin layer impact against steel projectile is simulated using 3D Lagrangian solver in ANSYS/AUTODYN® software. The target face, including the gelatin block, has a cross-section dimension of 200 × 200 mm. Owing to symmetry and to reduce the computation time, only quarter geometry is modelled. The geometry of the quarter target is shown in the Fig. 1a. Thickness of SiC ceramic and UD-UHMWPE fibre composite is considered 8 and 10 mm, respectively, as shown in Fig. 1b.

Thickness of the gelatin block is fixed to 150 mm in the present analysis. The projectile features a conical nose and measures 33.8 mm in length and 6 mm in diameter, identical to the core of 7.62 mm armour-piercing rounds. The UD-UHMWPE fibre composite is constructed using a sublimate approach to simulate the delamination between the layers. The entire UD-UHMWPE fibre composite is divided into 5 sub-laminates of thickness 2 mm each.

A phenomenological model developed by Johnson-Holmquist (JH-1) implemented in AUTODYN® library is used to describe the behaviour of the brittle material against impact loading. The model includes pressure-dependent strength, strain

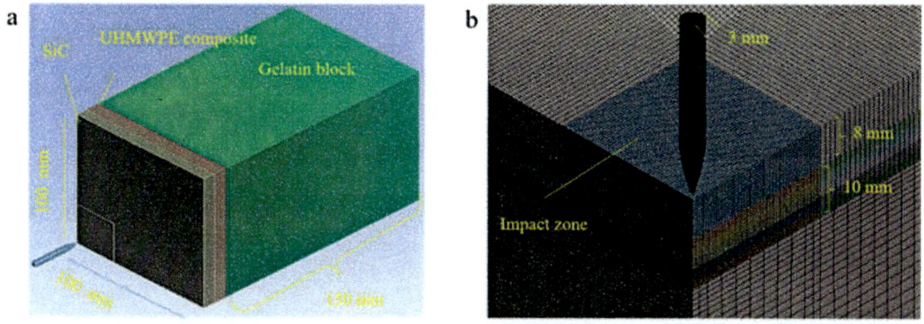

Fig. 1 **a** Schematic of quarter geometry of the model; **b** Finite element model

rate, and material damage under ballistic loading. Material parameters of SiC were taken from the open literature [14]. UD-UHMWPE composite laminates are treated as macroscopic homogeneous materials and modelled as a non-linear orthotropic material with orthotropic hardening and energy-based softening [15]. Gelatin is modelled as viscoelastic material and follows the Mie-Grüneisen form of the EOS and Johnson–Cook failure model. The material properties of the gelatin are listed in the Table 1. The projectile was modelled using the ANSYS/AUTODYN® material library 4340 steel, represented by linear EOS and the Johnson–Cook strength and failure model.

A stress-based failure criterion, governed by Eq. 1 is used to predict the delamination of the layers at their interfaces.

$$\left(\frac{\sigma_n}{S_n}\right)^m + \left(\frac{\sigma_s}{S_s}\right)^n \geq 1 \tag{1}$$

where the m and n are constants set to 1, and σ_n and σ_s are the interface normal and shear stresses, respectively, and S_n and S_s are interlaminar normal and shear strengths. Table 2 shows the interface properties of the different layers. The interface between gelatin and UHMWP is constrained to simulate the strapping of armour in real practice.

An eight-node Lagrangian linear element is utilized to discretize the entire model, as shown in Fig. 1b. The geometry of the SiC and UHMWPE composite near impact

Table 1 Material properties of the gelatin block [10]

Parameters	Value
Density (ρ), kg/m^3	1030
Viscoelasticity	
Instantaneous shear modulus (G_0), MPa	0.214
Shear modulus (G), MPa	0.158
Viscoelastic decay constant, (β), s^{-1}	0.00087
Mie-Gruneisen EOS	
Gruneisen coefficient	0.17
C_1, m/s^{-1}	1553
S_1	1.93
S_2, s.m^{-1}	0
Johnson–Cook failure model	
D_1	-0.13549
D_2	0.6015
D_3	0.25892
D_4	0.030127
D_5	0
Melting temperature, ^{0}C	20
Reference strain rate	0.001

Table 2 Interface properties of the ceramic-composite targets [14]

Contact layer	Target layer	Normal strength, MPa	Shear strength, MPa
SiC	UHMWPE	65	45
UHMWPE fibre composite	UHMWPE	5.75	7.85

zone was modelled with mesh element size of 0.5 and 1 mm in the outer region. The gelatin elements varied in size from 1 mm near the contact zone to 8 mm at the end. The projectile is discretized with 0.3 mm-sized elements. The total number of mesh elements in the numerical model is 529,823. A global geometric failure strain of 1.2 is implemented in the numerical model to avoid the convergence issue that arises due to highly distorted elements. Mass of the deleted elements was preserved in the analysis. No boundary condition is imposed on the target.

3 Results

3.1 Numerical Model Validation

Nguyen et al. [16] reported the ballistic limit of 394 m/s for a UHMWPE fibre composite when impacted against a 20 mm fragmented, segmented projectile (FSP). In order to verify the correct implementation of the composites material model in the ANSYS/AUTODYN®, a full-scale numerical analysis of the UHMWP target is performed at an impact velocity in the range of 350–450 m/s, and an error of 12% reduction in the ballistic limit was found. This error is due to several reasons—simplified use of the sub-laminate approach, variation in strain rate-dependent properties etc. The ceramic and projectile material properties are taken from the AUTODYN® library, hence not validated explicitly.

3.2 Damage Progress in Composite Target

The ballistic evolution of the ceramic-composite armour was determined at an impact velocity of 500 m/s. For impact velocities greater than 550 m/s, penetration of the projectile into the gelatin block was observed; therefore, the impact velocity was restricted to 500 m/s in the present study to prevent target perforation, and quantify the cavity formation and pressure value in the gelatin block more accurately. Figure 2 shows the variation in impact velocity of a steel projectile over time. The series of events, depicted in Fig. 2, occur during the course of projectile penetration into target. It is observed that the bullet's kinetic energy is significantly reduced on fracture of SiC composite at 0.056 ms. The velocity reduces further as a result of layer debonding and laminate deformation. At 0.15 ms, the impact velocity was decreased to 35 m/s,

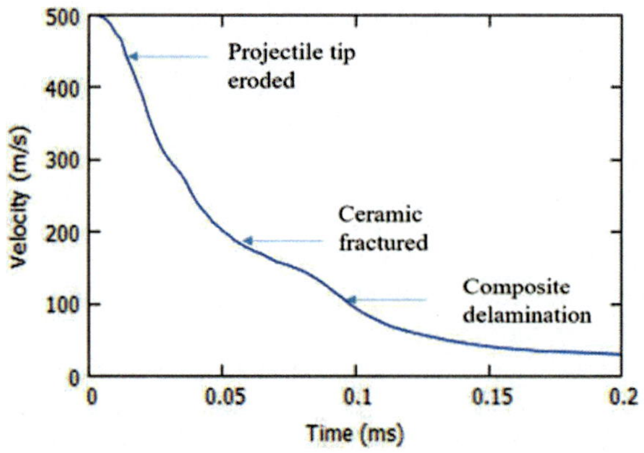

Fig.2 Variation of impact velocity with time for gelatin backed ceramic-composite target

and between 0.15 and 0.2 ms, velocity is nearly constant due the sufficient stiffness of UHMWPE fibre composite.

The damage sequence in the target material at the different time step is shown in the Fig. 3. The gelatin was included in the analysis, however, not represented in Fig. 3. When the projectile impacts the target, the tip of the projectile erodes, and the projectile surface looks like a mushroom, as depicted in Fig. 3. Due to high compressive stress below the projectile, formation of conoid is visible in the SiC ceramic. The eroding projectile continues to penetrate the volume of the damaged ceramic. As a result of the conoid positioned between the projectile and the backing plate, momentum may be dispersed throughout the large surface area of the backing plate. Continuous bulging and bending of the composite backing were observed. The rapid deceleration of the projectile can be attributed to its continued engagement with a ceramic conoid supported by a composite backing of moderate thickness.

3.3 *Deformation of Gelatin Block*

Figure 4 illustrates the cavity development or distorted shape of the gelatin block at three distinct times. At time, $t = 0.046$ ms, the cavity depth measured from the top face of the gelatin was the least; the deformation of gelatin increased to 6.37 mm at $t = 0.094$ ms. The maximum deformation of 11.65 mm is measured in the gelatin block before the velocity of the projectile reduces to 31 m/s at $t = 0.19$ ms. The impact pressure related to momentum transferred to the human tissue is also determined. The peak pressure of 14.2 MPa was observed near the cavity region; the pressure gradually decreases from the depth of the cavity site to the rear face of the gelatin. According to NIJ 0101.04 (17), the maximum depth of the cavity of 40 mm is permissible for hard armour impacted against ballistic projectile. At a distance of 40 mm from the top face of the gelatin block, a pressure value of 6.48 MPa was measured.

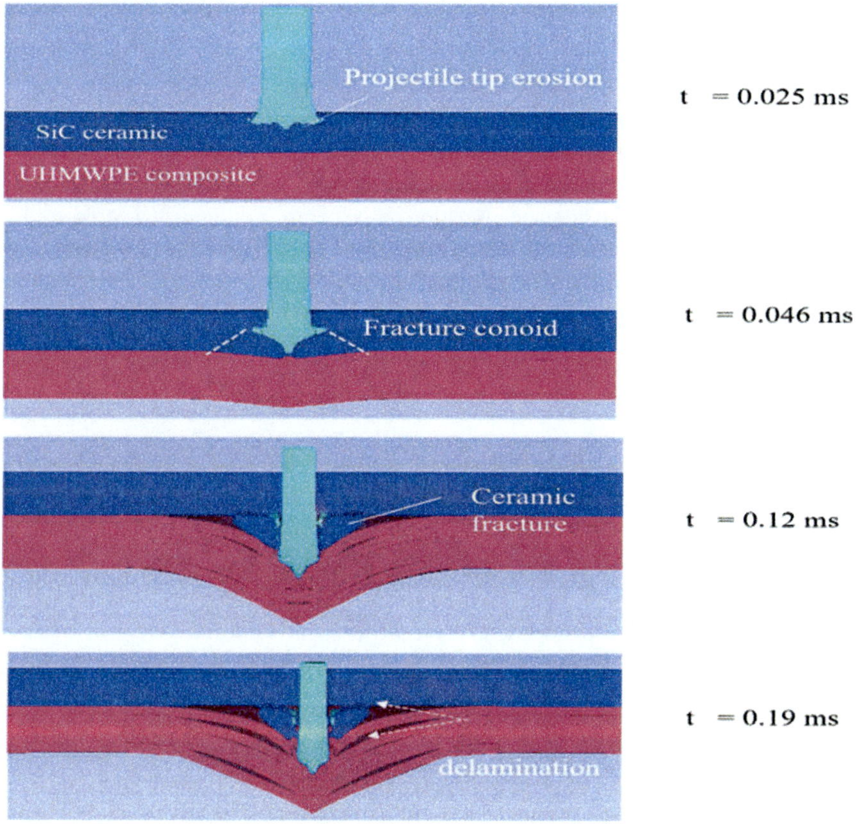

Fig. 3 Sequence of damage progress in the ceramic-composite target, the gelatin block is not shown in the figure, however, it included in the numerical analysis

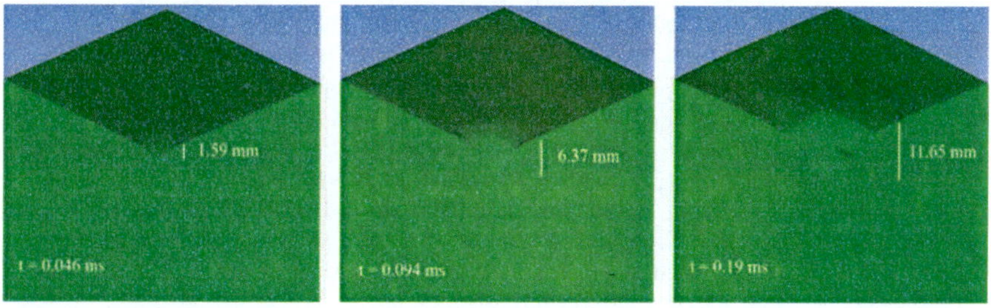

Fig.4 Cavity development in gelatin block at various time intervals

4 Conclusion

The ballistic impact response of a gelatin block protected by a target comprised of silicon carbide ceramic and unidirectional ultrahigh molecular weight polyethylene fibre (UD-UHMWPE) composite is numerically studied in order to evaluate the behind armour blunt trauma. The model is analysed using a 3D Lagrangian solver in ANSYS/AUTODYN®. The damage evolution in the ceramic composite is predicted against steel projectile at an impact velocity of 500 m/s. The transverse displacement in ballistic gelatin and the pressure values of the bullet's impact were measured to characterize the behaviour of a bullet striking a ceramic-composite target.

References

1. Cannon L (2001) Behind armour blunt trauma–an emerging problem. J R Army Med Corps 147(1):87–96. https://doi.org/10.1136/jramc-147-01-09
2. Carr DJ, Horsfall I, Malbon C (2016) Is behind armour blunt trauma a real threat to users of body armour? A systematic review. J R Army Med Corps 162(1):8–11. https://doi.org/10.1136/jramc-2013-000161
3. Shen W, Niu Y, Bykanova L, Laurence P, Link N (2010) Characterizing the interaction among bullet, body armor, and human and surrogate targets. J Biomech Eng 132(12). https://doi.org/10.1115/1.4002699
4. Wang Y, Shi X, Chen A, Xu C (2015) The experimental and numerical investigation of pistol bullet penetrating soft tissue simulant. Forensic Sci Int 249:271–279. https://doi.org/10.1016/J.FORSCIINT.2015.02.013
5. Carr DJ, Stevenson T, Mahoney PF (2018) The use of gelatine in wound ballistics research. Int J Legal Med 132(6):1659–1664. https://doi.org/10.1007/S00414-018-1831-7
6. Rahbek DB, Johnsen BB (2019) Fragmentation of an armour piercing projectile after impact on composite covered alumina tiles. Int J Impact Eng 133:103332. https://doi.org/10.1016/j.ijimpeng.2019.103332
7. Mawkhlieng U, Majumdar A, Laha A (2019) A review of fibrous materials for soft body armour applications. RSC Adv 10(2):1066–1086. https://doi.org/10.1039/c9ra06447h
8. Mishra VD, Mishra A, Singh A, Verma L, Rajesh G (2022) Ballistic impact performance of UHMWP fabric impregnated with shear thickening fluid nanocomposite. Compos Struct 281:11499 1. https://doi.org/10.1016/J.COMPSTRUCT.2021.114991
9. Zhang D, Sun Y, Chen L, Zhang S, Pan N (2014) Influence of fabric structure and thickness on the ballistic impact behavior of Ultrahigh molecular weight polyethylene composite laminate. Mater Des 1980–2015(54):315–322. https://doi.org/10.1016/J.MATDES.2013.08.074
10. Yoon GH, Mo JS, Kim KH, Yoon CH, Lim NH (2015) Investigation of bullet penetration in ballistic gelatin via finite element simulation and experiment. J Mech Sci Technol 29(9):3747–3759. https://doi.org/10.1007/s12206-015-0821-7
11. Wen Y, Xu C, Jin Y, Batra RC (2017) Rifle bullet penetration into ballistic gelatin. J Mech Behav Biomed Mater 67:40–50. https://doi.org/10.1016/J.JMBBM.2016.11.021
12. Gilson L, Rabet L, Imad A, Coghe F (2020) Experimental and numerical assessment of non-penetrating impacts on a composite protection and ballistic gelatine. Int J Impact Eng 136:103417. https://doi.org/10.1016/J.IJIMPENG.2019.103417
13. Batra RC, Pydah A (2020) Impact analysis of PEEK/ceramic/gelatin composite for finding behind the armor trauma. Compos Struct 237:111863. https://doi.org/10.1016/J.COMPSTRUCT.2020.111863

14. Shen Z, Hu D, Yang G, Han X (2019) Ballistic reliability study on SiC/UHMWPE composite armor against armor-piercing bullet. Compos Struct 213(April):209–219. https://doi.org/10.1016/j.compstruct.2019.01.078
15. Nguyen LH, Lässig TR, Ryan S, Riedel W, Mouritz AP, Orifici AC (2015) Numerical modelling of ultra-high molecular weight polyethylene composite under impact loading. Proc Eng 103:436–443. https://doi.org/10.1016/j.proeng.2015.04.043
16. Nguyen LH, Ryan S, Cimpoeru SJ, Mouritz AP, Orifici AC (2016) The efficiency of ultra-high molecular weight polyethylene composite against fragment impact. Exp Mech 56(4):595–605. https://doi.org/10.1007/s11340-015-0051-z

Investigation of the Effect of Ceramic/glass Fibre Sandwich Composite Subjected to Projectile Impact

Chalichemala Lalan Krishna, Jagannadham Bharath, Thota Sriharsha, Pritam Ghosh, and Stanley Samlal

Abstract Ceramic tiles are often preferred to prevent projectiles from penetrating lightweight armour systems. This study investigates the effect of projectile impact on the glass fibre-reinforced composite (GFRP) with ceramic sheet sandwich panels. In the experiments, using a single-stage gas gun, 15 mm hemispherical and conical projectiles were impacted on 140 × 140 mm sandwich plates at velocities of about 80 m/s. Four shots for each hemispherical and conical projectile were fired on, 4-layer GFRP layer, 2-layers of GFRP with ceramic core and 4-layers of GFRP with ceramic sheet totalling 32 samples. Perforated panels were investigated for damage using C-scan. The effect of adding the ceramic sheet was evident in the energy absorption of the laminate. In comparison between glass fibre laminate and glass fibre/ceramic sandwich panels, it was observed that the ballistic limit was enhanced by 10% with the addition of a ceramic sheet in the laminate. Further, it also shows that adding GFRP can improve the ballistic resistance of the proposed GFRP/ceramic composite sandwich.

Keywords Ceramic Cored · GFRP · Ceramic cored sandwich structure · Projectile Impact · Damage · Ballistic limit

Nomenclature

G	Glass fibre
Co	Ceramic fibre
P1	G/CO/G
P2	G/G/Co/G/G
P3	G/G
P4	G/G/G/G

C. L. Krishna · J. Bharath · T. Sriharsha · P. Ghosh (✉) · S. Samlal
SIMCRASH Centre, Department of Aeronautical Engineering, Hindustan Institute of Technology and Science, Padur, Kelambakkam, Chennai 603103, India
e-mail: pritamrajghosh@gmail.com

R. Velmurugan et al. (eds.), *Dynamic Behavior of Soft and Hard Materials Volume 1*, Springer Proceedings in Materials 34, https://doi.org/10.1007/978-981-99-6030-9_34

1 Introduction

Due to numerous influencing factors and the evolution of ballistic threats, studies based on ceramic-based lightweight armour have been on the rise for decades [1, 2]. Non-metallic multi-component armour systems, which typically consist of a ceramic plate and a fibre-reinforced polymer (FRP) laminate, have achieved extensive application as protective structures in many engineering sectors [3]. As a result, it is critical to do extensive research into the projectile impact performance of ceramic/FRP laminate composites.

An experimental study is one of the most widely used approaches for investigating the projectile impact performance of ceramic/FRP composite constructions. Savio and Madhu [4] studied the ballistic efficiency of ceramic material in projectile impact tests to determine the influence of projectile velocity, backing material and tile thickness. Evci and Gülgeç [5] studied the energy absorption and fracture behaviour of the ceramic/E-glass composite laminates reinforced with Boron carbide ceramic through an impact test and determined that Boron carbide ceramic had better fracture behaviour than Alumina ceramic. Alebooyeh et al. [6] worked on finite element simulation with a 3D lagrangian model to examine the impact of the ceramic plate's thickness on the ballistic performance of the Al2O3-SiC/UHMWPE composite laminate.

Researchers have extensively investigated the impact response and energy absorption capability of various engineering structures. The energy absorption, failure mode and ballistic limit velocity of the projectile-plate impact system are significantly affected by different nose shapes of the projectile. The projectile kinetic energy is absorbed mainly through the plastic deformation and shear expansion of the face sheets, the fracture damage of the solidified epoxy resin and ceramic prisms, and the macroscopically bending deformation of the whole structure.

The high-strength ceramic prisms inserted in the pyramidal core can significantly enhance the ballistic resistance of the structure, resulting from the drastic erosion effects of the ceramic prism on the ballistic projectile [7]. The high hardness and strength of ceramic materials are exploited to design armour systems, but their low toughness and brittleness limit the ballistic performance. Hence, ceramics are joined with either metallic or long fibre-reinforced composite backing plates to improve their ballistic performance [8]. Ceramic materials are always used in combination with ductile backing material. Due to this, ceramic material's ballistic efficiency is sometimes different and changes concerning the configuration, type and properties of the backing material used.

Several concerns still require more consideration about the ballistic performance of the ceramic/FRP laminate composite structure, despite the significant contributions made by researchers in this area. No studies provide a thorough description of how the behaviour of a woven GFRP laminate with ceramic sheets is affected. The present work aims to get a deeper insight into the influence of the ballistic performance of a ceramic/GFRP woven composite sandwich structure. The effect of

GFRP face sheet on ballistic limit velocity, energy absorption and delamination area are experimentally evaluated.

2 Experimental

The GFRP composite comprising woven glass fibres reinforced with epoxy resin was fabricated using vacuum bagging. The laminate with a dimension of 140 × 140 mm with thickness (P1 = P2 = 3 ±0.3, P3 = P4 = 1 ± 0.3 mm) were fabricated with [0/90]s stacking sequence. Sandwich laminates were prepared using plain-weave E-Glass fibre having Tensile strength of 3445 MPa, Compressive strength of 1080 MPa, Young's modulus of 76 GPa, Density of 2580 kg/m^3 and Supercera Ceramic Paper having Tensile Strength of 0.039 MPa, Density of 240 kg/m^3. The Araldite LY556 epoxy resin and Aradur HY951 hardener are used as matrix (ratio 100/10 g) in the lay-up process, followed by sealing the vacuum bag using a tacky tape, applying 2 bar vacuum for compaction and allowing a 24 h-curing period after that. The laminates were then demolded and post-cured in an oven at 80°C for 8 h. The fibre volume fraction of the laminates was ≈ 54%.

2.1 Tensile Testing

The sandwich specimens were machined to obtain different specimen geometries for impact and tension testing. The tensile specimens were flat with gage length 150mm and gage width 25 mm. The laminates machined for tension test were used to carry out the test at 2 mm/min, according to ASTM D3039. Five samples were used to determine the tensile strengths (*XT and YT*), elastic modulus (*EX and EY*) and Poisson's ratio (υxy) were obtained. Since the E-Glass fabric used in this research was a balanced plain weave, the 'x' and 'y' in-plane properties were considered to be identical.

2.2 Projectile Impact Testing

The projectile impact tests were performed at the SIMCRASH Centre, Hindustan Institute of Technology and Science, India. The impact tests were conducted using a single-stage gas gun with a 5-litre pressure vessel, connected to 2 m long barrel and the pressure release was controlled using solenoid valve as shown in Fig. 1. The sandwich laminates were impacted by Conical and Hemispherical projectiles as shown in Fig. 2 which were made of high-strength steel approximately weighing 20 g with a nominal velocity ranging between 40 to 75 m/s. The distance between

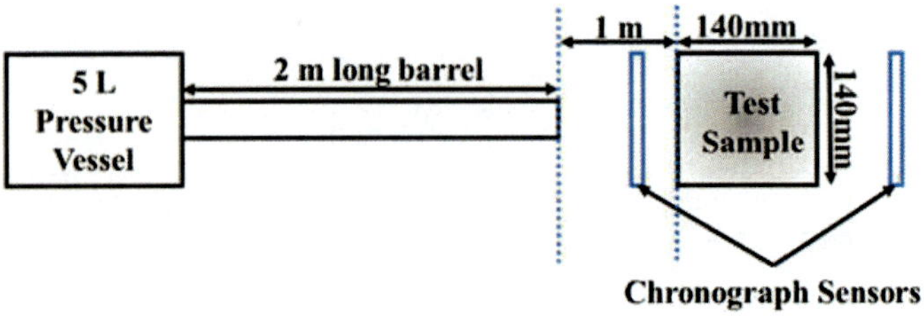

Fig. 1 Schematic representation of projectile impact test setup

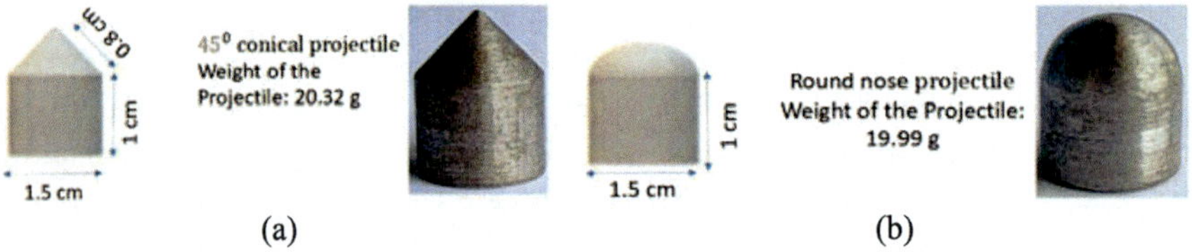

Fig. 2 **a** Conical projectile with 45° **b** Spherical projectile

the specimen and barrel was 1 m, and initial and residual impact velocities were measured by chronograph sensors.

The plate configuration considered for the study is (a) 2-layer GFRP (b) 4-layer GFRP layer (c) a ceramic sheet, (d) 2-layers of GFRP with ceramic core and (e) 4-layers of GFRP with ceramic core. Since the 2-layer GFRP and Ceramic Sheet showed no evident change at most of the residual velocities, the results for these two configurations were not reported in the manuscript.

2.3 Ultrasonic Testing

Non-destructive testing (C -Scan ultrasonic testing) was employed to determine the delamination area of the samples at the rebounding velocity for the four proposed configurations. Ultrasonic Testing (UT) with high-frequency sound waves between 0.5 and 15 MHz was used to determine the interlaminar damage area for the proposed composite configuration. The transmission method was utilised to determine the delamination dimensions using an immersion UT system. The transmitting and receiving probes were utilised, which were placed on the opposite side at equal distances from the specimen. In the un-defected region, the sound waves travel through and return without much diffraction; in the damaged area, they get scattered, resulting in lower sound wave energy at the receiving transducer. A colour gradient image was obtained based on the strength of sound wave energy, depicting the damaged zone with a different colour gradient.

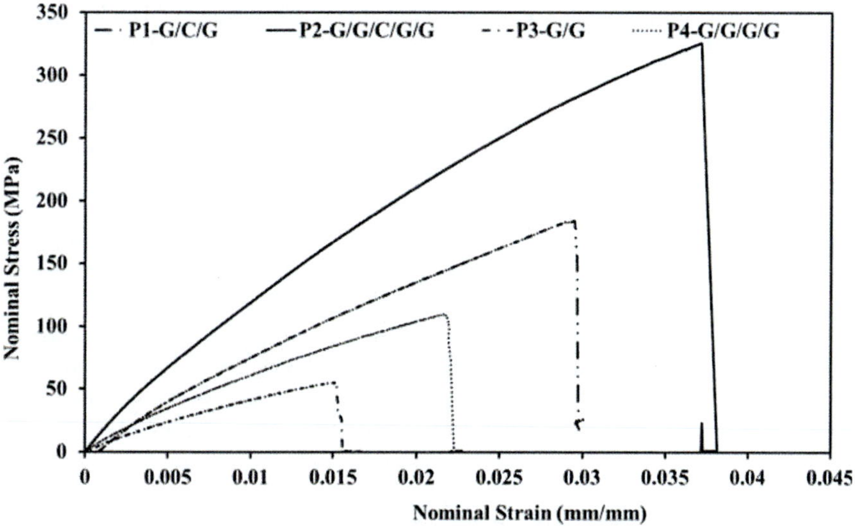

Fig. 3 Stress–strain representation for different laminate configurations

3 Experimental Results

3.1 Tensile Testing Results

One commonly used method to assess the mechanical characteristics of materials, such as GFRP laminates, is through tensile testing. GFRP laminates consist of glass fibres and a matrix, typically composed of a thermosetting resin, and are classified as composite materials. Figure 3 represent the nominal stress–strain curve for the tensile testing of the specimen. It is evident that P2 samples produced the best result compared to neat, P3 and P4. The introduction of ceramic layers in composite laminates resulted in an enhancement of tensile strength. The maximum tensile strength for P1, P2, P3, P4 were 184.2 MPa, 325.8 MPa, 55.1 MPa and 109.8 MPa respectively.

3.2 Projectile Impact Testing Results

The residual velocity of the projectiles in this series of tests together with the reduction of incident velocity for the hemispherical and conical projectile is shown in Tables 1 and 2 respectively. The results show that all the composite laminates with ceramic layers have reduced the residual velocity relative to the neat GFRP samples. There was a 40.9% enhancement in ballistic limit for P2 samples when compared to P4 samples whereas the ballistic limit was enhanced by 17.87% for P1 samples when compared to P3 samples. Thus, it can be concluded that the residual energy absorption increases on the addition of ceramic layer in the composite laminates. The ballistic limit for P2 was obtained experimentally, where the projectile perforated the laminate but it got stuck in the laminate which is clearly evident in Fig. 4.

The results show that all the composite laminates with ceramic layer have reduced the residual velocity relative to the neat GFRP samples. There was a 32.9% enhancement in ballistic limit for P2 samples when compared to P4 samples whereas the ballistic limit was enhanced by 11.97% for P1 samples when compared to P3 samples. Adding a GFRP layer improved the ballistic resistance of a GFRP/ceramic composite laminate. GFRP laminate had better mechanical properties than ceramic paper leading to enhanced mechanical properties of the combined sandwich structure. The GFRP laminates absorbed the initial impact energy through fibre failure and matrix damage, preventing projectile penetration to the sandwich structure. Also, it can be concluded that the residual energy absorption increases on the addition of GFRP layer in the composite laminates when impacted by the conical projectile.

Table 1 Residual velocity and energy comparison for hemispherical projectile

Specimen	Initial velocity (m/s)	Residual velocity (m/s)	Initial energy (J)	Residual energy (J)	Ballistic limit (m/s)
P1–1	72.3	55.9	52.25	31.23	45.85
P1–2	57.2	35	32.70	12.24	45.24
P1–3	50.1	23.2	25.09	5.38	44.40
P1–4	46.7	10.4	21.80	1.08	45.53
P2–1	75.4	Rebound	56.82	–	68.2
P2–2	70.1	Rebound	49.12	–	
P2–3	68.9	Rebound	47.45	–	
P2–4	68.2	Ballistic Limit	46.49	–	
P3–1	74.6	65.4	55.62	42.75	35.89
P3–2	67.2	55.6	45.14	30.90	37.74
P3–3	60.9	48.3	37.07	23.32	37.09
P3–4	51.3	34.5	26.30	11.90	37.97
P4–1	75.1	63.4	56.37	40.18	40.25
P4–2	40.28	Ballistic Limit	16.23	-	40.28
P4–3	60.2	45.7	36.22	20.87	39.19
P4–4	55.3	38.2	30.57	14.59	39.99

Table 2 Residual velocity and energy comparison for conical projectile

Specimen	Initial velocity (m/s)	Residual velocity (m/s)	Initial energy (J)	Residual energy (J)	Ballistic limit (m/s)
P1–1	75.4	62.7	56.82	39.29	41.88
P1–2	63.7	48.7	40.56	23.71	41.06
P1–3	53.1	33.8	28.18	11.42	40.95
P1–4	47.1	22.5	22.17	5.06	41.38
P2–1	75.8	50.1	57.43	25.09	56.88
P2–2	69.3	41.5	48.00	17.21	55.50
P2–3	64.2	30.6	41.20	9.36	56.44
P2–4	55.1	Rebound	30.34	–	–
P3–1	72.6	63.5	52.68	40.30	35.19
P3–2	68.4	57.2	46.76	32.70	37.51
P3–3	62.1	50.7	38.54	25.69	35.86
P3–4	55.4	41.3	30.68	17.05	36.93
P4–1	75.1	65.1	56.37	42.36	75.1
P4–2	73.4	62.4	53.85	38.92	73.4
P4–3	65.7	53.9	43.14	29.04	65.7
P4–4	59.8	46.8	35.74	21.89	59.8

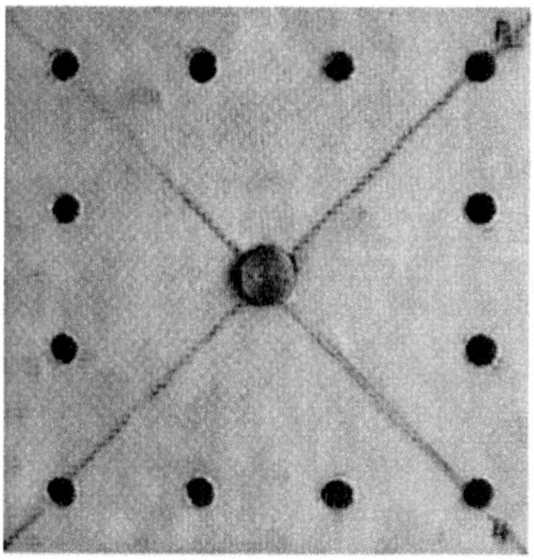

(a)

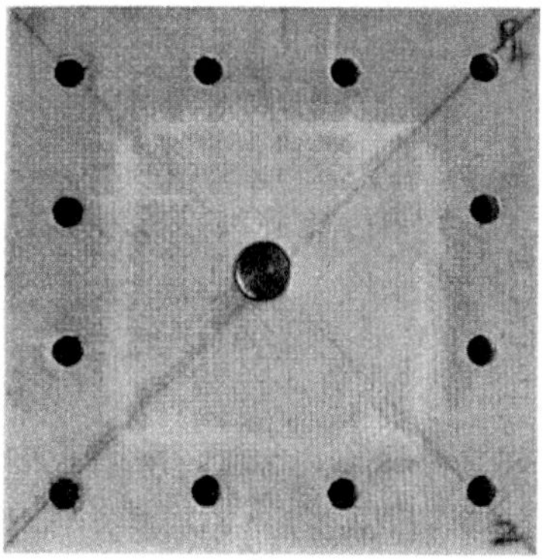

(b)

Fig. 4 **a** ballistic limit for P2 sample at 68.2 m/s **b** ballistic limit for P4 sample at 40.28 m/s

3.3 *Ultrasonic Testing Results*

The energy absorbed in the form of delamination was more in the P2 sample as compared to the P4 sample since the plate was made of only 2-layers of GFRP in the P4 sample whereas in the P2 sample, the laminate was made of 2 GFRP layers

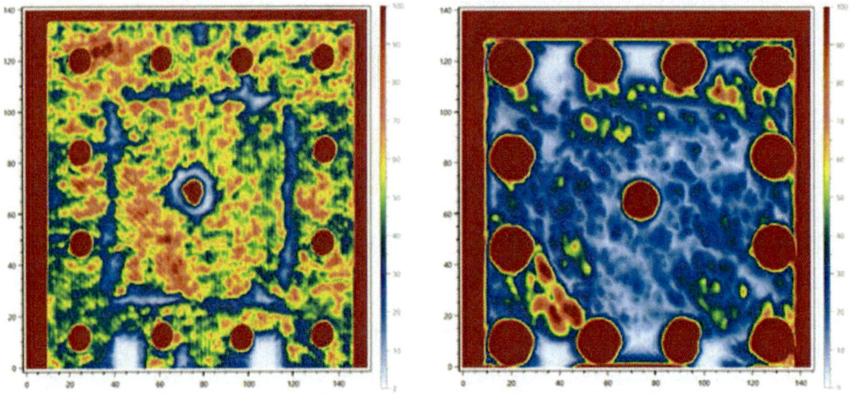

Fig. 5 **a** C scan results for P2 sample at 68.2 m/s **b** C scan results for P4 sample at 40.28 m/s

sandwiched between a ceramic layer resulting in comparatively greater thickness. The delamination area obtained from the ultrasonic C scan is represented in Fig. 5. The delamination area of p4 is 21mm*21mm that is 441 mm^2 whereas P2 is 16 mm*16 mm that is 256 mm^2.

4 Conclusion

In this study, projectile impact penetration tests were performed using a light gas gun on six different composite sandwich panel configurations. The penetration and dynamic out-plane response of the composite samples were investigated based on the residual velocity and ultrasonic C-scan technique to determine the ballistic limit and delaminated area, respectively. The prevalent failure mechanisms of the six sandwich panel layouts were also experimentally investigated. The following conclusions were drawn:

An enhancement of tensile strength was observed on addition of ceramic paper layers in GFRP laminates. On adding ceramic to P3 sample, tensile strength increased from 55.1 MPa to 184.2 MPa(P1 sample).

For hemispherical projectile, there was a 40.9% enhancement in ballistic limit for P2 samples when compared to P4 samples whereas the ballistic limit was enhanced by 17.87% for P1 samples when compared to P3 samples.

For conical projectile there was a 32.9% enhancement in ballistic limit for P2 samples when compared to P4 samples whereas the ballistic limit was enhanced by 11.97% for P1 samples when compared to P3 samples.

References

1. Zhang Y, Dong H, Liang K, Huang Y (2021) Impact simulation and ballistic analysis of B4C composite armour based on target plate tests. Ceramics Int 47(7, Part A):10035–10049. https://doi.org/10.1016/j.ceramint.2020.12.150
2. Wu S, Xu Z, Hu C, Zou X, He X (2022) Numerical simulation study of ballistic performance of Al_2O_3/aramid-carbon hybrid FRP laminate composite structures subject to impact loading. Ceram Int 48(5):6423–6435. https://doi.org/10.1016/j.ceramint.2021.11.186
3. Shanazari H, Liaghat G, Feli S, Hadavinia H (2017) Analytical and experimental study of high-velocity impact on ceramic/nanocomposite targets. J Compos Mater 51(27):3743–3756. https://doi.org/10.1177/0021998317692658
4. Savio SG, Madhu V (2018) Ballistic performance evaluation of ceramic tiles with respect to projectile velocity against hard steel projectile using DOP test. Int J Impact Eng 113:161–167. https://doi.org/10.1016/j.ijimpeng.2017.11.020
5. Evci C, Gülgeç M (2014) Effective damage mechanisms and performance evaluation of ceramic composite armors subjected to impact loading. J Compos Mater 48(26):3215–3236. https://doi.org/10.1177/0021998313508594
6. Alebooyeh M, Baharvandi HR, Aghanajafi C (2016) Two-dimensional FE simulation of impact loading on alumina matrix nanocomposite reinforced by dyneema® HB25 laminates. J Mech 33(1):1–11. https://doi.org/10.1017/jmech.2016.26
7. Ni CY, Li YC, Xin FX, Jin F, Lu TJ (2013) Ballistic resistance of hybrid-cored sandwich plates: numerical and experimental assessment. Compos A Appl Sci Manuf 46:69–79. https://doi.org/10.1016/j.compositesa.2012.07.019
8. Serjouei A, Gour G, Zhang X, Idapalapati S, Tan GEB (2017) On improving ballistic limit of bi-layer ceramic–metal armor. Int J Impact Eng 105:54–67. https://doi.org/10.1016/j.ijimpeng.2016.09.015

Effect of Hybrid Fibers on the Mechanical Properties and Impact Behavior of Concrete Panels

Beri Ashok Kumar Reddy, Partheepan Ganesan, M. Chinnapandi, and R. Velmurugan

Abstract Structures of high importance such as multi storied buildings, military structures, bunkers, pressure vessels and nuclear containments have been a target during intentional terrorist attacks, war or by accidental explosion. In this research, experimental investigations were conducted to find the mechanical properties as well as impact response behavior of square concrete targets of size 450 × 450 mm having an average target compressive strength of around 48 MPa. Impact force on concrete targets was applied by ogive nosed projectiles of two different masses. The effect of the hybridization of microfibers and macro fibers on the impact behavior of plain concrete and reinforced concrete targets as well as mechanical properties of the concretes were studied. Several concrete panels of three different thicknesses namely 100 mm, 80 mm and 60 mm were casted and tested for the present study. The test results showed that the hybrid fibers in the plain and reinforced concrete lead to lesser crater areas and reduce the damage in the back face by around 40% when compared to those panels having no fibers and reinforcements.

Keywords Mechanical properties · Projectile impact · Concrete target · Hybrid fibers · Ballistic limit

B. Ashok Kumar Reddy (✉) · P. Ganesan
Department of Civil Engineering, MVGR College of Engineering, Vizianagaram, AP 535005, India
e-mail: beri.kumar@gmail.com

M. Chinnapandi · R. Velmurugan
Department of Aerospace Engineering, Indian Institute of Technology Madras, Madras, TN 600036, India
e-mail: ramanv@iitm.ac.in

R. Velmurugan et al. (eds.), *Dynamic Behavior of Soft and Hard Materials Volume 1*, Springer Proceedings in Materials 34, https://doi.org/10.1007/978-981-99-6030-9_35

1 Introduction

Concrete ranks first amongst the most commonly used building materials all over the world. Although concrete is incredibly strong in compression, it has intrinsic weaknesses in tensile strength that cause brittle fracture, as is well known. This is obviously very undesirable for any building material. In order to prevent brittle failure and increase tensile strength, researchers have developed many types of tensile reinforcements that can be used in any sort of structural design. Usually Concrete structures are typically designed to last 100 years, under quasi-static loading; however, when exposed to sudden or impact loads, the behavior of such structures differ dramatically from that under static loading conditions. This is due to concrete's low energy absorption capacity, which causes fragmentation. To overcome this serious concern, researchers developed fiber reinforced concrete, which contains thin, short-length fibers distributed throughout the concrete mix. As a result, a variety of natural and synthetic fibers, including steel, glass and polypropylene, are currently employed in concrete to enhance concrete's energy absorption capacity. Moreover, the blending of different types of fibers (hybridization) has been proven to have the ability to improve concrete's impact and mechanical properties.

Varma and Raju [1] concluded from their experimental study that while the addition of polypropylene and steel fibers in the concrete significantly improved the flexural strength and split tensile strength of the concrete but did not help very much in enhancement of elastic modulus and compressive strength of the concrete. Alhozaimy et al. [2] observed from their study that polypropylene fibers were not significant in improving the mechanical properties such as flexural strength and compressive strength but helped in enhancing the crack initiation and impact resistance of the concrete. Li et al. [3] discovered that concrete specimens reinforced with hooked end steel fibers had improved mechanical properties when compared to specimens reinforced with straight steel fibers, however polypropylene fibers, when blended with hooked end steel fibers, the efficiency of polypropylene in improving characteristics is insignificant. Similarly, a host of other researchers [4–12] have used hybrid fiber reinforced concrete made of different fibers in enhancing the mechanical properties of concrete.

Dancygier and Yankelevsky [13] observed in their study that high-strength concrete by virtue of its increased brittleness leads to larger crater dimensions compared to the normal strength concrete. Steel fibers and steel reinforcements in high-strength concrete helped in reducing the crater dimensions, penetration depth and improved the concrete targets impact strength. However, the difference is not significant to the response of normal strength concrete reinforced with fibers. Zhang et al. [14] noted that the strength of concrete does not matter much in reducing the crater diameters when fiber reinforced concrete is used. On the other hand, concrete reinforced with steel fibers or polypropylene fibers alone under-performed in reducing the crater diameter when compared to the concrete reinforced with the combination of steel and polypropylene fibers. Kojima [15] observed that the impact resistance of concrete panels reinforced on both the front and back faces was lower to slabs reinforced on only one face. Furthermore, the use of steel lining to the rear face helped in preventing scabbing or perforation effect.

Abdel-Kader and Fouda [16, 17] used concrete targets with steel lining at front and rear faces so that secondary shrapnel effects caused by crater spalling or scabbing can be effectively prevented. Concrete targets reinforced with rear steel plate lining and/or GFRP lining have prevented perforation resistance, however, it did not help in minimizing penetration depth and spalling craters at front side. Zhang et al. [18] observed that the impact response of the target had improved when it is subjected to projectile impact due to the bond that exists between the ribs present in the reinforcement bars and the concrete in the reinforced concrete panels.

Sangho et al. [19] noted from their experiment on projectile impact on reinforced concrete targets that the diameter and yield strength of the reinforcing bars had a minor effect on impact response of the specimens, but lesser spacing between the reinforcement bars definitely helped in improving the impact resistance because of the greater confinement. Rajput et al. [20–22] noticed that, because of concrete confinement by the reinforcements, small cracks are formed in reinforced concrete targets due to very slow propagation of pressure wave. On the contrary, in plain concrete targets, the cracks are wide and long, originating from the target center and resulting in concrete splitting. Kristoffersen et al. [23] found through their experimental and numerical study that the size of scabbing damage was more than the spall crater damage, and both the scab and spall diameters were increasing when the grade of concrete increased.

Brown et al. [24, 25] performed projectile impact tests to study the impact behavior of concrete panels of various characteristic strengths, thicknesses and reinforcement types. They noted that with increasing in compressive strength of the target, the loss in concrete mass due to projectile impact increased. Also, when the target thickness is increased, the failure pattern changed from perforating to scabbing. Jamnam et al. [26] noticed that depending on their thickness and fiber content, steel fiber reinforced concrete panels fail in different ways under various impact loads with various geometries and muzzle velocities.

Almusallam et al. [27, 28] found that hybrid fibrous concrete prepared with three different types of fibers minimizes crack development and also greatly reduces the size of the damaged region, when subjected to projectile impact compared to concrete prepared with one or two types of fibers. Ueno et al. [29] revealed that polypropylene fiber reinforced concrete slabs significantly reduced the scabbing damage caused by the projectile impact when compared to a plain concrete slab.

Xu et al. [30] studied the effects of impact velocity and concrete compressive strength on the perforation performance of large, thick reinforced concrete slabs and came to the conclusion that, the residual velocity and residual thickness increased with increasing initial impact velocity, and the ballistic limit increased as the concrete's compressive strength increased, but the residual thickness decreased.

The majority of ballistic tests were conducted on concrete targets reinforced with either steel or polymer fibers; however, very few works were reported on targets made of concrete using a combination of steel, glass and polypropylene fibers. In the current work, the effectiveness of hybrid fibers with fixed proportions of steel, polypropylene and glass fibers on the mechanical characteristics and impact response of concrete panels with compressive strengths of 48 MPa of different thicknesses was investigated. The impact reaction parameters such as perforation, scabbing and spalling areas, and penetration depth were measured and compared.

2 Experimental Work

2.1 Materials

Ordinary Portland cement available in local market is used for conducting experimental investigation. The cement was of 53 grade meeting the requirements specified in IS 12269 (2013) [31]. The fine aggregate used was natural river sand belonging to Zone III having a fineness modulus value of 2.6. The coarser aggregate of 10 mm size obtained from a nearby quarry was used. SP 430 superplasticizer was mixed with potable water to maintain workability. The reinforcement bars used are of 8 mm in diameters of Fe500 Grade (Yield strength 500 MPa).

In the present study, 3 different types of fibers were used. The hooked end steel fibers used to have an aspect ratio (L/D) 50 conforming to ASTM A820. Fibrillated multidimensional polypropylene fibers of diameter 35–40 μm made of 100% virgin polypropylene home-polymer conforming to ASTM C-1116 are also used in addition to Alkali resistant glass fibers of diameter 10–13 μm.

2.2 Mix Design

Two different mixes namely plain cement concrete (M1) and hybrid fibrous composite concrete (HFCC) (M2) of grade M40 are designed using IS 10262 (2019) [32]. For HFCC, the steel, polypropylene and glass fibers are taken in fixed quantity respectively as 20 kg, 10 kg and 10 kg per m^3 of concrete. Table 1 shows the mix proportions used.

Table 1 Mix proportions for 1 m^3 of plain concrete (M1) and hybrid fiber concrete (M2)

Mix type	Cement (kg)	Sand (kg)	Coarse aggregate (kg)	Admixture (liter)	Water (liter)	W/C ratio	SF (kg)	GF (kg)	PPF (kg)	f_{ck} (N/mm^2)
M1	401.7	915.29	1017.37	2.0085	160.68	0.4	0	0	0	47.2
M2	401.7	915.29	1017.37	2.0085	160.68	0.4	20	10	10	50.9

SF, GF and PPF respectively stand for Steel, Glass and Polypropylene fibers; f_{ck}: Characteristic compressive strength of concrete

2.3 Test Specimens

Mechanical properties of concrete such as compressive strength, split tensile strength and flexural strength were evaluated using the required numbers of standard test specimens viz cube, cylinder and prism respectively as per the specification recommended in IS 516 (1959) [33].

Impact resistance subjected to projectile impact was studied using four different configurations of concrete target specimens described as follows:

a. Plain cement concrete (M1)

Plain concrete with a mean target compressive strength of 48 MPa is designed according to IS 10262 (2019).

b. Reinforced cement concrete (M1R)

Target specimens were casted using reinforced cement concrete in which 8 mm diameter Fe500 reinforcements were provided in the rear side of the specimen at a clear cover of 15 mm in orthogonal directions at a uniform of spacing of 80 mm as shown in Fig. 1.

c. Hybrid fiber concrete (M2)

Specimens were also casted using hybrid fibrous composite concrete by adding steel, polypropylene and glass fibers in fixed quantities of 20 kg, 10 kg and 10 kg per m3 to the plain concrete.

d. Reinforced hybrid fiber concrete (M2R)

Specimens were casted by adding an orthogonal reinforcement mesh of 8 mm diameter at a spacing of 80 mm at the rear side of the targets of the category c as above.

Before casting After casting

Fig. 1 Casting of concrete panels

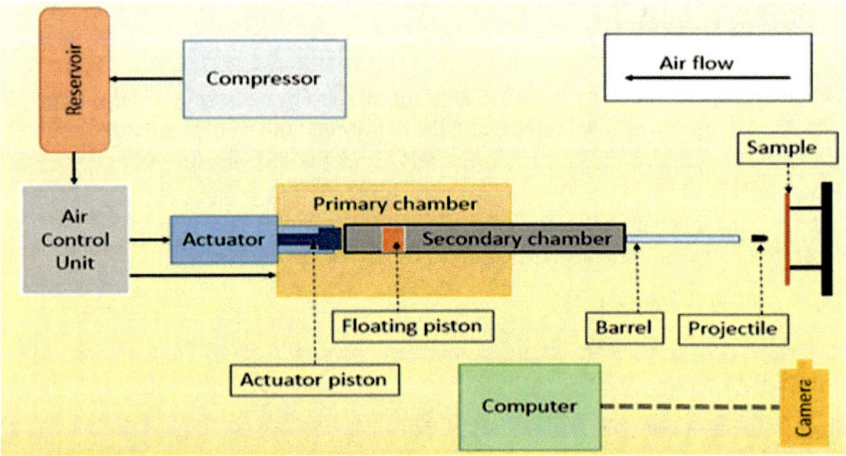

Fig.2 Air gun setup-schematic representation

2.4 *Test Setup*

Standard test setup was used to determine the compressive strength and flexural strength as per IS 516 (1959) and for split tensile strength, the procedure as per IS 5816 (1999) [34] were followed.

The projectile impact tests were performed using a pneumatic gun. It includes a long cylindrical barrel (length 9 m and diameter 25 mm), an air booster and compressor. In the present study, the projectiles struck the target normal to the target at the center of its face. Two projectiles of diameters of 23.30 mm were prepared from steel of Fe500 Grade weighing 1 kg and 1/2 kg having a length of 320 mm and 150 mm respectively (Fig. 2).

3 Results and Discussions

3.1 *Mechanical Properties*

The compressive strength of concrete with addition of fibers (M2) has increased when compared to plain cement concrete (M1) as shown in Fig. 3. The hybrid fibers led to an increase in the compressive strength over plain concrete up to 7.75% at 7 days and up to 7.8% at 28 days.

It can be observed from Fig. 4 that addition of hybrid fibers resulted in significant increase in the splitting tensile strength of concrete. The failure pattern revealed that, unlike normal concrete, HFCC specimens were not separated out after the first fracture. This improvement can be attributed to the enhanced ductile behavior

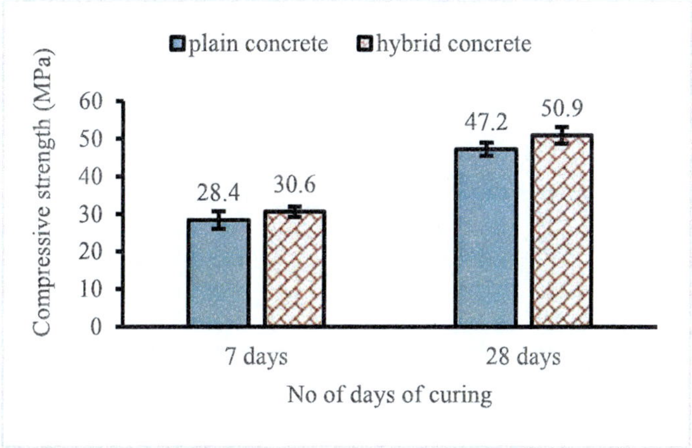

Fig. 3 Comparison of mean compressive strength

provided by the fiber. The % increase of splitting tensile strength of hybrid fiber concrete was 43.8% at 7 days and 43% at 28 days in comparison to plain concrete.

Figure 5 shows the flexural strength test results of PCC and HFCC targets. It is evident from the figure that fiber addition had beneficial effect on the flexural strengths of hybrid fibrous concrete. The percentage increase of flexural strength of hybrid concrete was 9.5% at 7 days and 10.9% at 28 days when compared with the plain concrete.

For the purpose of comparison, the mechanical properties of different concrete are summarized in the following Table 2.

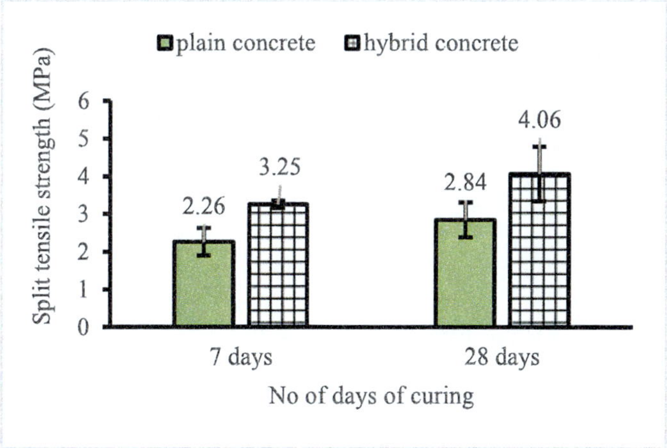

Fig. 4 Comparison of split tensile strength

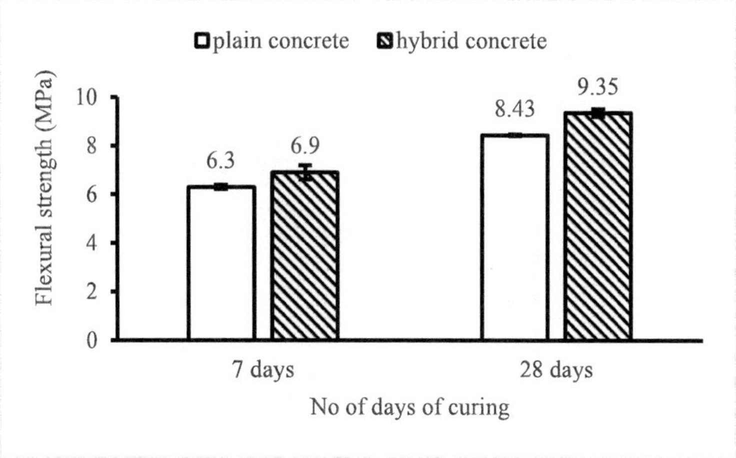

Fig. 5 Comparison of average flexural strength

Table 2 Mechanical properties of plain concrete (M1) and hybrid fiber concrete (M2)

Type of mix	No of days of curing	Average compressive strength (MPa)	Standard deviation (MPa)	Average tensile strength (MPa)	Standard deviation (MPa)	Average flexural strength (MPa)	Standard deviation (MPa)
M1	7	28.4	2.33	2.26	0.362	6.3	0.100
	28	47.2	1.73	2.84	0.461	8.43	0.046
M2	7	30.6	1.35	3.25	0.095	6.9	0.280
	28	50.9	2.19	4.06	0.728	9.35	0.015

It is observed from Fig. 6 that the peak load and displacements in the flexure test of concrete are increased with the addition of hybrid fibers. Thus showing the improved energy absorption capacity. This is due to the fact that micro and macro fibers helped in restricting the crack widening.

3.2 *Impact Response*

The impact responses such as the nature of failure, the damages in the face receiving the impact and its back face and the depth of penetration of projectile (the depth of penetration was measured as the distance from the top level of the surface receiving the impact to the deep most point in the damage) were observed for the different concrete targets impacted by projectile. The average crater diameter of damaged area is calculated based on the measurement of diameters at four different places as shown in Fig. 7.

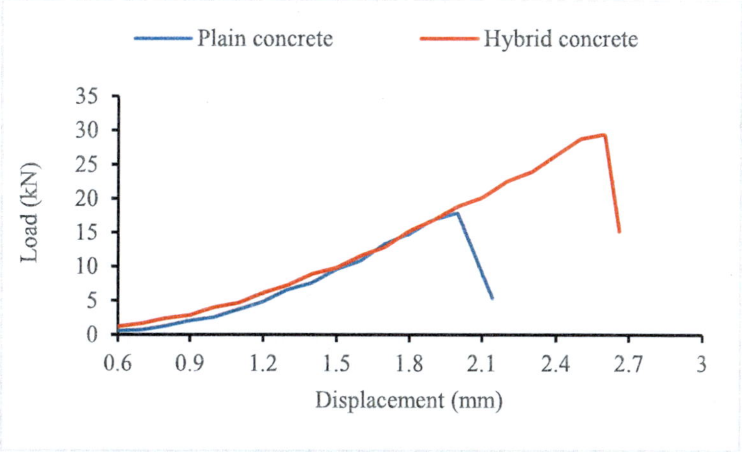

Fig. 6 Comparison of load displacement curve (flexure test)

Fig. 7 Equivalent diameter measurement of affected area after perforation

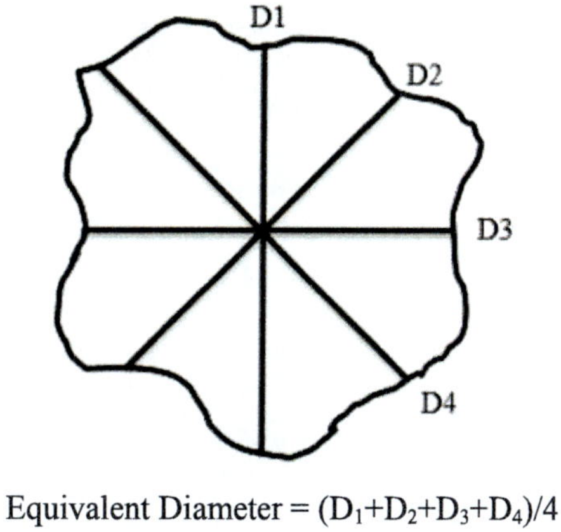

3.2.1 Damage

When a concrete panel is impacted with projectile, very high stress is generated at the projectile contact point leading to the formation damage both at the impact receiving surface and the rear face as well. For instance, the impact response of different types of concrete panels of 80 mm thick impacted by 1 kg ogive nosed projectile is shown in Table 3. It is observed that the plain concrete specimen (M1) gets completely fragmented in the absence of steel reinforcement and even in the presence of hybrid fibers (M2) visible cracks are found both in the front and rear surfaces. Whereas, when the reinforcements are provided in the concrete slab, there are no visible cracks as these reinforcement bonded together with concrete helped them keep intact. This can be observed from the following Table 3 for the specimens in the reinforced category (M1R and M2R).

Table 3 Impact test response of different types of concrete targets of 80 mm thick impacted by 1 kg projectile

Target type	Front face	Back face	Impact test parameters and response
Plain concrete (M1)	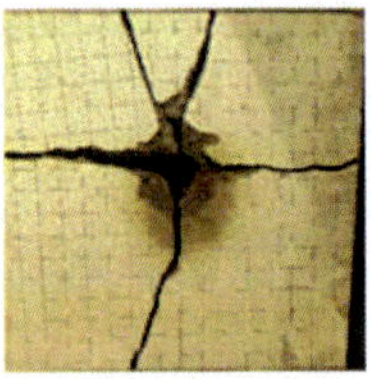	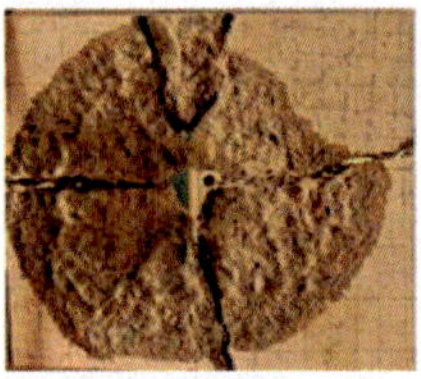	Impact velocity 103 m/s Crater dia in front face = 91 mm Crater dia in back face = 338 mm Damage type = fragmented % loss of mass = 8
Hybrid fiber concrete (M2)		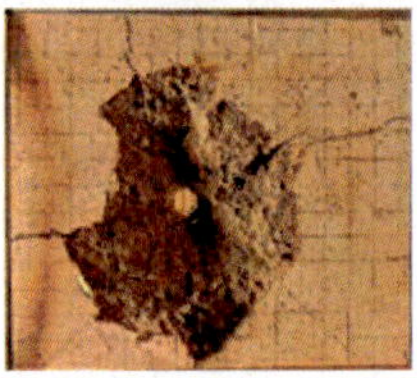	Impact velocity 112 m/s Crater dia in front face = 105 mm Crater dia in back face = 254 mm Damage type = penetration with spalling % loss of mass = 5.2
Reinforced Concrete (M1R)	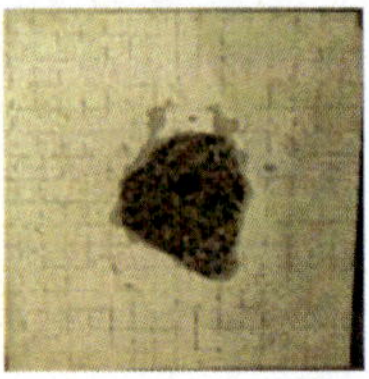	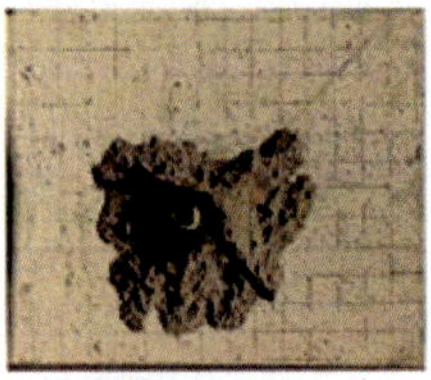	Impact velocity 102 m/s Crater dia in front face = 142 mm Crater dia in back face = 192 mm Damage type = penetration with spalling % loss of mass = 3.4
Reinforced hybrid concrete (M2R)			Impact velocity 108 m/s Crater dia in front face = 112 mm Crater dia in back face = 165 mm Damage type = penetration with spalling % loss of mass = 2.7

3.2.2 Crater Diameter

It is observed from the number of tests that the average crater diameter at the back face is high in the case of plain concrete specimen (M1). However, for the specimens made of reinforced concrete together with hybrid fibers (M2R) had least average crater diameter at the back face. As a consequence of these observations, it is possible to conclude that the hybrid fibers efficiently reduced the extent of the damage while also preventing the concrete from splitting.

3.2.3 Loss of Mass

Mass of the fallen concrete from both the faces of the specimen due to projectile impact has been quantified by the percentage of mass loss from the concrete specimens after testing. It is noted from the observation that the percentage of loss of mass was high in case of plain concrete specimen and was least in the case of reinforced hybrid fiber concrete due to the fact that the reinforcements in conjunction with hybrid fibers held together the concrete by bonding.

A comparison of failure patterns of plain and reinforced concrete slabs with hybrid fibrous and reinforced hybrid fibrous concretes indicated that plain concrete exhibited thick radial cracks and failed by perforation into many pieces, however, the reinforced concrete prevented radial cracking and splitting. Similar observations were made also for the specimens impacted with 1/2 kg projectile. For the purpose of comparison, the average back spalling diameter and percentage of mass loss for the different concrete specimens impacted by projectiles having mass of 1/2 and 1 kg are presented in Figs. 8 and 9 respectively.

4 Conclusions

- Hybridization of macro fibers (hooked end steel fibers) and microfibers (glass and polypropylene fibers) into the plain concrete helped significantly in improving the split tensile strength of concrete by over 43% compared to that of plain concrete specimens. Whereas it moderately improved the compressive strength and flexural strength of concrete by 8–10% compared to those values of plain concrete specimens.
- Hybridization also helped in increasing the energy absorption characteristics in flexural tensile test by increasing the peak load as well as the corresponding displacements. It increased the ductility of the specimen slightly.
- Plain concrete specimen when impacted by projectile gets completely fragmented in the absence of steel reinforcement and wide cracks are found both in the front and rear surfaces.

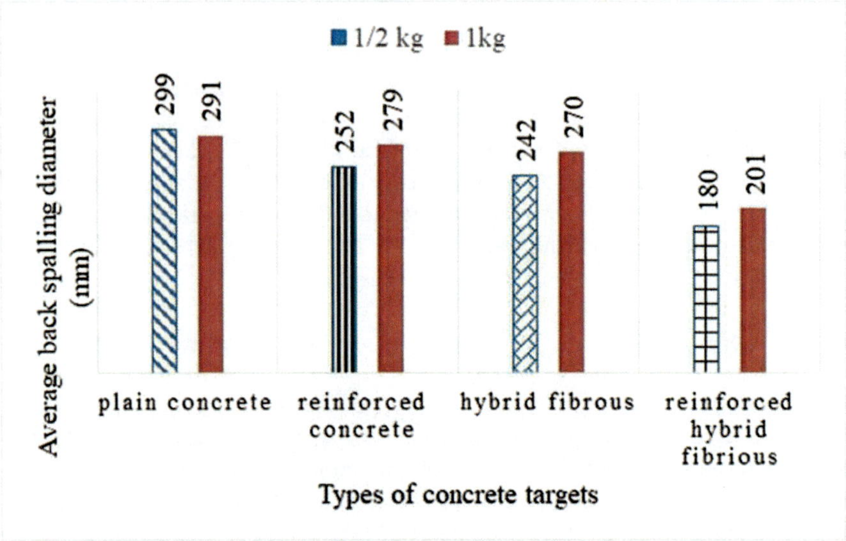

Fig. 8 Back spalling diameter for different concrete targets (80 mm thick)

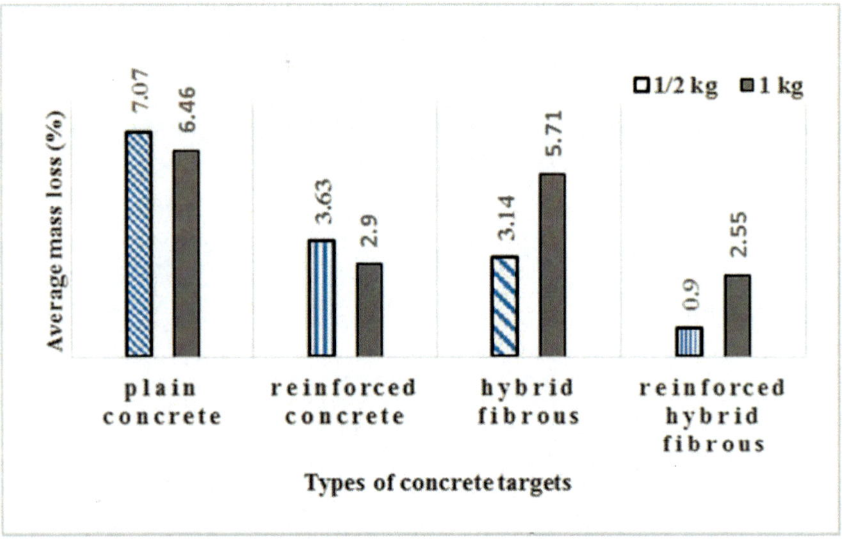

Fig. 9 Percentage of mass loss for different concrete targets (80 mm thick)

- The hybrid fibers efficiently reduced the extent of the damage while also preventing the concrete from splitting. The back face average crater diameter is reduced by more than 40% by the incorporation of hybrid fibers in conjunction with reinforcement bars. Also the loss of mass of concrete as result of projectile impact is reduced in reinforced hybrid fiber specimen. Hybrid fibers have definitely improved the mechanical properties as well as impact response of the concrete specimens.

The work can be extended further for finite element simulation and numerical modeling of projectile impact on hybrid fiber reinforced concrete targets.

Acknowledgements The authors gratefully acknowledge funding received for this current work from Science and Engineering Research Board, Government of India under the scheme Teachers Associateship for Research Excellence (TARE) through the project TAR/2018/000472.

References

1. Varma SMS, Raju PM (2021) Mechanical properties of hybrid polypropylene-steel fibre-reinforced concrete composite. In: IOP conference series: material science engineering, vol 1025(1). pp 012022
2. Alhozaimy AM, Soroushian P, Mirza F (1996) Mechanical properties of polypropylene fiber reinforced concrete and the effects of pozzolanic materials. Cem Conc Comp 18(2):85–92
3. Li B, Chi Y, Xu L, Shi Y, Li C (2018) Experimental investigation on the flexural behavior of steel-polypropylene hybrid fiber reinforced concrete. Const Buil Mat 191:80–94
4. Naaman AE (2003) Engineered steel fibers with optimal properties for reinforcement of cement composites. J Adv Conc Tech 1(3):241–252
5. Tassew ST, Lubell AS (2014) Mechanical properties of glass fiber reinforced ceramic concrete. Const Buil Mat 51:215–224
6. Liu J, Jia Y, Wang J (2019) Experimental study on mechanical and durability properties of glass and polypropylene fiber reinforced concrete. Fib Poly 20(9):1900–1908
7. Abbass W, Khan MI, Mourad S (2018) Evaluation of mechanical properties of steel fiber reinforced concrete with different strengths of concrete. Const Buil Mat 168:556–569
8. Nia AA, Hedayatian M, Nili M, Sabet VA (2012) An experimental and numerical study on how steel and polypropylene fibers affect the impact resistance in fiber-reinforced concrete. Int J Imp Engg 46:62–73
9. Huang L, Chi Y, Xu L, Chen P, Zhang A (2016) Local bond performance of rebar embedded in steel-polypropylene hybrid fiber reinforced concrete under monotonic and cyclic loading. Cons Build Mat 103:77–92
10. Ahmed AA, Jia Y (2019) Effect of using hybrid polypropylene and glass fibre on the mechanical properties and permeability of concrete. Mat 12(22):3786
11. Banthia N, Majdzadeh F, Wu J, Bindiganavile V (2014) Fiber synergy in hybrid fiber reinforced concrete (HyFRC) in flexure and direct shear. Cem Conc Comp 48:91–97
12. Hussain, Ali B, Akhtar T, Jameel MS, Raza SS (2020) Comparison of mechanical properties of concrete and design thickness of pavement with different types of fiber-reinforcements (steel, glass, and polypropylene). Case Stu Cons Mat 13:e00429
13. Dancygier AN, Yankelevsky DZ (1996) High strength concrete response to hard projectile impact. Int J Imp Engg 18(6):583–599
14. Zhang MH, Sharif MSH, Lu G (2007) Impact resistance of high-strength fibre reinforced concrete. Mag of Conc Res 59(3):199–210

15. Kojima (1991) An experimental study on local behaviour of reinforced concrete slabs to missile impact. Nuc Engg Des 130(2):121–132
16. Abdel-Kader M, Fouda A (2014) Effect of reinforcement on the response of concrete panels to impact of hard projectiles. Int J Imp Engg 63:1–17
17. Abdel-Kader M, Fouda A (2019) Improving the resistance of concrete panels to hard projectile impact. Int J Prot Strus 10(4):510–538
18. Zhang X, Wu H, Zhang S, Huang F (2020) Projectile penetration of reinforced concrete considering the effect of steel reinforcement: experimental study and theoretical analysis. Int J Imp Engg 144:103653
19. Sangho LEE, Chunghyeon KIM, Yongjae YU, Jae-Yeol CHO (2021) Effect of reinforcing steel on the impact resistance of reinforced concrete panel subjected to hard-projectile impact. Int J Imp Engg 148:103762
20. Rajput MA, Iqbal P (2017) Bhargava, Experimental and numerical study of concrete targets under high rate of loading. Proc Engg 173:130–137
21. Rajput MA, Iqbal NK (2017) Gupta, Ballistic performances of concrete targets subjected to long projectile impact. Thin-Walled Struc 126:171–181
22. Rajput R, Jakka BM, Pruthvik MA (2016) Iqbal, Experimental study of plain and reinforced concrete targets subjected to impact loading. Proc Engg 144:124–131
23. Kristoffersen M, Toreskås OL, Dey S, Børvik T (2021) Ballistic perforation resistance of thin concrete slabs impacted by ogive-nose steel projectiles. Int J Imp Engg 156:103957
24. Brown JL, Howard IL, Barnes AT (2021) Response of concrete elements subjected to impact by fragments with varying aspect ratios. J Mater Civ Engg 33(4):04021031
25. Brown JL, Howard IL, Woodson BG (2019) Influence of compressive strength, fiber reinforcement, and thickness on spall and breach performance of concrete elements impacted with high-aspect-ratio fragments. Int J Imp Engg 133:103342
26. Jamnam S, Maho B, Techaphatthanakon A, Sonoda Y, Yoo DY, Sukontasukkul P (2020) Steel fiber reinforced concrete panels subjected to impact projectiles with different caliber sizes and muzzle energies. Case Stud Construct Mater 13:e00360
27. Almusallam TH, Siddiqui NA, Iqbal RA, Abbas H (2013) Response of hybrid-fiber reinforced concrete slabs to hard projectile impact. Int J Impact Eng 58:17–30
28. Almusallam TH, Abadel AA, Al-Salloum YA, Siddiqui NA, Abbas H (2015) Effectiveness of hybrid-fibers in improving the impact resistance of RC slabs. Int J Impact Eng 81:61–73
29. Ueno H, Beppu M, Ogawa A (2017) A method for evaluating the local failure of short polypropylene fiber-reinforced concrete plates subjected to high-velocity impact with a steel projectile. Int J Impact Eng 105:68–79
30. Xu X, Ma T, Ning J (2019) Failure mechanism of reinforced concrete subjected to projectile impact loading. Eng Fail Anal 96:468–483
31. IS 12269 (2013) Ordinary Portland cement, 53 grade—specification, Bureau of Indian Standards, New Delhi
32. IS 10262 (2019) Concrete mix proportioning—guidelines, Bureau of Indian Standards, New Delhi
33. IS 516 (1959) Methods of tests for strength of concrete, Bureau of Indian Standards, New Delhi, 2004
34. IS 5816 (1999) Splitting tensile strength of concrete-method of test, Bureau of Indian Standards, New Delhi, 2004

A Visco-Damage-Plastic Model to Analyze the Impact Loading Response in Synthetic Rock

Sachin Kumar, Arghya Das, and Gaurav Tiwari

Abstract In the present study, a coupled damage-plastic constitutive model is proposed for capturing the rate-dependent response of intact rocks under dynamic loading conditions. A thermo-mechanically consistent continuum damage model is utilized for this study. The strain rate effect is introduced through viscoplasticity by enhancing the flow rule using the Perzyna type overstressed function. A fully implicit stress integration scheme is used for the numerical implementation of the model. The model response is calibrated based on the in-house dynamic experiments on synthetic rock specimens under Split Hopkinson Pressure Bar (SHPB). The model qualitatively predicts the variation in dynamic response, especially the peak stress and post-peak softening for different strain rates in the range of 50–150/s.

Keywords Damage plasticity · Perzyna viscoplasticity · Impact load · Synthetic rock · Constitutive model type your

1 Introduction

Several rock engineering and geophysical activities like rock blasting, excavation, pile driving, etc., involve high (dynamic) loading rate conditions [1, 2]. In order to analyze those engineering problems, it is essential to understand the strain rate-dependent response of rocks and model the observed response. For the modelling of mechanical and fracture behaviour of rock-like brittle materials, several pure damage models were used [3–5]. Although these models were quite successful, the plasticity and frictional behaviour of rocks were not included. Due to the absence of the ability to capture the frictional dissipation during the failure process of rock-like materials, the applicability of pure damage models is questionable [6, 7]. These plasticity and frictional effects are involved by coupling the damage with plasticity, as observed in many coupled damage-plastic models [8–10]. However, in several damage-plastic

S. Kumar · A. Das (✉) · G. Tiwari
Department of Civil Engineering, Indian Institute of Technology (IIT) Kanpur, Kanpur, Uttar Pradesh 208016, India
e-mail: arghya@iitk.ac.in

R. Velmurugan et al. (eds.), *Dynamic Behavior of Soft and Hard Materials Volume 1*, Springer Proceedings in Materials 34, https://doi.org/10.1007/978-981-99-6030-9_36

models, a weak coupling was found that resulted in the requirement of multiple loading surfaces [11]. This need for separate loading surfaces for plasticity and damage can be avoided by strong coupling of damage and plasticity (either by calculating the damage variable as a function of plastic strain or by adopting a damage-based model with the evolution law of plastic strain) [12, 13]. In literature, many coupled damage-plastic models are proposed to capture the mechanical response of rocks [14–16]. These models describe the interaction between damage and plasticity to capture the reduction in strength and stiffness observed during inelastic deformation. However, very few coupled damage-plastic models exist that can predict the rate and pressure-dependent behaviour of rock-like materials [17, 18].

In the present study, a combined yield failure function is selected from the existing literature [19] to capture the rate-dependent mechanical response of rock-like material subjected to impact loading. Further, a simple rate-dependent enhancement is incorporated through a Perzyna type overstressed function [20] to account for different impact loading rate effects.

2 Model Description

A coupled damage-plastic model is adopted from a previous study [19–21]. The experimental analysis is performed using impact loading through Split Hopkinson Pressure Bar analysis. Since one-dimensional loading is applied and a corresponding response is captured, constitutive equations for 1D stress–strain analysis are presented here for brevity. However, such 1D formulation is not a limitation of the model since the original damage model is already established in multiaxial stress condition and Perzyna viscoplasticity was also implemented in multiaxial stress space. In this study, the objective is to capture the constitutive response and not solve boundary value problems. Therefore, complete tensor representations of stress ($\boldsymbol{\sigma}$), strain ($\boldsymbol{\varepsilon}$) and stiffness are not presented.

$$\sigma = (1 - D)E\varepsilon^e \tag{1}$$

In Eq. 1, D represents the continuum damage variable and E is the Young's modulus (initial tangent modulus).

The combined damage-yield loading function is given by

$$y = \frac{1}{2}\frac{\sigma^2}{E(1 - D)^2} - F(D) \leq 0 \tag{2}$$

Here $F(D)$ is a function that controls the rate of dissipation due to damage and plasticity [19]. $F(D)$ is defined by the expression

$$F(D) = \frac{f_c^2}{2E}\left\{\frac{E + E_{pc}(1-D)^{n_c}[\ln(1+D)]^{m_c}}{E(1-D) + E_{pc}(1-D)^{n_c}[\ln(1+D)]^{m_c}}\right\}^2 \tag{3}$$

where f_c is the uniaxial compressive stress of the material at static loading condition, E_{pc} is peak tangent stiffness (it controls the softening behaviour of stress–strain response), m_c is a material parameter which governs the ductility of the stress–strain curve and n_c regulates the tail response of the model.

Flow rules are adopted from Nguyen et al. [21]

$$\Delta\varepsilon^p = 2\Delta\lambda\frac{\sin^2\omega}{\sigma} \tag{4}$$

$$\Delta D = 2\Delta\lambda\frac{\cos^2\omega}{F(D)} \tag{5}$$

Here, ω is the parameter governing the coupling between damage and plasticity.

It is important to note that the model considers incremental total strain ($\Delta\varepsilon$) decomposition into elastic ($\Delta\varepsilon^e$) and plastic strain ($\Delta\varepsilon^p$), i.e. $\Delta\varepsilon = \Delta\varepsilon^e + \Delta\varepsilon^p$ but not in the cumulative form.

The constitutive equations can be integrated using an implicit algorithm (return mapping) to trace the complete stress–strain response. In such algorithms, the updated stress is given by

$$\sigma_i^{n+1} = \sigma_i^{trial} - \left(1 - D_i^{n+1}\right)E\Delta\varepsilon^p - \frac{\sigma_i^{n+1}}{\left(1 - D_i^{n+1}\right)}\Delta D \tag{6}$$

Here, the subscript 'i' indicates i th integration point where the stress and state variables are computed. The superscript $n+1$ corresponds to the state of stress and other state variables at the time step $t + \Delta t$. σ_i^{trial} represents the elastic trial stress, i.e. $(\sigma_i^n + (1-D)E\Delta\varepsilon)$.

3 Rate-Dependent Model

The proposed model is extended to capture the rate-dependent response of rock-like materials based on a viscousplasticity. The flow rule is enhanced using an overstress function proposed by Perzyna. Therefore, the flow rules are modified to define the viscoplastic strain and damage evolution as follows

$$\Delta\varepsilon^{vp} = \eta\langle\varphi(y)\rangle\Delta t\frac{2\sin^2\omega}{\sigma} \tag{7}$$

$$\Delta D = \eta\langle\varphi(y)\rangle\Delta t\frac{2\cos^2\omega}{F(D)} \tag{8}$$

where η is the viscosity parameter, Δt is incremental time and $\varphi(y)$ is the overstressed function. The Macaulay brackets $\langle \ldots \rangle$ indicate zero plastic or damage dissipation during elastic loading condition. The time increment Δt is defined such that axial strain rate becomes $\dot{\varepsilon} = \Delta\varepsilon/\Delta t$. The overstressed function $\varphi(y)$ is expressed as

$$\varphi(y) = (y)^m \tag{9}$$

where m is the dimensionless parameter.

In order to perform stress integration, a fully implicit stress-return algorithm is developed following the stress residual method proposed by Etse and Willam [22]. Combining Eqs. 6, 7 and 8, the following stress residual (r_i) can be formed:

$$r_i = \sigma_i - \begin{pmatrix} \sigma_i^{trial} - \left(1 - D_i^{n+1}\right)E\eta\langle\varphi(y)\rangle\Delta t\dfrac{2\sin^2\omega}{\sigma_i^{n+1}} \\ -\dfrac{\sigma_i^{n+1}}{\left(1 - D_i^{n+1}\right)}\eta\langle\varphi(y)\rangle\Delta t\dfrac{2\cos^2\omega}{F(D)} \end{pmatrix} \tag{10}$$

A Newton–Raphson formulation is adopted to minimize the residual stress and obtain the updated stress state and state variables for the Perzyna type viscoplastic damage model formulation.

4 Calibration and Validation

In this section, the proposed rate-dependent model is calibrated and validated with the experimental results of impact testing of rock-like material (dental plaster) under different strain rate ranges. The details of sample preparation and experimental studies through the Split Hopkinson Pressure Bar compression test of synthetic rock specimens are taken from Kumar et al. [23]. Impact loading with four different strain rates, 61 s^{-1}, 83 s^{-1}, 119 s^{-1} and 130 s^{-1} was applied on the cubic specimens. The fracture pattern indicates tensile dominated cracks at low strain rate conditions while influence of shear crack is more in the high strain rate conditions. A few typical fractured rock images are given in Fig. 1.

Notice that the formulation of the Perzyna type strain rate model is such that the parameters rely on a reference strain rate response. In the present study, the lowest strain rate response, at 61 s^{-1}, is considered as a reference. All the rate-independent model (solving $\Delta\lambda$ using consistency model) parameters E, E_{pc}, n_c, m_c, ω, are calibrated with experimental results of impact tests at the lowest strain rate (61 s^{-1}). The rate dependent parameter η is calibrated comparing the responses of rate-independent and dependent model (using Perzyna formulation) at the lowest strain rate (61 s^{-1}). Nevertheless, for damage analysis, f_c is obtained from the static UCS test. Figure 2a describes the selection of E (secant modulus of synthetic specimen at 61 s^{-1}) and E_{pc} for the proposed model. Considering the brittleness in the response, the coupling

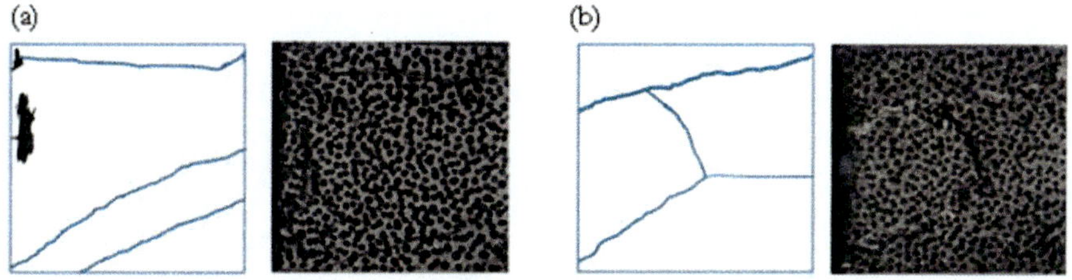

Fig. 1 Schematic and original fracture pattern of rock-like specimens. **a** At a strain rate of 61 s-1. **b** At a strain rate of 130 s-1

angle is assumed to be $\omega = 0$ indicating elastic damage since the static unconfined compression test shows elastic brittle response without any hardening or softening. In the case of the 3-dimensional formulation of this proposed model, these parameters will change due to the effect of Poisson's ratio. The calibrated parameters are listed in Table 1. Figure 2b compares the response of the rate-independent and proposed rate-dependent models with the same calibration parameters. With increasing strain rate, the peak strength increases.

Figure 3 demonstrates a comparison between the experimental results and model response at the constitutive level under one-dimensional impact loading at different

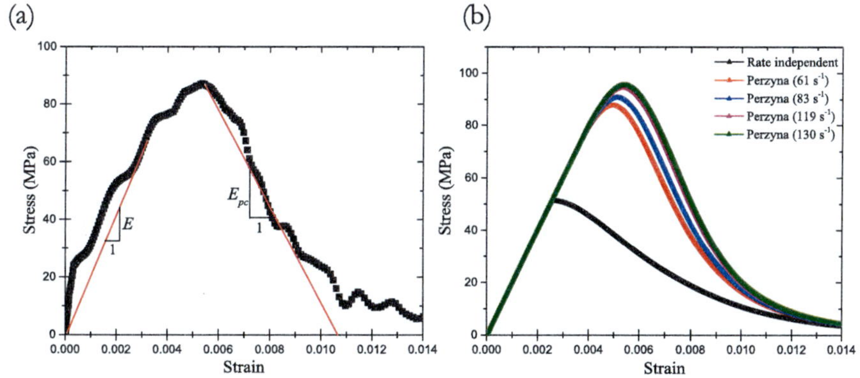

Fig. 2 **a** Description of secant modulus (E) and peak tangent stiffness (E_{pc}) used for calibration from the experimental results at strain rate of 61 s-1; **b** Model response under different strain rate conditions

Table 1 Calibrated parameters of the proposed model

Parameter	Value	Parameter	Value
f_c	51.2 MPa	m_c	1.4
E	20.3 GPa	m	2.77
E_{pc}	17.2 GPa	η	0.003
n_c	0.2	ω	0

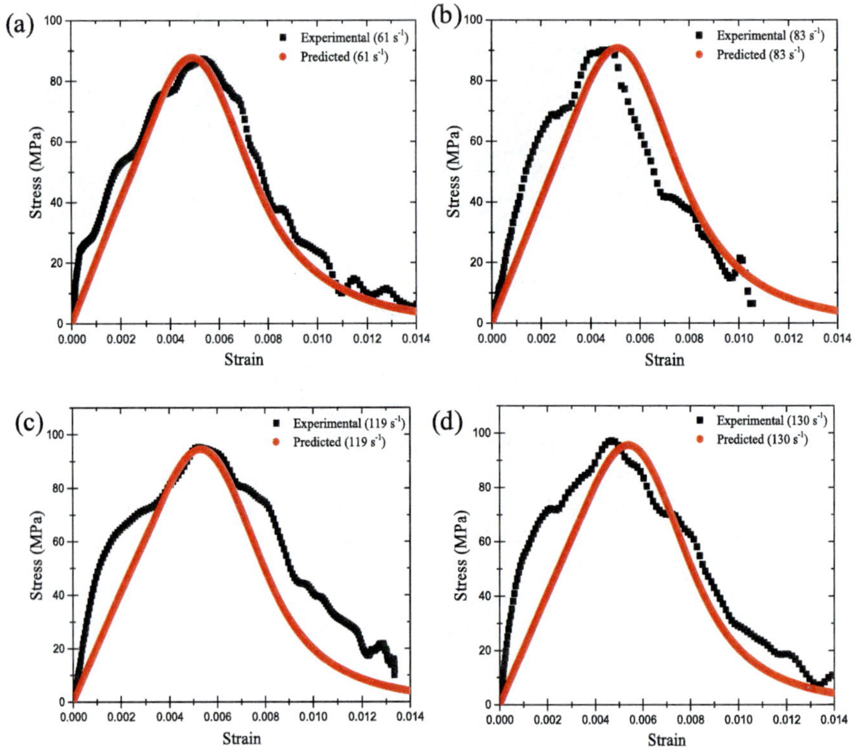

Fig. 3 Comparison on experimental and predicted stress–strain response. **a** 61 s^{-1} **b** 83 s^{-1} **c** 119 s^{-1} **d** 130 s^{-1}

strain rates. These responses are obtained by only changing Δt to incorporate the strain rate effects while other calibration parameters remain the same. The predicted response shows a qualitative agreement against the experimental plots. In addition, the change in the peak stress due to increasing impact loading rate is well-captured results in this model (Fig. 4). In rate-dependent model, the strain rate ($\dot{\varepsilon} = \Delta\varepsilon/\Delta t$) can be varied in two ways, (a) while maintaining a constant displacement/velocity and changing the value of Δt, and (b) keeping the value of Δt fixed while varying the displacement/velocity at the boundary of the specimens. Hence, in the present work, we keep the displacement fixed and varied the Δt to capture the rate enhancement.

It is important to note that the pre-peak response (possibly elastic response) shows increasing stiffness with increasing strain rate. Since the present model assumes a constant linear elasticity, such a response is not captured. On the other hand, the post-peak response shows undulation due to structural influence during fracture propagation. Modelling such phenomena requires solving the boundary value problem.

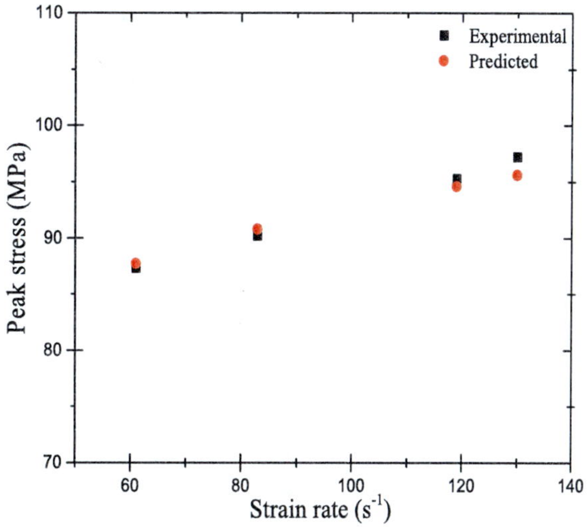

Fig. 4 Experimental and predicted peak stress of synthetic rock specimens at different strain rates

5 Conclusions

This study proposed a coupled damage-plastic constitutive model to predict the (impact and/or dynamic) strain rate-dependent mechanical response of rock-like specimens. A simple Perzyna type overstressed function is implemented into an existing coupled damage-plastic model for the rate enhancement in stress–strain response. Unlike previous studies where only low strain rate response is modelled using Perzyna type viscoplasticity, the proposed model successfully captured the stress–strain response of rock-like specimens under impact load. The modelling approach demonstrates that model calibration based on any impact loading response (possibly the slowest strain rate) can easily predict other high impact responses as well. However, the experiments also demonstrate that dynamic strain rate alters the elastic stiffness of the rock, and the relationship between modulus and strain rates is highly chaotic; thus, ad-hoc viscoelasticity can be incorporated in the proposed model to obtain better agreement with the experimental observation. Ongoing research by the authors is aimed at formulating the model in multiaxial stress space and solving boundary value problem on dynamic fracture in rocks.

References

1. Zhang QB, Zhao J (2014) A review of dynamic experimental techniques and mechanical behaviour of rock materials. Rock Mech Rock Eng 47:1411–1478. https://doi.org/10.1007/s00603-013-0463-y
2. Xia K, Yao W (2015) Dynamic rock tests using split Hopkinson (Kolsky) bar system e a review. J Rock Mech Geotech Eng 7:27–59. https://doi.org/10.1016/j.jrmge.2014.07.008

3. Benvenuti E, Borino G, Tralli A (2002) A thermodynamically consistent nonlocal formulation for damaging materials. Eur J Mech A Solids 21:535–553
4. Comi C (2001) A non-local model with tension and compression damage mechanisms. Eur J Mech—A/Solids 20:1–22
5. Borino G, Failla B, Parrinello F (2003) A symmetric nonlocal damage theory. Int J Solids Struct 40:3621–3645
6. Perdikaris PC, Romeo A (1995) Size effect on fracture energy of concrete and stability issues in three-point bending fracture toughness testing. ACI Mater J 92:483–496
7. Reinhardt HW, Cornelissen HAW, Hordijk DA (1986) Tensile tests and failure analysis of concrete. J Struct Engng 112:2462–2477
8. Abu Al-Rub RK, Kim S-M (2010) Computational applications of a coupled plasticity-damage constitutive model for simulating plain concrete fracture. Eng Fract Mech 77:1577–1603
9. Luccioni B, Oller S, Danesi R (1996) Coupled plastic-damaged model. Comput Methods Appl Mech Engng 129:81–89
10. Salari MR, Saeb S, Willam KJ, Patchet SJ, Carrasco RC (2004) A coupled elastoplastic damage model for geomaterials. Comput Methods Appl Mech Engng 193:2625–2643
11. Einav I, Houlsby GT, Nguyen GD (2007) Coupled damage and plasticity models derived from energy and dissipation potentials. Int J Solids Struct 44:2487–508
12. Grassl P (2009) On a damage-plasticity approach to model concrete failure. Proc ICE Engng Comput Mech 162:221–31
13. Faria R, Oliver J, Cervera M (1998) A strain-based plastic viscous-damage model for massive concrete structures. Int J Solids Struct 35:1533–1558
14. Chen L, Wang C, Liu J, Wang J, Jia Y et al (2015) Damage and plastic deformation modelling of Beishan granite under compressive stress conditions. Rock Mech Rock Eng 48:1623–1633
15. Chiarelli A, Shao J, Hoteit N (2003) Modelling of elastoplastic damage behaviour of claystone. Int J Plast 19:23–45
16. Unteregger D, Fuchs B, Hofstetter G (2015) A damage plasticity model for different types of intact rock. Int J Rock Mech Min Sci 80:402–411
17. Mukherjee T, Nguyen GD, Mir A, Bui HH, Shen L, El-zein A et al (2017) Capturing pressure- and rate-dependent behaviour of rocks using a new damage-plasticity model. Int J Impact Eng 110:208–218. https://doi.org/10.1016/j.ijimpeng.2017.01.006
18. Yang L, Wang G, Zhao G, Shen L (2020) A rate—and pressure-dependent damage-plasticity constitutive model for rock. Int J Rock Mech Min Sci 133:104394
19. Nguyen GD, Houlsby GT (2008) A coupled damage–plasticity model for concrete based on thermodynamic principles : Part I : model formulation and parameter identification. Int J Numer Anal Meth Geomech 32:353–389. https://doi.org/10.1002/nag
20. Perzyna P (1966) Fundamental problems in viscoplasticity. Adv Appl Mech 9:243–377
21. Nguyen GD, Einav I, Guiamatsia I (2012) On the partition of fracture energy in constitutive modelling of quasi-brittle materials. Eng Fract Mech 79:225–244. https://doi.org/10.1016/j.engfracmech.2011.11.002
22. Etse G, Willam K (1999) Failure analysis of elastoviscoplastic material models. J Eng Mech 125(1):60–69
23. Kumar S, Tiwari G, Parameswaran V, Das A (2022), Rate-dependent mechanical behavior of jointed rock with an impersistent joint under different infill conditions. J Rock Mech Geotech Eng

Development of a Simplified Numerical Model to Characterize the Ballistic Behaviour of Woven Fabrics

Charitha Mudi and Amit Shaw

Abstract In computational simulations of ballistic impact on woven fabric, the idealization of fabric structures is very important. In the available literature, several attempts have been made using commercial finite element packages. Those studies typically result in either overly simplified models or fully discretized 3D models that are prohibitively computationally intensive. The present study provides a simplified finite element model of plain-woven fabric, which is developed based on characterizing the behaviour in the pull-out test. The warp and weft yarns are modelled independently with straight elastic bars by 1D truss elements, representing the straightened crimped yarns. The orthogonally arranged weft and warp yarns are connected through viscoelastic interaction at crossovers. The required interaction parameters of the viscoelastic system at crossover points are obtained using the pull-out test data. The obtained results of the simplified model in terms of the deformation history of the fabric target and residual velocity history of the projectile are in reasonable congruence with the experimental and other numerical observations reported in the literature. The proposed idealization is computationally efficient and may be further explored to model more complicated weave patterns, for which 3D modelling, so far, is the only viable option.

Keywords Viscoelastic interaction · Inter-yarn spring-dashpot system · Kevlar fabric · Abaqus

1 Introduction

Woven fabrics comprised of high-strength fibres offer exceptional mechanical properties such as high tenacity, modulus, and stiffness, making them suitable for ballistic applications. Notably, the aramid fibres such as Kevlar possess additional properties such as being lightweight and flexible, making them ideal over conventional heavy metals in manufacturing body armour and bulletproof vests. Over the last few

C. Mudi · A. Shaw (✉)
Department of Civil Engineering, Indian Institute of Technology, Kharagpur 721302, India
e-mail: abshaw@civil.iitkgp.ac.in

R. Velmurugan et al. (eds.), *Dynamic Behavior of Soft and Hard Materials Volume 1*,
Springer Proceedings in Materials 34, https://doi.org/10.1007/978-981-99-6030-9_37

decades, investigations on woven fabric's impact and penetration resistance against ballistic impact have emerged as a research trend. Several studies indicated that various factors such as weave architecture, geometry, and material characteristics of fabric [3] play important roles in altering the ballistic performance of woven fabrics. Brisco and Motamedi [4], Bazhenov [5], and Tan et al. [7] concluded that the interfacial friction between the yarns and fabric-projectile affects woven fabrics' energy dissipation and absorption capacity. However, the experimental studies could not provide much insight into how friction affects the energy absorption mechanisms [3–7]. Fabric can absorb more energy without undergoing severe failure when the yarns are able to dissipate the force exerted by the projectile to neighbouring yarns. Estimating the inter-yarn friction is important in numerical simulation to investigate the realistic ballistic response of a fabric. Sebastian et al. [1] addressed this by obtaining the yarn friction beforehand through the yarn pull-out test. The pull-out study would certainly help determine woven fabric's actual performance by estimating a substantial friction value. Numerous numerical techniques are employed to better understand fabric deformation and failure. Attempts are made by developing 3D solid elements to represent the actual weave geometry of the fabric architecture and are found advantageous as the results obtained are remarkable [7–10]. However, such models require many elements due to the full discretization of yarns and are thus computationally intensive. Therefore, the idealization of woven fabric for the numerical modelling of weave architecture, along with other essential features, has become a great challenge to researchers. The yarn-level representation of woven fabric using 2D membrane or shell elements has then come into practice [12–16]. Nevertheless, some fabric features, such as unravelling, slippage of yarns, and friction between yarns, are not addressed properly. Some studies put effort into developing idealized fabric modelling using one-dimensional elements. That led to using a 1D linear bar or truss elements to resemble the geometry of woven fabric and trying to incorporate all the necessary features. The yarns at crossovers are tied together or provided with an inter-yarn spring [7], whereas the inter-yarn friction is given as general contact to avoid computational complexities. Such shortcomings are a great task to modify and improve the idealization of woven fabric. Shim et al. [12], Johnson et al. [15], and Das et al. [17] reported that the 1D truss elements provide reasonable accuracy. However, the incorporation of inter-yarn friction in such models is still a point of concern.

The present study aims to develop an efficient way of incorporating inter-yarn friction and the effect of weave architecture in a fabric model where yarns are idealized by 1D truss elements. Herein, the single-layer plain-woven fabric subjected to impact with a spherical tip projectile is modelled using commercial FEA software ABAQUS®. The yarns are modelled with 1D linear bar or wire elements representing the straightened crimped yarns. The warp and fill yarns are modelled individually and then arranged orthogonally to allow them to join but not in direct contact. An inter-yarn spring-dashpot is provided at every crossover point between each warp and fill, facilitating the inter-yarn friction and viscoelastic nature of yarns. The required parameters for the spring-dashpot system are derived from a yarn pull-out test. The spherical projectile is modelled as a rigid body using solid elements. The computed

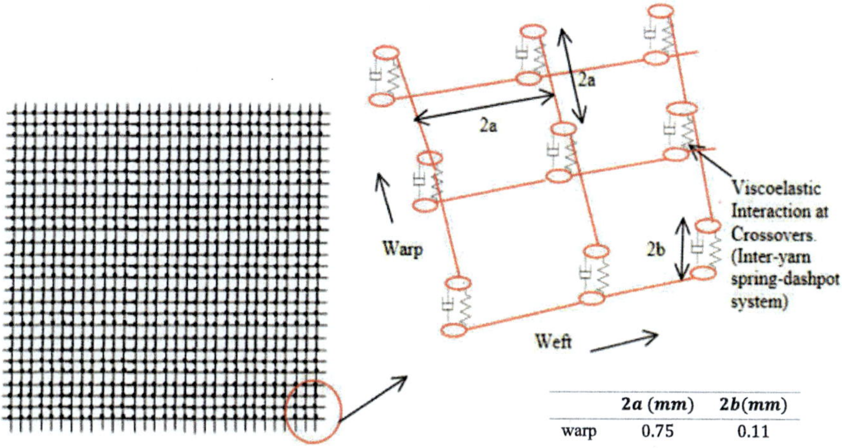

	$2a$ (mm)	$2b(mm)$
warp	0.75	0.11
weft	0.63	0.11

Fig. 1 Geometry details of plain-woven fabric

results are compared with the numerical results given by Das et al. [17]. The results obtained from the present study are in good agreement.

2 Computational Model

2.1 Geometry Details of Woven Fabric

The plain-woven Kevlar fabric of square size 120 × 120 mm is considered for the present study. The Kevlar of K-706 style contains KM2 600 denier yarns with a yarn density of 34 yarns per inch. The woven fabric has a density of 1440 kg/m3. Figure 1 illustrates the schematic representation of the fabric cross-section.

2.2 FEA Model Description

Improvements to simplified models, such as yarn-level resolution where the woven fabric is represented by a network of bar elements, are actively developed since intricate models incorporating full discretization of yarns are computationally intensive to simulate comprehensive fabric armour systems.

In the present study, a simple and effective yarn-level-based model is developed using FEA code ABAQUS® to examine the ballistic performance of the plain-woven fabric. The network of linear 1D bar elements is used to model the yarns. The warp and fill yarns are individually modelled and slid orthogonal to each other. The orthogonally arranged yarns are kept in contact but not joined directly at crossovers. The nodes of the warp and weft yarns are maintained 0.11 mm apart in the thickness direction. The yarns are brought into contact at the crossover points by a spring-dashpot system, which also aids the viscoelastic nature of the yarns as they interact. An average coefficient of friction of 0.2 is given. The corresponding dashpot coefficient is computed using the equivalent viscous damping concept [22]. The values of the equivalent viscous damping for different friction coefficients are shown in Table 3.

The spring coefficient value is obtained as 5.5 N/mm by performing a linear fit to the corresponding curve region in a pull-out load–displacement plot where the crimped yarns become straightened. Figures 1 and 2 depict the proposed idealization of the plain-woven fabric model. The fabric is considered to be an elastic continuum; beyond the ultimate strength, the fully damaged elements are assumed to be removed from the model.

The fracture energy-based failure criterion is employed for the yarn elements. Despite the fact that it does not expressly reflect yarn failure, it prevents yarns from becoming perfectly brittle. The spherical tip projectile is modelled as rigid material using 3D solid elements. A general hard contact algorithm provided by ABAQUS® Explicit has been used to describe the projectile–fabric interaction. According to the stated literature in [17], impact velocities are assumed to be in the range of 35–110 m/s. Tables 1 and 2 provide information on Kevlar fabric and projectile details.

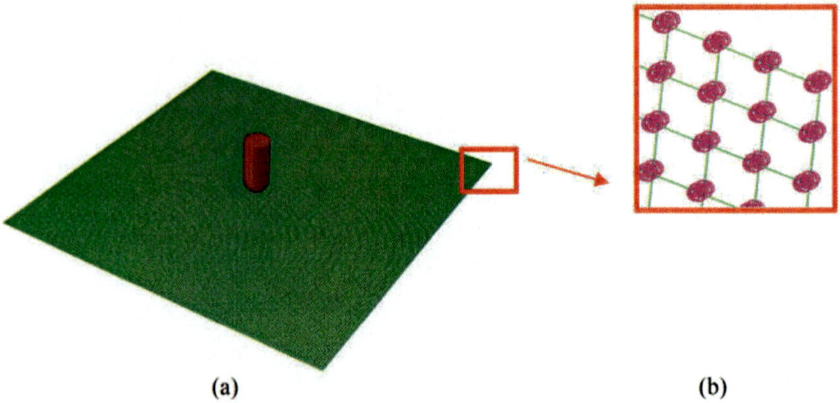

Fig. 2 **a** Numerical model of plain-woven fabric; **b** Inter-yarn spring-dashpot at crossovers

Table 1 Kevlar fabric material properties [17]

Young's modulus (MPa)	Ultimate strength (MPa)
82,600	2929

Table 2 Projectile details

Length (mm)	Diameter (mm)	Mass (g)
12	9	7.5

Table 3 Viscous damping coefficient values obtained from coulomb friction

Coefficient of friction	Equivalent viscous damping $(N - s/mm)$
0.1	5.5e-8
0.2	1.10e-7
0.3	1.656e-6
0.4	2.209e-5
0.5	2.761e-4
0.6	3.313e-4

2.3 *Validation of Results*

A series of simulations are performed with varying initial velocities to explore the critical velocity, where the fabric is able to absorb the total impact energy exerted by the projectile. The results revealed that the projectile is penetrated into the fabric at an initial velocity of 48 m/s for a spherical tip projectile, below which the projectile is bounced off.

Das et al. [17] performed a numerical study on woven fabric modelled with 1D linear elements. For validation of the present model, a few representative cases with spherical projectile are considered. The deformation time history plots are shown in Fig. 3. The results from the current model revealed some minor discrepancies, such as less deformation comparatively. This may be ascribed to the inter-yarn spring-dashpot system, which restrains the lateral movement of yarns slightly and would result in less fabric deformation. The deformed shape of the fabric at 0.4 ms for impact velocity 55 m/s and for different boundary conditions are shown in Fig. 4. The computed impact velocity vs. residual velocity curve is compared with that of given in [17] in Fig. 5.

Next, the simulations of fully clamped fabric subjected to a projectile impact with 55 m/s are performed with different friction coefficients ranging from 0.1 to 0.6. Corresponding equivalent viscous damping coefficients are given in Table 3. The energy absorption in friction with different friction coefficients is shown in Fig. 6. Generally, when a projectile strikes a target, the fabric fails mostly because the yarns are pushed apart, and the yarns break at the impact centre. However, when no appropriate inter-yarn friction is facilitated in the fabric model, it would result in early fabric failure or some fictitious fabric response. Therefore, the inter-yarn friction coefficient must be incorporated carefully while modelling the fabric because primary and secondary yarns participate in energy dissipation. In this work, the rise in fabric energy absorption is observed for friction coefficients ranging from 0 to

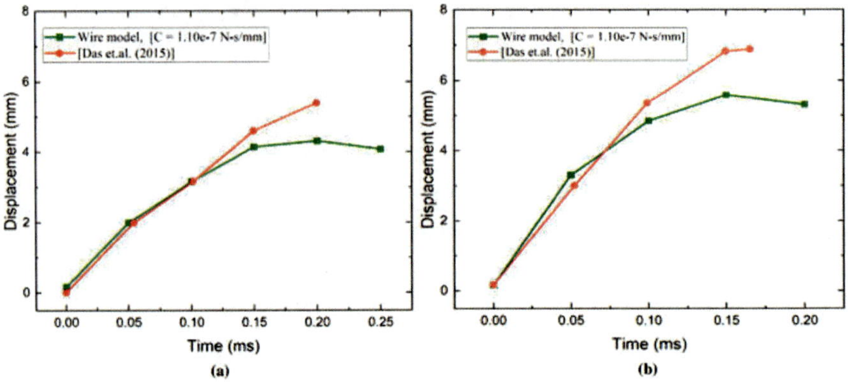

Fig. 3 The deformation time histories of plain-woven fabric **a** 38 m/s; **b** 63 m/s

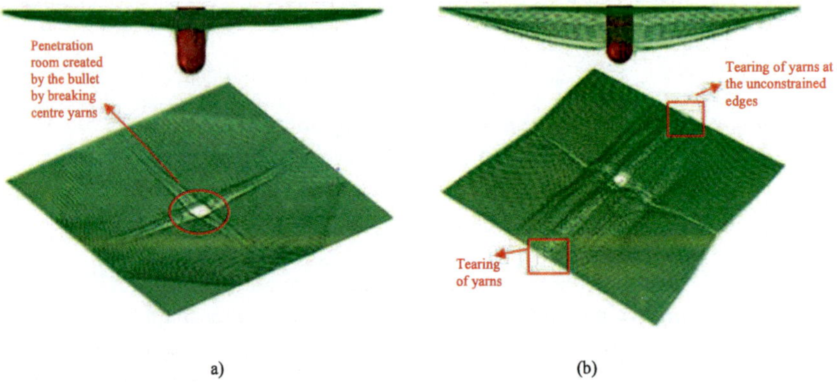

Fig. 4 Deformed configurations for critical initial velocity 55 m/s at 0.4 ms. **a** fully clamped target; **b** two opposite sides are clamped

0.4, after which it drops. When the projectile hits the target, the lateral movement of the warp and weft yarns is primarily restrained at greater and optimum inter-yarn friction. At the optimum friction coefficient, the fabric fails mainly owing to yarn failure at the impact point. It results in excellent energy dissipation between the yarns. As demonstrated in Fig. 6, the progressive increase in energy absorption is absorbed until the coefficient of friction reaches 0.4. Beyond this threshold, it has a contrary effect on energy absorption. A fabric with a very high friction coefficient would behave as an overly stiff material, resulting in false ballistic fabric behaviour.

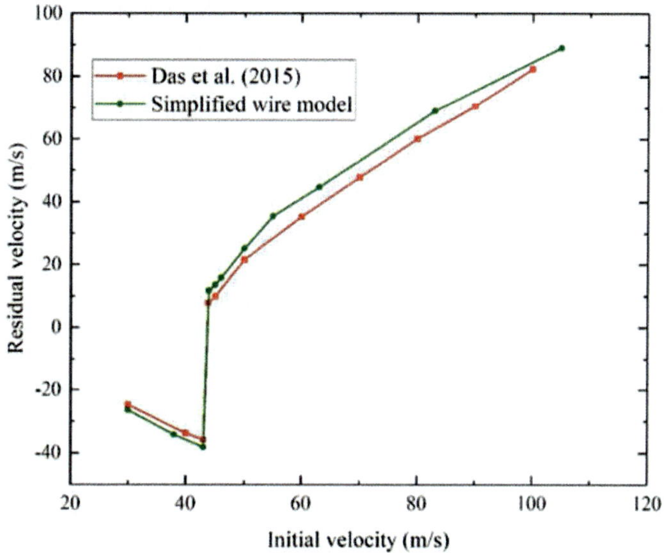

Fig. 5 Residual velocity time history of the projectile

Fig. 6 Energy absorption with coefficient of friction

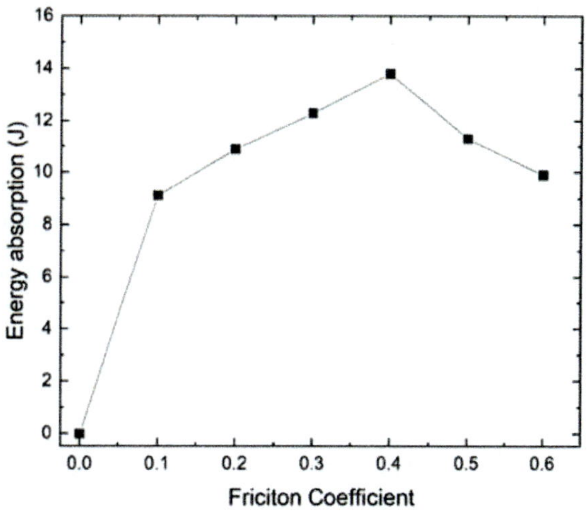

3 Closure

The objective has been to develop an efficient, simple numerical model to examine the ballistic behaviour of woven fabrics at low-velocity impacts. The simplification is made according to the fabric and yarn behaviour in the pull-out test. The pull-out test helps in estimating the inter-yarn friction. The ballistic response of single-layer woven fabric, when impacted with a rigid projectile is modelled using ABAQUS® explicit. In this study, the warp and fill yarns are individually modelled using a one-dimensional linear bar or wire elements. The warp and fill yarns are joined by each other by inter-yarn spring-dashpot, incorporating the friction with the corresponding dashpot coefficient. It is observed that inter-yarn spring-dashpot substantially affects the yarn impact performance in terms of yarn deformation path and failure. This work significantly shows how friction governs the energy absorption capabilities of a fabric. The results obtained are in good agreement with the reported numerical results. The yarns which are left unconstrained have a greater impact on performance sensitivity when the projectile strikes the fabric. The accuracy in results provides encouragement in suggesting the yarn resolution models with reduced computational cost and rapid analysis. The higher frictions show a reverse effect in fabric energy absorption mechanisms; a friction coefficient of 0.4 is found as an optimal value.

References

1. Sebastian SARD, Bailey AI, Briscoe BJ, et al. (1986) Effect of a softening agent on yarn pull–out force of a plain weave fabric. Text Res Inst 56:604–611. 10.1177%2F004051758605601003
2. Shim VPW, Tan VBC, Tay TE (1995) Modelling deformation and damage characteristics of woven fabric under small projectile impact. Int J Impact Eng 16.4:585–605. https://doi.org/10.1016/0734-743X(94)00063-3
3. Cunniff PM, Ting J (1999) Development of a numerical model to characterize the ballistic behaviour of fabrics. In: Proceedings of the 18th international symposium of ballistics, pp 822–28
4. Briscoe BJ, Motamedi F (1992) The ballistic impact characteristics of aramid fabrics: the influence of interface friction. Wear 158(1–2):229–247. https://doi.org/10.1016/0043-1648(92)90041-6
5. Bazhenov S (1997) Dissipation of energy by bulletproof aramid fabric. J Mater Sci 32.15:4167–73. https://doi.org/10.1023/A:1018674528993
6. Duan Y, Keefe M, Bogetti TA, Cheeseman BA (2005) Modeling the role of friction during ballistic impact of a high-strength plain-weave fabric. Compos Struct 68.3:331–37. https://doi.org/10.1016/j.compstruct.2004.03.026
7. Tan VBC, Ching TW (2006) Computational simulation of fabric armour subjected to ballistic impacts. Int J Impact Eng 32.11:1737–51. https://doi.org/10.1016/j.ijimpeng.2005.05.006
8. Shockey DA, Erlich DC, Simons JW (1999) Lightweight fragment barriers for commercial aircraft. In: Proceedings of the 18th international symposium of ballistics, pp 1192–99
9. Blankenhorn G, Schweizerhof K, Finckh H (2003) Improved numerical investigations of a projectile impact on a textile structure. In: 4th European LS-DYNA users conference, vol 49.0. pp 7–14
10. Billon HH, Robinson DJ (2001) Models for the ballistic impact of fabric armour. Int J Impact Eng 25(4):411–422. https://doi.org/10.1016/S0734-743X(00)00049-X
11. Duan Y, Keefe M, Bogetti TA, Cheeseman BA, Powers B (2006) A numerical investigation of the influence of friction on energy absorption by a high-strength fabric subjected to ballistic impact. Int J Impact Eng 32.8:1299–1312. https://doi.org/10.1016/j.ijimpeng.2004.11.005
12. Simons JW, Erlich DC, Shockey DA (2001) Finite element design model for ballistic response of woven fabrics. In: Proceedings of the 19th international symposium of ballistics, pp 1415–22
13. Brueggert M, Tanov RR (2002) An LS-DYNA user defined material model for loosely woven fabric with non-orthogonal varying weft and warp angle. In: Proceedings of the 7th international LS-DYNA users conference, pp 8–13

14. Ala T, Ivanov I (2002) Computational miro-mechanical model of flexible woven fabric for finite element impact simulation. Int J Numer Methods in Eng 53.6:1259–76. https://doi.org/10.1002/nme.321
15. Johnson GR, Beissel SR, Cunniff PM (1999) A computational model for fabrics subjected to ballistic impact. In:18th International symposium on ballistics, 2 January 1999, pp 962–69
16. Lim CT, Shim VPW, Ng YH (2003) Finite-element modeling of the ballistic impact of fabric armor. Int J Impact Eng 28(1):13–23. https://doi.org/10.1016/S0734-743X(02)00031-3
17. Sanhita D, Jagan S, Shaw A, Pal A (2015) Determination of inter-yarn friction and its effect on ballistic response of para-aramid woven fabric under low velocity impact. Compos Struct 120:129–40. https://doi.org/10.1016/j.compstruct.2014.09.063
18. Gaurav N, Gillespie JW (2012) Ballistic impact modeling of woven fabrics considering yarn strength, friction, projectile impact location, and fabric boundary condition effects. Compos Struct 94.12:3624–34. https://doi.org/10.1016/j.compstruct.2012.05.030
19. Gaurav N, Gillespie JW (2013) Yarn pull-out behavior of plain woven kevlar fabrics: effect of yarn sizing, pullout rate, and fabric pre-tension. Compos Struct 101:215–24. https://doi.org/10.1016/j.compstruct.2013.02.018
20. Rodríguez Millán M, Moreno CE, Marco M, Santiuste C, Miguélez H (2016) Numerical analysis of the ballistic behaviour of kevlar® composite under impact of double-nosed stepped cylindrical projectiles. J Reinforced Plastics and Compos 35.2:124–37. https://doi.org/10.1177/0731684415608004
21. Sagar I, Yerramalli CS, Guha A, Mishra S (2021) Effect of material properties on ballistic energy absorption of woven fabrics subjected to different levels of inter-yarn friction. Composite Struct 266. https://doi.org/10.1016/j.compstruct.2021.113824
22. Stutts, Daniel S (2013) Equivalent viscous damping. 2.7:11–13

Low-Velocity Impact Response of Nano-Silica Reinforced Aluminum/PU/GFRP Laminates

M. Vijayan, V. Selladurai, Vishnu Vijay Kumar, G. Balaganesan, and K. Marimuthu

Abstract The bonding of metal sheets and Fiber-reinforced composites is called Fiber Metal Laminates (FML). In the current work, Polyurethane (PU) foam is inserted as a core material in FML, and E-Glass fibers enclose it. The exterior layers are of aluminum alloy 2024 sheets bonded by nanofiller dispersed in an epoxy resin matrix. The nano-silica is used as nanofillers dispersed in the epoxy resin matrix. Nano-silica of 1wt.% is used with epoxy for fabricating the NFML (Nano Fiber Metal Laminate). Laminates are fabricated by hand lay-up and compression molding at 180 MPa at 70 °C. The laminates are tested under low-velocity impact test using a drop weight impactor at different energy levels of 60 J, 80 J, and 90 J for a specimen size of 150 mm x 150 mm. The failure modes of aluminum sheet, PU Foam, and Composite laminate are studied. The results are compared between FML with and without nano-silica. The effect of nano-silica becomes significant in the laminate from 90 J impact energy. Below that, the pristine FML has a high impact force and withstands the impact load effectively. The failure sign of dent depth, delamination area on the rear side, and peak impact load are measured.

Keywords Nano-silica · PU foam · Drop weight test · Adhesive bonding · FML

M. Vijayan (✉) · K. Marimuthu
Department of Mechanical Engineering, Coimbatore Institute of Technology, Coimbatore 641014, India
e-mail: vijayan.m@cit.edu.in; aerovijay97@gmail.com

V. Selladurai
Department of Management Studies, Coimbatore Institute of Technology, Coimbatore 641014, India

V. Vijay Kumar · G. Balaganesan
Indian Institute of Technology Madras, Chennai 600036, India

V. Vijay Kumar
Department of Mechanical Engineering, National University of Singapore, Singapore 117575, Singapore

R. Velmurugan et al. (eds.), *Dynamic Behavior of Soft and Hard Materials Volume 1*,
Springer Proceedings in Materials 34, https://doi.org/10.1007/978-981-99-6030-9_38

1 Introduction

Composite materials are used in various fields like aerospace, construction, communication, military, and so on because of their high specific strength, increased design flexibility, thermal expansion properties, and good resistance to fatigue and corrosion [1–3]. The Fiber Metal Laminate (FML) is an emerging hybrid composite material class of thin metal layers bonded together with composite prepreg. They were first developed in the 1980s at the Delft University of Technology, Netherlands [4]. FML is fabricated by sandwiching fibers and metals under pressure, bonded by an epoxy resin [5]. It combines both the advantages of composite and metal. The PU foam is incorporated into the FML layer to improve the energy-absorbing capacity of the FML.

Akshaj Kumar et al. [6] discuss the sandwich structure material for low-velocity impact response having mild steel as skin material bonded to thermoplastics/PU foam as core material in drop weight test. The effects of a projectile impact test on a sandwich panel made of CSM/WRM/Foam/Epoxy laminate are discussed and concluded, indicating that thin foam does not improve energy absorption or panel stiffness [7]. Balaganesan et al. [8] studied the energy absorption characteristic of PU foam, silicon carbide inserts, and silicon carbide (SiC) plate bonded to GFRP composite laminate. SiC inserts and plates are combined as the front layer of the composite laminate to improve energy absorption. Target impact testing at low-velocity is done using a drop mass setup and a mild steel spherical nosed impactor with fixed boundary conditions by Akshaj Kumar et al. [9]. Low-velocity impact test was conducted by impactors with various geometries on rigid PU foam blocks at velocities ranging from 2 to 4 m/s employing a lumped mass approach and formulating based on finite deformation [10]. Wang et al. [11] conducted experimental and numerical analysis approach to the foam-core sandwich panel under low-velocity impacts. PU foam-cored panels with laminated face sheets of plain weave carbon fabric were subjected to low-velocity impacts with hemispherical steel impactors of varying diameters and energies. S. Dariushi and M. Sadighi [12] studied a novel hybrid material using Glass–Epoxy/Aluminum (GE/AL) laminate face sheets and two types of crushable PU and un-crushable EVA 180 foam cores tested under three-point bending. Similar specimens with Glass–Epoxy (GE) faces were created for the comparison and tested under identical conditions.

The effect of Nano-graphene (GnP) on impact properties and reducing the extent of damage was studied by Haq et al. [13]. GnP platelets were used to fabricate sandwich structures made of woven glass fabric and PVC foam. Their impact behavior was assessed by referencing pristine samples. Zhang et al. [14]used a PU-filled pyramidal lattice core sandwich panel under low-velocity impact and discovered that the impactor's contact time with the sandwich specimens is shorter. The impact peak load is slightly higher for the foam-filled specimens. For the study on the impact response of high-density flexible PU-based foam, they are loaded along with a crushing up of the foam and irreversible heating [15]. Only limited works on the reinforcement of Nano-silica filler in Al/PU/GFRP laminates are available in the literature.

This paper investigates the energy absorption of Al/GFRP/PU foam with and without nano-silica under low-velocity impact. PU foam serves as the core material and is attached to an outer layer of aluminum sheet to increase energy absorption. Drop mass setup is used to conduct low-velocity impact. Impact testing of laminates in fixed boundary conditions is performed using a mild steel spherical nose impactor. The failure of GFRP, PU foam, and Aluminum layers are determined comparatively with and without nano-silica. The parameters like peak force, dent depth, and delamination area for the failed samples are compared to the FML laminate incorporated PU foam core with and without nano-silica.

2 Experimental Section

2.1 Specimen Preparation

The materials used in this study involve aluminum alloy of grade 2024 (0.4 mm thickness) sheet, woven rowing E-glass fiber of 610 g/m^2 with a dimension of 150 × 150 × 15 mm, and PU foam of 450 kg/m^3 density. Epoxy resin (Araldite LY556) and hardener (Aradur HY951) are used for bonding purposes. The nano-silica filler of 17 nm is used as secondary reinforcement, dispersed in the epoxy resin using a mechanical stirrer for 2 h to obtain a homogeneous distribution. Before lamination, the aluminum sheet is pre-treated with a chemical for better adhesion properties. The specimen is fabricated by the hand-layup method. The composite is fabricated by laying an aluminum sheet over the wax-coated mold and placing three woven rowing glass fibers, and PU forms alternatively as AL/WRM3/PU/WRM3/AL configuration. Epoxy resin is used for bonding aluminum, glass fiber, and PU foam together under compression. The steel roller is used to wet the glass fiber and remove excess resin while laminating [16, 17]. The components of the hybrid sandwich specimen are the woven roving mat (WRM) glass/epoxy composite, along with PU foam as the core composite laminate. Aluminum sheets of 150 × 150 × 0.4 mm were used to create the samples, along with epoxy resin and one weight percent of nano-silica.

The sandwich specimen made from aluminum alloy sheet as a skin material, bonded with GFRP laminate and PU foam as core material is shown in Fig. 1. Two aluminum sheets and three E-glass fiber mats were bonded with PU foam on either side. After cured in room temperature, the specimen edges are trimmed.

2.2 Drop Weight Impact Test

A drop weight tower was used to evaluate the specimens under low-velocity impact conditions. The specimens were struck orthogonally with an impactor that accelerates freely in gravity through a guide. The mild steel cylindrical impactor had a 16 mm

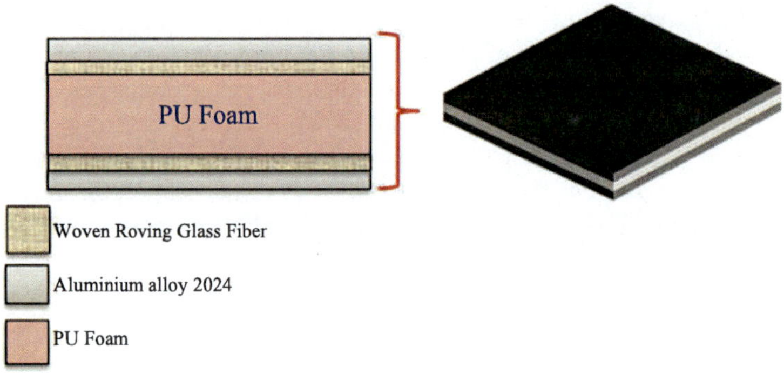

Fig. 1 AL/GFRP/PU Foam laminate

Fig. 2 Photograph showing the drop weight setup and the impactor

diameter hemispherical snout at the end. The impactor weighing 17.5 kg was dropped freely under various energy levels of 60 J, 80 J, and 90 J during the experiment. These energy levels are obtained by dropping the mass from a height of 0.35 m, 0.47 m, and 0.52 m respectively. A high-speed camera was employed to monitor the impactor velocity before and after the impact. The experimental setup for drop mass impact loading is shown in Fig. 2.

2.3 Energy Absorbed by Laminate

The energy absorbed by the FML during low-velocity impact is studied. An analytical model was used based on the idea of kinetic energy (KE) absorption [9]. During impact, the laminate absorbs incident energy from the impactor through a variety of energy-absorbing and damage mechanisms [18].

The Kinetic Energy lost by the impactor

$$E_L = E_i - E_f \tag{1}$$

where E_f is final K.E of impactor and E_i is the initial K.E of the impactor.

During impact, the energy gets absorbed and distributed in various mechanisms. The energy lost by impactor is absorbed by aluminum sheet, PU foam, and nano-composite laminate and is given by

$$E_L = E_{\text{Al}} + E_{\text{PU}} + E_{\text{GFRP}} \tag{2}$$

where, E_{Al} denotes the energy absorbed by aluminum sheet, E_{PU} energy absorbed by PU foam-core layer, and E_{GFRP} energy absorbed by composites layer.

Energy absorbed by aluminum layer is a combination of plastic strain deformation, E_{ps} [19] and petaling, E_{pt} [20],

$$E_{\text{Al}} = E_{\text{ps}} + E_{\text{pt}} \tag{3}$$

Energy absorbed by the PU foam layer is a combination of core crushing, E_c [21] and to shear plugging, E_s and is given by

$$E_{\text{PU}} = E_c + E_s \tag{4}$$

Energy absorbed by the GFRP laminate is given by

$$E_{\text{GFRP}} = E_d + E_t + E_{\text{del}} \tag{5}$$

where, E_{d} is energy absorbed by composite due to elastic deformation, E_t is energy due to tensile failure of fibers, and E_{del} is energy absorbed due to delamination [22].

3 Results and Discussions

After fabrication, the total laminate thickness is 20 ± 0.5 mm, with foam thickness 14.5 ± 0.4 mm. Figure 3 depicts the photograph of damaged specimens under various energy conditions of the impactor. For the 60 J impact energy level, the impactor perforates up to the PU foam layer, and the dent damage on the rear side aluminum layer is seen. Figure 3 shows the bulge formation on the rear side aluminum. The PU foam core arrests the impactor's further travel and acts as a protective layer. With an increased energy level of 80 J impact energy of the impactor, a single line crack initiation on the back of the layer is observed. It is an extension of the bulge, causing crack failure in the aluminum sheet. At 90 J impact energy, the extended crack in the

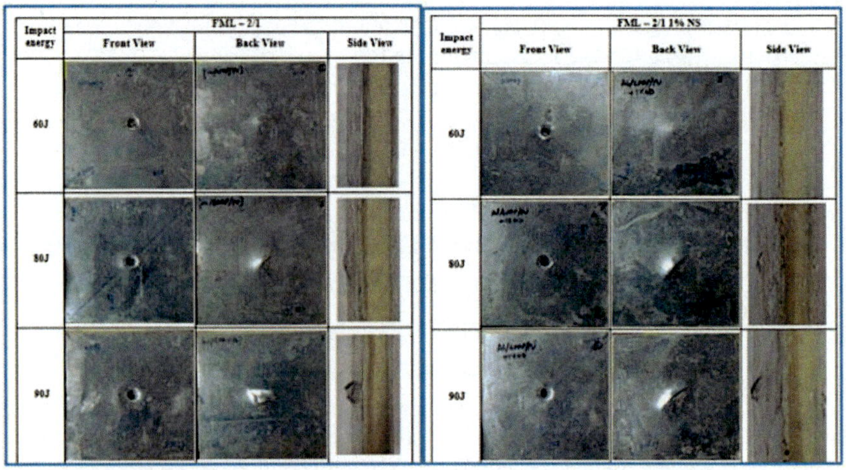

Fig. 3 Photograph showing the damaged specimens at various impact energies

rear side of FML is visible. Petaling is noticeable in the side view aspect. A complete perforation is observed when the energy levels increase beyond 90 J, which creates the petaling in the rear side aluminum sheet and causes a fiber breakage, thereby failing the material.

Only dent formation was on the impact side, and no cracks were observed on the rear side of the sample at 60 J. While increasing the impact energy to 80 J, dent depth increases, and crack initiation is observed. A complete perforation is occurring from 90 J impact energy and more. The FML without nano-silica materials failures is accumulated around the impacted area as failure breakage and petaling in rear side aluminum sheet. In nano-silica dispersed FML specimens, the impact energy is distributed among the matrix and fiber. The failure of the aluminum sheet on the rear side shows cracks formation only. The failed specimen discovered the maximum dent depth was 31.7 mm, and bulge formation was seen on the back.

The Impact Force vs time graphs of FML at 60, 80, and 90 J impact energies are depicted in Fig. 4. From the graph, it can be inferred that the effect of nano-silica is not effective up to 80 J the peak impact force and is maximum for pure FML/PU foam material, which suggests that for pristine FML, the nano-silica reinforcement is less significant below impact energy of 80 J under the low-velocity impact test. From 90 J impact energy, the level of the impact force increases for silica dispersed FML laminate. This proves that under high impactor velocities beyond 90 J, the nano-silica reinforcement comes into play and effectively helps to transfer the load into the matrix and fibers.

Figure 5a–d shows the high-speed camera image sequences during the low-velocity impact test. The drop mass impactor and the specimen are fixed using various clamps with a base fixture plate. The onset of experiments starts in Fig. 5a, and d shows the final impacted specimen. Crushing of PU foam and bouncing back of the

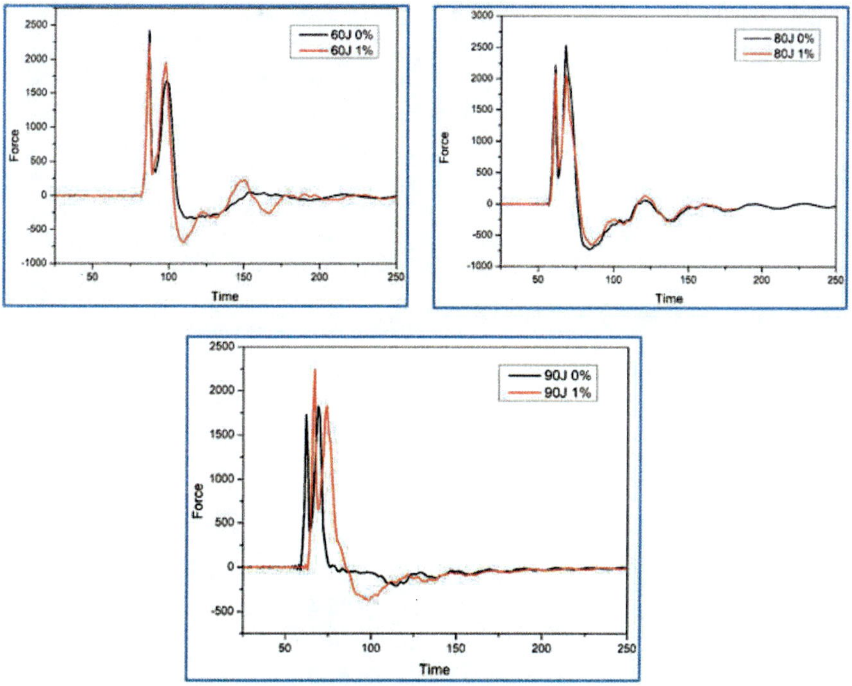

Fig. 4 Force vs time graphs of FML at 60, 80, and 90 J impact energies

impactor are observed during the final stage. The impactor's velocity is determined by closely computing the data obtained from the recordings. Using ImageJ software, the delamination area of the failed specimen is marked and computed (Fig. 6). On measuring the delamination area in the rear side of the specimen, with increasing impact energy levels, the damaged area also increases. In comparison, the FML with nano-silica shows a slightly higher delamination area under the same impactor energy. This shows that the nano-silica effectively transfers the impact energy and acts as a protective barrier to further damage to the specimen.

While comparing the dent depth, the specimen with nano-silica shows a lesser dent than pure FML/PU laminate. The damage to the sample increases with increasing energy levels. The maximum delamination area of 914mm^2 and 1086mm^2 was observed at 90 J impact energy for Pure FML, and 1wt% Nano-silica dispersed FML, respectively as shown in Table 1.

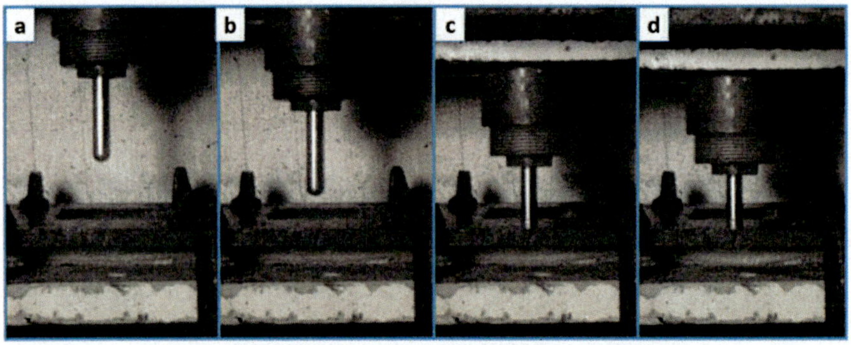

Fig. 5 High speed camera sequence image of drop weight impact

Fig. 6 Measuring the damage area using IMAGE J software

4 Conclusions

The low-velocity impact test is conducted on FML laminate with PU foam as core incorporated in a drop mass impact testing using mild steel hemispherical nose head projectile.

Based on the study, the following conclusions were drawn.

Table 1 Estimating the dent depth and delamination area under various impact energy levels

Specimen	Impact energy (J)	Peak force (N)	Dent depth (mm)	Delamination area (mm^2)
FML 2/1	60	2421	16.8	664
	80	2540	20.4	741
	90	1729	29.1	914
FML 2/1 1% NS	60	2243	17.4	719
	80	1977	19.9	891
	90	2241	27.5	1086

(1) The properties of the FML were enhanced by adding PU foam as core materials, withstanding up to 90 J impact energy without perforation.
(2) Failure of FML materials is Dent Formation, and Bulge failure was observed in rear side 80 J impact energy. For higher impact energy levels, crack initiation is observed on the Rear side of the aluminum sheet.
(3) Common failures like Fiber breakage, PU foam crushing, Delamination around the impact zone, and debonding between Aluminum and GFRP layers.
(4) The nano-silica inclusion in Epoxy enhances the sandwich composite material.

Acknowledgements The authors would like to thank the All-India Council for Technical Education (AICTE), New Delhi, India for the research grant under the RPS-NDF scheme (8-24/RIFD/RPS-NDF/Policy-1/2018-19).

Declaration of Conflicting Interest The authors declare that there is no conflict of interest.

Data Availability Statement The datasets generated during and/or analyzed during the current study are available from the corresponding author on reasonable request.

References

1. Vijay Kumar V, Rajendran S, Surendran S, Ramakrishna S (2022) Enhancing the properties of carbon fiber thermoplastic composite by nanofiber interleaving. In: 2022 IEEE International conference on nanoelectronics, nanophotonics, nanomaterials, nanobioscience and nanotechnology, pp 1–4
2. Vijay Kumar V, Ramakrishna S, Kong Yoong JL, Esmaeely Neisiany R, Surendran S, Balaganesan G (2019) Electrospun nanofiber interleaving in fiber reinforced composites—recent trends. Mater Des Process Commun 1(1):e24
3. Kumar VV, Balaganesan G, Lee JKY, Neisiany RE, Surendran S, Ramakrishna S (2019) A review of recent advances in nanoengineered polymer composites. Polymers (Basel) 11(4)
4. Vlot A, Vogelesang LB, De Vries TJ (1999) Towards application of fibre metal laminates in large aircraft. Aircr Eng Aerosp Technol 71(6):558–570
5. Nandhini N, Vijayan M, Rajesh Kumar MA, Kumaresan G, Siva Shanmugam N, Dhinakaran V (2021) Numerical investigation on aluminum alloys under medium velocity impact of projectiles. In: Advances in materials research select proceedings of ICAMR 2019. Springer Nature Singapore, Singapore, pp 1113–1121

6. Akshaj Kumar V, Surya P, Pandit MK (2017) Low velocity impact response of composite/sandwich structures. Key Eng Mater 725:127–131
7. Velmurugan R, Ganesh Babu M, Gupta NK (2006) Projectile impact on sandwich panels. Int J Crashworthiness 11(2):153–164
8. Balaganesan G, Akshaj Kumar V, Khan VC, Srinivasan SM (2017) Energy-absorbing capacity of polyurethane/SiC/glass-epoxy laminates under impact loading. J Eng Mater Technol Trans ASME 139(2):1–9
9. Akshaj Kumar V, Khan VC, Balaganesan G, Sivakumar MS (2017) Low velocity impact behavior of SiC/PU/GFRP laminates. Key Eng Mater 725:122–126
10. Shim VPW, Tu ZH, Lim CT (2000) Two-dimensional response of crushable polyurethane foam to low velocity impact. Int J Impact Eng 24(6):703–731
11. Wang J, Waas AM, Wang H (2013) Experimental and numerical study on the low-velocity impact behavior of foam-core sandwich panels. Compos Struct 96:298–311
12. Dariushi S, Sadighi M (2013) A study on flexural properties of sandwich structures with fiber/metal laminate face sheets. Appl Compos Mater 20(5):839–855
13. Haq M, Umer R, Khomenko A, Loos AC, Drzal LT (2013) Manufacturing and impact behavior of sandwich composites with embedded graphene platelets. ICCM International conference on composite materials, vol 2013, pp 2435–2440
14. Zhang G, Wang B, Ma L, Wu L, Pan S, Yang J (2014) Energy absorption and low velocity impact response of polyurethane foam filled pyramidal lattice core sandwich panels. Compos Struct 108(1):304–310
15. Zaretsky E, Asaf Z, Ran E, Aizik F (2012) Impact response of high density flexible polyurethane foam. Int J Impact Eng 39(1):1–7
16. Vijayan M, Selladurai V, Balaganesan G, Suganya Priyadharshini G (2022) Comprehensive characterization of AA 2024T3 fiber metal laminate with nanosilica-reinforced epoxy based polymeric composite panel for lightweight applications. Polym Compos 43(11):8274–8296. https://doi.org/10.1002/pc.v43.11, https://doi.org/10.1002/pc.26998
17. Vijayan M, Selladurai V, Kumar VV (2023) Investigating the influence of nano-silica on low-velocity impact behavior of aluminium-glass fiber sandwich laminate. Silicon 15(11):4845–4859. https://doi.org/10.1007/s12633-023-02391-w
18. Akil Hazizan M, Cantwell WJ (2002) The low velocity impact response of foam-based sandwich structures. Compos Part B Eng 33(3):193–204
19. Zhang D, Zhang X, Luo Y, Wang Q (2018) Experimental study on drop-weight impact response of basalt fiber aluminum laminates (BFMLs). Adv Mater Sci Eng 2018
20. Ahmadi H, Liaghat GH, Sabouri H, Bidkhouri E (2013) Investigation on the high velocity impact properties of glass-reinforced fiber metal laminates. J Compos Mater 47(13):1605–1615
21. Hosseini M, Khalili SMR (2013) Analytical prediction of indentation and low-velocity impact responses of fully backed composite sandwich plates. J Solid Mech 15(3):278–289
22. Balaganesan G, Velmurugan R, Srinivasan M, Gupta NK, Kanny K (2014) Energy absorption and ballistic limit of nanocomposite laminates subjected to impact loading. Int J Impact Eng 74:57–66

Elastoplastic Design of a Dome Structure for Water Impact Loads

Mohammed Afzal Rafiq, Shijo Xavier, A. P. Beena, and Sirajudeen Ahamed

Abstract Dome structures of re-entry modules are generally subjected to very high impact loads during landing. Design of such structures is governed by the displacement limits imposed by envelope constraints and high strains developed during impact. The constraints in payload mass too pose challenges in the design and calls for adopting an elasto-plastic design methodology, wherein significant yielding needs to be permitted under the ultimate loads. This paper discusses the design methodology of the dome structure subjected to water impact loads. The impact loads are estimated by simulating the worst-case conditions at touchdown of the structure, considering dynamic amplification imparted due to excitation of the fundamental mode of the structure. Finite Element (FE) simulations with material and geometric non-linearities are employed to estimate the plastic strains and displacements on the structure. Loading and unloading curves are used to study the residual displacements and to assess the integrity of the structure. FE based optimization is carried out with constraints on plastic strain as well as total displacements on the structure to arrive at a minimum mass configuration for the dome.

Keywords Elastoplastic design · Dome structure · Water impact · Re-entry module

1 Introduction

Dome structures of typical re-entry modules are subjected to heavy loads at the time of final splashdown. Design of such a structure for water impact load is constrained by limits on the displacements and strains on the hardware, as well as the requirement of achieving minimum mass. Configuration of dome for the dynamic water impact loads requires certain simplifications and definition of equivalent static loads. This paper presents the methodology for design of such a structure.

M. A. Rafiq (✉) · S. Xavier · A. P. Beena · S. Ahamed
Structural Engineering Entity, Vikram Sarabhai Space Centre, Thiruvananthapuram 695022, India
e-mail: mohammedafzalrafiq@gmail.com

R. Velmurugan et al. (eds.), *Dynamic Behavior of Soft and Hard Materials Volume 1*, Springer Proceedings in Materials 34, https://doi.org/10.1007/978-981-99-6030-9_39

2 Design Methodology

Water impact landing of the re-entry module is a highly dynamic event. Impact load on the module decelerates it from its defined touchdown velocity. The pressure distribution over the dome with respect to time is estimated from a coupled Eulerian–Lagrangian finite element analysis assuming the structure to be rigid. This is estimated for various touchdown conditions with different impact angles, touchdown velocity, etc. Critical pressure distributions at various time instants are identified to define the design loads for the structure. Normalized pressure distribution over the dome at the instant of maximum deceleration for a 12° side impact is shown in Fig. 1. Over this pressure distribution, a dynamic amplification factor is considered to account for the flexibility of the structure. Once the design loads are defined, these equivalent static pressure distributions can be used to configure the structure.

A critical constraint on design is posed by the physically available gap between the outer mould line and the inner structure. This available gap should accommodate the thickness of the dome as well as its displacement during impact. To achieve this stringent requirement with minimum mass, the dome is configured as a gridded structure with flanged ribs as shown in Fig. 2. Cross section of a typical flanged rib is shown in Fig. 3. The higher moment of inertia provided by flanging of the ribs ensure that the buckling limits for the ribs under bending loads are higher than its yield point. Optimum flanged isogrid configuration of the central dome for the defined external pressure is worked out based on the basic theory available in Isogrid Design Handbook [1].

Since the buckling stress is higher than yield, the design can allow yielding of material for ultimate loads. Therefore, an elasto-plastic analysis with material and geometric non-linearity is called for to estimate the actual strains and displacements developed on the structure under water impact load.

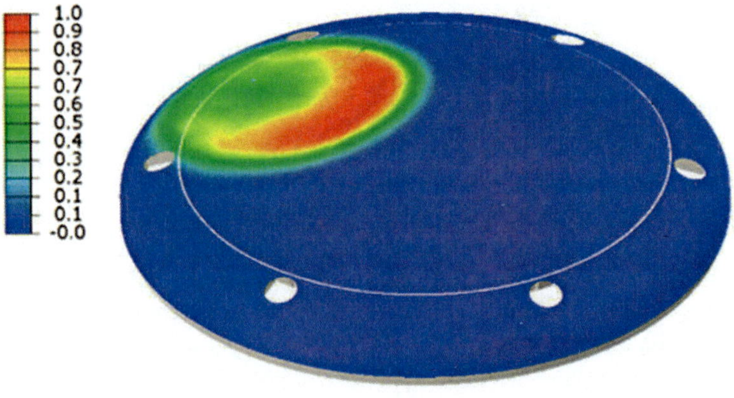

Fig. 1 Normalized pressure distribution on the dome

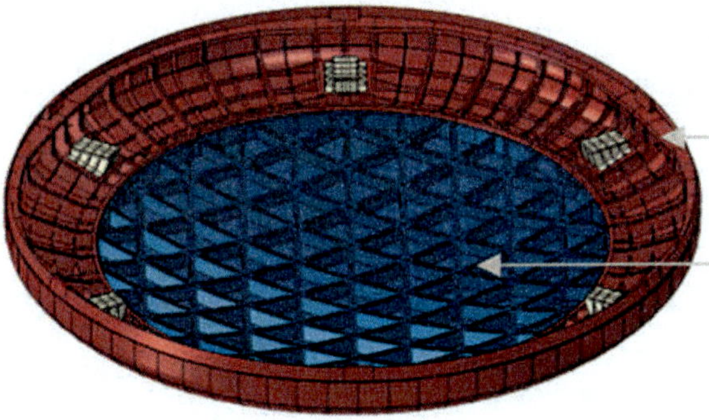

Fig. 2 Construction of the dome structure

Fig. 3 Cross section of a typical flanged rib

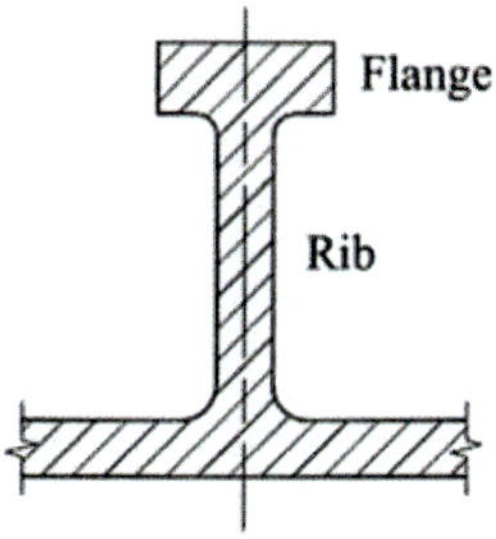

3 FE Simulations and Results

The dome structure is modelled using a combination of quad/tri shell and hex/wedge 3D elements. The structure is attached to six longerons through which it interfaces with the re-entry module inner structure. Interfaces between the longeron and inner structure are provided with a fixed boundary condition. This boundary condition is a reasonable assumption for obtaining a simplified model for preliminary analysis and design iterations. Pressure distribution corresponding to 12° water impact is applied on the outer surface of the dome. FE model of the structure is shown in Fig. 4.

True stress–strain curve corresponding to the dome material is used for the finite element simulation. This curve is generated using the Ramberg–Osgood model [2], with the tested values of the material's Young's modulus, yield strength, ultimate strength, and ultimate strain. True stress–strain curve for the material is shown in Fig. 5. Ultimate strain for the material is only 1%. This corresponds to the properties achievable in realizing the heavy forgings required for the dome.

FE analysis is carried out with geometric and material non-linearity. Displacement plot of the dome is shown in Fig. 6. Equivalent plastic strains developed on a rib in impact bay are shown in Fig. 7. Equivalent plastic strains developed on the central dome are shown in Fig. 8. It can be seen that significant yielding has taken place in the structure.

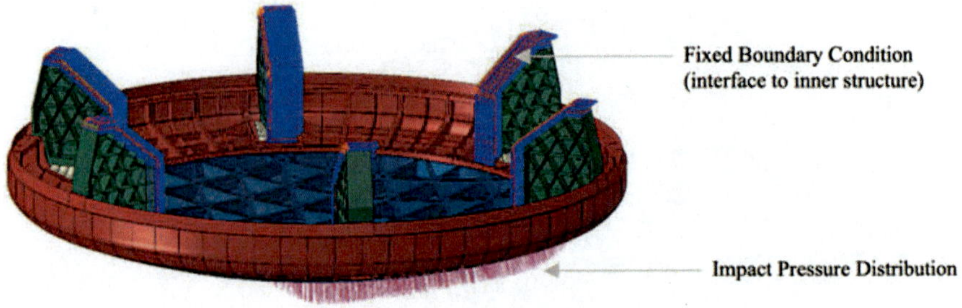

Fig. 4 Finite element model of dome structure

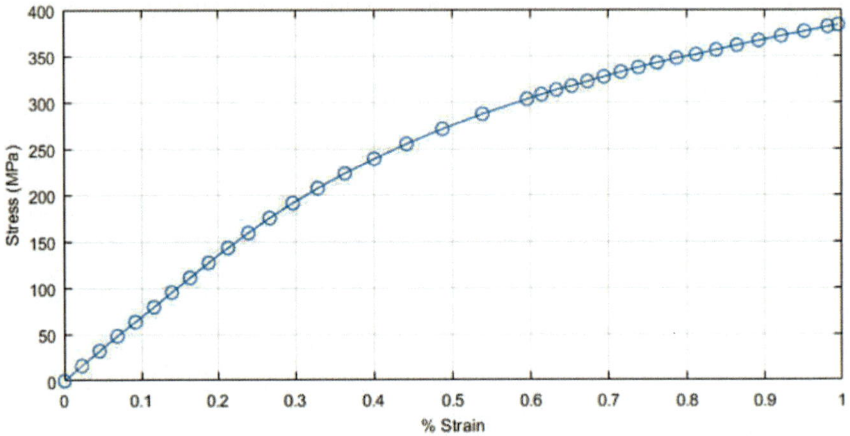

Fig. 5 True stress–strain curve for the dome material

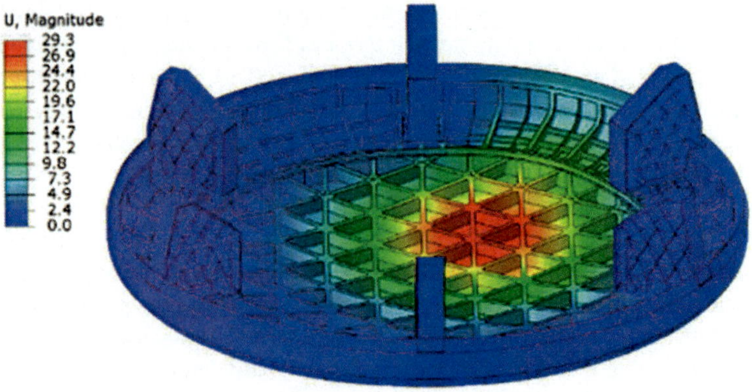

Fig. 6 Displacement plot for 12° water impact

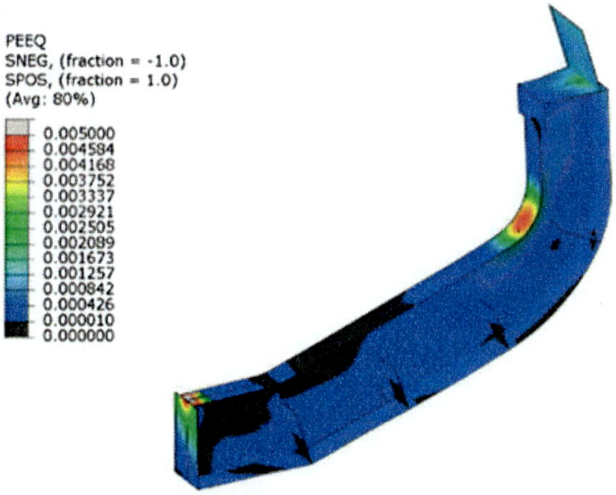

Fig. 7 Equivalent plastic strains on an impact bay rib

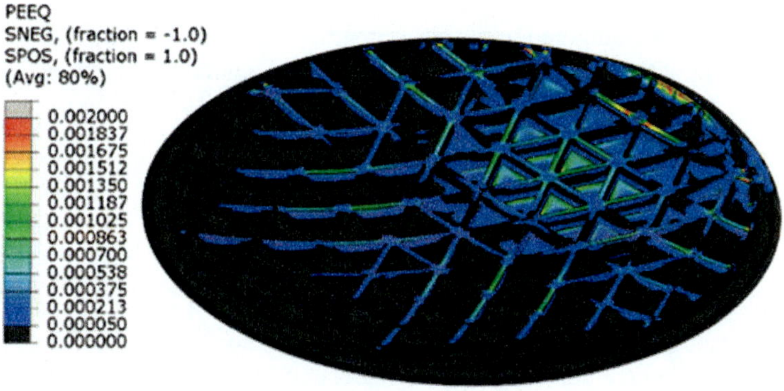

Fig. 8 Equivalent plastic strains on the central dome

For comparison, the analysis is repeated with linear material and non-linear geometry. Load–displacement curves at the maximum displacement point of the dome structure for non-linear and linear material are shown in Fig. 9. It can be seen that the structure undergoes significant yielding with an increase of 17% in the maximum displacement as compared to displacement with linear material. 13% of the maximum displacement remains in the structure as residual after unloading. Load–displacement behaviour until buckling is estimated by a collapse load analysis. As can be seen from the plot, a margin of 0.2 is available against collapse.

It can be seen that, an elasto-plastic analysis is crucial for design of such a dome with stringent constraints on displacements as well as mass. Parametric size-optimization is carried out on the above simplified boundary model to arrive at a minimum mass solution to achieve the specified displacement and strain constraints.

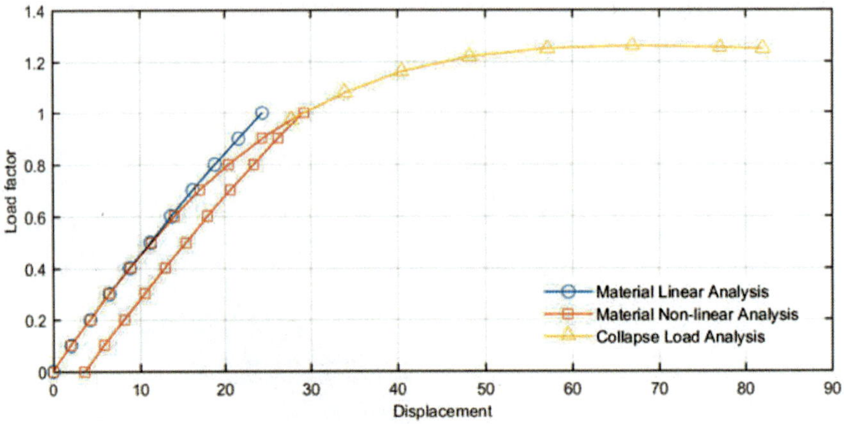

Fig. 9 Load–displacement curve for maximum displacement point on dome

Once the structure is configured in the above method, integrated inertia relief analysis is carried out with the full module to simulate actual boundary condition for the structure.

4 Conclusion

A methodology is established for design of dome structures for re-entry modules subjected to water impact loads. Flanged rib configuration adopted for design is effective in achieving the objective of allowing material yielding without structural instability. The local model of the structure with simplified boundary condition is capable of predicting the strain levels on the structure with reasonable accuracy compared to the integrated model. Runtime for the model with simplified boundary is reasonable enough to make FE based optimization procedures amenable.

Acknowledgements The authors acknowledge Shri. Nirmal Ramachandran and Shri. AG Balamurali from dynamics team at VSSC for carrying out water impact dynamic simulations and providing the pressure distributions as an input for design. The contributions of Shri. Yash Mittal and Shri. Sabin Kurian in the integrated inertia relief simulations of the dome with the re-entry module inner structure is also acknowledged.

References

1. Meyer RR, Harwood OP, Harmon MB, Orlando JI (1973) Isogrid design handbook, NASA CR-124075. McDonnell Douglas Astronautics Corporation
2. Ramberg W, Osgood WR (1943) Description of stress-strain curves by three parameters, Technical note No. 902. National Advisory Committee for Aeronautics, Washington, D.C., USA

An Investigation of Damage and Residual Strength of Aircraft Composite Structure Caused by Foreign Object Debris Impact

M. M. Ansari, K. Sharad, J. Dhayanidhi, and S. Manju

Abstract Fiber-reinforced polymer composite components are required to be resilient against foreign object damage caused by debris impact. The FOD impact induced damage is a probable threat in the composite components of aircraft. The induced damage influences the compressive strength of laminate and thus load transfer capability. The adequacy of structural components for design and repair can be assessed based on investigation of probable FOD impact induced damage. This paper presents an investigation of FOD impact induced damage in the composite component of aircraft. Cohesive zone model is used to predict delamination. Impact induced damage variation at different impact energies is discussed.

Keywords Low velocity impact · FOD impact · CFRP · Cohesive zone model

1 Introduction

Aircraft composite structural components experiences different loads over the span of its service. The laminated composite is more vulnerable under impact load. The occurrence of foreign object damage (FOD) due to impact in the composite structure is the major concern in the aviation industry [1, 2]. Especially, high velocity impact due to foreign objects creates damages in the composite components in the form of matrix failure, delamination and fiber fractures [3–5]. The propagation of these failure modes under the service life load may lead to major failure. Subsequent to these failure modes occurrence, the load-carrying capacity as well as stiffness of composite component worsens further that may cause safety issues. For these

M. M. Ansari (✉)
Structural Engineering, CSIR-Central Building Research Institute Roorkee, Roorkee 247667, India
e-mail: mmansari@cbri.res.in

K. Sharad · J. Dhayanidhi · S. Manju
Structural Technological Division, CSIR-National Aerospace Laboratories Bangalore, Bangalore 560017, India

R. Velmurugan et al. (eds.), *Dynamic Behavior of Soft and Hard Materials Volume 1*, Springer Proceedings in Materials 34, https://doi.org/10.1007/978-981-99-6030-9_40

reasons, FOD induced damage in the aircraft composite components needs to be studied at the design stage.

A significant amount of work has been reported over the years addressing the investigation of impact behavior of GFRP [6, 7] and CFRP [8, 9]. Cui et al. [10] has investigated impact induced damage in IM7 carbon composite at varied impact velocity ranging from 21 to 157 m/s. The delamination was observed to be prominent failure mode for low to medium velocity range. Dorey and Sidey [11] experimented the ballistic response of 3 mm thick carbon/epoxy composite laminates using 6 mm diameter steel projectile. The projectile was fired from an air gun with impact speed up to 300 m/s. The numerical model is now being used to capture failure modes and their propagation over short period of time effectively under the impact phenomenon. Also, 3D model with solid elements are needed to study the stress distribution through the thickness of laminate under failure occurrence [12].

The current study focuses on the analysis of impact induced damage assessment in the carbon fiber composite (CFC) which were fabricated from unidirectional carbon prepreg (HTS40 12K300) fiber/epoxy resin [13] with stacking sequence of [0/90]2 s under the range of impact energies. A symmetric CFC plate of size 150 mm × 150 mm was cut from CFC laminate, casted and cured in autoclave [13] which is impacted with 50 gm spherical projectile with varied impact energy. Failure propagation and modes of damages are discussed.

2 Constitutive Material Model

The failure in composite is categorized in four modes such as tension/compression of fiber and matrix. Quadratic stress-based Hashin failure criteria [14, 15], popular in the industry, is implemented to capture these modes. The initiation of failure is as per Eqs. (1)–(4) below.

Fiber tension failure ($\hat{\sigma}_{11} > 0$);

$$F_f^t = \left(\frac{\hat{\sigma}_{11}}{X_T}\right)^2 - \left(\frac{\hat{\sigma}_{12}}{S_{12}}\right)^2 + \left(\frac{\hat{\sigma}_{13}}{S_{13}}\right)^2 \geq 1 \tag{1}$$

Fiber compression failure ($\hat{\sigma}_{11} < 0$);

$$F_f^t = \left(\frac{\hat{\sigma}_{11}}{X_C}\right)^2 \geq 1 \tag{2}$$

Matrix tension failure ($\hat{\sigma}_{22} + \hat{\sigma}_{33} \geq 0$);

$$F_m^t = \left(\frac{\hat{\sigma}_{22} + \hat{\sigma}_{33}}{Y_T}\right)^2 + \frac{1}{S_{23}^2}\left(\hat{\sigma}_{23}^2 - \hat{\sigma}_{22}\hat{\sigma}_{33}\right) + \left(\frac{\hat{\sigma}_{12}}{S_{12}}\right)^2 + \left(\frac{\hat{\sigma}_{13}}{S_{13}}\right)^2 \geq 1 \tag{3}$$

Table 1 Material properties of CFC and cohesive zone

Density (kg/m^3)	1600
Stiffness properties (GPa)	$E_{11} = 150, E_{22} = E_{33} = 10.3$ $G_{12} = G_{13} = 6, G_{23} = 3.7$
Strength properties (MPa)	$X_T = 2537, X_C = 1580, Y_T = 82, Y_C = 236, S_{12} =$ $90, S_{23} = 40$
Poisson's ratio	$\upsilon 12 = \upsilon 13 = 0.3, \upsilon 23 = 0.4$
Cohesive properties	
Elastic modulus (GPa/mm)	Knn = 1373.3, Kss = Ktt = 493.3
Inter-laminar strength (MPa)	$t_n^0 = 62.3, t_s^0 = t_t^0 = 92.3,$
Inter-laminar fracture toughness (J/m^2)	$G_{IC} = 280, G_{IIC} = G_{IIIC} = 790$ $\beta = 1.45$

Matrix compression failure ($\hat{\sigma}_{22} + \hat{\sigma}_{33} < 0$);

$$F_m^C = \left(\frac{\hat{\sigma}_{22} + \hat{\sigma}_{33}}{2S_{23}}\right)^2 + \frac{\hat{\sigma}_{22} + \hat{\sigma}_{33}}{Y_C}\left[\left(\frac{Y_C}{2S_{23}}\right)^2 - 1\right] + \frac{1}{S_{23}^2}\left(\hat{\sigma}_{23}^2 - \hat{\sigma}_{22}\hat{\sigma}_{33}\right) + \left(\frac{\hat{\sigma}_{12}}{S_{12}}\right)^2 + \left(\frac{\hat{\sigma}_{13}}{S_{13}}\right)^2 \geq 1 \quad (4)$$

where $\hat{\sigma}_{ij}$ (i, j = 1,2,3) are the effective stress tensor, X_k and Y_k (k = T,C) longitudinal and transverse strength of lamina. After failure initiation, the damage evolution progresses exponentially. The damage failure initiation and evolution are initiated using VUMAT subroutine. Delamination of composite is implemented through surface-to-surface cohesive zone modeling (CZM). Material properties of CFC and cohesive zones are given in Table 1 [13]. CZM is used to quantify the delamination in the laminate which is one of the major failure modes under impact loading [7]. CZM is based on traction separation law that comprises the evolution of delamination based on fracture energies in all the three direction such as normal and two shear and thus effective in capturing the delamination [15]. CZM descriptions is available in details in [17]. Also, surface to surface CZM is computationally economical as compared to element based CZM.

3 Numerical Model

A 3D FE model is made using ABAQUS/CAE. A projectile with 15 mm diameter hemispherical tip is modeled as rigid body. The projectile is constrained to translate in z direction only. A circular composite plate of diameter 75 mm consists of 8 plies $(0/90)_{2s}$ which makes total target thickness of 2 mm. Fine mesh of size 1 mm × 1 mm is applied near the impact region in the target plate to capture damage

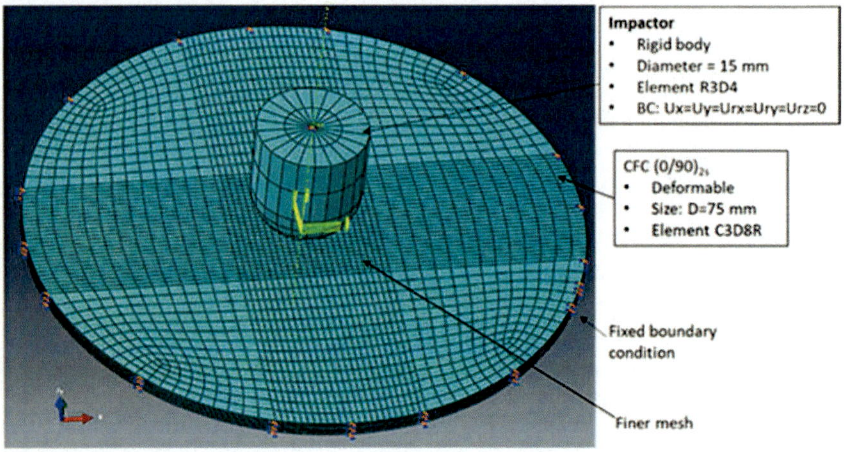

Fig. 1 Finite element model of CFC and projectile

and deformation accurately, whereas coarser elements of size 2 mm x 2 mm are used through biasing from impact region to outer edge. Composite plate is having encastre boundary condition at edge as shown in Fig. 1.

4 Results and Discussions

This paper aims to present a numerical model to demonstrate impact induced damage in the composite structure of aircraft. The details of materials and target are taken from literature [13] to validate the numerical model. The structural response of target plate in terms of contact force and deformation is given below. The time histories of contact force in the target plate at different energy levels of impact have been presented in Fig. 2. The predicted contact forces are having good correlation with the test results in terms of time variation and occurring of peak. The maximum contact force of 3400N is predicted as compared to 3460 N test result for 7.34 J impact energy. For 11 J impact energy, 4070N peak force is predicted with test value of 3800 N. The predicted contact force shows error of 0.0001% and 0.007% for 7.34 J and 11 J, respectively. The increase in contact force is observed to be high up to 1 ms and then a sudden drop as shown in Fig. 2a. This may be due to first occurrence of delamination in the target. In general, more oscillation in force–time history is observed near peak, which shows the progression of delamination.

The force response of target plate with deflection has also been predicted and validated with test data as shown in Fig. 3. Force–deflection history attributes the stiffness of the target. The prediction of stiffness degradation in the target plate is also having similar trend as in test. The composite plate shows maximum deflection about 4 mm against 3.9 mm test results, see Fig. 3a. Considering the permanent deformation

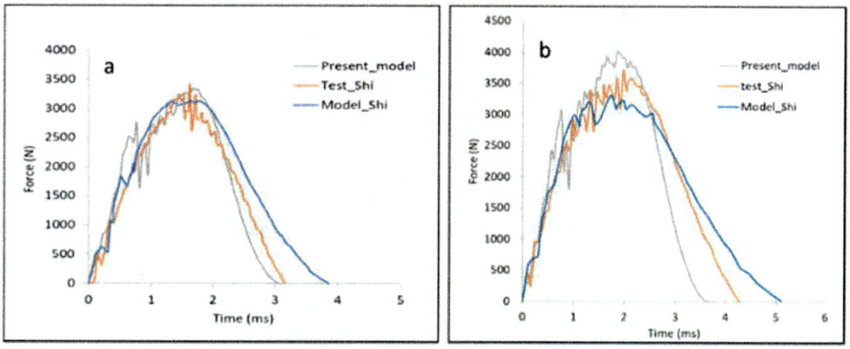

Fig. 2 Contact force–time histories as predicted from FE model and experimentation **a** 7.34 J, **b** 11 J

of target plate, its value is predicted to be 2.2 mm as compared to 2.5 mm in tested specimen, in case of 7.34 J impact energy.

In the transverse loading of laminated plate, delamination is predominant which generally starts from outer plies [16]. Damage occurrence in the target plate due to impact is also predicted and validated with the test results [16]. Composite plate of size 45 mm × 67.5 mm is impacted by 2.28 kg projectile of diameter 12.5 mm. Composite plate consists of 12 plies of $(0_3/90_3)$s. Surface-based cohesive zone model [17] is applied in between 0° and 90° plies to predict the delamination. Material details for target, projectile, and cohesive surface are taken from [16]. Delamination in bottom 0/90 interface is critical as compared to top (0/90) interface which is plotted in Fig. 4. The variation of peanut-shaped delamination with impact energy is predicted which shows close correlation with test results. The predicted value of height and width of peanut-shaped delamination is also being compared with test delamination in Table 2. In general, the predicted value of delamination is slightly conservative with respect to test.

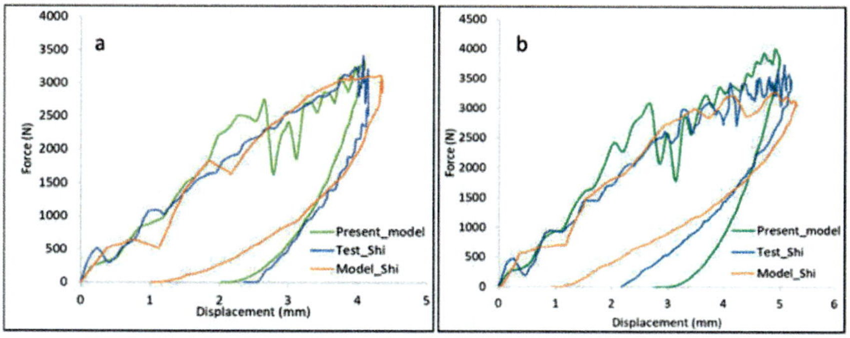

Fig. 3 Numerical prediction of contact force–displacement histories as predicted from FE model and experimentation **a** 7.34 J, **b** 11 J

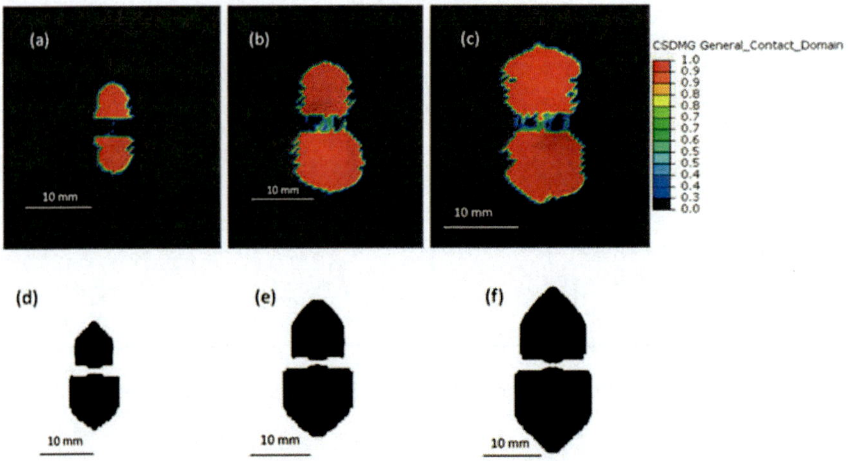

Fig. 4 Total delamination in composite plate with present model (**a, b, c**) and compared with test delamination [16] (**d, e, f**) at impact energy **a, d** 1 J,(**b, e** 3 J and **c, f** 7 J

Table 2 Delamination in target at different impact energy

Impact Energy	1 J		3 J		7.35 J	
Delamination height/width	H	W	H	W	H	W
Present model	6.2	6	14	13.2	20.4	18.2
Test [16]	5.96	5.20	14.44	12.33	20.58	15.73

Contact force peak is also compared with test data for different impact energies of projectile as shown in Fig. 5. The predicted values show a close relation with test results.

5 Conclusions

The present work is aimed to present a simple model to predict damages due to FOD to the composite structural component of aircraft. Since, the composite structure of aircraft is more susceptible to FOD threat under service life, the adequacy of structural components at design ultimate load needs to be assessed based on investigation of probable FOD impact induced damage. Cohesive zone model is used in the present 3D FE model to predict delamination. Structural responses of composite structure in terms of contact force, force–deflection, as well as delamination are predicted and validated with test results at different impact energies. The predicted results show good agreement with test results [13]. At low energy impact, peanut-shaped delamination is predominant failure mode, and the same is predicted reasonably

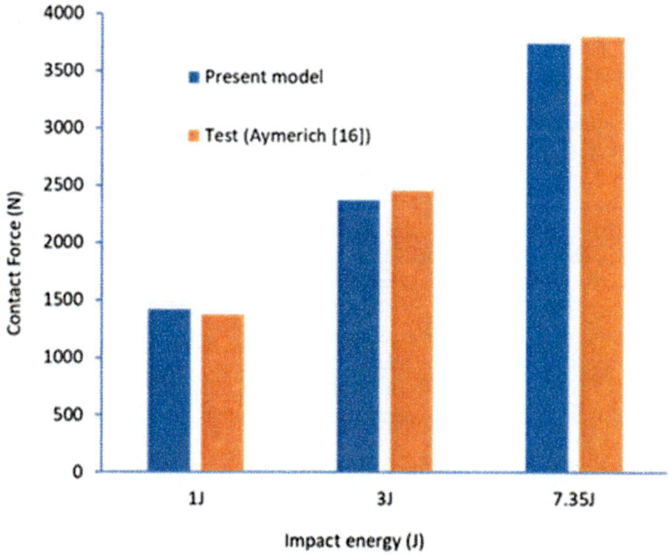

Fig. 5 Validation of contact force at different energies

well in line with test data [16]. The predicted delamination can be used further in the design and repair assessment of composite component.

Acknowledgements The authors gratefully acknowledge Dr. S. Raja, Head of the division, STTD, CSIR-NAL Bangalore, for kind support.

References

1. Duo P, Liu J, Dini D, Golshan M, Korsunsky AM et al (2007) Evaluation and analysis of residual stresses due to foreign object damage. Mech Mater 39:199–211
2. Presbya MJ, Mansoura R, Manigandana K et al (2019) Characterization and simulation of foreign object damage in curved and flat SiC/SiC ceramic matrix composites. Ceram Int 45(2A):2635–2643
3. Ashwin RK, Xiaodong X, David N, Michael RW, Stephen RH (2019) Experimental investigation of high velocity oblique impact and residual tensile strength of carbon/epoxy laminates. Compos Sci Tech 182:107772
4. Abrate S (1991) Impact on laminated composites. Appl Mech Rev 44(4):155–190
5. Cantwell WJ, Morton J (1989) Comparison of the low and high velocity impact response of CFRP. Composites 20:545–551
6. López-Puente J, Zaera R, Navarro C (2008) Experimental and numerical analysis of normal and oblique ballistic impacts on thin carbon/epoxy woven laminates. Composites A 39:374–387
7. Ansari MM, Chakrabarti A (2016) Impact behavior of FRP composite plate under low to hyper velocity impact. Compos Part B: Eng 95:462–474
8. Kim H, Welch DA, Kedward KT (2003) Experimental investigation of high velocity ice impacts on woven carbon/epoxy composites panels. Compos A 34:25–41

9. Shetty H, Sethuram D, Rammohan B, Budarapu PR (2020) Low-velocity impact studies on GFRP and hybrid composite structures. Int J Adv Eng Sci Appl Math 12:125–141
10. Cui H, Thomson D, Secondary S, Petrinic N (2019) A critical study on impact damage simulation of IM7/8552 composite laminate plate. Int J Impact Eng 127:100–109
11. Dorey G, Sidey GR (1974) Residual strength of CFRP laminates after ballistic impact. In: Proceedings of mechanical properties high rates strain conference. Oxford, pp 344–351
12. Seo H, Hundley J, Hahn HT, Yang JM (2010) Numerical simulation of glass-fiber-reinforced aluminum laminates with diverse impact damage. AIAA J 48:676–687
13. Shi Y, Swait T, Soutis C (2012) Modelling damage evolution in composite laminates subjected to low velocity impact. Compos Struct 94:2902–2913
14. Hashin Z (1980) Failure criteria for unidirectional fiber composites. J Appl Mech 47(2):329–334
15. Zhou J, Wen P, Wang S (2020) Numerical investigation on the repeated low-velocity impact behavior of composite laminates. Compos B 185:107771
16. Francesconi L, Aymerich F (2017) Numerical simulation of the effect of stitching on the delamination resistance of laminated composites subjected to low-velocity impact. Compos Struct 159:110–120
17. ABAQUS User manual 2012

Understanding the Impact of Functionalization of Foam on Electrical and Mechanical Properties of Epoxy Composites

S. Semion Kingslee, Chillu Naresh, Guvvala Nagaraju, Ramachandran Velmurugan, Ramanujam Sarathi, and Rengaswamy Jayaganthan

Abstract Syntactic foam mixed with epoxy thermoset polymer offers low density, exceptional mechanical properties, high thermal conductivity, and light weight due to which these composites are often used in thermal insulation, submarine, and aerospace applications. In this work, surface of the filler is modified by the silane-coupling agent in order to make it compatible with the organic host and prepare a composite with homogeneously distributed filler material. Impact test shows better energy absorption with silanized samples, and the effect of functionalization on the filler has been studied with respect to its electrical and mechanical properties. Dielectric constant of silanized composites has shown reduced value in comparison with the unsilanized specimens.

Keywords Hollow glass microspheres · Dielectric properties · Silanization · Thermosets

S. Semion Kingslee (✉) · R. Jayaganthan
Department of Engineering Design, Indian Institute of Technology Madras, Chennai 600036, India
e-mail: ed20s200@smail.iitm.ac.in

C. Naresh · G. Nagaraju · R. Sarathi
Department of Electrical Engineering, Indian Institute of Technology Madras, Chennai 600036, India

R. Velmurugan
Department of Aerospace Engineering, Indian Institute of Technology Madras, Chennai 600036, India

R. Velmurugan et al. (eds.), *Dynamic Behavior of Soft and Hard Materials Volume 1*, Springer Proceedings in Materials 34, https://doi.org/10.1007/978-981-99-6030-9_41

1 Introduction

Hollow glass microspheres embedded in polymer matrices are gaining popularity due to their superior compressive properties, low moisture absorption, and low density [1]. Syntactic foams are material systems that consist of a matrix mixed with hollow particles [2]. Hollow glass microspheres are made of outer stiff glass and inner inert gas, which results in some unique properties such as light weight and low dielectric constant [3]. The high-dimensional stability and load-bearing capacity of syntactic foams, combined with their thermal stability, make them enticing in aerospace, automotive, and structural applications. Further, because of its widespread use in the aerospace and transportation industries, epoxy thermoset polymer is the most used base matrix material in syntactic foam research. Epoxy resin and its composites are the most commonly used insulating materials in electrical power applications, such as high voltage bushings, as a spacer in gas insulated systems (GIS), dry type transformers, and printed circuit boards for electronics applications. The glass spheres with high strength due to their close porosity in foam will aid in reducing the overall weight of the material. Interest in utilizing the low density of syntactic foams in other applications such as aero-space structures and sports equipment has necessitated characterizing these materials for tensile loading and studying various parameters affecting their properties [4].

Daniel Paul et al. studied the mechanical properties of syntactic foams and discovered that tensile values decreased as the HGM volume fraction increased [5]. This is because the HGM and matrix are not properly bonded. It was also noted that the HGM-epoxy interfacial characteristics can improve tensile properties. Mohammed Imran et al. studied the impact of hollow glass microspheres on the compressive strength and storage modulus of CNT modified epoxy resin [6]. The compressive strength of CNT/Epoxy composites decreases as HGM volume increases. According to research, HGM plays an essential function in composite reinforcement by enhancing the damping qualities of epoxy. Yung and colleagues investigated the effect of increasing filler loading on dielectric characteristics [3]. The dielectric constant is shown to decrease as the HGM content increases. Mechanical parameters such as the coefficient of linear expansion also improved. Thermal conductivity, on the other hand, exhibited decreased behavior due to the presence of an interior inert gas in HGM. Vasanth Chakravarthy et al. [7] investigated the impact properties of vinyl ester matrix syntactic foams mixed with hollow microspheres. In comparison with vinyl ester resin, the impact strength of syntactic foams is observed to be lower. Tensile, impact, hardness, and dielectric properties of HGM-epoxy syntactic foams are investigated in this study. The effect of functionalization on all of these properties has been studied. The sample preparation method and its influence on microstructure have also been discussed.

2 Experimental Studies

2.1 Materials

The resin system, hollow glass particles, and hardener are the constituents used in the composites fabrication process. The base matrix material is epoxy resin (Araldite CY-205). It is a DGEBA (diglycidyl ether of bisphenol A)-based resin. HY951, and a triethylenetetramine (TETA)-based compound was used as a curing agent. Silane coupling agent in the form of (3-Glycidyloxypropyl) trimethoxy silane is used for surface functionalization which is supplied by Sigma-Aldrich. The hollow glass particles, also known as glass micro balloons (GMB), were obtained from 3 M Corporation and were of the K15 type.

The silane coupling agent 3-Glycidyloxy propyl trimethoxy silane (GPS) supplied by Sigma-Aldrich was used for surface functionalization. This coupling agent can organically bond inorganic materials (such as glass and fillers) and organic polymers together, acting as a promoter of adhesion, crosslinking agent, and surface modifier. It is a colorless, low viscosity liquid with a faint odor of ketones, and aromatic hydrocarbons. Hydroxyl groups on the particle surface are required for the functionalization process. The epoxy group has the ability to react with various resins, and the silicon hydroxyl group formed after hydrolysis of the trimethoxyl group can condense with the hydroxyl group on the surface of the inorganic material, forming a molecular bridge between the inorganic material and the resin. A coupling agent must have functional groups at the head to interact with the filler and at the tail to interact with the polymer. Surface functionalization serves two purposes: it adds relevant chemical groups to enable effective filler dispersion, and it improves the compatibility of the filler and the host material. The surface of the HGM powder was pre-treated with coupling agent at a concentration of 3 wt% of HGM powder (Fig. 1).

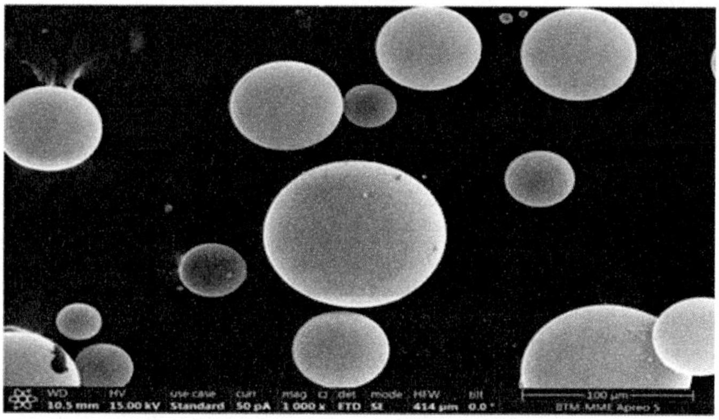

Fig. 1 SEM picture of HGM used

2.2 Sample Preparation

The schematic representation of sample preparation is shown in Fig. 2. Initially, the required amount of HGM powder is heated at 110 °C for 3 h to remove any traces of moisture. Filler loadings of 1%, 3%, and 5% are weighed and used for preparing the samples. Pure epoxy resin sample is used as reference specimen. A 3 wt% silane coupling agent is used for specimen preparation (Table 1).

The heated filler is then mixed with ethanol, and the combined mixture is stirred for 30 min in an ultrasonic bath. Following that, the hollow particle dispersed ethanol was shear mixed with epoxy resin for 8 h. Following that, a high-frequency sonication process (500 W, 20 kHz) is performed for 1 h in an ice-filled bath. Later, the mixture was placed in an oven to remove any excess ethanol. The necessary amount of hardener is added and thoroughly mixed until uniform dispersion is achieved. Upon degassing the mixture, it is then poured into a mold with the required dimensions for the various tests. The mixture is allowed to cure for 24 h at room temperature. The nomenclature used for sample identification for various tests is represented in Table 2.

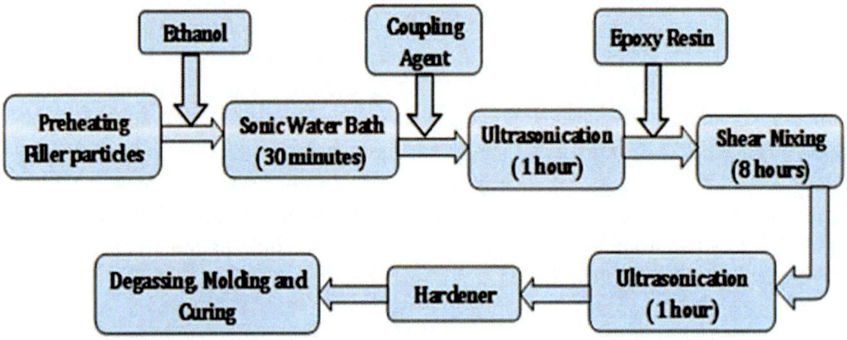

Fig. 2 Schematic diagram of sample preparation

Table 1 Properties of HGM used

Property	Value
Density	0.15 gm/cc
Thermal conductivity	0.055 W/mk
Particle size	115 microns
Color	White
Dielectric constant	1.45 at 100 MHz
pH	9.5
Packing factor	0.6

Table.2. Nomenclature of samples

Sample designation	Sample composition
E	Pure epoxy
1F	1 wt% HGM in epoxy
1F-Si	Silanized 1 wt% HGM in epoxy
3F	3 wt% HGM in epoxy
3F-Si	Silanized 3 wt% HGM in epoxy
5F	5 wt% HGM in epoxy
5F-Si	Silanized 5 wt% HGM in epoxy

2.3 Microstructure and FTIR Analysis

The microstructure of polymer composites is determined entirely by the dispersion and distribution of nanoparticles in a polymer matrix. The investigation of filler dispersion in polymer composites is critical, and it is evaluated using field emission scanning electron microscopy (FESEM (Model: Hitachi (S-4800)). FTIR analysis has been carried out to comprehend the presence of extra functional groups that were brought by the addition of silane coupling agent. A Cary 630 FTIR type instrument, made by Santa Clara in the USA, was used for FTIR analysis.

2.4 Mechanical Characterization

The impact test was carried out using a Frank Bacon Baldwin bi-model Izod impact tester. The laminate was cut into test samples with dimensions of 63.5 × 12.8 mm. The specimen was clamped in the vertical plane as a cantilever beam and impacted by a swinging pendulum at the unsupported end in the Izod test [8]. To determine the surface hardness of samples, Vickers hardness test (The FALCON 603 hardness tester) was carried out at room temperature by applying a load of 0.98 N on the samples at ten different places, and then averaged for hardness value. The Vickers hardness test is carried out using a diamond indenter and a hardness scale of HV0.1 with a dwell period of 10 s.

The tensile test measures the force required to break a sample and the amount by which the sample stretches or elongates to that breaking point. This measurement utilizes a machine with a gripping force of 20 KN. To investigate the effect of filler on resistance to fracture under applied stress, samples are conventionally tested with a universal testing machine (KALPAK UTM) of 20 kN loading capacity and 2 mm/min displacement rate at room temperature. In accordance with the ASTM D-638 (Type IV) standard procedure, dumbbell-shaped tensile specimens measuring 115 × 6 × 3 mm and 33 mm in gauge length were prepared for this study. For accuracy and dependability, tests were conducted on four samples with the same concentration and averaged.

2.5 *Electrical Characterization*

Dielectric relaxation spectroscopy (DRS) is used to study the changes in dielectric properties of prepared composites. The broadband dielectric/impedance spectroscopy analyzer (Novocontrol technologies) was used to examine the properties. Variations in the dielectric behavior of unsilanized and silanized samples were investigated at room temperature in the frequency range of 0.1–1 kHz. The diameter and thickness of the sample were 20 mm and 1 mm, respectively. Because the permittivity of hollow glass microspheres is greater than that of the polymer matrix, the nanocomposite with different base constituents is expected to improve effective dielectric constant.

3 Results and Discussion

3.1 *Microstructure Analysis and FTIR*

Figure 3a–f show SEM micrographs of composites with various filler concentrations. The microstructure of the epoxy and epoxy/HGM composite samples was examined using SEM. Sample images were obtained from the fractured edges. A random distribution of filler particles was found in the 1 wt. percent filler loaded specimen, as illustrated in Fig. 3b. Fillers are shown by little hollow particles in Fig. 3b. The silanized foam specimen in Fig. 3c shows the coupling agent around the filler particles, resulting in uniform dispersion. Agglomeration occurs further with the addition of filler particles, resulting in larger particles as shown in Fig. 3d. The presence of a coupling agent up to a specified filler concentration resulted in uniform filler dispersion and distribution in the matrix.

Increased filler addition, on the other hand, leads in a reduction in inter filler distance and hence shattering of numerous hollow particles, as seen in Fig. 3d. Since the filler is a porous material, presence of voids in the samples is unavoidable. At low filler loadings, presence of additional functional groups can be noticed in the Fig. 3e. For 3F sample, matrix cracking phenomenon is the reason for the failure criterion. Microspheres can rupture at higher filler loadings, which is caused by rapid shear mixing involved in sample preparation process. Interaction of silane coupling agent with the filler particles has been characterized through FTIR spectroscopy. The characterization regarding the presence of additional functional groups for the modified surface is analyzed now through Fourier transform infra-red (FTIR) spectra with a resolution of 0.4 cm^{-1} in the wavelength range of 500–4000 cm^{-1}. The FTIR spectra of unmodified and modified samples are shown in Fig. 4. The symmetrical and asymmetrical vibration peaks of Si–O (802 cm^{-1} and 1078 cm^{-1}) are present in the FTIR spectra of silane-added particles. The asymmetric vibration of the $-CH_2$ groups can be observed at 2925 cm^{-1} indicating the presence of organic molecular chains on the surface of HGM after the coupling agent treatment. The presence of

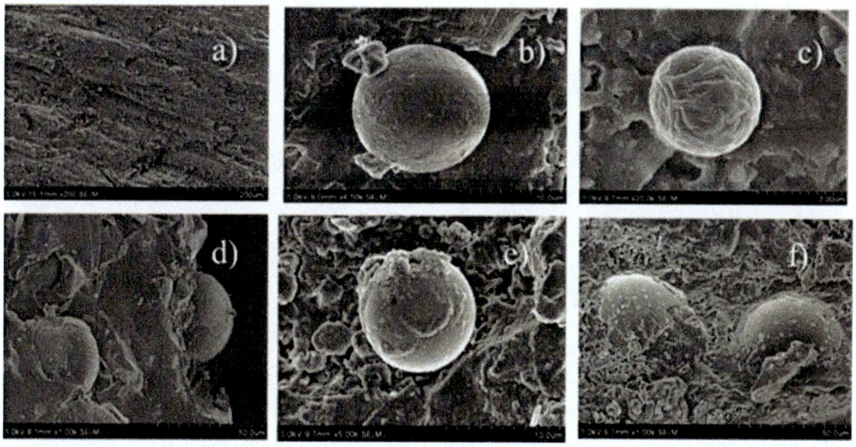

Fig. 3 SEM micrographs of fractured samples: **a** Pure epoxy, **b** 1F, **c** 1F-Si, **d** 3F, **e** 3F-Si and **f** 5F

alkyl groups on the modified HGM surface proves that the coupling agent molecules are successfully connected to the surface of the HGM. Further narrowing of the peaks between wavenumbers 2750 cm^{-1} and 3050 cm^{-1} can be attributed to presence of the additional functional groups that absorb the energy corresponding to the functional group.

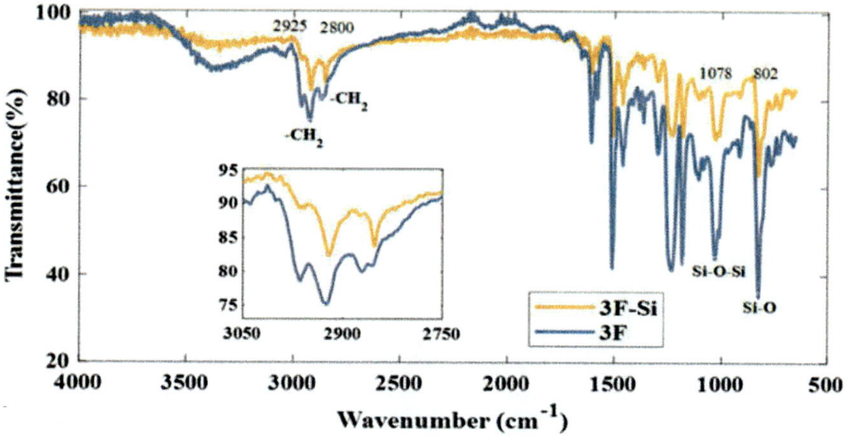

Fig.4 FTIR spectra of 3F and 3F-Si samples

3.2 Mechanical Characterization

3.2.1 Impact Strength and Hardness

Average impact strength measurements were obtained for three samples of each specimen arrangement. The variation in impact toughness of samples with respect to sample configuration is shown in Fig. 5.

$$\text{Toughness} = (\text{Energy} \cdot \text{absorbed})/(\text{Width} \times \text{Thickness})$$

The impact strength has decreased slightly at lower filler loadings particularly for unsilanized 1 and 3 wt% samples. The impact strength of the functionalized sample improved for 3 wt% specimen after silanization, but decreased drastically at 5 wt%. This could be attributed to unequal filler dispersion at larger filler loadings. The functionalization method has improved the impact toughness of up to 3 wt% samples. The increased impact properties after functionalization are due to the large surface area of the fillers and the interaction of the silane functional group with the epoxy matrix. Furthermore, spherical morphology is more conducive to homogenous filler dispersion than irregular shapes [15]. The Vickers hardness test is carried out using a diamond indenter and a hardness scale of HV0.1 with a dwell period of 10 s.

Figure 6 depicts the variation of hardness with regard to sample configuration. The initial hardness diminishes with the addition of filler, whereas the hardness value of the silanized specimen (1F-Si) improves. This could be because, with a minimal amount of filler added, the composite becomes light weight and its hardness value drops. With the addition of more fillers, the hardness value improved dramatically and showed a higher value than sample E (pure epoxy) alone.

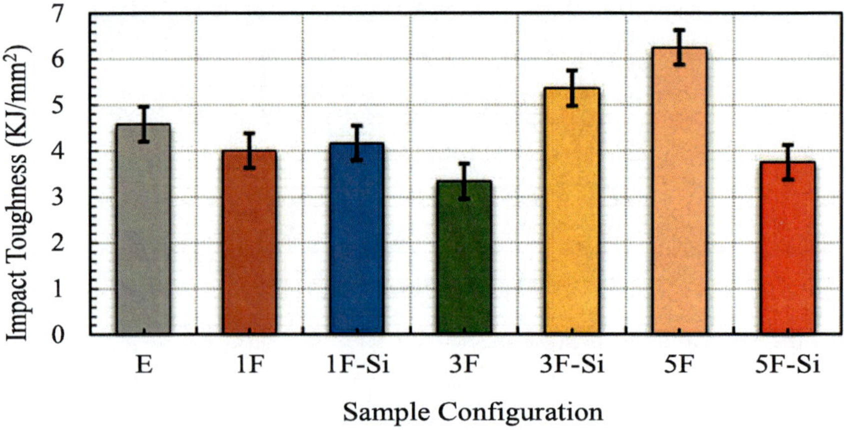

Fig. 5 Variation in impact toughness of samples

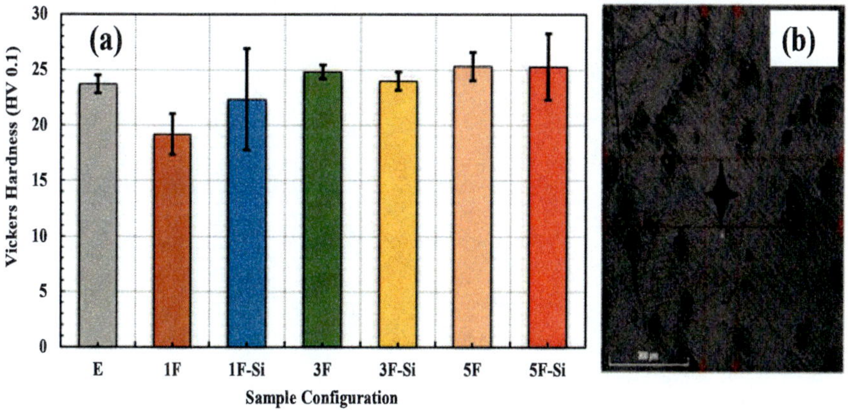

Fig. 6 Variation in hardness of samples with respect to sample configuration

3.2.2 Tensile Test

The force required to break a specimen and the extent to which the specimen stretches or elongates to that breaking point are both measured in this tensile test. In this measurement, the KALPAK UTM machine KIC-2–0200-C with a gripping load of 20 KN is used. Each sample configuration's specimens are dog bone shaped to ensure that failure occurs within the gauge length and not in the grip region. Tensile experiment was carried out in accordance with the ASTM D 638 standard. Figure 7 shows a stress–strain map of composite samples. The area under the stress–strain diagram reflects the energy absorbed by the specimen. As shown in the picture, the pure epoxy E sample has the maximum area under the stress strain curve, which is equivalent to the other mechanical tests. Due to the nature of the filler, adding HGM filler has greatly reduced the area under the curve, but silanization at low filler loadings has led to much of the property enhancement.

The tensile parameters such as load at maximum tensile stress and ultimate tensile strength parameters are presented in Table 3. It is noticed that 3F samples improved significantly after silanization, which is consistent with other mechanical testing. The tensile properties of 5F sample are significantly reduced, which is much enhanced after the silanization procedure. Uniform dispersion of fillers appears to be an issue with higher filler weight percentages. As a result, 5F samples have demonstrated substantially reduced characteristics; however, the enhancement after silanization is noteworthy. The initial decrement of UTS for 1F and 3F samples are because the presence of HGM, which makes the matrix more porous and the microstructure uneven. The interfacial region present in the composite has become strong with addition of more fillers in case of 5F sample. For 5F sample, the increase in the ultimate tensile strength is attributed to the effective mixing and dispersion of filler in the matrix, which leads to increased interfacial adhesion between matrix and filler. Increased volume of HGM spheres in the matrix restrict the fracture flow and

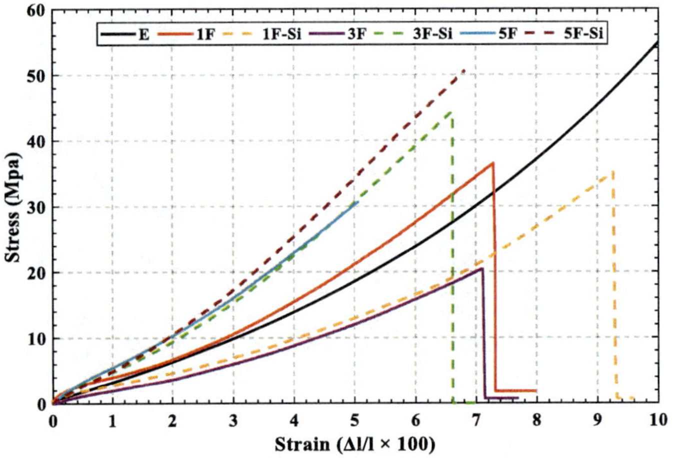

Fig. 7 Tensile stress–strain curve of samples

Table 3 Tensile properties of samples

Sample	Load at maximum tensile stress (N)	Ultimate tensile strength, S_u (MPa)
E	1135.429	43.007
1F	875.435	36.474
1F-Si	844.651	35.198
3F	489.431	20.395
3F-Si	1064.846	44.371
5F	738.673	30.774
5F-Si	1213.526	50.561

contributes to the increase in UTS. This is in accordance with the SEM images where more interfacial regions can be noticed in the 5F samples.

3.3 *Dielectric Relaxation Spectroscopy*

The variation in dielectric constant of the samples with respect to frequency is shown in Fig. 8a and b. The permittivity of the filler has considerably influenced the dielectric behavior of the nanocomposite at low frequencies. The permittivity decreases over the measured frequency range, regardless of filler concentration or silane coupling agent inclusion. The presence of more orientable dipoles that can align according to the applied field results in increased permittivity at lower frequencies. In general, atomic orientation and interfacial polarization influence material dielectric characteristics. Polarization (dipole alignment) is caused by the addition of filler to an applied electric

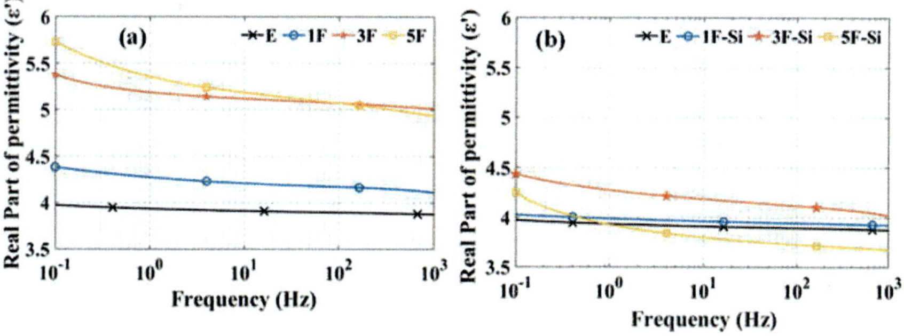

Fig. 8 Variation in relative permittivity of samples with respect to frequency (a) Unsilanized samples (b) Silanized samples

field. It has been discovered that as the concentration of filler increases, permittivity is increased.

The number of orientable dipoles between the filler and the polymer matrix increases as the filler % increases. The presence of fillers changes the functional groups and dipole alignment (polarization) in the direction of the applied electric field [9, 10]. The enhanced polarizability caused by filler inclusion results in increased permittivity for sample 5F when compared to other unsilanized specimens.

On the other hand, the addition of a silane coupling agent can modify the molecular structure of silanized polymeric materials by a variety of internal mechanisms. Silanization can enhance the bonding in insulating materials and even stop the ionization process. Ruptures can be suppressed in either the main chain or the side branch, preventing the creation of free radicals (tiny molecules) during the sample preparation process. Suppression of these free radicals prevents the creation of various functional groups by avoiding many mechanisms such as crosslinking, chain scission, gas generation, and oxidation. In the literature, the hydroxyl group (–OH) and the carbonyl group ($C = O$) were reported as reaction products for epoxy resin [11, 12]. Because of the absence of polymeric chains, crosslinking of broken polymeric chains, and small radicles, the polarizability of silanized composites is decreased [13, 14]. As a result of reduced polarizability, silanized samples have a lower dielectric constant than unsilanized specimens.

4 Conclusions

The polymer composites with incorporation of functionalized foams are investigated for its mechanical and electrical properties in the present work. The conclusions accrued based on the study are the following:

(1) Microstructural properties demonstrated that the HGM particles were well distributed during silanization, whereas larger filler concentrations resulted in foam microsphere fracture.
(2) Silanized specimens have slightly higher impact hardness than unsilanized specimens. The addition of fillers and silanization, on the other hand, increases the Vickers hardness.
(3) FTIR spectra of HGM modified samples indicated the presence of additional functional groups
(4) Except for 1 wt% specimen, the silanized samples outperform their unsilanized counterparts in terms of tensile parameters such as load at maximum tensile stress, Young's modulus, and ultimate tensile strength.
5) Regardless of silanization or filler concentration, relative permittivity decreases with frequency. Filler addition increased polarizability; however, silane addition decreased the dielectric constant.

References

1. Gupta N, Pinisetty D, Shunmugasamy VC (2013) Reinforced polymer matrix syntactic foams. Springer International Publishing, Cham. https://doi.org/10.1007/978-3-319-01243-8
2. Paul D, Velmurugan R (2018) Analysis of the specific properties of glass microballoon-epoxy syntactic foams under tensile and flexural loads. Mater Today Proc 5:16956–16962. https://doi.org/10.1016/j.matpr.2018.04.099
3. Yung K, Zhu B, Yue T, Xie C (2009) Preparation and properties of hollow glass microsphere-filled epoxy-matrix composites. Compos Sci Technol 69:260–264. https://doi.org/10.1016/j.compscitech.2008.10.014
4. Gupta N, Nagorny R (2006) Tensile properties of glass microballoon-epoxy resin syntactic foams. J Appl Polym Sci 102:1254–1261. https://doi.org/10.1002/app.23548
5. Paul D, Velmurugan R, Jayaganthan R, Gupta NK, Manzhirov AV (2018) Analysis of syntactic foam—GFRP sandwich composites for flexural loads. J Phys Conf Ser 991:12064. https://doi.org/10.1088/1742-6596/991/1/012064
6. Imran M, Rahaman A, Pal S (2019) Effect of low concentration hollow glass microspheres on mechanical and thermomechanical properties of epoxy composites. Polym Compos 40:3493–3499. https://doi.org/10.1002/pc.25211
7. Shunmugasamy VC, Anantharaman H, Pinisetty D, Gupta N (2015) Unnotched Izod impact characterization of glass hollow particle/vinyl ester syntactic foams. J Compos Mater 49:185–197. https://doi.org/10.1177/0021998313515290
8. Cantwell WJ, Morton J (1991) The impact resistance of composite materials—a review. Composites 22:347–362. https://doi.org/10.1016/0010-4361(91)90549-V
9. Singha S, Thomas M (2008) Dielectric properties of epoxy nanocomposites. IEEE Trans Dielectr Electr Insul 15:12–23. https://doi.org/10.1109/T-DEI.2008.4446732
10. Eloundou JP (2002) Dipolar relaxations in an epoxy–amine system. Eur Polym J 38:431–438. https://doi.org/10.1016/S0014-3057(01)00200-2
11. Gao Y, Du BX (2012) Effect of gamma-ray irradiation on permittivity and dielectric loss of polymer insulating materials. In: International conference on high voltage engineering and application. IEEE, pp 229–232. https://doi.org/10.1109/ICHVE.2012.6357027
12. Gao Y, Du BX, Ma ZL, Zhu XH (2010) Decay behavior of surface charge on gamma-ray irradiated epoxy resin. In: 2010 10th IEEE International conference on solid dielectrics. IEEE, pp 1–4. https://doi.org/10.1109/ICSD.2010.5568073

13. Banford HM, Fouracre RA (1999) Nuclear technology and ageing. IEEE Electr Insul Mag 15:19–27. https://doi.org/10.1109/57.793826
14. Dodd SJ (2003) A deterministic model for the growth of non-conducting electrical tree structures. J Phys D Appl Phys 36:129–141. https://doi.org/10.1088/0022-3727/36/2/309
15. Kim HS, Khamis MA (2001) Fracture and impact behaviours of hollow micro-sphere/epoxy resin composites. Compos Part A Appl Sci Manuf 32:1311–1317. https://doi.org/10.1016/S1359-835X(01)00098-7

Perforation of AA-2024 Aluminium Targets Subjected to Impact by Spherical Aluminium Projectiles

Mithilesh Kumar Dewangan, Davinder Kumar, Vikas Bhardwaj, Manjeet, Sohan Lal, Prince Sharma, and Rajesh Kumari

Abstract All spacecraft and protective structures are prone to impacts by projectiles of various shapes and sizes. The perforation process of 5 mm thin AA 2024-T4 sheets subjected to different spherical projectiles of AA 2014-T652 with 6 mm and 8 mm diameters was studied experimentally and numerically. The influence of impact conditions such as impact velocity and size of spherical projectiles on the failure evolution was observed. The projectiles are impacted at high velocities ranging from 1400 to 1850 m/s using a 40 mm calibre propellant smooth-bore gun. High-speed videography was deployed to capture the projectile-target interaction and measure impact and residual velocities. A 2D axisymmetric FE model was used to determine the ballistic limit and study the damage evolution. The Johnson–Cook plasticity and failure model was used to model target and projectile materials' high strain rate behaviour. The damage profile of AA 2024-T4 sheets and residual velocities of the projectile are in good correlation with the experimental results.

Keywords Ballistic impact · FEA · AA-2024 thin sheets · Deformable projectiles

1 Introduction

Modern spacecraft and weapon systems require lightweight materials for technological advancements and enhanced protection against impacting projectiles. It is essential to comprehend the attributes of the projectile-target interaction mechanism, which depends on several parameters such as shape, size, incident velocity of the projectile and target material. Several researchers have studied the ballistic behaviour of thin and thick sheets when subjected to impact by rigid projectiles [1–7]. Chen et al. [8] studied the behaviour and damage morphologies for harder steel targets when subjected to cylindrical soft steel projectiles by increasing the incident velocity. The failure of projectiles by mushrooming, petalling and plugging phenomena during

M. K. Dewangan · D. Kumar · V. Bhardwaj · Manjeet · S. Lal · P. Sharma (✉) · R. Kumari
Ballistic Evaluation and Hypervelocity Impact Group, Terminal Ballistics Research Laboratory, Sector 30, Chandigarh, India
e-mail: prince@tbrl.drdo.in

R. Velmurugan et al. (eds.), *Dynamic Behavior of Soft and Hard Materials Volume 1*, Springer Proceedings in Materials 34, https://doi.org/10.1007/978-981-99-6030-9_42

perforation is discussed in detail. Iqbal et al. [9] studied the perforation of 12 mm thick Weldox 460 E steel and Aluminium 1100-H12 targets when subjected to normal and oblique impact by ogive-nosed and conical steel projectiles. Vijayan et al. [10] investigated the effect of the apex angle of the projectile in the perforation mechanism of thin aluminium targets. Sharma et al. [11] investigated the ballistic penetration of soft and hard steel projectiles on Aluminium AA2014-T652 targets. The Johnson–Cook material properties for the AA2014-T652 using experimental analysis are evaluated, and the ballistic perforation of these plates using numerical and experimental techniques was studied further. Cheng et al. [12] studied the multiple ballistic impacts on 2024-T4 aluminium alloys by multiple steel projectiles using numerical and experimental studies. The Cowper Symmonds model parameters are evaluated by conducting experiments and are used as input parameters for numerical study. Chen et al. [13] studied the strain rate sensitivity of AA2024-T351 at different temperature ranges during quasi-static and dynamic loading conditions. Very few pieces of literature are focused on the studies where both projectile and the target plate are deformable during the interaction phenomena. Most of the research was focused on the perforation of rigid projectiles on deformable targets, and an enticing sector in this field, where both projectile and target plate are deformable, needs to be more explored. One such instance is the fragment impact produced when space debris strikes Whipple shields. Another illustration is the deployment of steel fragments generated by warheads to pierce steel targets. So, the present research work examines the behaviour of Aluminium AA2024-T4 target plates when subjected to spherical Aluminium AA2014-T652 projectiles. Spherical projectiles of 6 mm and 8 mm diameter were impacted on 5 mm thick target plates at 1400–1850 m/s.

The study highlights deformable projectiles' damage and perforation capabilities on a deformable plate. Furthermore, a series of numerical simulations using AUTODYN is also presented. A 2D axisymmetric model was used to define the ballistic penetration of these aluminium projectiles in the target plate.

2 Experimental Details

2.1 Experimental Setup

A smooth-bore 40 mm propellant gun was used to project the aluminium spherical projectiles in the target plate. As the projectile diameter is very small compared to the propellant gun's bore, a two-piece nylon sabot was used to launch the projectile. The complete assembly of the projectile sabot and cartridge is shown in Fig. 1b. To arrest the forward movement of the sabot, a sabot trapper is placed at a distance of 3 m from the gun barrel to ensure sufficient sabot opening. The velocities are recorded using aluminium foil-sensors and a clock set at 5 m from the muzzle end, as shown in Fig. 1a. The target was at 5.5 m from the muzzle end of gun. High-speed videography is used to capture the projectile-target interaction at 20,000 FPS and

Fig. 1 **a** Test setup, **b** projectile, sabot and cartridge

was placed in a camera shelter 10 m away from the target. The incident and residual velocities are recorded with velocity screens and with high-speed videography by calculating frames.

2.2 *Target and Projectile*

The target plates are made of Aluminium AA2024T4 thin sheets having thickness of 5 mm. The spherical-shaped Aluminium AA2014-T652 projectiles are propelled onto the target plates. Two different projectiles of diameter 6 mm and 8 mm are used and are engraved inside the sabot to be propelled from the 40 mm smooth-bore propellant gun. The amount of propellant is varied inside the cartridge in order to achieve different velocity ranges and to study the behaviour of the perforation phenomena.

3 Numerical Setup

3.1 Constitutive and Failure Model

The commercial finite element code AUTODYN is used to simulate the interaction of aluminium projectile and target plates. A 2D axisymmetric model is used to define the ballistic penetration. The target plates are defined using the Johnson–Cook plasticity and damage model. Johnson–Cook proposed a phenomenological model to predict the isotropic material behaviour subjected to different dynamic loadings. The model shows the equivalent stress (σ) as a function of equivalent plastic strain (ε), strain rate $\varepsilon^{\cdot}$ and homologous temperature (T^*).

$$\sigma = \left(A + B\varepsilon^n\right)\left(1 + C\ln\dot{\varepsilon}^*\right)\left(1 - T^{*m}\right) \tag{1}$$

A, B, C, n and *m* are material parameters which are evaluated experimentally and are used in the numerical studies. Parameter *A* is the quasi-yield strength, *B* and *n* are strain hardening parameters, *C* is the strain rate hardening parameter, and m is thermal softening exponent. The material properties are taken from the data available from the literature works [11–13] and are given in Table 1 for constitutive plasticity model.

The constitutive damage model proposed by Johnson–Cook is applied to define the fracture behaviour of metals under dynamic loading conditions. The model represents the fracture strain of the material as a function of stress triaxiality, strain rate and temperature.

$$\varepsilon_f = \left[D_1 + D_2 e^{D_3\overline{\sigma}^*}\right]\left[1 + D_4 In\dot{\int}^p\right]\left[1 + D_5 T^{*m}\right]. \tag{2}$$

Here, D_1, D_2, D_3 are stress triaxiality parameters where as D_4 is strain rate parameter and thermal sensitivity parameter is defined by D_5. The first part in right side highlights the effect of stress triaxiality which is expressed as the ratio of hydrostatic to deviatoric stress components. Similarly, second and third terms predict the effect of strain rate and temperature, respectively. The material properties are taken from the data available from the literature works [11–13] and are given in Table 2 for damage model.

Table 1 Johnson–Cook constitutive model parameters for projectile and target plate

Material	Density (kg/m^3)	A (MPa)	B (MPa)	*n*	*C*	*m*	Melting temperature (K)
AA2024-T4	27875	291.497	588.21	0.501	0.08	2.242	1220
AA2014-T652	2850	453	453	0.5948	0.013	1.08	775

Table 2 Johnson–Cook damage model parameters for projectile and target plate

Material	D_1	D_2	D_3	D_4	D_5
AA2024-T4	0.31	0.045	−1.7	0.005	0
AA2014-T652	0.087	2.68	−8.01	0.029	5.332

3.2 Finite Element Model

As the loading and geometry is simple and axisymmetric in nature, a 2D axisymmetric model is used to analyse the ballistic phenomena. This will considerably decrease the computational time and enable the study of a quick prediction for this phenomenon. With the limitation in experiments, a series of numerical simulations were conducted for determining the ballistic limit (complete perforation) numerically. And the FE model results are compared with the experimental results for both 6 and 8 mm projectile perforations. A 2D plate is modelled having thickness 5 mm and length 50 mm as shown in Fig. 2. The spherical ball is also modelled as half projectile and is given a minimal distance away from the target plate, so that there should not be any pre-interaction effects. The ball is given the initial incident velocities and is propelled in one direction for interaction with the target plate. A fixed boundary condition is provided in the top for restricting the motion of the plate. As the actual target plate will have quite large size, this will predict an equivalent results. Based on mesh convergence studies, 50 elements in through the thickness direction is selected and 100 elements are selected in the radial direction. Also a fine mesh is provided up to 10 mm in radial direction which is quite above the radius of the projectiles having diameters 6 and 8 mm. Also, shock waves are observed on the target plate using numerical simulation, as it should not touch the top most layer during this interaction phenomena, otherwise effect of boundary and selection of length of the plate will be affected. This is evaluated by applying a gauge at the top location of the plate near the boundary and pressure values are studied. The studies concludes that the projectile usually perforates before the shock wave travels to the boundary in this arrangement which will suffice the real experimental scenario.

4 Results and Discussions

4.1 Ballistic Perforation of 5 mm Thin AA2024T4 Sheets by 6 mm Spherical Projectiles

The residual velocities are measured for the impacting projectiles using both numerical and experimental techniques. The residual velocities predicted by the numerical model are within 10% deviation from the experimental values which is a good correlation for a simple axisymmetric model. The numerical ballistic limit is 1075 m/s

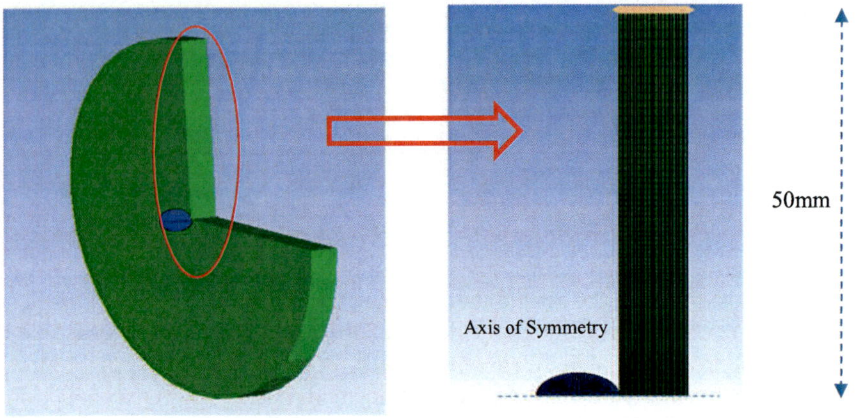

Fig. 2 2D axisymmetric finite element model

which causes complete perforation. Figure 3 represents the residual velocities against impacting velocities for 6 mm projectile and velocity vs time graph for the incident velocity of 1461 m/s. All the residual velocities are recorded in a similar way and are plotted against the incident velocities as shown in Fig. 3. There is a dip in the residual velocities in the plot between 1400 and 1500 m/s which is observed. This is due to the extraction of more material from the target plate which resulted in further shift in the slope of velocity vs time, as can be observed in Fig. 3 in both experimental and numerical analysis. Figure 4 represents the interaction phenomena at different stages of perforation by 6 mm projectiles on the target plate using high-speed videography for experimental studies and numerical simulation results.

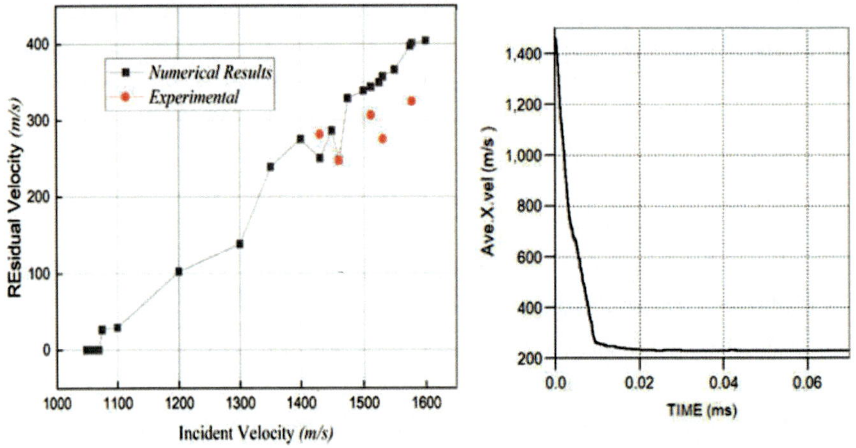

Fig. 3 Residual velocity vs. incident velocity highlighting numerical and experimental results and velocity vs time graph for incident velocity of 1461 m/s

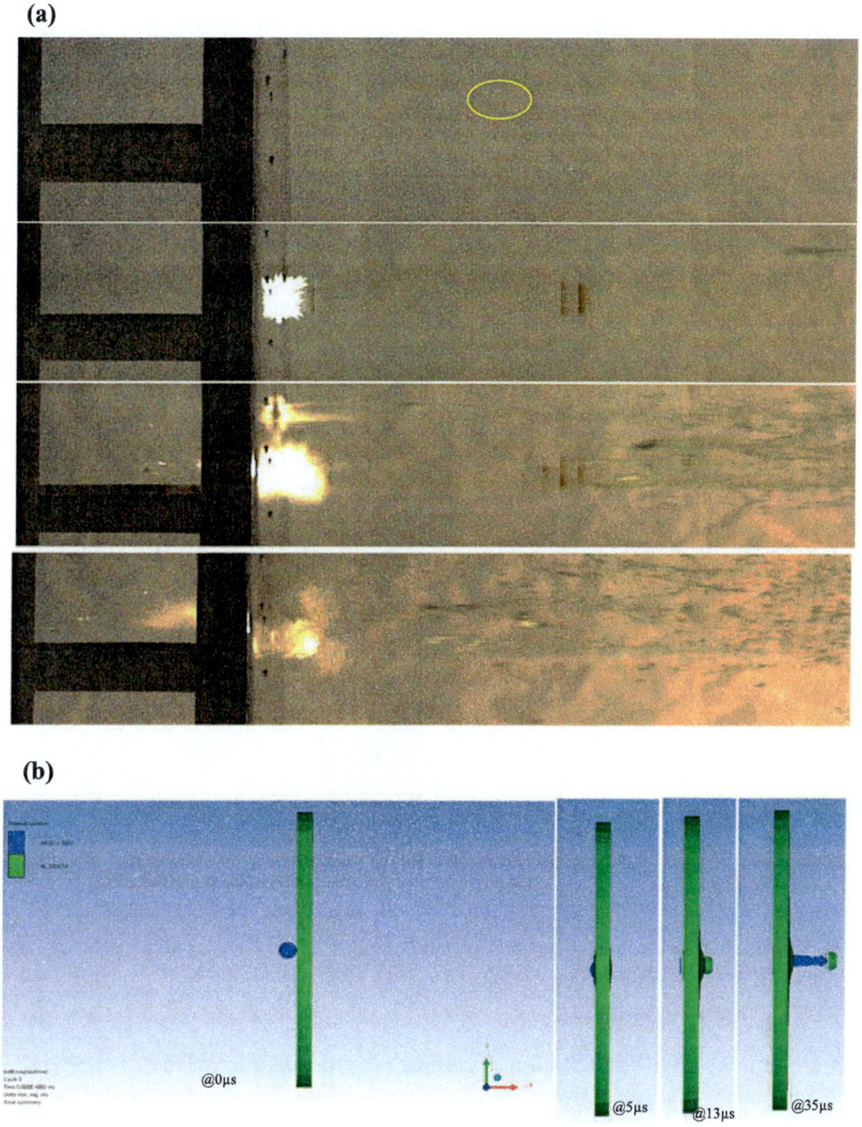

Fig. 4 Ballistic perforation phenomena in **a** experimental and **b** numerical studies for incident velocity of 1461 m/s

The plate perforation can be observed with lip formation in the front and rear of the target plate, and a similar result is observed by numerical analysis as shown in Fig. 5. With the increase in the velocity, the damage caused inside the plate is more and an extra amount of plug is detached from the target plate. This results in the variation in the residual velocities as observed in numerical studies. The front face

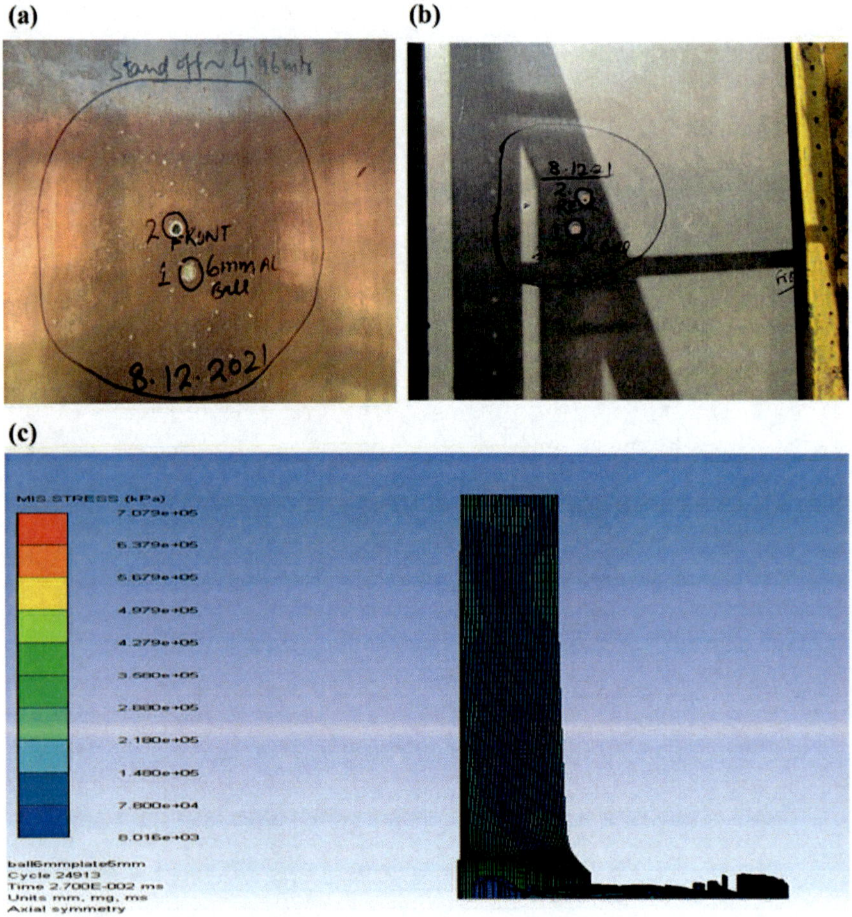

Fig. 5 Perforation of target plate (front and rear) and Von-misses stress for target plate subjected by 6 mm projectile at 1461 m/s incident velocity

signature shows higher lip formation as compared to the rear face of the target plate, and the damage is more of divergent-convergent-divergent type as predicted from Fig. 5.

4.2 Ballistic Perforation of 5 mm Thin AA2024T4 Sheets by 8 mm Spherical Projectiles.

The similar studies are conducted for the 8 mm projectile perforation in experiments and numerical studies. The numerical ballistic limit is observed at the velocity 950 m/

s complete perforation. The residual velocities are recorded with screen recorders as well as high-speed videography (as in the previous studies for 6 mm) and are compared with the numerical results. The numerical residual velocities are close to the experimental results which can be observed from the Fig. 6a and have nearly 10% variations which is a good correlation for understanding the material behaviour during perforation phenomena. All the experimental velocities are evaluated numerically as well for more and close predictions for the ballistic phenomena of the present case. The perforation phenomena by 8 mm projectiles for incident velocity of 1648 m/s is shown in Fig. 6b and are compared with the numerical results. The study highlights that the 8 mm projectiles will have more back face deformation as compared to the front face, and the damage profile is more of convergent–divergent type. This may be accounted due to the higher diameter of the projectile which causes more failure and expansion of the material.

5 Conclusions

The perforation of AA2024T4 thin sheets when subjected to impact by aluminium projectiles of 6 mm and 8 mm is studied. The lip formation at the rear face is observed for both the cases, but a higher formation is observed for 8 mm projectiles as compared to 6 mm projectiles. The numerical results are very close to the experimental results.

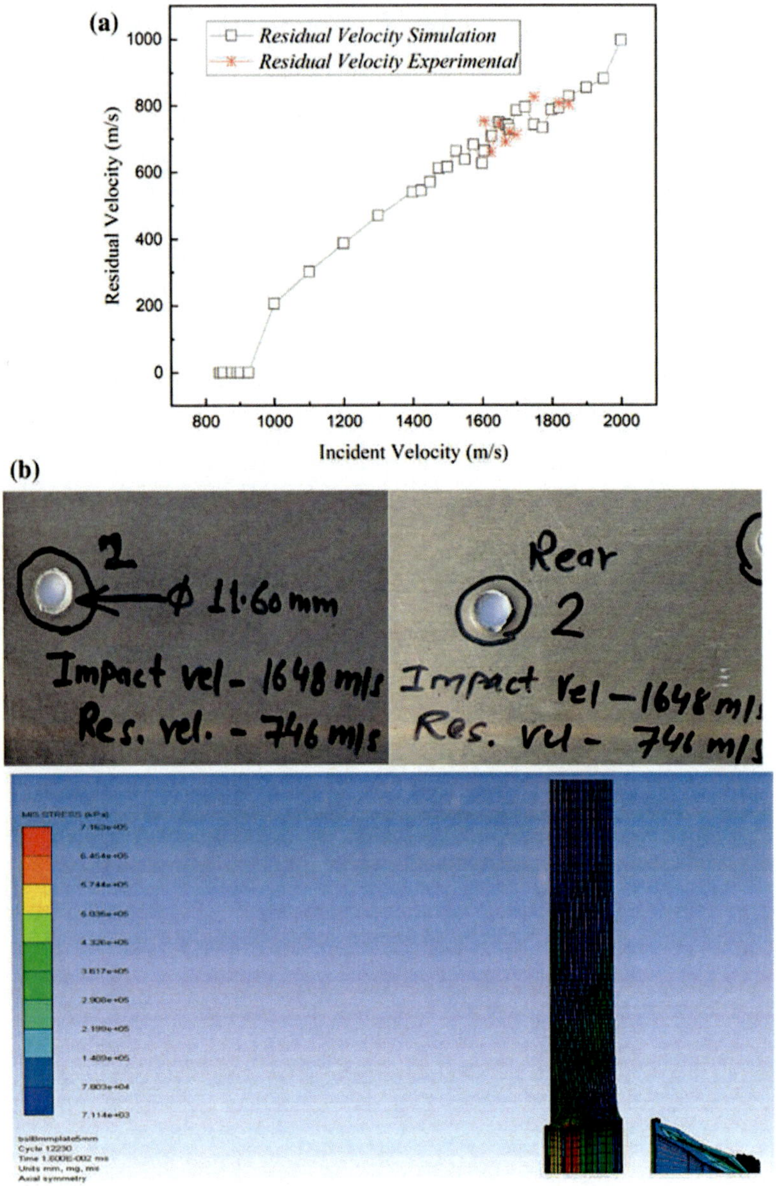

Fig. 6 **a** Residual velocities against incident velocities for 8 mm aluminium projectiles and **b** perforation at the 1648 m/s velocity

Acknowledgements The authors are grateful to the director TBRL Chandigarh, for providing this opportunity for the research and the staff and officers of BEHI Group for carrying out the experimental studies.

References

1. Borvik T, Langseth M, Hopperstad OS, Malo KA (2001) Perforation of 12 mm thick steel plates by 20 mm diameter projectiles with flat, hemispherical and conical noses—Part I: experimental study. Int J Impact Eng 27:19–35. https://doi.org/10.1016/S0734-743X(01)00034-3
2. Szymczyk M, Sumelka W, Łodygowski T (2018) Numerical investigation on ballistic resistance of aluminium multi-layered panels impacted by improvised projectiles. Arch Appl Mech 88:51–63. https://doi.org/10.1007/s00419-017-1247-8
3. Wen K, Chen XW, Chi RQ, Lu YG (2020) Analysis on the fragmentation pattern of sphere hypervelocity impacting on thin plate. Int J Impact Eng 146:103721. https://doi.org/10.1016/j.ijimpeng.2020.103721
4. Jones N, Kee J (2012) Impact perforation of aluminium alloy plates. Int J Impact Eng 48:46–53. https://doi.org/10.1016/j.ijimpeng.2011.05.007
5. Deng Y, Zhang Y, Xiao X, Hu A, Wu H, Xiong J (2020) Experimental and numerical study on the ballistic impact behavior of 6061-T651 aluminum alloy thick plates against blunt-nosed projectile. Int J Impact Eng 2020:103659. https://doi.org/10.1016/j.ijimpeng.2020.103659
6. Mohammad Z, Gupta KP, Baqi A (2020) Experimental and numerical investigations on the behavior of thin metallic plate targets subjected to ballistic impact. Int J Impact Eng 146:103717. https://doi.org/10.1016/j.ijimpeng.2020.103717
7. Børvik T, Olovsson L, Dey S, Langseth M (2011) Normal and oblique impact of small arms bullets on AA6082-T4 aluminium protective plates. Int J Impact Eng 38:577–589. https://doi.org/10.1016/j.ijimpeng.2011.02.001
8. Chen XW, Chen G, Zhang FJ (2008) Deformation and failure modes of soft steel projectiles impacting harder steel targets at increasing velocity. Exp Mech 48:335–354. https://doi.org/10.1007/s11340-007-9110-4
9. Iqbal MA, Chakrabarti A, Beniwal S, Gupta NK (2010) 3D numerical simulations of sharp nosed projectile impact on ductile targets. Int J Impact Eng 37:185–195. https://doi.org/10.1016/j.ijimpeng.2009.09.008
10. Vijayan V, Hegde S, Gupta NK (2017) Deformation and ballistic performance of conical aluminum projectiles impacting thin aluminum targets: Influence of apex angle. Int J Impact Eng 110:39–46. https://doi.org/10.1016/j.ijimpeng.2017.05.009
11. Sharma P, Chandel P, Bhardwaj V, Singh M, Mahajan P (2018) Ballistic impact response of high strength aluminium alloy 2014–T652 subjected to rigid and deformable projectiles. Thin-Walled Struct 126:205–219. https://doi.org/10.1016/j.tws.2017.05.014
12. Cheng JC, Zhao SP, Fan D, Chai HW, Ye SJ, Li C et al (2021) Multiple ballistic impacts on 2024–T4 aluminum alloy by spheres: Experiments and modelling. J Mater Sci Technol 94:164–174. https://doi.org/10.1016/j.jmst.2021.04.012
13. Chen G, Lu L, Ren C, Ge X (2018) Temperature dependent negative to positive strain rate sensitivity and compression behavior for 2024–T351 aluminum alloy. J Alloys Compd 765:569–585. https://doi.org/10.1016/j.jallcom.2018.06.196

Analysis of High-Speed Impact Behavior of Al 2024 Alloy Using Machine Learning Techniques

S. Siri, Navya Gara, R. Velmurugan, and R. Jayaganthan

Abstract The large deformation of Al alloys subjected to high-speed impact may lead to catastrophic failure of aerospace structural components fabricated using these materials. The present work is focused to analyze dynamic behavior of Al2024 alloy subjected to very high impact velocities using FEA software (LS DYNA) along with machine learning (ML) techniques. The transient impact behavior of the alloy was estimated using modified Johnson–Cook visco-plastic model for a strain rate range of 100–3000/s. The residual velocities of Al 2024 subjected to high-speed impact were estimated through FEA, analytical routes, and with experimental studies. In the present work, five machine learning (ML) algorithms, such as K-Nearest Neighbor (KNN), Decision Trees (DT), Random Forests (RF), Support Vector Regression (SVR), and Extreme Gradient Boosting (XGB) algorithms are implemented to predict the crashworthiness of structures constituting Al2024 alloy. The procedures used for optimization of hyper parameters in each of the above ML models are discussed. The comparative analysis of ML models was made for its predictive accuracy in estimating the dynamic behavior of Al 2024 alloy based on their R2 scores, mean squared error, and mean absolute error. Extreme Gradient Boosting has performed better for the crashworthiness predictions providing least mean squared errors and higher R2 scores compared to other models.

Keywords High-speed impact behavior · 2024 alloy · Random forest · Support vector · Decision tree · Machine learning

S. Siri · N. Gara · R. Jayaganthan (✉)
Department of Engineering Design, Indian Institute of Technology Madras, Chennai 600036, India
e-mail: edjay@iitm.ac.in

N. Gara · R. Velmurugan
Department of Aerospace Engineering, Indian Institute of Technology Madras, Chennai 600036, India

R. Velmurugan et al. (eds.), *Dynamic Behavior of Soft and Hard Materials Volume 1*, Springer Proceedings in Materials 34, https://doi.org/10.1007/978-981-99-6030-9_43

1 Introduction

The crashworthiness of aluminum alloys used as structural components in the aerospace, defense industries is of paramount concern for its reliable performance in actual service conditions. Particularly, Al2024 alloy is preferred for its high fatigue resistance, fracture toughness, and its high strength-to-weight ratio [1]. During the functional role of components, these may be subjected to high pressures, varied temperatures, and sometimes to debri or bird impact that may lead to detrimental effects and thus leading to failure [2]. In order to protect the components constituting Al alloys, it is essential to understand the collision impact and its effect at different velocities on this alloy. However, performing the tests at that high strain rates may be expensive owing to the destructive testing which is quite essential in the estimation. Hence, there is a need to adopt the cheapest way of analyzing the high strain rate behavior of the alloy. Finite element modeling (FEM) plays a vital role to simulate the experimental testing condition of Al alloy. The analysis of material behavior in the form of stress strain curves, force–displacement curves, fringe plots, etc., provide us with a sense of the component's material constitutive response [3]. However, this involves huge generation of input/output parametric data and computational effort especially with the post processing of the data obtained through the numerical modeling. Thus, handling such huge data could sometimes lead us with the missing of partial datasets. With more number of iterations, it becomes computationally intensive to handle and the corresponding storage issues [4]. To curb this shortcoming and with the advancement in the computational engineering, machine learning (ML) is one of the merging tools that aid in predicting the response based on training the model with the available data without any further generation of data of the similar kind.

The use of ML models is gaining momentum in recent times for predicting the effects of various engineering applications such as fatigue life [5], corrosion properties [6], and fatigue crack growth [7] by feeding the input parameters for training the ML model. This present work focuses to predict the ballistic impact behavior of the Al2024 alloy for different impact velocities, projectile nose shapes, projectile dimensions—diameter and length—and for varied target thicknesses. The residual velocities from the already available experimental literature and simulated data for distinct geometrical parameters were utilized in developing the different ML models. Five ML models such as Random Forest Regression (RF), Decision Tree Regression (DT), Support Vector Regression (SVR), K-Nearest Neighbor Regression (KNN), and Extreme Gradient Boosting (XGBOOST) were utilized to predict the residual velocities. The optimization of the parameters was carried out to estimate the dynamic behavior of material as output parameter. These were further compared, and the best algorithm for its predicting crashworthiness of Al 2024 alloy is discussed in the present work.

Table 1 Parameters for pre-processing data

Input parameters	Parameter 1	Parameter 2
Nose shape of projectile	Blunt	Hemispherical
Diameter of projectile [D] (mm)	12.7	10.5
Length of projectile [L] (mm)	40.25	40
Target thickness[T] (mm)	2	4
Impact velocity range [I] (m/s)	Ballistic limit- 900	Ballistic limit- 500

2 Methodology and Modeling

2.1 Data Collection

The dynamic behavior of Al 2024 alloy under ballistic impact with respect to the residual velocities was estimated using the results as obtained from the literature [1] and is given in Table 1. Datasets were collected from the different sources, and reliable 500 data points were selected to train the models for obtaining better accuracy in prediction. The geometric variation in the ballistic impact process was considered as the independent input variables while the residual velocities were considered as the dependent output variable. The histograms representing the data point distributions with respect to the impact velocity, nose shape (25% blunt and 75% hemispherical), diameter of the projectile, and the target thickness are as shown in Fig. 1.

2.2 Data Pre-Processing

The categorical data is further encoded into a readable format that can be fed into the ML models. Standardization is performed on the entire dataset which limits the range of variables, so that the variables can be compared on the common grounds. The formula for standardization is as given in Eq. (1):

$$X' = \frac{X - \mu}{\sigma} \tag{1}$$

where X', X, μ, σ are the new scaled value, feature value, mean, and standard deviation of the feature values, respectively. Table 2 presents the scaled dataset used in developing models. The dataset is split into train (75%—335points) and test (25%—111points), where the model has been developed on the train dataset. The test dataset is used to verify the developed model performing on the unseen data. The performance was analyzed using metrics such as mean squared error, mean absolute error, and R2 score.

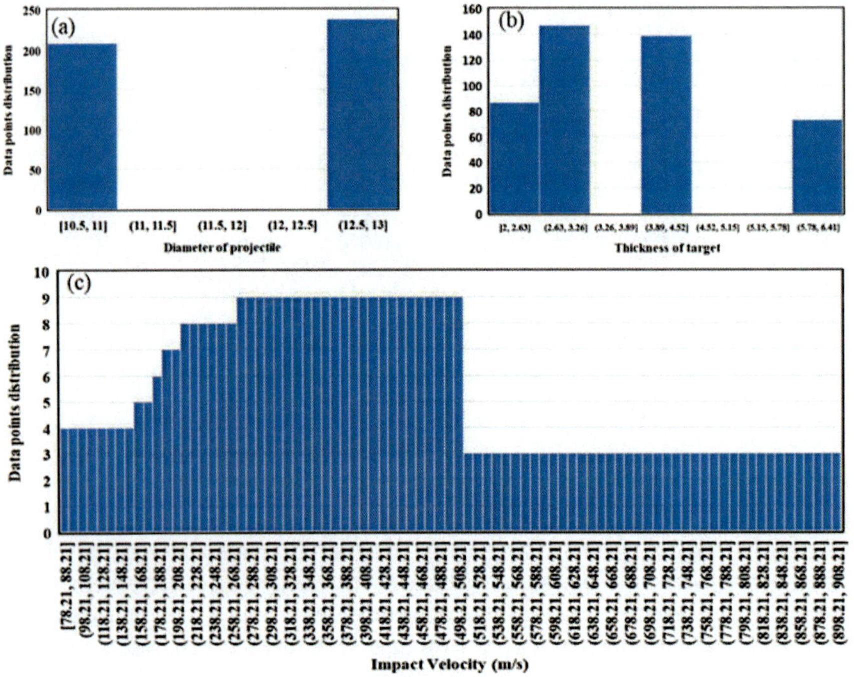

Fig. 1 Data points distribution: **a** Diameter of projectile, **b** thickness of target and **c** impact velocity

Table 2 Scaled dataset

Nose shape	Diameter [D] (mm)	Length [L] (mm)	Target thickness [T] (mm)	Impact velocity [I] (m/s)
−0.572172	−1.069687	−1.069687	−0.806390	−0.979380
−0.572172	−1.069687	−1.069687	−0.806390	−1.029069
−0.572172	0.934853	0.934853	−0.076877	−0.640045
1.747726	−1.069687	−1.069687	−1.535904	−0.688522
1.747726	−1.069687	−1.069687	−1.535904	−1.415669

2.3 Development of ML Models

The five models were developed by utilizing pre-defined python libraries such as (a) NumPy—for defining the scientific calculations; (b) Pandas—for managing and analysis of data points; and (c) Matplotlib—for plotting the graphs obtained from the ML models. The selection of models was based on literature [5–7] for the desired accuracy in the prediction. The impact dynamic behavior that needs to be predicted was compared and contrasted from the five models considered.

Table 3 Random forest hyper parameters

Hyper parameters	Value
n_estimators	400
max_features	sqrt
max_depth	None
min_samples_split	2
min_samples_leaf	1
Bootstrap	False

2.3.1 Random Forest Regression

A large number of individual decision trees that make a class prediction form the random forest. The accuracy in the prediction depends upon the degree of correlation between the classes utilized. The lesser the correlation, the more is the accuracy (by reducing the errors) and thus, to attain the lower correlation, boot strap aggression and feature randomness were adopted. Furthermore, the several datasets that aggregate as the multiple trees were trained and the average of predictions by each tree was taken as the prediction of the model. The hyper parameters of Random Forest Regression were tuned using random search CV. The best hyper parameters estimated for this algorithm using the dataset are as given in Table 3.

2.3.2 Decision Tree Regression

The model was based on the predictive binary sets by breaking down into numerous subsets such as (a) root nodes—representing entire population of data, (b) decision nodes—a subset data, and (c) leaf node—decision on the target variable. CART algorithm is used to build a tree. A set of questions in the form of True/False were verified at each sub node, and a decision was derived at the leaf nodes. Further the nodes were split, based on the minimization of mean squared errors on the child nodes. The minimization function for splitting the nodes is as given in Eqs. (2) and (3). Mean square Average (MSA) @node A=

$$\frac{1}{N_A} \cdot \Sigma_A \left(Y - \hat{Y}_A\right)^2 + \frac{1}{N_B} \cdot \Sigma_B \left(Y - \hat{Y}_B\right)^2 \tag{2}$$

$$\hat{Y}_A = \frac{1}{N_A} \cdot \Sigma_A Y_i \; and \; \hat{Y}_B = \frac{1}{N_B} \cdot \Sigma_B Y_i \tag{3}$$

where Y is the predicted value in corresponding child node, $\hat{Y}_A$ is the average of the predictions in the child node A, and $\hat{Y}_B$ is the average of predictions in the child Node B.

The obtained decision of the leaf node was calculated for its standard deviation [S(T,X)] which is used to evaluate the homogeneity of the decision as the final output

Table 4 Decision tree hyper parameters

Hyper parameters	Value
max_depth	7
max_features	auto
max_leaf_nodes	None
min_samples_leaf	3
min_weight_fraction_leaf	0.1
Splitter	best

Table 5 Support Vector hyper parameters

Hyper parameters	Value
C	1000
Epsilon	0.01
Gamma	1

variable given in Eq. (4).This S(T,X) is reduced further by splitting the dataset known as standard deviation reduction [SDR(T,X)] as given in Eq. (5). Table 4 presents the hyper parameters utilized in this prediction using GridSearch CV.

$$S(T, X) = \sum_{c \in X} P(c)S(c) \tag{4}$$

$$\mathrm{SDR}(T, X) = S(T) - S(T, X) \tag{5}$$

2.3.3 Support Vector Regression

The model prediction is based on the discrete values of the input and output parameters. A hyper plane is selected where many data points would fall. The data points on either set of planes would represent the support vectors. Unlike the remaining models that minimizes the error between the actual and predicted, this model finds the best fit line within the threshold range (plane to boundary line) where maximum of the points lies. This model is the best choice for a less noise data. The hyper tuning of the parameters was made on GridSearch CV as given in Table 5.

2.3.4 K-Nearest Neighbor Regression

KNN or K-Nearest Neighbors could be used for both regression and classification of data, with a feature to predict the value of hidden data points. The model determines the new datasets that are close to the trained datasets by using the average distance value of decided K-nearest neighbor. The Euclidean distance average method was

Table 6 K-Nearest Neighbor hyper parameters

Hyper parameters	Value
Best leaf_size	9
Best p	2
Best n_neighbors	2

used as given in Eq. (6).

$$d = \sqrt{\sum_{i=1}^{k} (x_i - y_i)^2} \tag{6}$$

where x_i and y_i are the coordinates of the new and existing data points, respectively. The parameter 'K' (number of neighbors) was tuned for the hyper parameters, and a value of 2 is obtained with the best least square error. Table 6 presents the hyper parameters utilized in this prediction using Grid Search CV.

2.3.5 Extreme Gradient Boosting

Extreme Gradient Boosting or XGBOOST is an optimized Gradient Boosting algorithm which enables to perform efficiently for most datasets. The algorithm can be represented by objective function as given in Eq. (7):

$$L = \sum_{i} l(y_a, y_p) + \Sigma_k \beta(f_k) \tag{7}$$

where L is the loss function depicting the difference between the actual and the predicted values and $\beta(f_k)$ called the regularization term, which makes the algorithm robust. This term can be further represented by Eq. (8).

$$\beta(f_k) = \gamma T + 0.5 \cdot \lambda \cdot \omega^2 \tag{8}$$

where T represents the number of trees, γ represents the pruning index, ω represents the leaf weight in the tree model, and λ is the scaling factor for those weights. All these terms were hyper parameters for the XGB algorithm and were tuned using Grid Search CV as given in Table 7.

Table 7 XGB hyper parameters

Hyper parameters	Value
Booster	Gb tree
Lambda	0.5
Learning rate	0.3
Maximum depth	2
Number of estimators	100

3 Model Evaluation

3.1 Model Validation

The obtained developed models were evaluated with the test dataset, to understand the accuracy of the predicted residual velocities to those obtained experimentally. Performance of the model was measured by three metrics (a) R2 scores—variance in the predictions explained by the dataset by keeping it minimum, (b) Mean squared error (MSE)—Proximity of regression line to the set of data points by nearing it to 1, and (c) Mean absolute error (MAE)—magnitude of difference between the predicted and observed true values by keeping it minimum. Thus,

$$\mathrm{MSE} = \frac{1}{\mathrm{N}}\left[\sum(\mathrm{y_a} - \mathrm{y_p})\right]^2 \tag{9}$$

$$MAE = \frac{1}{N}\Sigma\left|y_a - y_p\right| \tag{10}$$

$$R_2 = 1 - \frac{\sum(y_a - y_p)^2}{\sum(y_a - y_m)^2} \tag{11}$$

where y_a is the experimental value, y_p is the predicted value, and y_m is mean of the experimental values.

From Fig. 2a, it could be observed that the error in the MAE and MSE was small compared to the R2 scores for all the ML models considered. Further, the decision tree algorithm showed a maximum error than the other regression models utilized for predicting the residual velocities. From Fig. 2b, R2 Score for train and test dataset was higher with XGB algorithm which makes it superior for this dataset. Since the residuals of the preceding regression tree were trained to the succeeding regression tree with the residuals shrinking to zero thus leading to the patterns in the data which are identified. However, there is higher chance of overfitting when higher number of trees are trained. In our present study, R2 score for both the test and train dataset was comparable, and hence, model is not overfitted. For both the algorithms, R2 scores for train and test datasets were similar and R2 scores of XGB were slightly higher than RF, which indicates that XGB performs better.

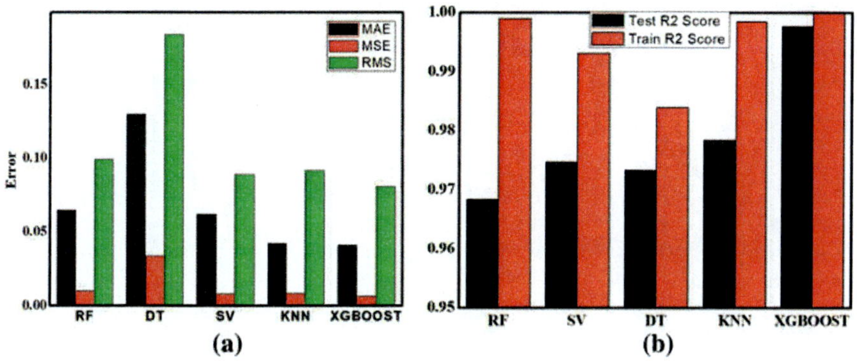

Fig. 2 **a** Error plots, **b** R2 scores for different ML models

3.2 Predictions Using Machine Learning Models

Predictions on the experimental impact data were carried out and comparison between the actual and predicted data is shown in Fig. 3a. Y test represents the experimental residual velocity in the test set while Y_predict represents the residual velocities predicted by the ML algorithms. From Fig. 3, it could be observed that the KNN and XGBOOST plots were approximately linear. Additionally, Fig. 2a shows that these models have the least MSE and MSA values. These models outperformed the others with minimum error values and have comparable R2 scores. For the dataset considered in the present study, XGBOOST has the better scores and better predictions compared to KNN algorithms; thus, XGBOOST is best fit algorithm in predicting the high-speed impact behavior of Al 2024 alloy. To understand the predictive accuracy of the ML algorithms for the dataset considered in the study, the experimental impact velocity and residual velocities were plotted along with predicted residual velocities as shown in Fig. 3b. The red data points are the experimental residual velocities, and the green data points are the residual velocities predicted by the models.

3.3 Feature Importance Analysis

Feature importance analysis was carried out to estimate the relevant and irrelevant features (inputs) that are essential in predicting the target variable using ML model. A score/rank is calculated for all the input parameters required for the prediction and the feature with a higher score has a larger effect on the predictions by ML model. In the present study, the features considered were impact velocity, target thickness, projectile nose shape, and projectile diameter. Figure 4 shows the result of the feature importance analysis. It could be observed that the most influencing

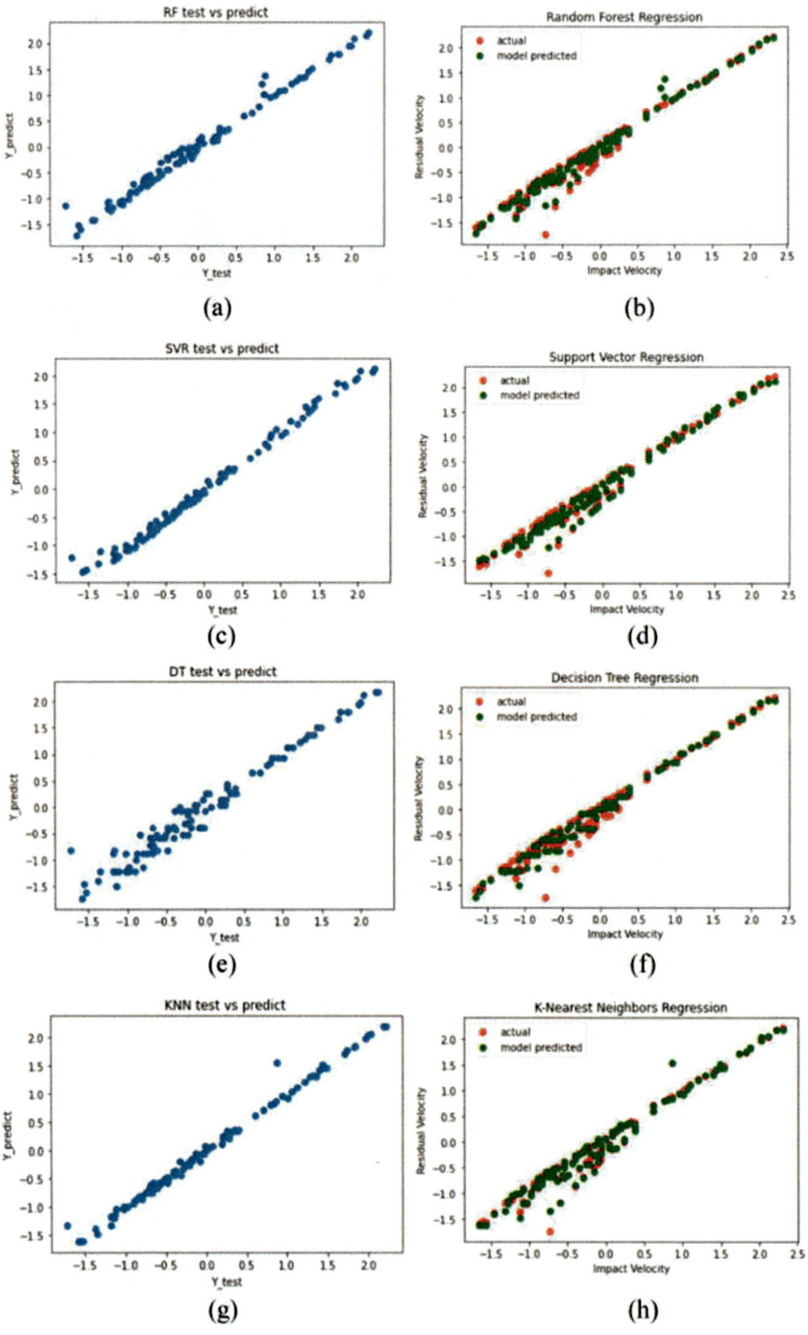

Fig. 3 **a, c, e, g, i**—Actual test data v/s predicted; **b, d, f, h, j**—predicted residual velocities for impact velocities

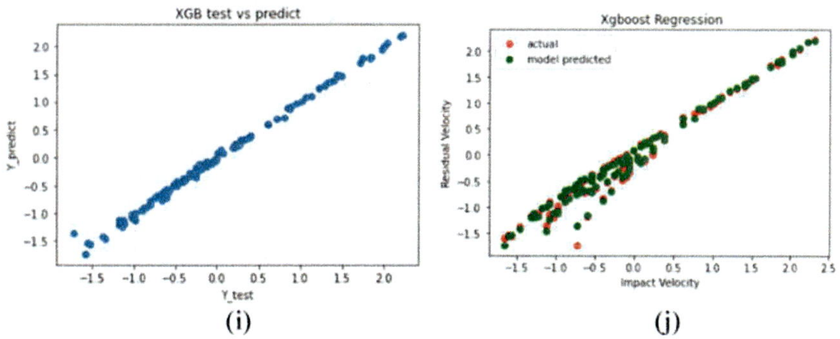

Fig. 3 (continued)

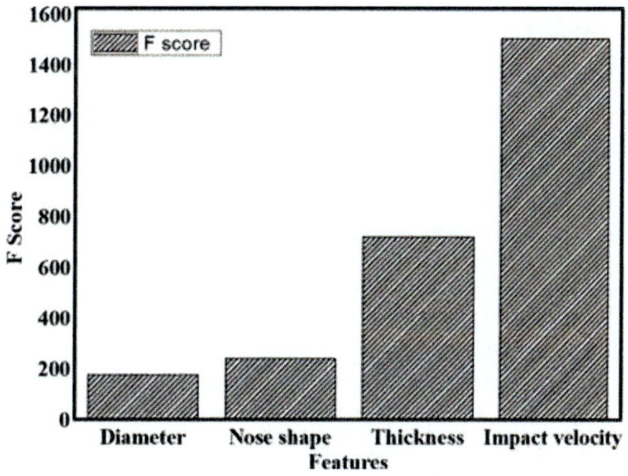

Fig. 4 F-scores for the different features

feature is the impact velocity as it is the testing condition that has direct influence on the component behavior. However, the remaining features were the other conditions whose importance is analyzed by feature importance analysis for the chosen dataset. ML models are capable of identifying the influence of independent parameters appropriately along with predicting accurate experimental results.

4 Conclusions

Analysis of high velocity impact behavior of Al 2024 alloy was carried out by using five different ML models using the experimental data reported in the literature. The following are the conclusions based on the present study.

- It was observed that the XGB algorithm had led to a best R2 score with least mean squared error and least mean absolute error in predicting the impact behavior of Al 2024 alloy.
- In the feature importance analysis, impact velocity, thickness, nose shape, and diameter are identified as important parameters for predicting the impact behavior of Al 2024 alloy with impact velocity being the most influencing feature.

References

1. Gara N, Ramachandran V, Rengaswamy J (2021) Analytical and FEM analyses of high-speed impact behaviour of Al 2024 alloy. Aerospace 8:281
2. Navya G, Jayaganthan R, Velmurugan R, Gupta NK (2022) Finite element analysis of tensile behaviour of glass fibre composites under varying strain rates. Thin-Walled Struct 172:108916
3. Lei XD, Wu XQ, Zhang Z, Xiao KL, Wang YW, Huang CG (2021) A machine learning model for predicting the ballistic impact resistance of unidirectional fiber-reinforced composite plate. Sci Rep 11:6503
4. Wadagbalkar P, Liu GR (2021) Real-time prediction of projectile penetration to laminates by training machine learning models with finite element solver as the trainer. Defence Technol 17(1):147–160
5. Konda N, Verma R, Jayaganthan R (2021) Estimation of high cycle fatigue life of additively manufactured Ti6Al4V using data analytics. In: Procedia Engineering. Pre-Publish Press
6. Mythreyi OV, Rohith Srinivaas M, Kumar TA, Jayaganthan R (2021) Machine-learning-based prediction of corrosion behavior in additively manufactured Inconel 718. Data 6(8)
7. Raja A, Chukka ST, Jayaganthan R (2020) Prediction of fatigue crack growth behaviour in ultrafine grained Al 2014 alloy using machine learning. Metals 10(10):1–13

Ballistic Resistance of Finite Plate Targets Against 7.62 NATO AP Ammunition

Zaid Mohammad, J. Venkatesan, and Pramod Kumar Gupta

Abstract In the present study, 12 mm thick single and in-contact multi-layered target configurations composed of Armox 500 T and mild steel materials have been investigated against 7.62 NATO AP projectiles. The effect of projectile–target surface interaction properties on the ballistic resistance was studied by varying the tangential friction coefficient between projectile and target. Finite element analysis was carried out in ABAQUS/Explicit software. The numerical study was verified from previously published ballistic study on the same materials. JC constitutive model was used to simulate the material behavior of the target, whereas the projectile body was modeled as analytical rigid instance. The observed impact behavior of different target configurations was compared in terms of velocity drop in the projectile incident velocity, deformations, and strain energy absorption in the target configurations.

Keywords Ballistic impact · Mild steel target · Armox 500 T · Contact friction · Energy dissipation

1 Introduction

The mechanics involved in the large deformation impact problems lead to the evaluation of complex solutions using advanced numerical techniques. However, the accuracy of these numerical simulations is affected with the geometrical, material, and interaction properties of striker as well as target, resulting into inaccurate ballistic results. Previous research works have investigated the impact resistance of the target in terms of strain energy absorption in the deformation of the target during impact

Z. Mohammad
Department of Civil Engineering, ZHCET, Aligarh Muslim University, Aligarh 202002, India

J. Venkatesan (✉)
CSIR, Structural Engineering Research Center, Chennai 600113, India
e-mail: jsvenkat@serc.res.in

P. K. Gupta
Department of Civil Engineering, Indian Institute of Technology Roorkee, Roorkee 247667, India

R. Velmurugan et al. (eds.), *Dynamic Behavior of Soft and Hard Materials Volume 1*,
Springer Proceedings in Materials 34, https://doi.org/10.1007/978-981-99-6030-9_44

[1–8]. Moreover, experimental investigations of the influence of contact friction on the impact process are difficult to achieve due to very short duration of the perforation process [9].

Friction dissipation studies conducted by Krafft [10] showed that about 3% of the kinetic energy was lost due to friction between the projectile and target interface. Woodward [11] concluded that the frictional effects significantly influence the ballistic resistance of the target. Sutter and Ranc [12] found that the temperature at the interface could increase up to 1100 °C at higher impact velocities. Holmen et al. [13] and Rosenberg and Vayig [14] reported that the contact friction could reduce the residual velocity by 20%. Mohammad et al. [15] studied the effect of surface contact friction on the ballistic resistance of thin aluminum plate target against ogive nosed projectile. It was observed that the ballistic limit of the target was increased with increase in the tangential frictional coefficient values. The strain energy absorbed in frictional dissipation was about 17.7% and 33.7% at 0.05 and 0.1 friction coefficient, respectively.

Mohammad et al. [4] presented an elaborative literature review of the studies investigating the ballistic performance of single and multi-layered metallic targets impacted with different nose shaped projectiles. It was observed that the geometrical and material properties of the projectile as well as target play an important role in the impact behavior. Teng et al. [16] concluded that multi-layered targets against blunt nose projectile showed 25% higher impact resistance than single layer targets of same thickness. Zhou and Stronge [17] found that the multi-layered targets of stainless-steel material performed better than the single layer targets against blunt projectiles at inclined impact. Flores-Johnson et al. [18] investigated single and multi-layered steel targets against 7.62 APM2 projectiles and found that single layer targets exhibited higher impact resistance than double-layer targets. Senthil and Iqbal [19] showed higher ballistic resistance of monolithic targets than multi-layered aluminum targets against ogive nosed penetrators. Wei et al. [20] investigated the impact performance of single and multi-layered steel targets, against steel projectile of blunt nose shape. It was concluded that the single targets showed higher ballistic performance than the multi-layered targets. Senthil and Iqbal [21] studied the impact performance of single and three-layer configurations of mild steel, 7075-T651 aluminum, and Armox steel plate targets against 7.62 AP bullets. The layered targets were configured to observe the suitability against high as well as low strength and ductility under impact. It was determined that the best configuration against the impact would be with the front layer of low ductility and high strength and rear layer having high ductility and low strength.

Previous research studies suggest that the contact friction between projectile and target surfaces significantly influences the damage morphology and overall impact resistance of thin and thick targets. Moreover, frictional resistance in multi-layered targets was found to be more profound than the single targets due to enhanced interaction time duration in layered targets. Thus, in this study, numerical analyses have been performed to assess the effect of contact friction on the different ballistic parameters of 12 mm thick single as well as layered targets of commercial mild steel and Armox 500 T steel material against 7.62 mm AP ammunition. The observed impact behavior of different target configurations was discussed in terms of velocity drop in the incident velocity of the projectile, deformations, and kinetic energy dissipation in the targets.

2 Materials and Methods

Three-dimensional finite element modeling of 7.62 mm AP projectile, 12 mm thick single and double-layered plate targets of 6 mm individual thickness, was carried out in ABAQUS/CAE. Both target configurations of the mild steel and Armox 500 T steel square plates of 200 × 200 mm size were modeled as a deformable instance, whereas the projectile core was modeled as analytical rigid instance. The targets were then discretized using C3D8R brick elements with variable mesh size to reduce the analysis time. The mesh convergence study was also conducted and the element size in central contact region was considered to be 0.2 mm with unit aspect ratio. The element size was increased up to 5 mm in the non-contact region using transition wedge elements (C3D6R) of varying size to reduce distortion in the mesh (see Fig. 1).

Pre-defined incidence velocity was given to the projectile, and the outer periphery of plate targets was kept fixed. The surface interaction properties between projectile and target during perforation were applied using kinematic contact algorithm. To investigate the effect of contact friction on the ballistic performance, the tangential friction coefficient value was taken as 0 (frictionless), 0.01 and 0.02. However, the value of friction coefficient between the inner contact layers of the double-layered target was kept constant as 0.5, and the interaction between the layers was modeled with general contact algorithm [21].

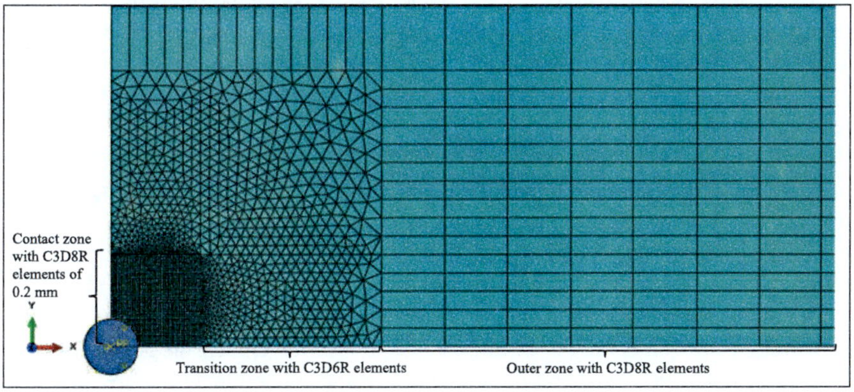

Fig. 1 Finite element model of target and projectile and element types used in mesh discretization

Table 1 Ballistic parameters obtained from experimental and numerical studies

Target material	Impact velocity (ms^{-1})	Experimental residual velocity (ms^{-1}) [21]	Numerical residual velocity(ms^{-1})
Mild steel	818.0	661.5	661.6
Armox 500 T steel	823.6	334.3	332.1

The material behavior of both target configurations was modeled using JC strength and damage material model with Hillerborg's fracture energy criterion [4]. Johnson–Cook strength model parameters for both mild steel and Armox 500 T steel materials such as Young's modulus of elasticity, Poisson's ratio, yield stress, and strain hardening parameters were obtained from uniaxial tensile tests. The remaining material damage parameters were undertaken from previously published characterization studies on the same material [21].

The validation of the numerical modeling was done by conducting impact simulations on 12 mm thick mild steel and 8 mm thick Armox steel plate targets of 200 × 200 mm size. Both targets were impacted with 7.62 mm projectile core at 818.0 ms^{-1} and 823.6 ms^{-1}, respectively [21]. The residual velocities obtained from the finite element study were then compared with the previously obtained experimental values and were found in close corroboration (see Table 1).

3 Results and Discussion

Ballistic impact simulations of monolithic as well as layered mild steel and Armox 500 T steel plate targets against 7.62 mm NATO projectiles were performed in ABAQUS/Explicit solver. The effect of contact friction between the projectile and target surface was observed on the ballistic performance of the targets. The impact parameters such as velocity drop, central deformation, damage morphology, and energy absorption in the target was ascertained and compared within the different target configurations.

Both monolithic as well as layered target models were impacted with incident velocity of 1000 ms^{-1} and the residual velocities after the perforation of the target were obtained. The percentage velocity drop in the various targets has been displayed in Fig. 2. Mild steel targets exhibited considerably less ballistic resistance in comparison with Armox steel targets. Monolithic as well as layered mild steel targets showed only 13.2 to 14.2% velocity drop.

However, Armox steel targets exhibited 59.7 to 88.8% velocity drop in the incident velocity. It was observed that the impact resistance of single and multi-layered targets of both mild steel and Armox 500 T steel was increased with increase in the contact friction coefficient. However, this increase in the impact performance was observed to more prominent in Armox steel targets as compared with mild steel

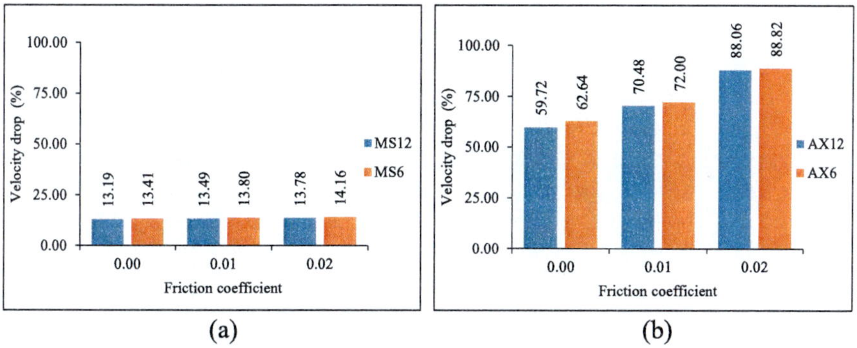

Fig. 2 Percentage velocity drop in the incident velocity in **a** single (MS12) and double-layered (MS6) mild steel targets, **b** single (AX12) and double-layered (AX6) Armox 500 T steel targets, impacted at 1000 ms^{-1}

target configuration. Moreover, the layered target configurations showed marginally higher velocity drop than the monolithic targets of same material.

The perforation in single and multi-layered targets at 1500 μs is shown in Figs. 3 and 4. It could be observed that the central deflection in the targets has increased with increase in the frictional coefficient. The layered in-contact targets showed more central deflection as compared to monolithic targets due to layered configuration. There was approximately 7.8% higher deflection observed in mild steel layered targets than their monolithic counterparts. However, layered Armox steel targets exhibited about 42.4% higher central deflection in comparison with monolithic targets of same material.

During the impact process, the local deformation in the mild steel targets was observed to have more distinct crater and petal formation on the impact and distal sides, respectively, pertaining to its more ductile flow behavior in comparison with Armox steel targets. However, the Armox steel targets showed higher deformation than the mild steel targets due to higher ballistic limit, reaching to the incident velocity of the projectile.

Previous studies have discussed the damage morphology and related the ballistic resistance capacity of the targets in terms of strain energy dissipation in different deformation modes [4]. Mohammad et al. [15] have described the energy balance in the deformation of target during impact process. The elapsed energy of the projectile in the impact processes was observed to be dissipated in elastic and plastic deformation as parts of internal work done, while the external work done in the target constitutes frictional and viscous dissipation in the target material. The energy distribution in various target configurations has been described in Fig. 5. It was observed that about 90 to 95% of the projectile kinetic energy has been dissipated in plastic deformation of the target. Moreover, it was observed to be marginally increased with increase in the value of contact friction coefficient. However, the energy absorbed in frictional dissipation was increased by 5.6% and 15.1% in mild steel and Armox steel targets, respectively. The amount of total plastic strain energy absorbed in the

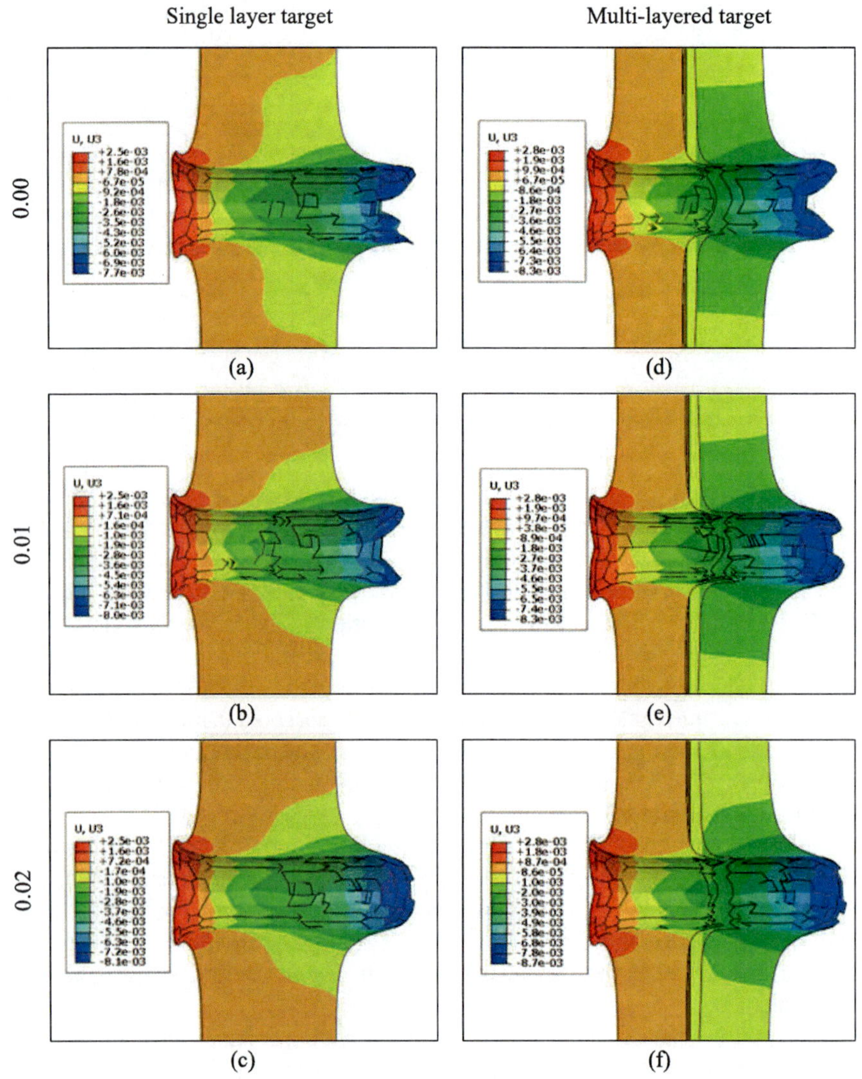

Fig. 3 Deformation in mild steel single and multi-layered targets impacted at various friction coefficients (at 1500 μs)

multi-layered targets was further divided and evaluated in the individual target layers (see Fig. 6). It was found that the rear target layer dissipated more energy than the front target layer, irrespective of the type of target material. Also, the plastic strain energy in front layer of the target was marginally reduced with increase in the friction coefficient, while the rear layer has shown the opposite behavior.

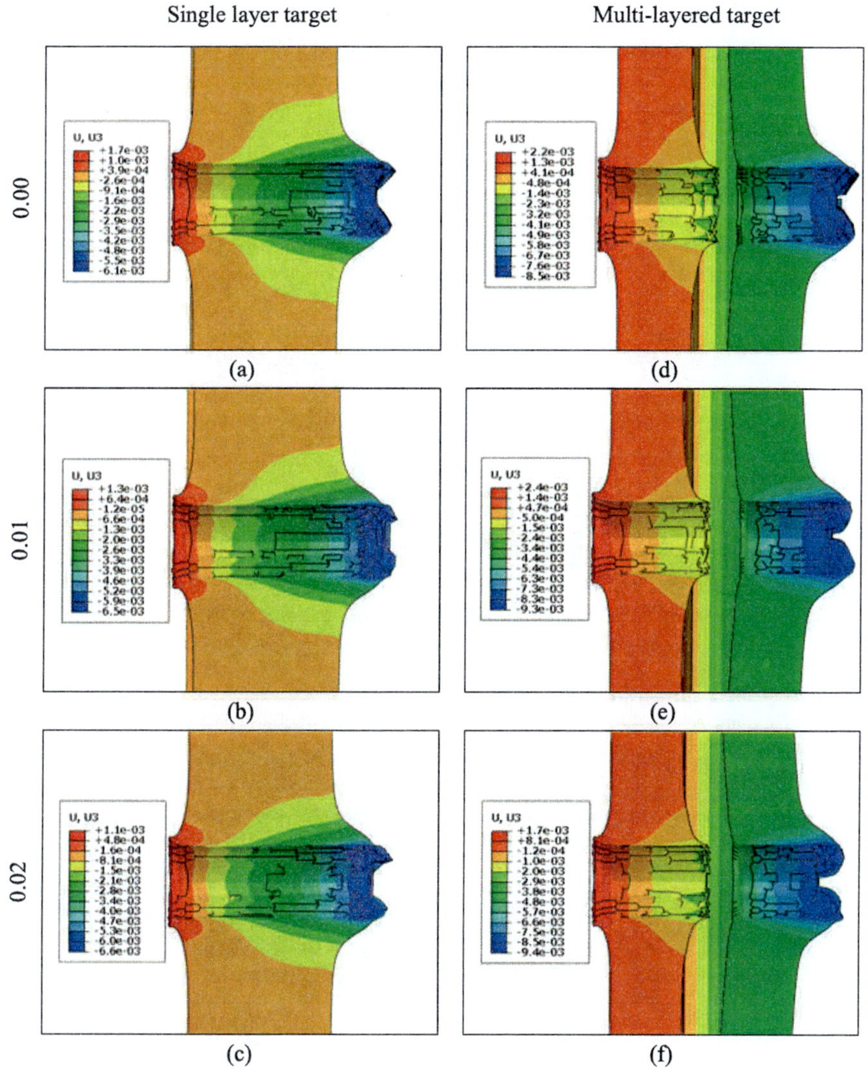

Fig. 4 Deformation in Armox 500 T steel single and multi-layered targets impacted at various friction coefficients (at 1500 μs)

4 Conclusions

The present numerical study has investigated the impact performance of 12 mm thick single and double-layered plate targets of 6 mm individual thickness against 7.62 mm AP projectiles. The effect of contact friction between the target and projectile surface with different friction coefficient values, viz. 0.00, 0.01, and 0.02 was studied on the overall impact performance. Both single as well as multi-layered target configurations

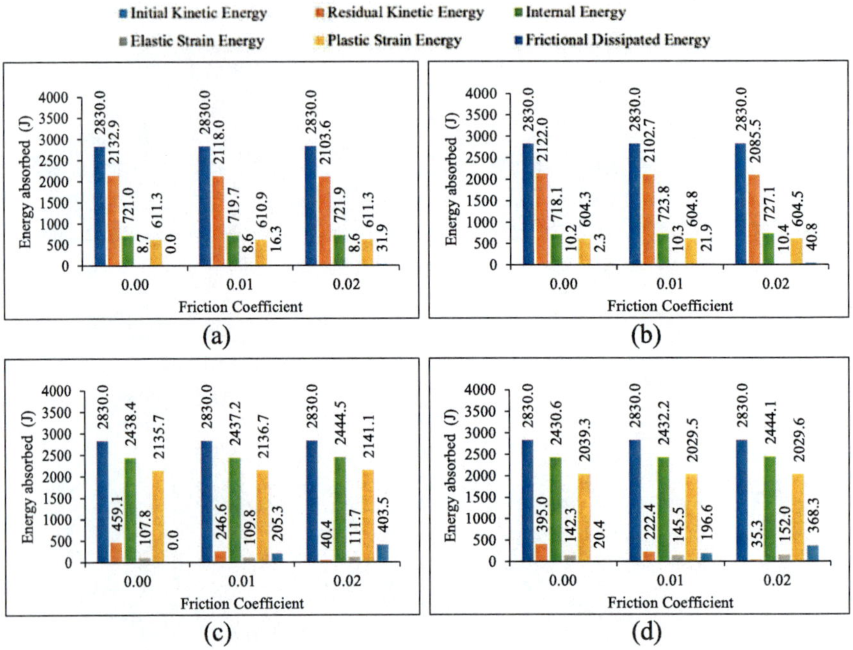

Fig. 5 Energy dissipation in **a** MS12, **b** MS6, **c** AX12, and AX6 targets impacted at incident velocity of 1000 ms^{-1} with different values of tangential friction coefficient

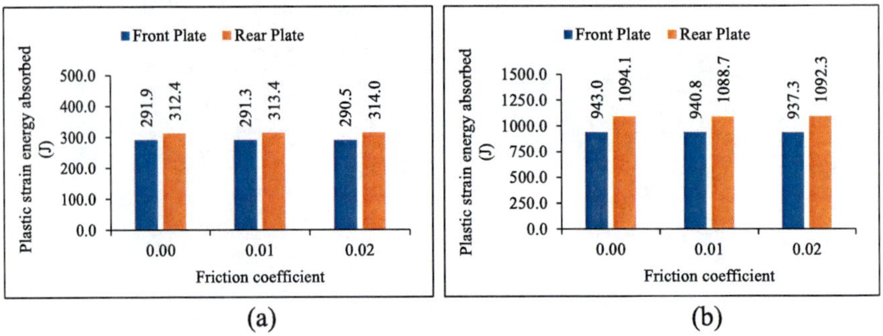

Fig. 6 Plastic strain energy dissipation in front and rear layers of **a** MS6 and **b** AX6 targets at incident velocity of 1000 ms^{-1} with different values of tangential friction coefficient

were impacted with incident velocity of 1000 ms^{-1}, and the impact resistance of the targets was discussed in terms of velocity drop, central deformation, damage morphology, and energy dissipation in the target. The ballistic resistance of both single and multi-layered targets was found to be significantly higher than the mild steel targets, and about 88.8% of velocity drop in Armox steel targets was observed.

The ballistic resistance of both mild steel and Armox 500 T steel target configurations was increased with increase in the contact friction coefficient. The layered target configurations showed marginally higher velocity drop than the monolithic targets of same material. The layered Armox steel targets exhibited about 42.4% higher central deflection in comparison with monolithic targets, whereas mild steel targets showed very insignificant difference. Further, it was observed that a large portion of kinetic energy of the projectile was absorbed in the plastic work done in the target material. Also, a marginal increase in the energy value was observed with increase in coefficient of contact friction, while the energy in frictional dissipation was found to be increased by 5.6% and 15.1% in mild steel and Armox steel targets, respectively. Moreover, the rear target layer dissipated more energy than the front target layer, irrespective of the type of target material.

References

1. Mohammad Z, Gupta PK, Iqbal MA, Baqi A (2017) Energy absorption in metallic targets subjected to oblique impact. Procedia Eng 173:145–152
2. Gupta PK, Iqbal MA, Mohammad Z (2017) Energy dissipation in plastic deformation of thin aluminium targets subjected to projectile impact. Int J Impact Eng 110:85–96
3. Gupta PK, Iqbal MA, Mohammad Z, Baqi A, Gupta NK (2018) Energy absorption in thin metallic targets subjected to oblique projectile impact: a numerical study. Thin-Walled Struct 126:58–67
4. Mohammad Z, Gupta PK, Baqi A (2020) Experimental and numerical investigations on the behavior of thin metallic plate targets subjected to ballistic impact. Int J Impact Eng 146:103717
5. Mohammad Z, Gupta PK, Baqi A, Iqbal MA (2021) Ballistic performance of monolithic and double layered thin-metallic hemispherical shells at normal and oblique impact. Thin-Walled Struct 159:107257
6. Mohammad Z, Gupta PK, Baqi A, Iqbal MA (2021) Energy dissipation characteristics of single curvature metallic shells subjected to ballistic impact. Eur J Mech A Solids 89:104279
7. Jayaraman V, Mohammad Z, Iqbal MA, Damage characteristics of semi-infinite aluminium targets subjected to normal and oblique projectile impact. Mater Today Proc (in press)
8. Khan MK, Iqbal MA (2022) Failure and fragmentation of ceramic target with varying geometric configuration under ballistic impact. Ceram Int. https://doi.org/10.1016/j.ceramint.2022.05.297
9. Ben-Dor G, Dubinski A, Elperin T (2007) Localized interaction models with non-contact friction for rigid penetrating impactors. Int J Solid Struct 44:2593–2607
10. Krafft JM (1955) Surface friction in ballistic penetration. J App Phys 26:1248–1253
11. Woodward RL (1978) The penetration of metal targets by conical projectiles. Int J Mech Sci 20:349–359
12. Sutter G, Ranc N (2010) Flash temperature measurement during dry friction at high sliding speed. Wear 268:1237–1242
13. Holmen JK, Hopperstad OS, Borvik T (2017) Influence of yield-surface on simulation of ballistic impact. Int J Impact Eng 108:136–146
14. Rosenberg Z, Vayig Y (2021) On the friction effect in the perforation of metallic plates by rigid projectiles. Int J Impact Eng 149:103794
15. Mohammad Z, Venkatesan J, Gupta PK (2022) On the contact behavior of aluminum plate targets subjected to ogive nosed projectile impact. Mater Today Proc 56:420–424. https://doi.org/10.1016/j.matpr.2022.01.390
16. Teng XQ, Dey S, Borvik T, Wierzbicki T (2007) Protection performance of double-layered metal shields against projectile impact. J Mech Mater Struct 2:1309–1330

17. Zhou DW, Stronge WJ (2008) Ballistic limit for oblique impact of thin sandwich panels and spaced plates. Int J Impact Eng 35:1339–1354
18. Flores-Johnson EA, Saleh M, Edwards L (2011), Ballistic performance of multi-layered metallic plates impacted by a 7.62-mm APM2 projectile. Int J Impact Eng 38:1022–1032
19. Senthil K, Iqbal MA (2013) Effect of projectile diameter on ballistic resistance and failure mechanism of single and layered aluminum plates. Theoret Appl Fract Mech 67:53–54
20. Wei Z, Yunfei D, Sheng CZ, Gang W (2012) Experimental investigation on the ballistic performance of monolithic and layered metal plates subjected to impact by blunt rigid projectiles. Int J Impact Eng 49:115–129
21. Senthil K, Iqbal MA (2021) Prediction of superior target layer configuration of armour steel, mild steel and aluminium 7075-T651 alloy against 7.62 AP projectile. Structures 29:2106–2119

Effect of Time Step Scale Factor Value on the Low Velocity Impact Numerical Simulation Results in LS-DYNA

Mahesh, Shivank Kumar, Kalyan Kumar Singh, and Prashant Rawat

Abstract The major focus of this investigation is to analyze the effect of the time step scale factor (TSSFAC) parameter value on the numerical simulation results obtained by LS-DYNA for glass fiber reinforced polymer (GFRP) composite laminates subjected to low velocity impact. The TSSFAC is a parameter in the CONTROL_TIMESTEP keycard. The numerical simulations are conducted for two impact velocities 1.5 m/s and 4 m/s. At 1.5 m/s impact velocity, the impactor rebounded, whereas at an impact velocity 4 m/s, the impactor perforated the GFRP composite laminate. The TSSFAC values considered for the numerical simulation are 0.001, 0.01, 0.1, 0.5, 0.6, 0.7, and 0.9. The TSSFAC values affected both numerical simulation accuracy and numerical simulation time. However, the extent of influence of TSSFAC is high for simulation time compared to the numerical simulation accuracy. The simulation time increased as the TSSFAC value decreased. It was found that the TSSFAC value 0.6 yielded minimum simulation time compared to other TSSFAC values. Further, the best correlation between experimental and simulation results is found for 0.6 TSSFAC value.

Keywords GFRP · LVI · Numerical simulation · LS-DYNA · TSSFAC

1 Introduction

Fiber reinforced polymer (FRP) composite laminates show high strength and stiffness along with good resistance to fatigue loading and chemical attacks. Further, the strength properties of FRP composite laminates can be manipulated according to the required direction. Due to these advantages, FRP composite laminates are extensively used in various industries such as aerospace, marine, civil construction,

Mahesh · K. K. Singh
Indian Institute of Technology (Indian School of Mines), Dhanbad 826004, India

S. Kumar · P. Rawat (✉)
Indian Institute of Technology Madras, Chennai 600036, India
e-mail: prashant.rawat@iitm.ac.in

R. Velmurugan et al. (eds.), *Dynamic Behavior of Soft and Hard Materials Volume 1*,
Springer Proceedings in Materials 34, https://doi.org/10.1007/978-981-99-6030-9_45

and automobiles as primary and secondary structural members [1]. The primary limitation of the FRP composite laminates is their weak strength properties across thickness direction. Due to this, the FRP composite laminates become highly susceptible to impact loading. If the impact loading is low, then the vulnerability of FRP composite laminates further increases because the low velocity impact (LVI) creates barely visible impact damage (BVID) [2].

The barely visible impact damage is not visible to the naked eye and hard to detect during a non-destructive inspection. Furthermore, the BVID significantly reduces the residual strength of the FRP composite laminates, particularly the compressive strength. The severe reduction in the strength properties led to the sudden and catastrophic failure of FRP composite structures. Thus, understanding the behavior of FRP composite laminates under LVI is necessary. However, understanding the impact behavior of FRP composite laminates is quite difficult because the impact loading occurs for a very short time and creates a three-dimensional stress state in FRP composite laminates. The compressive stresses are generated at impacting face side and tensile stresses are generated at the non-impacting face side. Moreover, the FRP composite laminates are heterogeneous and anisotropic, which further adds to the complexity of the stress distribution in the FRP composite laminates under impact loading. Thus, finite element numerical simulation is vital in establishing the damage behavior and stress distribution in FRP composite laminates under LVI [3].

Rawat et al. [4–7] conducted a series of numerical simulations on GFRP composite laminates subjected to LVI. Rawat et al. [4] used the LS-DYNA to study the behavior of GFRP composite laminates under oblique impact. Numerical simulation was conducted for four oblique impact angles 0°, 15°, 30°, and 45°. In another work, Rawat et al. [5] conducted the LVI numerical simulation on GFRP composite laminates for different impact energies. Rawat et al. [6] studied the influence of impactor shape on the damaged area in GFRP composite laminates under LVI. Four different impactor nose shapes hemispherical, spherical, ogival, and flat were considered for the numerical simulation. Rawat et al. [7] used the LS-DYNA to perform the numerical simulation on symmetric and asymmetric GFRP composite laminates with pre-crack subjected to LVI.

In a series of numerical simulation works, Mahesh et al. [8–10] used the LS-DYNA to carry out the LVI numerical simulation on GFRP composite laminates. Mahesh et al. [8] conducted the in-plane LVI numerical simulation on the GFRP composite laminates. In another study, Mahesh et al. [9] conducted a numerical simulation using LS-DYNA to study the influence of the area available within the circular boundary condition on the damage formation when GFRP laminates were subjected to LVI. In another work, Mahesh et al. [10] examined the effect of different edge constraining boundary conditions on the damage formation in GFRP composite laminates under LVI.

Menna et al. [11] conducted LVI numerical simulation on GFRP composite laminates. The MAT_059 material model was used to model the GFRP composite plate. It was observed that the numerical simulation results were sensitive to the through thickness, Young's modulus, shear moduli, and tensile strength values. However, the numerical simulation results were not affected by the variation in compressive strength values. Fragrassa et al. [12] conducted the parametric analysis for the material model MAT_054 to achieve accurate correlation at a minimum cost. For parametric optimization ALPH (Coefficient in stress–strain behavior of material along the longitudinal, transverse shear direction), BETA (Tensile fiber mode coefficient in Chang–Chang failure model), SOFT (element softening), FBRT (Fiber tensile strength reduction factor), and YCFAC (Reduction factor of compressive strength) parameters were considered. The ALPH value was varied from 0 to 1. The BETA value considered was 0 and 1. If the BETA value was equal to 0, then the shear stress elements were eliminated; whereas if the BETA value was equal to 1, then the maximum shear stress criterion was considered. The SOFT values varied from 0 to 1. The FBRT value was dependent on the tensile strength of the laminate. It was concluded that explicit parameters were important in establishing a good numerical simulation. The importance of these explicit parameter values increased as the number of layers in the laminate increased. From the numerical simulation investigation, following observations were made (i) In the 2-shell model, the BETA and FBRT parameters were important; and (ii) In the 4-shell mode, the ALPH value should be minimum, the BETA value should be intermediate, and the SOFT and FBRT should be high.

Many review works available in the literature which highlight the importance of numerical simulation of LVI on the FRP composite laminates [13–16]. From the literature survey, it is found that most of the numerical simulation investigations focused on the optimization of primary parameter values such as material models or on developing the new materials models. To the best of author's knowledge, no work provides detailed information about the secondary parameters associated with different keycards in the LS-DYNA. Thus, this numerical simulation investigation tries to fill this research gap. This numerical investigation aims to determine the optimum time step scale factor (TSSFAC) value for better numerical simulation results. The numerical simulation is conducted using the LS-DYNA commercially available finite element simulation software. The obtained numerical simulation results were validated with experimental results from the literature [17].

2 Numerical Modeling

The numerical simulation of GFRP composite laminates under LVI impact is conducted using LS-DYNA finite element simulation software for two impact velocities of 1.5 m/s and 4 m/s. Various material cards and keycards from the LS-DYNA are required to conduct the numerical simulation of LVI on GFRP composite laminates. Figure 1 shows the numerical simulation model of the GFRP composite plate and impactor. Solid elements are used to model both GFRP composite plate and impactor. Solid elements can sustain more distortion and damage. Further, they provide stress distribution in the thickness direction and accurate failure behavior. Thus, the impactor and GFRP plates are modeled using solid elements. The GFRP composite plate and impactor are meshed using the 4-node quadratic elements. The

MAT_020 (MAT_RIGID) material card is used to model the impactor. The MAT_020 considers the impactor as a rigid body with negligible deformation. The rigid body parts are bypassed from element processing, thus becoming a cost-effective approach.

The MAT_059 (MAT_COMPOSITE_FAILURE_SOLID_MODEL) material card is used to model the GFRP composite laminate. The MAT_059 material card functions on the modified Hashin failure criteria. Both ply axes and orientations are defined using AOPT and BETA parameters in the MAT_059 material card. The MAT_059 model is an enhanced version of MAT_022 and MAT_054/55. This material model considers the failure criteria in longitudinal and transverse directions under tensile and compression. Further, the model also considers the failure criteria in the laminate through thickness direction under compression and shear [18]. The contact between the GFRP composite plate and impactor is defined by the two way contact definition using the keycard CONTACT_AUTOMATIC_SURFACE_TO_SURFACE. The importance of defining the slave and master parts is not necessary. The contact among the plies is defined by the tied contact definition using the keycard CONTACT_TIED_SURFACE_TO_SURFACE. Here, the slave nodes move along

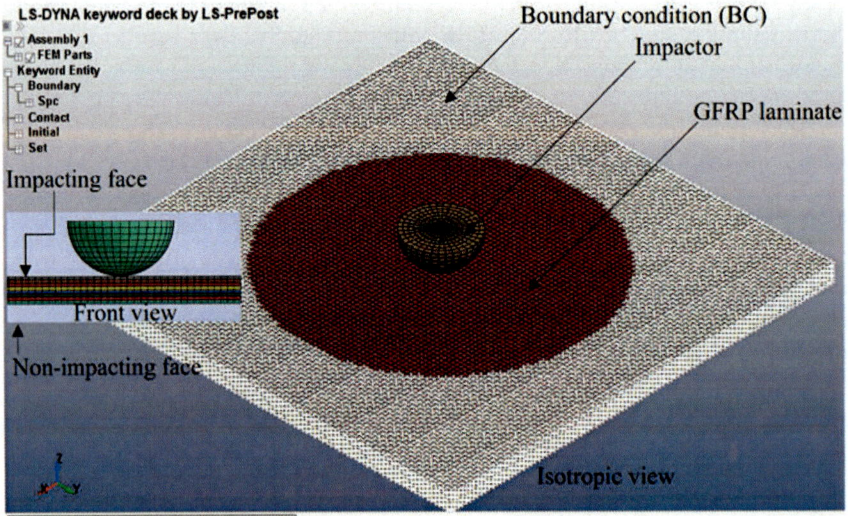

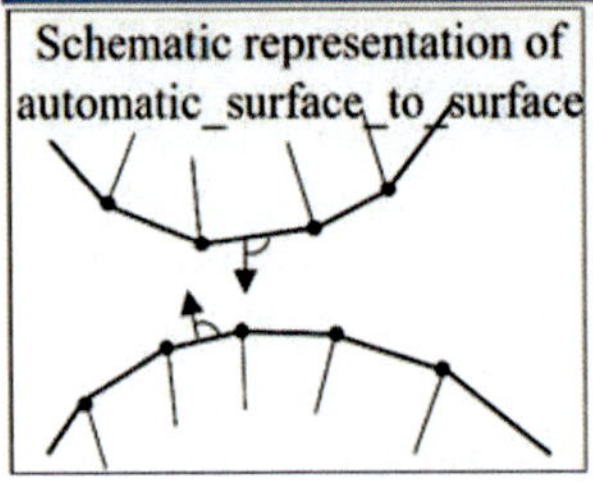

Part	Degree of Freedom					
	Translation constraint in direction			Rotational constraint about axis		
	X	Y	Z	X	Y	Z
Impactor	×	×	✓	×	×	×
BC	×	×	×	×	×	×

Fig. 1 Numerical simulation model of GFRP composite laminate and impactor

the master nodes. In this type of contact definition, the rotational degrees of freedom of the slave nodes are not constrained.

The CONTROL_TIMESTEP is used for the mass scaling of elements in LS-DYNA. The mass scaling of elements helps in achieving the larger time step. The mass scaling approach adds a non-physical mass to the structural elements during the numerical simulation. According to Eqs. 1 and 2, the time step value depends on the element size and the material density. Here, Eq. 1 is called the Courant–Freidrichs condition. The time step determines the total run time of the simulation. As the time step increases, the simulation time reduces and vice-versa. However, as the time step increases, the accuracy of numerical simulation reduces. Further, the choice of time step depends on the type of numerical simulation and the accuracy of numerical simulation results. In CONTROL_TIMESTEP, there is a parameter called time step scale factor (TSSFAC). The TSSFAC calculates the incremental time step (Eq. 1). The global time step is calculated using following equations

$$\Delta t = \text{scaling factor} \times \frac{e_{\min}}{c} \tag{1}$$

$$c = \sqrt{\frac{E}{\rho}} \tag{2}$$

$$\Delta t = \text{TSSFAC} \times 2\sqrt{\frac{2m_1 m_2}{(m_1 + m_2)k}} \tag{3}$$

$$\Delta t = \text{TSSFAC} \times \min(\Delta t_1, \ldots, \Delta t_n) \tag{4}$$

Δt—Time step,
k—Stiffness,
m—Spring mass,
$e_{\min}$—Smallest mesh element size,
c—Speed of the stress wave in the material,
E—Young's modulus of the material,
ρ—Density of the material.

3 Results and Discussion

Figures 2 and 3 show the force vs. time curve of GFRP composite laminates subjected to impact velocities 1.5 m/s and 4 m/s for different TSSFAC values. The nature of the force vs. time curve changed as the impact velocity increased from 1.5 m/s to 4 m/s. At an impact velocity of 1.5 m/s, fewer fluctuations on force vs. time curve are observed for GFRP composite laminate under impact velocity 1.5 m/s. However, the intensity of these fluctuations on force versus time curve increased for GFRP composite laminate impacted at 4 m/s. Both numerical and simulation results showed similar trends of force versus time curve.

It is evident from Figs. 2 and 3 that the numerical simulation results changed as the TSSFAC value changed. The formation of the maximum peak load (Fmax) and damage threshold force (Fd) are changed as the TSSFAC value changes. At an impact velocity of 1.5 m/s, the maximum peak load for TSSFAC values 0.9, 0.7, 0.6, 0.5, 0.1, and 0.01 are 5773 N, 5586 N, 5528 N, 5782 N, 5589 N, and 4614 N,

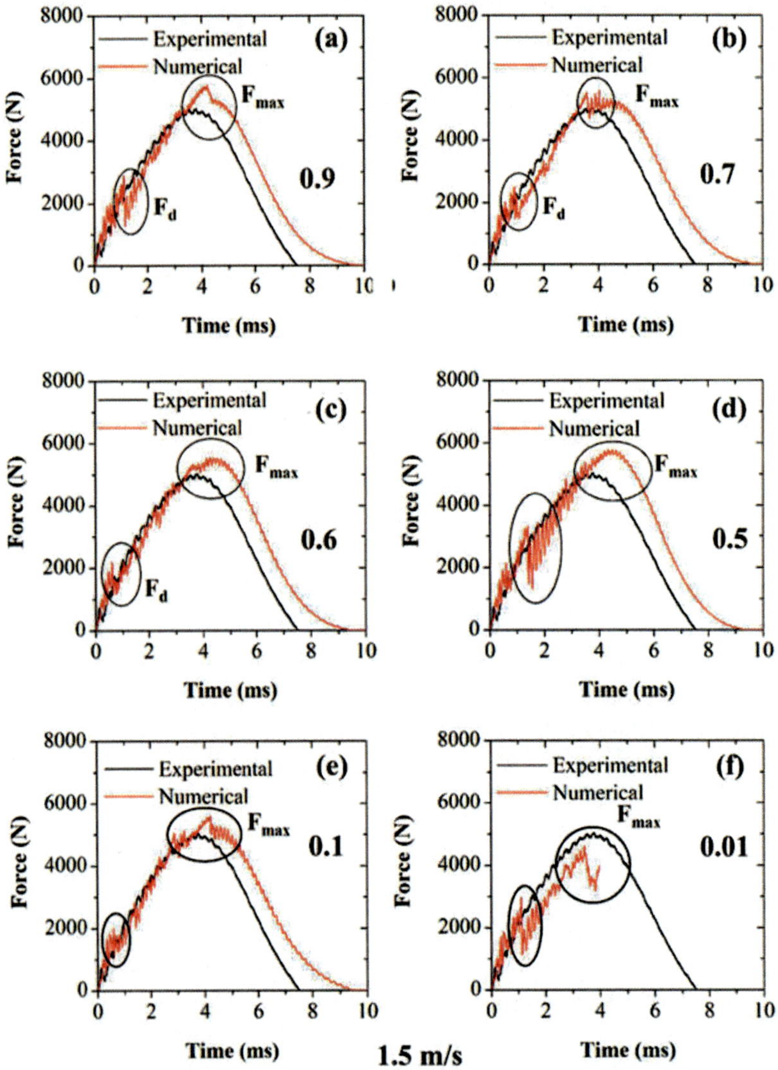

Fig. 2 Force versus time curves of GFRP composite laminates subjected to LVI at impact velocity 1.5 m/s for different TSSFAC values **a** 0.9, **b** 0.7, **c** 0.6, **d** 0.5, **e** 0.1, and **f** 0.01

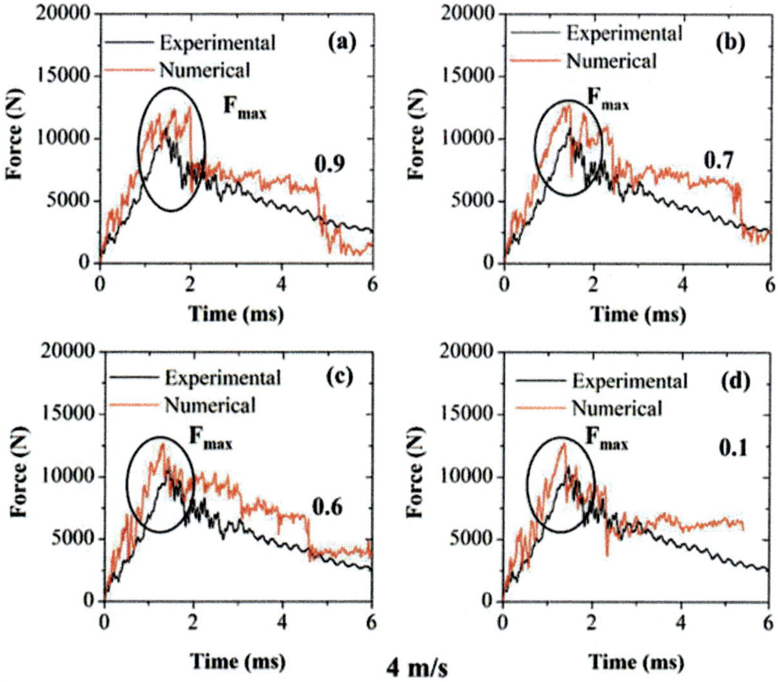

Fig. 3 Force versus time curves of GFRP composite laminates subjected to LVI at impact velocity 4 m/s for different TSSFAC values **a** 0.9, **b** 0.7, **c** 0.6, and **d** 0.1

respectively. At impact velocity 1.5 m/s, the corresponding time at maximum peak force for TSSFAC values 0.9, 0.7, 0.6, 0.5, 0.1, and 0.01 are 4.21 ms, 4.0399 ms, 4.4 ms, 4.49 ms, 4.25 ms, and 3.45 ms, respectively. At impact velocity 4 m/s, the corresponding time at maximum peak load for TSSFAC values 0.9, 0.7, 0.6, and 0.1 are 1.33 ms, 1.44 ms, 1.31 ms, and 1.37 ms, respectively. The damage threshold force for TSSFAC values 0.9, 0.7, 0.6, and 0.5 is 2890 N, 2484 N, 2178 N, and 3305 N, respectively. Similarly, the occurrence time of damage threshold force is 1.11 ms, 0.9399 ms, 0.6 ms, and 0.39, respectively.

The maximum peak force is considered for the comparison between experimental and numerical results. Thus, at impact velocity 1.5 m/s, the numerical simulation results showed 13%, 10%, 9%, 13%, and 10% more maximum peak load than the experimental maximum peak load for TSSFAC values 0.9, 0.7, 0.6, 0.5, and 0.1. However, for TSSFAC value 0.01, the numerical simulation maximum peak load is 9% less compared to the experimental peak load. At impact velocity 4 m/s, the numerical simulation maximum peak load value is 13%, 14%, 14%, and 14% more compared to the experimental results for TSSFAC values 0.9, 0.7, 0.6, and 0.1, respectively.

When closely observed, the correlation between numerical simulation and experimental results is highest for TSSFAC value 0.6 for both the impact velocities 1.5 m/

Table 1 Computational time at different TSSFAC value for GFRP composite laminates subjected to 1.5 m/s and 4 m/s impact velocities

TSSFAC value	Computational time (hrs)	
	At impact velocity 1.5 m/s	At impact velocity 4 m/s
0.001	4824.35	8683.85
0.01	587	936.81
0.1	77.95	120.2
0.5	22.23	51.5
0.6	14.21	36.71
0.7	7.83	22.9
0.9	7.43	13.35

s and 4 m/s. At impact velocity 1.5 m/s, the fluctuations on the force vs. time curve matched well with the experimental force vs. time for both the loading and unloading region of the force vs. time curve. A similar trend is observed for impact velocity 4 m/s. The rising region of the force vs. time curve is called the loading region, while the descending region of the force vs. time curve is called the unloading region.

Table 1 shows the computational time for impact velocities of 1.5 m/s and 4 m/s for different values of TSSFAC. Table 1 shows the dependency of simulation time on TSSFAC values. The simulation time increased as the TSSFAC values decreased and vice-versa. However, it is conclusive that the decrease in TSSFAC value may not show a better correlation with the experimental results. Further, neither increase in the TSSFAC value yield good correlation with the experimental results. A good correlation between the numerical and experimental is observed for TSSFAC value 0.6, with an error of 9% and 14% for the maximum peak load at impact velocity 1.5 m/s and 4 m/s, respectively. It is found that for TSSFAC values 0.9, 0.7, 0.6, and 0.5, the force vs. time curves showed the least fluctuations compared to the TSSFAC value of 0.5.

4 Conclusion

In this investigation, the numerical simulation of low velocity impact on GFRP composite laminates is conducted using the LS-DYNA finite element software. The major focus of this study is to analyze the influence of time step scale factor (TSSFAC) value on the numerical simulation results. The TSSFAC parameter value in *CONTROL_TIMESTEP* keycard is varied between 0.9 and 0.01. From the numerical simulation results, following conclusive points can be drawn.

- Both numerical simulation results and total simulation time are affected by the TSSFAC value.

- As the TSSFAC value decreased, the numerical simulation time increased or vice-versa. However, neither low nor high TSSFAC value assured a good correlation with the experimental results.
- The TSSFAC value affected the maximum peak load, time corresponding to the maximum peak load and damage threshold load. It also affected the nature of ascending and descending region of the force vs. time curve.
- For impact velocities 1.5 m/s and 4 m/s, a good correlation between numerical simulation results and experimental results in terms of maximum peak load, fluctuations, time corresponding to maximum peak load, and closeness of force vs. time curve is observed for the TSSFAC value 0.6.

References

1. Singh KK, Shindhe M (2022) Introduction to fibre reinforced polymer composite laminates. In: Singh KK, Shinde M (eds) Impact behavior of fibre reinforced laminates: fundamentals of low velocity impact and related literature on FRP. Springer Singapore, Singapore, pp 1–43. https://doi.org/10.1007/978-981-16-9439-4_1
2. Singh KK, Shindhe M (2022) Low velocity impact on fibre reinforced polymer composite laminates. In: Impact behavior of fibre reinforced laminates: fundamentals of low velocity impact and related literature on FRP. Springer Singapore, Singapore, pp 83–105. https://doi.org/10.1007/978-981-16-9439-4_3
3. Singh KK, Shindhe M (2022) Numerical analysis of low velocity impact and compression after impact on fibre reinforced composite laminates. In: Singh KK, Shinde M (eds) Impact behavior of fibre reinforced laminates: fundamentals of low velocity impact and related literature on FRP. Springer Singapore, Singapore, pp 265–302. https://doi.org/10.1007/978-981-16-9439-4_9
4. Rawat P, Singh NK, Singh KK, Agrhari N (2019) Influence of oblique impact on glass fiber-reinforced polymer composites: a numerical approach. In: Singh I, Bajpai PK, Panwar K (eds) Trends in materials engineering. Springer Singapore, Singapore, pp 77–86. https://doi.org/10.1007/978-981-13-9016-6_9
5. Rawat P, Singh NK, Singh KK, Kumar A (n.d.) Experimental and numerical approach to investigate damage tolerance in FRP composites subjected to transverse low-velocity impact
6. Rawat P, Singh KK, Singh NK (2017) Numerical investigation of damage area due to different shape of impactors at low velocity impact of GFRP laminate. Mater Today Proc 4:8731–8738. https://doi.org/10.1016/j.matpr.2017.07.222
7. Rawat P, Singh KK, Singh NK (2017) Numerical investigation of low-velocity impact in symmetric and asymmetric GFRP laminate with and without pre-crack. Adv Mater Proc 2:152–155. https://doi.org/10.5185/amp.2017/304
8. Mahesh, Singh KK (2020) In-plane low velocity impact behavior of GFRP laminate. In: Materials science forum. Trans Tech Publication, pp 257–263. https://doi.org/10.4028/www.scientific.net/MSF.978.257
9. Mahesh, Singh KK (2019) Influence of varying fully constrained circular boundary condition area on damage mechanism of GFRP laminate under low velocity impact loading. In: Narayanan RG, Joshi SN, Dixit US (eds) Advances in computational methods in manufacturing. Springer Singapore, Singapore, pp 1011–1021. https://doi.org/10.1007/978-981-32-9072-3_85
10. Mahesh, Singh KK (2019) Numerical simulation of GFRP laminate under low-velocity impact at different edge-constrained boundary conditions. In: Singh I, Bajpai PK, Panwar K (eds) Trends in materials engineering. Springer Singapore, Singapore, pp 87–95. https://doi.org/10.1007/978-981-13-9016-6_10

11. Menna C, Asprone D, Caprino G, Lopresto V, Prota A (2011) Numerical simulation of impact tests on GFRP composite laminates. Int J Impact Eng 38:677–685. https://doi.org/10.1016/J.IJIMPENG.2011.03.003
12. Fragassa C, Vannucchi de Camargo F, Pavlovic A, Minak G (2019) Explicit numerical modeling assessment of basalt reinforced composites for low-velocity impact. Compos Part B Eng 163:522–535. https://doi.org/10.1016/j.compositesb.2019.01.013
13. Andrew JJ, Srinivasan SM, Arockiarajan A, Dhakal HN (2019) Parameters influencing the impact response of fiber-reinforced polymer matrix composite materials: a critical review. Compos Struct 224:111007. https://doi.org/10.1016/j.compstruct.2019.111007
14. Bogenfeld R, Kreikemeier J, Wille T (2018) Review and benchmark study on the analysis of low-velocity impact on composite laminates. Eng Fail Anal 86:72–99. https://doi.org/10.1016/j.engfailanal.2017.12.019
15. Rawat P, Singh KK (2017) An impact behavior analysis of CNT-based fiber reinforced composites validated by LS-DYNA: a review. Polym Compos 38:175–184. https://doi.org/10.1002/pc.23573
16. Singh NK, Singh KK (2015) Review on impact analysis of FRP composites validated by LS-DYNA. Polym Compos 36:1786–1798. https://doi.org/10.1002/pc.23064
17. Singh KK (2022) Mahesh, Effect of ply position switching in quasi-isotropic glass fibre reinforced polymer composite subjected to low velocity impact. Int J Damage Mech 31:665–693. https://doi.org/10.1177/10567895211068176
18. Rabiee A, Ghasemnejad H (2022) Finite element modelling approach for progressive crushing of composite tubular absorbers in LS-DYNA: review and findings. J Compos Sci 6. https://doi.org/10.3390/jcs6010011

Design and Fabrication of a Single-Stage Gas Gun

Dhruv Narayan and Naresh Bhatnagar

Abstract Designing a gas gun system which could accelerate a projectile without using explosives is a challenge. There are no design handbooks available for this purpose. As it is a multidisciplinary activity, it requires knowledge of gas dynamics, machine design, material science, production engineering, etc. Manufacturability of the components also needs to be considered during the design process. This article is a walk through the design process of a single-stage gas gun focusing on various parameters of importance. Starting with the basic construction, mechanism and principal of working are discussed in detail. The effect of various design parameters on the performance of the gas gun is assessed analytically. Starting with an almost unrealistic ideal design, an approach is made toward the real design by gradually relaxing the assumptions. The results of analysis of the final design are correlated with the performance of an actual gas gun designed and fabricated at CoE, IIT Delhi. Description of this actual gas gun is also included.

Keywords Single-stage gas gun · Dual-diaphragm mechanism · Gas dynamics · Rarefaction · Projectile

1 Introduction

A gas gun is a device which can accelerate projectiles at speeds required for various scientific experiments to simulate fast-moving projectiles such as a bullet, fragments from a blast, and space debris.

A single-stage gas gun uses the potential energy of a compressed gas to accelerate the projectile. It has only one stage of energy transfer, i.e., from compressed gas to the projectile. Single-stage gas guns cannot fire projectiles with muzzle velocity more than 1500 m/s due to thermodynamic irreversibility and limited strength of pressurized chambers [1].

D. Narayan · N. Bhatnagar (✉)
Department of Mechanical Engineering, IIT Delhi, Delhi, India
e-mail: nareshb@mech.iitd.ac.in

R. Velmurugan et al. (eds.), *Dynamic Behavior of Soft and Hard Materials Volume 1*, Springer Proceedings in Materials 34, https://doi.org/10.1007/978-981-99-6030-9_46

Two-stage gas guns make use of an explosive charge. In the first stage, gas is compressed in a converging chamber using an explosive charge. In the second stage, potential energy of this compressed gas is transferred to the projectile. To achieve even higher muzzle speeds, multiple stages of gas compression could be used.

As the usage of an explosive charge is not allowed in academic institutions, single-stage gas gun is the only option available for academic research.

2 Literature Review

Woods [2] presented an approach to solve the Lagrange's ballistic problem. This approach considers gas pressure on both front and rear sides of the Projectile. VanderKlok et al. [3] made analysis of a gas gun with polytropic expansion assumption. Gupta et al. [4] have designed a dual-diaphragm rupture mechanism for a gas gun being used at BARC, India. Fowles et al. [5] noticed that if the ratio of mass of gas in the pressurized chamber to the mass of the projectile exceeds five, there is no significant increase in the performance of the gas gun. The research by Bourne [6] and Bourne and Stevens[7] is focused on designing of gas guns based on the theory articulated by Siegel [1].

3 Construction and Working

There are three major components in a single-stage gas gun—the gun-breech or pressurized chamber, the firing mechanism, and the barrel. The firing mechanism is installed between the pressurized chamber and the barrel. It is must to prevent any leakage of gas from the pressurized chamber.

Projectile under consideration, which is ~200 microns smaller in diameter than the bore of the barrel, is placed in the barrel. When the pressure in the gun-breech attains the required value for the given experiment, the firing mechanism is activated. Therefore, the compressed gas expands in the barrel. As a result, the projectile gets pushed forward.

3.1 Firing Mechanisms

As the projectile passes through the barrel, the compressed gas has to be released quickly, so that the maximum possible pressure acts on the projectile. Various mechanisms are used for rapid release of the gas.

3.1.1 Quick Acting Solenoid Valves

Solenoid valve could be used as a firing mechanism provided it has following characteristics-

- Low mechanical actuation time (microseconds),
- High operating pressure,
- Compatibility with the working gas.

3.1.2 Specialized Valves

Some specially designed valves are also being used as firing mechanisms as the one used in the gas guns supplied by Thiot Ingénierie, France [8].

3.1.3 Diaphragms

Diaphragm rupture is a cost effective and zero maintenance mechanism for releasing the compressed gas. Diaphragm of a given thickness, chosen as per the required firing pressure, ruptures and releases the compressed gas almost instantaneously as soon as the pressure inside the gun-breech reaches the desired value. For a better control, dual-diaphragm rupture mechanism is used which is further discussed in this article.

3.2 Assumptions

The following discussion starts with several assumptions to analyze the system in ideal conditions for maximum theoretical performance. The assumptions are relaxed progressively to direct the analysis toward real conditions.

i. Ideal gas assumption,
ii. Insulated barrel,
iii. Frictionless barrel,
iv. Presence of vacuum downstream,
v. Gun-breech and barrel have the same internal diameter,
vi. Effectively infinite breech length,
vii. Uniform pressure, temperature and density of gas:
    a. Polytropic expansion,
    b. Isentropic expansion,
    c. Constant driving pressure.

Other than above assumptions, the parameters given in Table 1 are also fixed based upon feasibility of the experiments and availability of the components.

Table 1 Parameters for the gas gun under consideration

Parameter	Value	Parameter	Value
Barrel length (L)	1200 mm	Chamber length (Lc)	400 mm
Barrel external diameter	130 mm	Chamber external diameter	88 mm
Barrel internal diameter (D1)	30 mm	Chamber internal diameter (D0)	60 mm
Projectile mass (m)	20 g	Working gas	He
Initial gun-breech pressure	0–100 bar	Initial gas temperature	300 K
Volume of compressed gas	1000 mL	Gas constant (R) in J/Kg-K	2077
Maximum projectile speed	500 m/s	Ratio of specific heats (γ)	1.66

4 Models for Prediction of Projectile Speed

The projectile speed could be predicted through various approaches. A summary of the approaches followed in this article is given in Table 2.

Table 2 A summary of assumptions made to analyze the behavior of a single-stage gas gun

Assumption	Actual	Analysis-I	Analysis-II	Analysis-III	Analysis-IV
Ideal gas	No	Yes	Yes	Yes	Yes
Insulated barrel	No	Yes	Yes	Yes	Yes
Frictionless barrel	No	Yes	Yes	Yes	Yes
Evacuated barrel	No	Yes	Yes	Yes	No
Ratio $\frac{D_1}{D_0}$	0.5	NA	1	(A,B) 1 (C,D) 0.5	0.5
Effectively infinite L_c	No	NA	Yes	(A,C)Yes (B,D) No	Yes
Volume of gun-breech	1000 mL	NA	1000 mL	NA	NA
Pressure behind projectile	Depends upon diaphragm rupture	Constant pressure	Uniform pressure (A) Isentropic expansion (B) Polytropic expansion	Non-uniform breech pressure, isentropic expansion	Non-uniform breech pressure, isentropic expansion

4.1 *Constant Driving Pressure*

This is the most ideal condition for a gas gun. The speed of projectile propelled by a constant driving pressure (Analysis-I) could be given as

$$u_p = \sqrt{\frac{2P_0AL}{m}}. \tag{1}$$

A detailed discussion and derivation of Eq. (1) is given in A.1.1. Although this analysis assumes the driving pressure to be constant; in practice, the driving pressure decreases as the projectile travels downstream as a result of which the acceleration of the projectile also decreases.

Therefore, this analysis overestimates the muzzle velocity for a single-stage gas gun. However, had this system been a propellant-based rifle, Eq. (1) would have been able to predict the velocity with better accuracy, e.g., for an AK-47 assault rifle, considering $P_0 = 3400$ bar [9], $D_1 = 7.62$ mm, $L = 40$ cm, and $m = 7.9$ g, velocity comes out to be 886.04 m/s.

4.2 *Thermodynamic Analysis*

In this section, state properties of the gas such that temperature, pressure, and density are assumed to be uniform throughout the volume, i.e., at a given instant of time, properties do not vary with spatial location. This assumption considers that during the expansion process, the compressed gas has enough time to redistribute the pressure. In order to improve the accuracy of the model, the assumption of uniform pressure and density is not considered in analysis given in Sect. 4.3.

4.2.1 Isentropic Expansion

In analysis II-A, the compressed gas behind the projectile is assumed to undergo isentropic expansion. The velocity of the projectile for such a gas gun could be given by-

$$u_p = \sqrt{\frac{2}{m} \cdot \frac{P_0V_0}{\gamma - 1} \cdot \left[1 - \left(\frac{V_0}{V_1}\right)^{\gamma-1}\right]} \tag{2}$$

A derivation of Eq. (2) is given in A.1.2 using Newton's second law of motion.

4.2.2 Polytropic Expansion

An analysis (Analysis II-B) similar to the one mentioned in A.1.2, was carried out for the polytropic expansion assumption. The predicted projectile speed is:

$$u_p = \sqrt{\frac{2}{m} \cdot \frac{P_0 V_0}{\eta - 1} \cdot \left[1 - \left(\frac{V_0}{V_1}\right)^{\eta-1}\right]} \tag{3}$$

The value of polytropic efficiency (η) is 1 for isotherms. For isentropes, η equals to γ. Value of η is taken to be 1.26 from the analysis by VanderKlok et al. [3].

4.2.3 Non-insulated Barrel

Heat transfer to the system reduces the polytropic efficiency of the expansion process. Interestingly, if the initial temperature of the gas is equal to the ambient temperature, heat flows inside the barrel during the expansion of gas. It reduces the pressure loss for a unit displacement of the projectile, thereby increasing the acceleration. Therefore, if the gas gun is kept at room temperature, the more be the thermal conductance of barrel and gun-breech, the more be the muzzle velocity, keeping all other parameters constant.

4.3 Wave Physics Inside the Barrel and Pressurized Gun-Breech

A limitation of the thermodynamic analysis is that it does not consider molecular weight of the driving gas. However, as per literature evidence, gases with lower molecular weight are more effective in gas guns [1]. Therefore, it is important to understand the Physics of waves inside the gun barrel and breech.

In this discussion, only the first four assumptions are valid, i.e., now the state properties are no longer uniform throughout the volume. Hence, the mechanism of projectile acceleration is slightly complicated as compression-rarefaction waves are introduced in the system as shown in Fig. 1. Also, the interference of these waves increases the complexity of this analysis.

4.3.1 Rarefaction Wave

When the firing mechanism is actuated, projectile starts experiencing a push force. As the projectile moves downstream by an infinitesimally small distance dx, driving pressure just behind it decreases by an infinitesimally small value dp due to slight

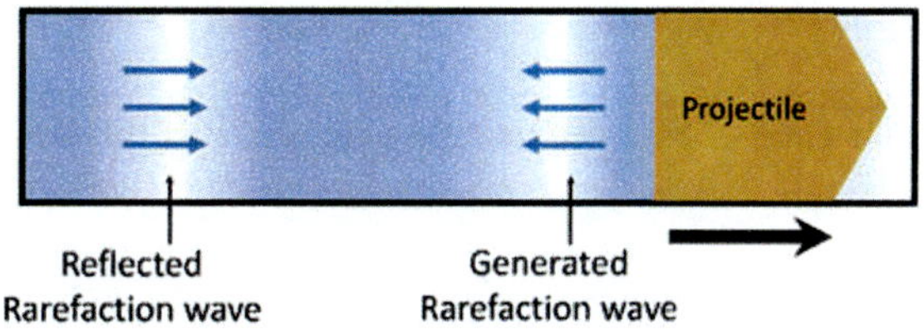

Fig. 1 Illustration of rarefaction waves behind the projectile

increase in the volume in which the compressed gas was contained. A gas element from upstream comes to fill this rarer region leaving a rarefaction at its original position. Thus, the rarefaction disturbance travels upstream with local acoustic speed.

4.3.2 Selection of Working Gas

As the projectile reaches Ma = 1, where the acoustic velocity corresponds to the region just behind the projectile, the gas from upstream is unable to fill the rare region behind the projectile. Therefore, the pressure behind projectile significantly reduces and so does the acceleration. Hence, initial acoustic velocity of gas should be as high as possible, which can be obtained by a gas having either high temperature or low molecular weight or both. For example, acoustic speed of Helium at room temperature is 1020 m/s while that of air is just 340 m/s.

Higher acoustic velocity implies that the rarefaction disturbance moves faster in the upstream direction. Thus, pressure behind the projectile changes its value from near zero to P at a faster rate. Therefore, the effective driving pressure is higher for higher acoustic velocity.

Moreover, in practice, the work done by the gas during expansion not only provides kinetic energy to the projectile, but it is also consumed to accelerate gas molecules inside the barrel. Dalton's law states that for the same pressure, all the gases will have the same number of moles in a given volume. Therefore, the heavier the gas, the more work it will store as its own kinetic energy.

There can be four different combinations of assumptions related to gun-breech. Table 3 clarifies these assumptions.

Table 3 Assumptions related to gun-breech

Variations in assumptions considered for analysis III		Ratio of barrel diameter to breech diameter	
		= 1	≠ 1
Breech length	Infinite	III-A	III-C
	Finite	III-B	III-D

4.3.3 Effectively Infinite Length Breech

Analysis III-A assumes that the length of barrel is designed such that the reflected rarefaction wave is unable to reach the projectile throughout its motion inside the barrel.

The relation between the driving pressure and projectile velocity is given by Eq. (4) below.

A detailed discussion along with derivations for infinitesimally small isentropic changes during expansion is given in A.2.1.

A limitation of this analysis is that the volume of compressed gas has not been considered. However, a thumb rule is suggested by Fowles et al. [5] according to which the mass of gas inside the chamber needs not be more than five times mass of the projectile.

4.3.4 Finitely Long Gun-Breech (Analysis III-B)

As illustrated in Fig. 1, the rarefaction wave while travelling upstream gets reflected by the other end of gun-breech and starts travelling downstream.

If the first reflected rarefaction wave reaches the projectile, an additional reduction in driving pressure takes place, reducing the muzzle velocity. On the other hand, if the projectile crosses the muzzle before the first reflected rarefaction could catch it, the gun-breech length is considered as effectively infinite.

Heybey [2] gave a relation in the projectile speed at this point where the rarefaction disturbance reaches the projectile for a given length of the breech. This analysis is detailed in A.2.1.

4.3.5 Unequal Breech and Barrel Sizes

Analysis III-C considers a gas gun in which the cross section area of the breech is more than that of the barrel. When the rarefaction wave travelling upstream reaches the transition point, relatively more amount of gas comes downstream to fill the rarefied region because of the larger breech diameter. Therefore, the gas downstream is effectively compressed.

$$\frac{P_0 A L}{m a_0^2} = \frac{2}{\gamma+1}\left\{\frac{\frac{2}{\gamma-1} - \frac{\gamma+1}{\gamma-1}\left[1 - \frac{u_p(\gamma-1)}{2a_0}\right]}{\left[1 - \frac{u_p(\gamma-1)}{2a_0}\right]^{\frac{\gamma+1}{\gamma-1}}} + 1\right\} \tag{4}$$

Hence, the rarefaction is partially reflected as a compression wave. This compression wave increases acceleration of the projectile resulting in slightly higher speed. A detailed analysis is given in A.2.3. Analysis III-D is beyond the scope of this article and could be performed by numerical methods suggested by Von Neumann and Richtmyer [10].

4.3.6 Non-evacuated Barrel

Analysis IV considers the presence of a gas at an initial pressure P_1 in front of the projectile. The pressure of this gas (P_f) increases as projectile velocity increases. This forward pressure tends to decelerate the projectile. A detailed discussion is given in A.2.4.

4.4 Effect of Friction in the Barrel

As the projectile and barrel have a clearance fit between them, only linear contact exists between barrel and projectile. If the barrel is machined smooth, role of friction inside the barrel is insignificant.

5 Design of Gas Gun

The required speed from the system is close to 500 m/s for medium velocity impact experiments. Accordingly, the design parameters were selected. Moreover, the projectile mass is limited to 20 g. As explained in Sect. 4.3.2, preferred working fluid is helium.

5.1 Design Parameters

5.1.1 Gun-Breech Design Pressure

To accelerate the projectile at speeds close to 500 m/s, a pressure of ~ 100 bar is required. However, for every experiment, this pressure could be controlled using dual-diaphragm rupture mechanism. The gun-breech has a factor of safety of 3 at 250 bar pressure.

5.1.2 Barrel Length

The effect of barrel length on projectile velocity keeps on reducing as the projectile moves down the barrel.

5.1.3 Barrel Diameter

The factor of safety of the above barrel for 300 bar pressure comes out to be ~ 4.

5.1.4 Gun-Breech Volume

According to the analysis by Fowles [5], a 100 g mass of gas is required for a 20 g projectile which corresponds to 250 mL of compressed helium at 250 bar and 300 K with ideal gas assumption.

5.1.5 Gun-Breech Length and Diameter

The effects of gun-breech length and diameter is explained in Sects. 4.3.4 and 4.3.5, respectively. The pressurized chamber for gun-breech was fabricated using schedule 60 XXS pipe, reducer, and spherical cap. It has a total volume of 1000 mL, so that the condition given by Fowles [5] is met at lower pressures also.

5.2 Design and Assembly

The flanges used were of class 2500. They have radial holes for inlet as shown in Fig. 2. Connecting hose was minimess hose with a rated pressure of 635 bar. The gun-breech was mounted on a slider assembly, so that it could easily be slid back for replacing the diaphragms.

Fig. 2 Single-stage gas gun designed and fabricated at CoE, IIT Delhi

5.3 Dual-Diaphragm Rupture Mechanism

While using a single diaphragm, user has minimal control over the burst pressure. Hence, dual-diaphragm mechanism is used. With this mechanism, the burst pressure could be controlled more precisely. One diaphragm (d_1) is used to close the pressurized chamber. As shown in Fig. 3, one surface of d_1 is acted by breech pressure p_1, other surface of d_1 is acted by p_2, the pressure in the intermediate chamber. The intermediate chamber is closed by another diaphragm d_2. One surface of d_2 is acted by p_2, and other surface is acted by atmospheric pressure where projectile is kept.

1. $p_1 > p_d > p_2, p_2$ is released for firing; d_1 ruptures first;
2. $p_1 > p_d > p_2, p_2$ is further increased for firing d_2 ruptures first;
3. $p_1 < p_2 < p_d, p_2$ is released; none will rupture;
4. $p_1 < p_2 < p_d, p_2$ is increased; only d_2 will rupture similar to a single diaphragm;

For example, if the rupture pressure (p_d) of a diaphragm is 100 bar, in single diaphragm rupture mechanism, the pressurized chamber could be held at 100 ± 5 bar.

The diaphragm always bursts catastrophically at its burst limit. It is not possible to hold the pressurized chamber at the burst pressure and let the pressure stabilize. However, with dual-diaphragm rupture mechanism, for the same diaphragms, rupture pressures of any value between 110 and 180 bar could be achieved reliably. The gun-breech could be held at the burst pressure for any duration.

6 Accessories

Various peripherals such as catcher plates, safety system, vacuum chamber, high-speed camera, and speed measurement system could be used with the gas gun depending upon the need of the experiments. Figure 4 shows a high-speed camera and lighting system for high-speed photography.

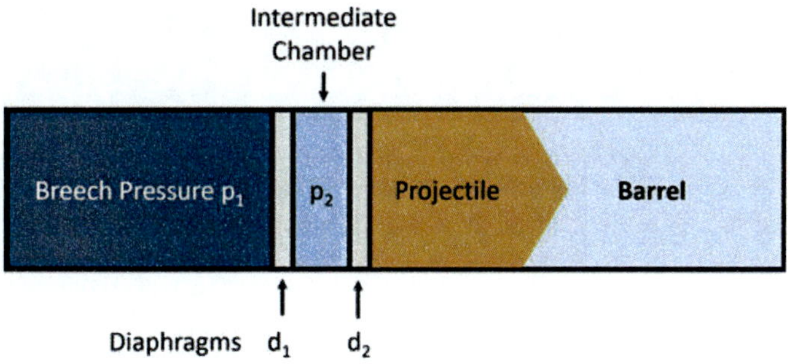

Fig. 3 Schematic of a dual-diaphragm rupture mechanism

Fig. 4 Photograph of high-speed camera

7 Results and Conclusion

The gas gun at IIT Delhi was used to accelerate 20 g wooden projectiles in order to assess its performance. Helium was used as the propelling gas, and a speed measurement device was used for recording the speed. A high-speed camera was also used to capture the flight of the projectile. A series of pictures captured at 28,500 frames per second are shown in Fig. 5.

Table 4 contains the data of these experiments. This data is also plotted in Fig. 6 with circular markers. A comparative analysis of all the approaches discussed in Sect. 4 has been performed and shown in Fig. 6. It is evident that 60–80 bar pressure of helium is sufficient to accelerate the projectile to 500 m/s.

Analysis III-A seems to be the closest to the experimental results. However, as the aerodynamic effects are not considered in this analysis, the speed of the projectile is lower than the predicted one. Moreover, the point at which speed is measured also plays a great role as the predicted velocity is the velocity at the muzzle of the gas gun. The actual velocity may reduce due to air drag when the projectile is flying

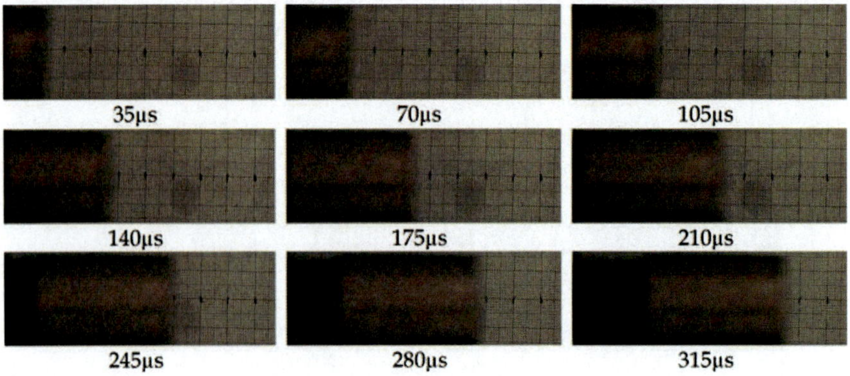

Fig. 5 Images of a wooden projectile obtained with high-speed camera

Table 4 Data obtained from experiments with the single-stage gas gun designed and fabricated at IIT Delhi

S. No.	Breech pressure (bar)	Velocity (m/s)
1	17	280.72
2	25	305.54
3	32	334.60
4	35	340.19
5	39	389.71

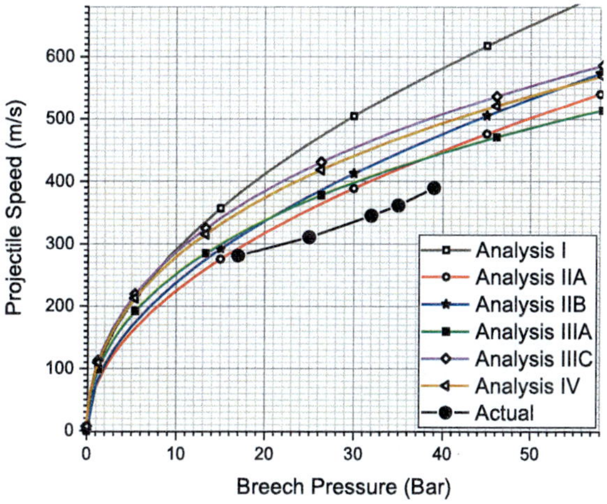

Fig. 6 Comparison of predicted and actual velocity

outside the barrel. Also, the experiments pertaining to other values of breech length, ratio D_1/D_0, and other parameters are beyond the scope of present research work.

Appendix-A. Derivations and Explanations

A.1. Thermodynamics

A.1.1. Constant Driving Pressure

From Newton's second law of motion,

$$p_0 \times A = m \times a_x \Rightarrow p_0 \times A = m \times u_p \frac{du_p}{dx}$$

$$\Rightarrow \int_0^L \frac{p_0 A}{m} dx = \int_0^v u_p du_p \Rightarrow \frac{p_0 A x_0}{m} = \frac{u_p^2}{2}$$

$$\Rightarrow u_p = \sqrt{\frac{2p_0 A x_0}{m}}$$

A.1.2. Isentropic Expansion

$$PV^{\gamma} = \text{constant}$$

$$mu_p \frac{\mathrm{d}u_p}{\mathrm{d}x} = A \cdot P_0 \left(\frac{V_0}{V}\right)^{\gamma}$$

$$\frac{1}{2} m u_p^2 = \int_0^L P_0 \left(\frac{V_0}{V}\right)^{\gamma} \cdot \mathrm{d}V$$

$$u_p = \sqrt{\frac{2}{m} \cdot \frac{P_0 V_0}{\gamma - 1} \cdot \left[1 - \left(\frac{V_0}{V_1}\right)^{\gamma - 1}\right]}$$

A.1.3. Polytropic Expansion

In polytropic expansion, polytropic coefficient η will be taken into account instead of γ.

$$PV^{\eta} = \text{constant}$$

$$u_p = \sqrt{\frac{2}{m} \cdot \frac{P_0 V_0}{\eta - 1} \cdot \left[1 - \left(\frac{V_0}{V_1}\right)^{\eta - 1}\right]}$$

A.2. Gas Dynamics

A.2.1. Infinitely Long Gun-Breech

Let D be the speed of a disturbance with respect to ground. At $t = 0$, as the gas particles travel with a speed u, the speed of disturbance with respect to the gas particles would be $D - u$. Also, let P be the pressure and ρ be the density of the gas at $t = 0$. At $t = \mathrm{dt}$, when the disturbance is past through an infinitesimal gas element of cross section area A, these quantities change to $u + \mathrm{d}u$, $P + \mathrm{d}P$, and $\rho + \mathrm{d}\rho$ (Fig. 7).

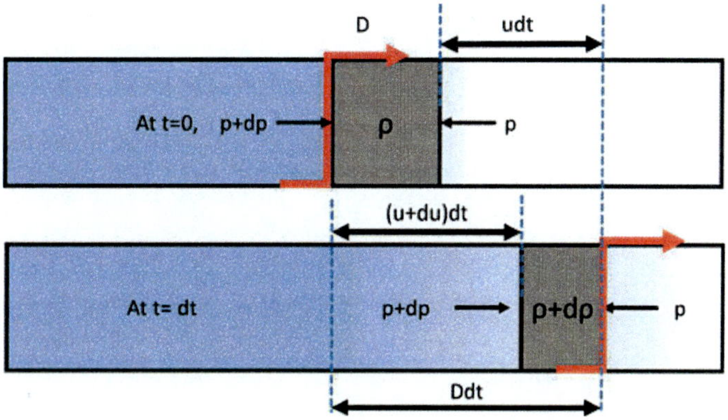

Fig. 7 Schematic of a pressure disturbance travelling through the gun barrel

Continuity equation, assuming 1D expansion could be written as

$$A(D-u)\rho \mathrm{d}t = A(D-u-\mathrm{d}u)\mathrm{d}t(\rho+\mathrm{d}\rho) \Rightarrow \mathrm{d}u = \frac{1}{\rho}(D-u)\mathrm{d}\rho$$

Momentum equation could be written:

$$A\mathrm{d}P = A\rho(D-u)\mathrm{d}t\frac{\mathrm{d}u}{\mathrm{d}t}$$

From above two relations:

$$\frac{\mathrm{d}P}{\mathrm{d}\rho} = (D-u)^2$$

For isentropic changes, acoustic speed:

$$a = \sqrt{\frac{\mathrm{d}P}{\mathrm{d}\rho}}$$

considering movement of disturbance in + x direction,

$$D - u = a$$

Now, the momentum equation will be:

$$A dP = A \cdot \rho \cdot a \cdot \mathrm{d}u$$

$$\mathrm{d}P = a \cdot \rho \cdot \mathrm{d}u$$

$$\Rightarrow \mathrm{d}P = z \cdot \mathrm{d}u$$

where z is the acoustic impedance of air.

For a gas gun with initial breech pressure P_0, the instantaneous speed u of gas particles will be

$$u = \int_{P_0}^{P} \frac{\mathrm{d}P}{\mathrm{d}\rho}$$

Considering isentropic expansion,

$$u = \frac{1}{a_0 \rho_0} \int_{P_0}^{P} \left(\frac{P_0}{P} \right)^{\frac{1+\gamma}{2\gamma}} \cdot \mathrm{d}P$$

$$u = 2\frac{a_0}{\gamma - 1} \left[1 - \left(\frac{P}{P_0} \right)^{\frac{\gamma-1}{2\gamma}} \right]$$

For an infinitely long barrel, P = 0

$$u_\infty = \frac{2a_0}{\gamma - 1}$$

If we keep on reducing the mass of the projectile, from the principle of conservation of energy, we should keep on achieving a higher and higher muzzle velocity, but it is not the case in practice because limiting velocity of the projectile is a function of initial values of propelling pressure and downstream pressure. As it is not possible to achieve absolute vacuum downstream, it is not possible to achieve the highest probable muzzle speed u_∞.

The pressure behind the projectile could be written as

$$P = P_0 \left(1 - \frac{u}{u_\infty} \right)^{\frac{2\gamma}{\gamma+1}}$$

Using Newton's second law to determine speed of the projectile (assuming u ~ u_p)

$$m u_p \frac{\mathrm{d}u_p}{\mathrm{d}x} = A P_0 \left[1 - \frac{u_p}{\frac{2}{\gamma-1} a_0} \right]^{\frac{2\gamma}{\gamma-1}}$$

Upon integrating,

$$\frac{P_0 AL}{ma_0^2} = \frac{2}{\gamma+1}\left\{\frac{\frac{2}{\gamma-1} - \frac{\gamma+1}{\gamma-1}\left[1 - \frac{u_p(\gamma-1)}{2a_0}\right]}{\left[1 - \frac{u_p(\gamma-1)}{2a_0}\right]^{\frac{\gamma+1}{\gamma-1}}} + 1\right\}$$

A.2.2. Finitely Long Gun-Breech (Analysis III-B)

The velocity of projectile at the instant when first reflected pulse reaches the projectile is related with chamber length L_c as:

$$\frac{P_0 AL_c}{ma_0^2} = \frac{2}{\gamma+1}\left\{\frac{1}{\left[1 - \frac{u_{p1}(\gamma-1)}{2a_0}\right]^{\frac{\gamma+1}{2(\gamma-1)}}} - 1\right\}$$

The propelling pressure reduces; therefore, projectile speed also reduces.

However, if the length of the chamber is designed such that x_{p1} (the distance travelled by the projectile before interacting the first reflected pulse) is more than length of the barrel, the velocity will be same as that in case of an infinitely long chamber.

$$x_{p1} \approx 2.5\frac{P_0 AL_c^2}{ma_0^2} > L$$

If $x_{p1} < L$, numerical methods suggested by Von Neumann and Richtmyer [10] have to be applied which is beyond the scope of this article.

A.2.3. Unequal Size of Gun-Breech and Barrel (Analysis III-C)

The increase in speed of projectile if the ratio of barrel diameter to breech diameter is reduced from 1 to 0 is ~ $0.5a_0$.

Hence, when the ratio is 0.5, the increase will be $0.25a_0$. However, this increase will be linear with respect to the projectile speed, i.e.,

$$If u_p < \frac{3a_0}{\gamma} \rightarrow \Delta u_p = \frac{\gamma u_p}{12}$$

A.2.4. Considering Atmospheric Pressure Downstream (Analysis IV)

The downstream pressure increases as projectile moves down the barrel.

$$P_f = P_1\left[1 + \frac{\gamma_1(\gamma_1+1)}{2}\cdot\left(\frac{u_p}{a_0}\right)^2\right]$$

Upon further investigation,

$$\frac{u_{p(P_1>0)}}{u_{p(P_1=0)}} \approx \sqrt{\frac{1-e^{-y}}{y}}$$

$$\text{such that } y = \frac{\gamma_1(\gamma_1+1)\cdot P_1\cdot A\cdot L}{ma_1^2}$$

If the value of limiting speed is more than u_p, above formulas could be used. In case limiting speed comes out to be less than u_p, limiting speed will be final speed of the projectile.

Appendix-B. List of Symbols Used

A	Projectile (or barrel) cross section area
a_0	Acoustic speed of gas inside breech (1020 m/s for helium)
a_1	Acoustic speed of gas present inside barrel (340 m/s for air)
a_x	Acceleration of the projectile
D	Speed of a rarefaction wave with respect to ground
D_1	Barrel diameter
D_0	Chamber diameter
L	Length of barrel
L_c	Breech length
m	Mass of projectile
η	Polytropic coefficient
P	Pressure of gas
P_0	Initial breech pressure
P_1	Initial pressure in front of the projectile
P_f	Instantaneous pressure in front of the projectile
u	Speed of air particles
u_p	Speed of projectile
u_{p1}	Speed of projectile when first reflected rarefaction wave reaches it
u_∞	Maximum speed of air in an infinitely long evacuated barrel without projectile
V_1	Volume of chamber plus barrel
V_0	Volume of chamber
X_{p1}	Position of projectile when first reflected rarefaction wave reaches it
z	Acoustic impedance
γ	Ratio of specific heats for gas filled in breech (1.67 for helium)
γ_1	Ratio of specific heats for gas present in barrel (1.4 for air)

ρ_0 Initial density of gas inside breech
ρ Density of gas

References

1. Seigel AE (1965) The theory of high speed guns. Advisory Group for Aerospace Research and Development Neuilly-Sur-Seine, France
2. Woods WA (1969) Paper 26: a solution to Lagrange's ballistic problem with gas on each side of the projectile. Proc Inst Mech Eng Conf Proc 184:76–82. https://doi.org/10.1243/PIME_CONF_1969_184_164_02
3. VanderKlok A, Stamm A, Dorer J, Hu E, Auvenshine M, Pereira JM et al (2018) An experimental investigation into the high velocity impact responses of S2-glass/SC15 epoxy composite panels with a gas gun. Int J Impact Eng 111:244–254. https://doi.org/10.1016/j.ijimpeng.2017.10.002
4. Gupta SC, Jyoti G, Suresh N, Sikka SK, Chidambaram R, Agarwal RG et al (1995) Gas-gun facility for shock wave research at BARC. Bhabha Atomic Research Centre
5. Fowles GR, Duvall GE, Asay J, Bellamy P, Feistmann F, Grady D et al (1970) Gas gun for impact studies. Rev Sci Instrum 41:984–996. https://doi.org/10.1063/1.1684739
6. Bourne NK (2003) A 50 mm bore gas gun for dynamic loading of materials and structures. Meas Sci Technol 14:273–278. https://doi.org/10.1088/0957-0233/14/3/304
7. Bourne NK, Stevens GS (2001) A gas gun for plane and shear loading of inert and explosive targets. Rev Sci Instrum 72:2214–2218. https://doi.org/10.1063/1.1359192
8. Ing T (n.d.) Single-stage gas guns design and manufacture. https://www.thiot-ingenierie.com/en/gas-guns-and-other-equipment/single-stage-gas-guns/. Accessed 26 May 2022
9. SAAMI (2015) Voluntary industry performance standards for pressure and velocity of centerfire Rifle ammunition for the use of commercial manufacturers
10. VonNeumann J, Richtmyer RD (1950) A method for the numerical calculation of hydrodynamic shocks. J Appl Phys 21:232–237. https://doi.org/10.1063/1.1699639

Finite-Element Analysis of Reinforced Concrete Target Against Blast Loading

Vaibhav Bhandari, Vimal Kumar, Arya Prakash Padhi, Anupam Chakrabarti, and Rajib Chowdhury

Abstract Civil and military infrastructure has been exposed to the potential of terrorist attacks using portable detonation devices and explosives; thus, the performance of common structural components and materials subjected to blast loading must be determined. To test the RCC slab's resistance to damage, numerical research was performed on it while subjected to near-field blast loading. TNT explosive was used to create an explosion on a 1.3 m × 1.0 m × 0.1 m target. The explosive was detonated from various standoff distances equating to 0.079–0.135 kg/m$^{1/3}$. In the modeling of concrete and steel, the concrete damage plasticity and Johnson–Cook plasticity models were utilized, respectively. The results demonstrated that the target slab's one-way bending and splitting resulted in a crater with clear visibility. The explosion pressure grows as the weight of the explosives increases, transforming localized failure into globalized failure. The greatest crater is 391 mm in diameter and is measured for 0.4 kg explosive weight at 0.1 m standoff distance. The smallest crater is 100 mm in size for 6.6 kg explosive weight at 0.3 m standoff distance. The safe standoff distance for non-nuclear explosives, such as 1 kg of TNT, is 0.5 m.

Keywords Finite-element modeling · Nonlinear dynamic analysis · Reinforced concrete slab

V. Bhandari (✉) · A. Chakrabarti · R. Chowdhury
Department of Civil Engineering, IIT Roorkee, Roorkee, India
e-mail: vaibhav.vibhz17@gmail.com

V. Kumar
Department of Civil Engineering, NIT Hamirpur, Hamirpur, India

A. P. Padhi
CTRANS, IIT Roorkee, Roorkee, India

R. Velmurugan et al. (eds.), *Dynamic Behavior of Soft and Hard Materials Volume 1*, Springer Proceedings in Materials 34, https://doi.org/10.1007/978-981-99-6030-9_47

1 Introduction and Literature Survey

During their service life, certain structures may be subjected to high loadings such as impact and explosive loading, which can result in an overload much greater than the structure's design load in a short period. This might result in significant damage to the reinforced concrete building, as well as numerous injuries and property losses. Previously, static and earthquake loads were the primary design considerations, but recently, more emphasis has been placed on designing structures of national and strategic significance that can withstand blast loads. The reason for such loadings might be unintentional, such as an internal explosion, or purposeful, such as a missile hit and explosion caused by a terrorist, or an explosion caused by mortar or shell in frontline regions. Intensive research has been undertaken in recent decades on the action, analysis, and design of reinforced structures to withstand short-term dynamic loading effects such as those caused by an air burst. The structural response is significantly influenced by short-duration, high-magnitude stress circumstances. The frequency of explosive loads may be much greater than those of normal loads. Furthermore, short-duration dynamic loads usually exhibit large spatial and temporal changes, resulting in structural stress gradients. As a result, the strain rate varies during the study. As a result, assessing structures made of fragile materials such as concrete that are subjected to blast loading becomes incredibly challenging. Additionally, the effects of shell impact or the detonation of high explosive (HE) rounds near concrete slab shelters on structural strength and stability are also frequently studied.

Xu [1] sought to design suitable/appropriate concrete spallation criteria capable of producing accurate concrete spallation findings. They carried out a near-field explosion and studied the influence of three-dimensional stress conditions. Lin [2] built a finite-element model in LS-DYNA to predict the structural response of steel-reinforced concrete panels under blast loading. The effect of element size on the concrete dynamic material model is investigated, and the model accounts for strain rate effects on concrete in tension and compression separately. Zhou [3] used a dynamic plastic damage model to model concrete material and predict the performance of both a conventional reinforced concrete slab and a high-strength steel fiber concrete slab exposed to blast loading in both experimental and computational simulations. The numerical results of the concrete slab simulation exposed to blast loading were compared to those of the weapon system division's explosive assessments. Wang [4] tested four 1000 × 1000 × 40 mm slabs against close-in blast loading using experimental and computational methods. For numerical investigations, a three-dimensional numerical model with air, explosion, concrete, and reinforcement was explored. The strain rate was also considered in the overall model. An appropriate erosion process was explored for material failure. The experimental and numerical results complimented each other. Schenker [5] conducted a full-scale field explosive test with aluminum foam on exposed and shielded slabs. Their primary goal was to investigate the dynamic response of aluminum foam to shock wave mitigation. The experimentally acquired results were utilized to validate their existing computer algorithms. The impacts of the explosion on buildings were thoroughly researched by Ngo [6]. The nature of blasts and the technique of blasting in free air have also been described. It goes into great depth into the dynamic response of various structural parts. It shows many methods for forecasting blast parameters. Kumar [7] investigated a reinforced concrete slab 1000 × 1000 × 100 mm against blast loading using numerical modeling and practical investigation. Finite-element simulations were carried out using the ABAQUS program, with the HJC model used to simulate concrete and the metal plasticity model used to model reinforcement. In reaction to slab against blast, standoff distance was shown to have a greater influence than blast weight. Nagy [8] subjected underground reinforced concrete buildings to blast loading. The FEM tool was used to analyze the entire system. A single model was devised and used to simulate the blast, blast wave propagation, and the interaction of buried structure and soil. Nebojsa [9] conducted an experimental test on six pre-stressed concrete slabs with varied reinforcing and prestressing. To prestress the slab, post-tensioning was used. Elsanadedy [10] investigated the impact of explosive loading on reinforced concrete circular columns that were already in place. Retrofitting was employed on existing structures against explosive loading, and the results were compared to prior findings obtained without retrofitting. Standoff distance and explosive weight were also investigated. Strengthening buildings using CFRP would be an effective technique to mitigate the effects of a modest explosion. Kong et al. [11] presented a finite-element model for the modeling of an aramid composite reinforced concrete slab. Their primary research goal was to investigate the composite slab's characteristics while taking into account the dynamic interfacial behavior of the RCC slab and aramid fiber. They got to the conclusion that aramid composite not only delays concrete fracture but also minimizes concrete spallation.

It is necessary to build a lot of field fortifications in the advanced regions quickly, including command posts, gun positions, and bunkers. Time is of the essence, thus to accelerate the pace of construction, the in situ building of these structures is avoided in the working area, and overhead protection is instead provided by precast movable reinforced concrete slabs. The field boundary condition of these portable structures is accurately represented by the experimental boundary condition of the slabs. So, the current study makes it possible to anticipate damage to the slabs in the field area in a way that is acceptable and practical. Only temporary protection is offered by merely acting as a support for the RC slabs' on two edges. The originality of the current study is that no previous research has been done on the impact of the field support condition (slab resting on its two opposite edges or flat slabs) on the performance of reinforced concrete slabs. The present study also uses the newly introduced failure model for concrete damage plasticity which has not been applied yet to blast loading cases [12].

The following paper is discussed in five sections. Section 1 briefs about the introduction and literature survey about the proposed study, Sect. 2 briefs about the experimental setup chosen for validation and for further study, Sect. 3 discusses material modeling used for numerical modeling, Sect. 4 discusses the result obtained, and Sect. 5 concludes the study.

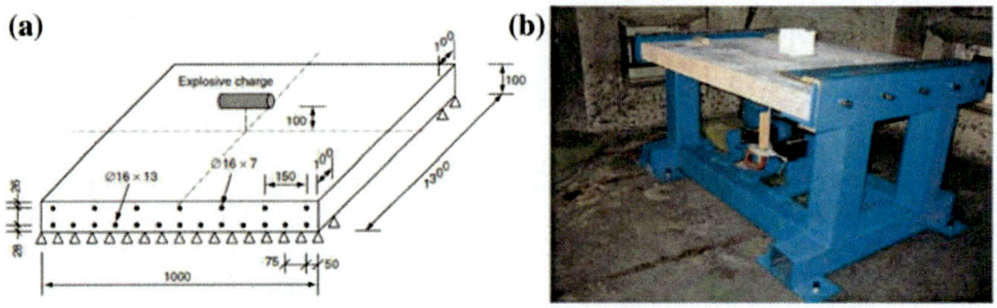

Fig. 1 **a** Schematic diagram; **b** experimental setup [3]

2 Experimental Setup

Figure 1 depicts the configuration of Zhou's [3] reinforced concrete slab, including its dimensions and boundary conditions, as well as the placement of explosives. It served as the foundation for our investigation. The length, breadth, and height of the reinforced concrete slab were 1300 × 1000 × 100 mm. In typical compression cylinder testing, the compressive strength of the slab was determined to be 50 MPA. For blast loading, composition B (60:40 RDX: TNT) and Powergel magnum explosives were employed. Reinforcement bars with 16 mm diameters were placed at the top and bottom of the slab, with 75 and 150 mm spacings, respectively. As seen in the illustration, the concrete slab was supported by a steel frame. We aimed to validate his results and extend our study with varying scaled distances.

3 Numerical Modeling in FEM

Numerical modeling of the slab was done in ABAQUS. A concrete damage plasticity model was used to model concrete of grade M50 whose property is given in Table 1. The properties used are that of the values in Abaqus. Fb0/fc0 is the ratio of initial equibiaxial compressive yield stress to initial uniaxial compressive yield stress. K is the ratio of the second stress invariant on the tensile meridian to that on the compressive meridian at initial yield for any given value of the pressure invariant p such that the maximum principal stress is negative. Steel bars were modeled using the Johnson–Cook model with a grade of steel 4340 whose properties are shared in Table 2. Blast was modeled using CONWEP. Dynamic explicit algorithm was used for the whole process. Concrete was modeled using three-dimensional solid continuum elements, and steel reinforcement was modeled using truss elements available in Abaqus.

3.1 Concrete Damage Plasticity

See Table 1.

Table 1 Concrete CDP parameters

Young Modulus (Mpa)	Poisson's ratio	Viscosity parameters	
19,700	0.19	0.001	
Dilation angle	Eccentricity	fb0/fc0	K
38	1	1.12	0.666
Compressive behavior	Compression damage		
Yield stress (kPa)	Crushing strain	Damage parameter	Crushing strain
15,000	0	0	0
20,197.804	0.0000747307	0	7.47307E−05
30,000;609	0.0000988479	0	9.88479E−05
40,303.781	0.000154123	0	0.000154123
50.007.692	0.000761538	0	0.000761538
40,236.090	0.002557559	0.195402	0.002557559
20,236.090	0.005675431	0.596382	0.005675431
5257.557	0.011733119	0.894865	0.011733119
Tensile behavior	Tension damage		
Damage parameter (kPa)	Cracking strain	Damage parameter	Cracking strain
1998.930	0	0	0
2842.000	0.00003333	0	0.00003333
1869.810	0.000160427	0.406411	0.000160427
862.723	0.000279763	0.69638	0.000279763
226.254	0.000684593	0.920389	0.000684593
56.576	0.00108673	0.980093	0.00108673

Table 2 Johnson–Cook parameters for steel

Density	7850 kg/m^3
Bulk modulus, K	159,000 Mpa
Shear modulus, G	81,800 Mpa
Basic yield stress, A	792 Mpa
Hardening constant, B	510 Mpa
Room temp	300 K
Strain rate constast, C	0.014
Thermal softening exponent, m	1.03
Strain rate $\dot{\varepsilon}$	1 s^{-1}

3.2 Johnson–Cook Model

It is a Mises plasticity model [13] which is used to model metals under a high-strain rate. It considers the effect of the hardening of strain in combination with the effect of strain rate and temperature. It is suitable for dynamic loading, as given in Eq. (1).

The model defines the yield stress as follows:

$$Y = \left(A + B\varepsilon_p^n\right)\left(1 + \log \varepsilon_p^* . C\right)\left(1 - T_H^m\right) \tag{1}$$

The first term in the above expression accounts for strain hardening. The second term accounts for the effect of strain rate and the last term accounts for temperature. Johnson–Cook dynamic failure was taken to model the damage initiation and evolution of steel under dynamic loading, as shown in Fig. 3.11. It was defined in the property module of steel. The fracture energy was taken as 0.001 J after going through the literature available.

3.3 CONWEP

The explosion is a quick release of energy that causes shock and pressure waves to occur. The pressure variation over duration is given below in Fig. 2.

Using the symmetry of the concrete slab about the two axes (x & y), a quarter slab with dimensions of 650 × 500 × 100 mm is modeled to decrease computational cost without sacrificing accuracy. The final model in Abaqus is shown in Fig. 3. Finer discretization has been made for precise capture of the resulting response. A

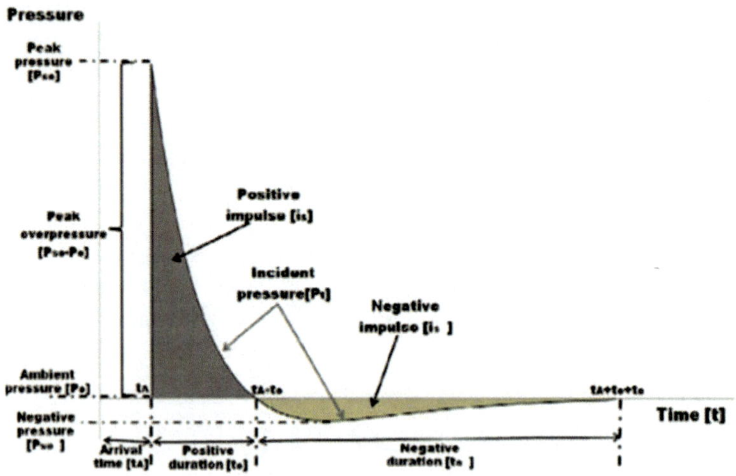

Fig. 2 Pressure versus time history of an ideal blast wave [14]

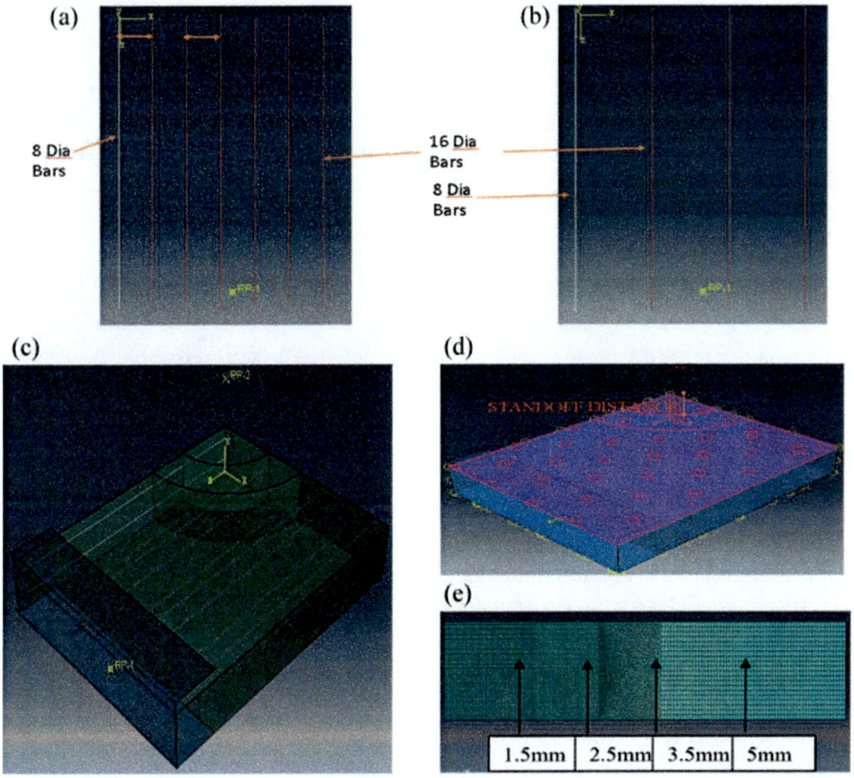

Fig. 3 **a** Lower reinforcement (16 Dia Bars Of 6 no. @75 mm C/C and 8 Dia Bars of 1 no.) **b** Upper reinforcement (16 Dia Bars of 3 no. @150 mm C/C and 8 Dia Bars of 1 no.) **c**, **d** Assembly of the model with standoff distance and mesh of the geometry, **e** Mesh refinement for concrete

total of 1,679,031 no. of three-dimensional continuum elements were created in the geometry which was proven to be sufficient for the mesh convergence of the study. The mesh convergence study has been shown in the Appendix 1.

3.4 Numerical Validation

The element erosion technique was adopted by writing two lines of code in the input file as follows:

$$* \text{ Concrete Failure, Type} = \text{Strain}$$
$$0.014, 0.0367, 0.99, 0.81,$$

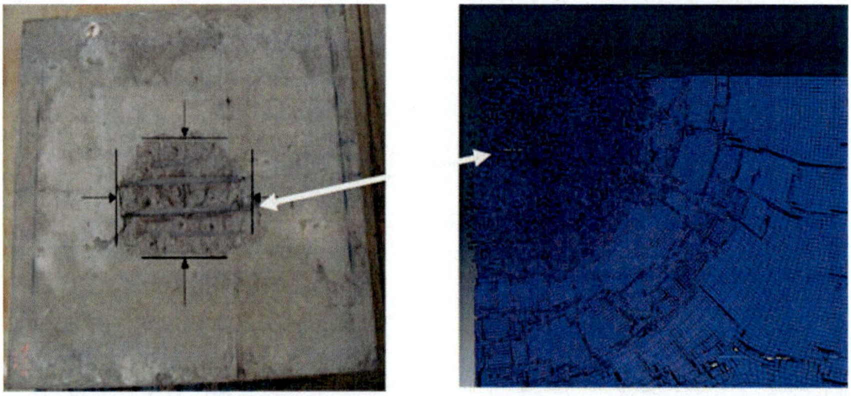

Fig. 4 Validation of experimental results

where 0.0014 is the damage parameter in tension and 0.00367 is the damage parameter under compression. 0.99 is the fracture strain in tension and 0.81 is the fracture strain in compression. The numerical model made up of constitutive models of CDP and JC was validated with the experimental results [3]. The damage shape obtained numerically complemented well the experimental results. The accuracy of the numerical model is also checked by comparing modeled peak pressure with that obtained from the empirical method in TM5-1300 [15]. The peak reflected pressure was within the 10% tolerance limit for most of the cases. The numerical results show the same spalling from the rear surface of concrete when compared to experimental results, as shown in Fig. 4, Table 3.

4 Results

The responses obtained for the 16-simulation performed have been summarized in Tables 4 and 5. Table 4 summarizes the behavior at various scaled distances for the same blast load at two distances. Table 5 summarizes the effect at the same blast-scaled distance. Out of 16 simulations performed the lowest pressure at the center of the slab was found for SN1 and as we move away from the center of the slab the pressure reduced. The red zone in tension damage shows that the majority of failure is due to tensile waves which lead to the spallation of concrete. The slab is unsafe. The crater formed is observed to be 101 mm and 105 mm in dimension in X and Z directions, respectively, as shown in Fig. 5.

The maximum crater for CS15 was detected, as indicated above. The highest pressure measured on the slab is 442.7 MPa, which is nearly the same as the reflected pressure of 445.8 MPa. When comparing damage to CS6, the region of full erosion

Table 3 Details of the case studies performed for numerical simulation

JOB ID	Simulation number (CS)	Weight of explosive in experiment* (Kg)	Weight of equivalent TNT explosive for simulation (Kg)	Standoff distance (mm)	Scaled distance (m/kg$^{1/3}$)
Case-1-400 gm	1	–	0.4	100	0.135
Case-1-500 gm	2	–	0.5	100	0.125
Case-1-545 gm	3	0.5	0.545**	100	0.122
Case-1-850 gm	4	–	0.85	100	0.105
Case-1-1000 gm	5	–	1	100	0.1
Case-1-2000 gm	6	–	2	100	0.079
Case-2-400 gm	7	–	0.4	500	0.676
Case-2-500 gm	8	–	0.5	500	0.628
Case-2-545 gm	9	–	0.545	500	0.61
Case-2-850 gm	10	–	0.85	500	0.527
Case-2-1000 gm	11	–	1	500	0.5
Case-2-2000 gm	12	–	2	500	0.397
Case-3-1360 gm	13	–	1.36	150	0.135
Case-3-10900 gm	14	–	10.9	300	0.135
Case-4-6600 gm	15	–	6.6	150	0.079
Case-4-52700 gm	16	–	52.7	300	0.079

* Explosive taken in the experiment is of composition 60RDX:40TNT
** Relative equivalence is taken as 1.09 [15]

increases. It demonstrates that the concrete has been entirely ruined. The damage indicates the creation of the crater and spalling. The crater generated measured 391 mm and 351 mm in the X and Z directions, respectively.

ABAQUS/EXPLICIT was used to run the dynamic nonlinear simulation. The simulation results were seen using a visualization feature of the finite-element program. The main goal of the preceding numerical approach is to investigate the behavior of the slab under various explosive weights and standoff distances. If the concrete reached the prescribed damage value and fracture strain, it was regarded to have failed and was eliminated from the computation.

However, during the simulation, it was discovered that removing an element from the concrete surface underpredicted the strength of the concrete, because concrete,

Table 4 Summary of the results obtained from the test of the first twelve simulations (CS1, CS2, CS3, CS4, CS5, CS6, CS7, CS8, CS9, CS10, CS11, and CS12)

Simulation number	Remark	Maximum pressure observed (Mpa)
CS1	Cracking with a little crater. Minor spallation on the back surface. The slab is hazardous	257
CS7	The slab is safe despite the little cracking	12.48
CS2	Moderate spallation at the rear surface. Slab is unsafe	279
CS8	Slight cracking at the rear surface, but no spalling and crater formation. The slab is safe	14.85
CS3	Spallation on the back surface is moderate. The slab is dangerous There is some cracking on the back surface, but no spalling or crater development. The slab is secure	289.14
CS9	The crater's diameter became larger than in SN1 and SN2. Moderate spallation off the back surface of the concrete. Cracks are seen to open. The slab is dangerous There is no spallation or crater development on the back surface. The slab is secure	18
CS4	Severe spallation with central bending is shown to be greater than in SN1 and SN2. The slab is dangerous	339.7
CS10	There is cracking but no spallation or crater development. The slab is referred to be safe	26.75
CS5	At the center, blast wave perforation is evident. Concrete disintegration at a variety of locations. The slab is dangerous	359.09
CS11	Flexural. The crack density grew. There is no spalling or crater development. The slab is secure	31.78
CS6	The region of the perforation blast wave grew. The area of failure grew larger	445.02
CS12	There is no evidence of spalling or cratering. Concrete disintegration with worldwide distortion. The slab is dangerous	43.5

when failed under loading, still takes the load, but due to element erosion, the concrete was removed from the analysis even before it could have taken more load [16]. As a consequence, two simulations were conducted for each scaled distance: one without element erosion, which allowed us to anticipate the peak pressure on the slab much more precisely, and one with element erosion, which provided us with a view of the damage process in concrete.

Table 5 Comparison of the results at the same scaled distance (SN13, SN14, SN15, SN16)

Simulation number	Remark	Maximum pressure (Mpa)
CS1	Cracking with a minor crater development. Spallation on the rear surface. There was no puncture caused by the explosion wave	257
CS13	Spalling crater formation. There was no puncture caused by the explosion wave	257.47
CS14	Perforation of a greater region by the blast wave. Target's total failure	257.5
CS6	Crater development was detected near the center fissures	445
CS15	The crater's diameter grew. Bending increased significantly	445.8
CS16	The concrete has completely eroded. The slab was entirely pierced by a planar blast	449.8

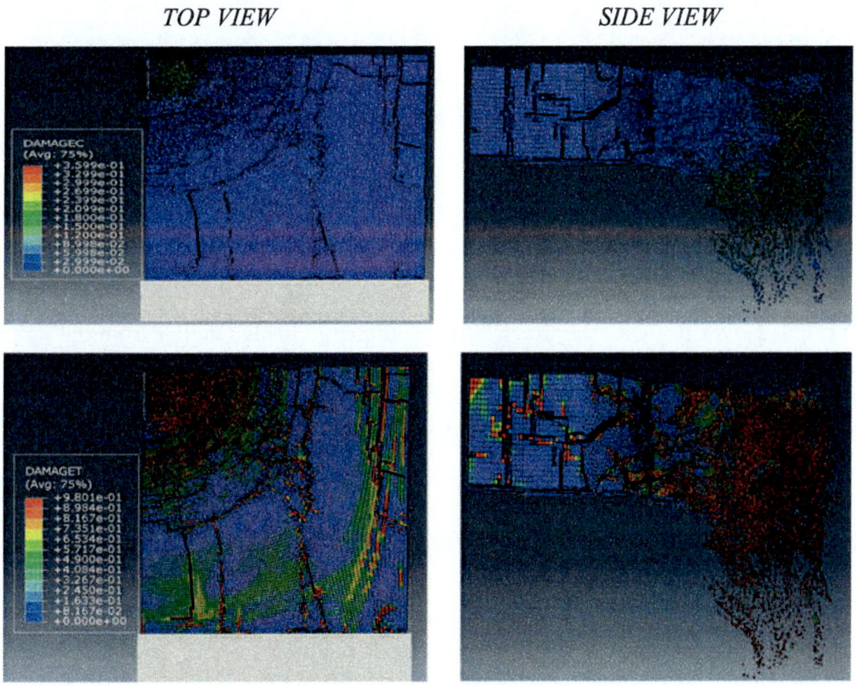

Fig. 5 Compression damage and tensile damage of CS1 simulation

A total of sixteen simulations were designed and carried out. The first 12 simulations were run with the effect of increasing scaled distances in mind. The first six simulations (CS1, CS2, CS3, CS4, CS5, CS6) were run with various blast loads at a fixed standoff distance of 0.1 m. The charge weight was gradually raised from 0.4 kg to 2 kg of explosive. The second set of six simulations (CS7, CS8, CS9, CS10,

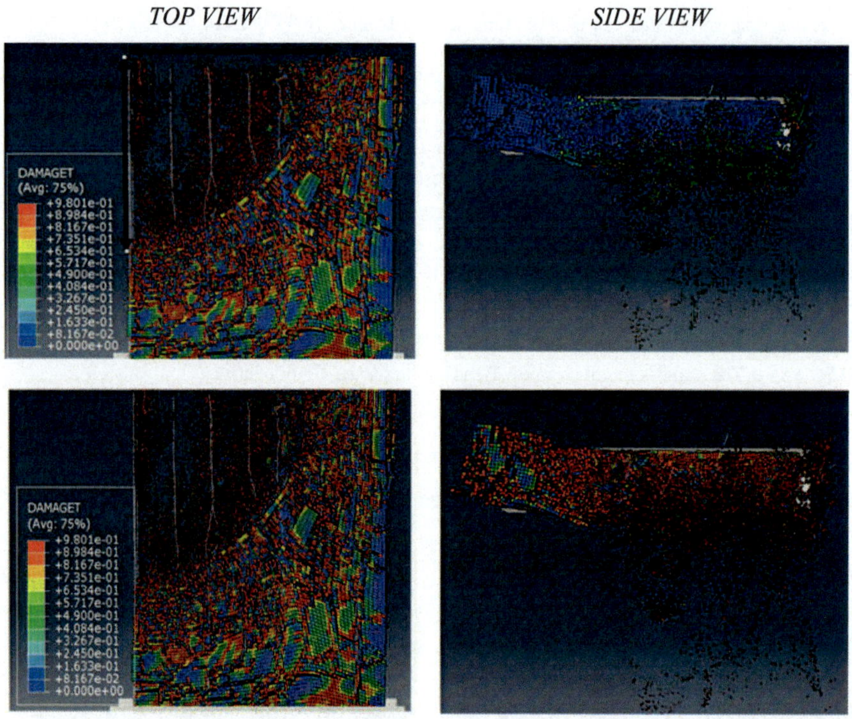

Fig. 6 Compression damage and tensile damage of CS15 simulation

CS11, CS12) was run with the previously taken charge loads ranging from 0.4 kg to 2 kg; however, the standoff distance was raised from 0.1 m to 0.5 m for all simulations. Overall, the simulated slab did not detach from the supports. The final four simulations (CS13, CS14, CS15, and CS16) were run at a constant scaled distance.

4.1 Variation of Blast Pressure Within the Target

The blast wave from a near-field explosion spreads spherically outwards, but the wave from a far-field explosion converts to a planar surface after traveling a distance. A pressure variation is recorded at the center (A), 0.150 m (B) 0.250 m (C) 0.350 m (D), and 0.500 m (E), as shown in Fig. 7. For CS6 and CS12 simulations, the blast pressure distribution is seen.

Figure 8 depicts the pressure distribution for a 2 kg explosion at a distance of 0.1 m. (CS6). It shows that there is a significant pressure difference between the center and the periphery of the slab. Figure 9 depicts the pressure distribution for 2 kg of explosive at 0.5 m. (CS12). When using CS12, the pressure fluctuation is quite modest compared to CS6.

5 Conclusion

The important observations are summarized below.

- The CDP constitutive model used here supports the experimental data effectively against such high-strain blast loading. The results of this study using the CDP constitutive model will aid researchers in better understanding the characteristics of reinforced concrete bunkers in field areas prone to mortar and rocket launcher explosions. As the explosive charge grows, the bending and one-way splitting of the slab, as well as localized damage in the shape of the crater, become more visible. By using correct failure properties for CDP, we can get the correct damaged shape of the concrete.
- The reduction in pressure is observed between 95 and 90% on increasing the standoff distance from 100 to 500 mm at the same explosive weight. The reduction in pressure is observed to be maximum for 0.4 kg explosive on increasing the standoff distance from 100 to 500 mm. The reduction in pressure is observed to

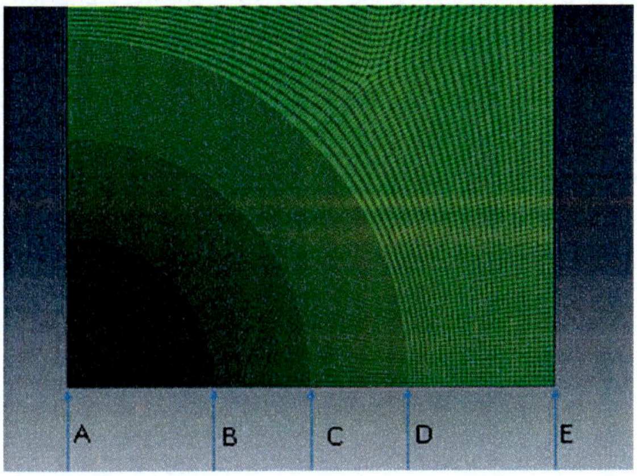

Fig. 7 Various locations for observing blast pressure distribution

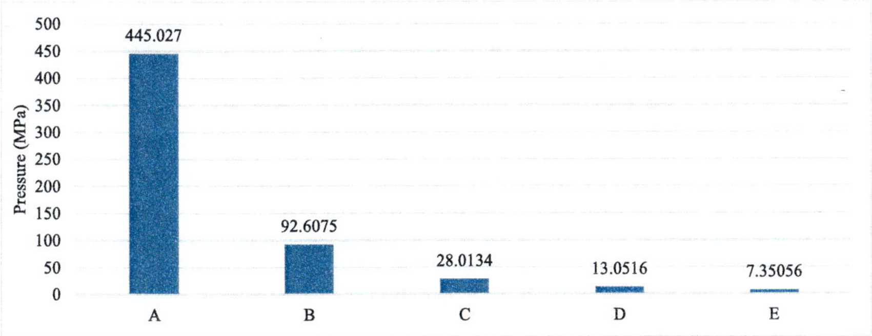

Fig. 8 Blast pressure distribution within the target (CS6)

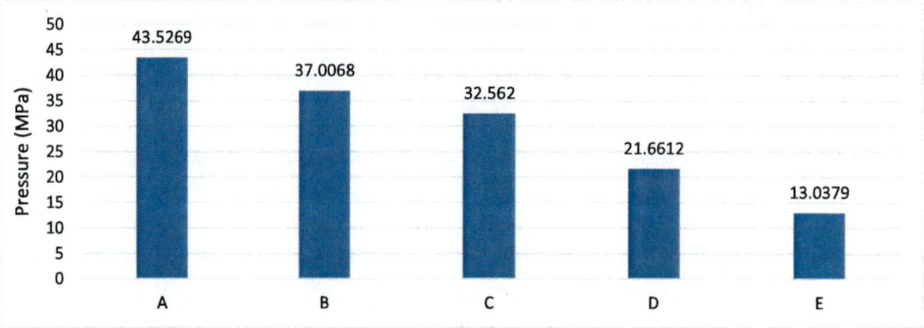

Fig. 9 Blast pressure distribution within the target (CS12)

be minimum for 2 kg of explosive on increasing the standoff distance from 100 to 500 mm.

- It was discovered that the material erosion approach underestimated the strength of concrete since the element was eroded from the study even before the concrete could have withstood greater pressure, implying that the material failure strain must be precisely defined. The fractures and damage were matched with the
- Compressive pressures were seen to be created on the front surface as soon as the shock wave struck it. The back surface of the concrete slab breaks owing to concrete spallation caused by tensile waves. It is advised that more studies in this arena be undertaken utilizing composite constructions with concrete since it may enhance the ductility of concrete, hence delaying the fracture of concrete.
- It was revealed that when the blast load is the same but the standoff lengths are different, the explosive closest to the slab has a higher effect on the slab in terms of localized collapse. As the standoff distance increased, the failure mode progressed from localized failure to global bending of the beam.
- The investigation also revealed that, for the same scaled distance, increasing the weight of the explosive along with the distance caused significant damage to the slab, increasing the deformation area and the localized failure region.
- The safe standoff distance for non-nuclear explosives, such as 1 kg of TNT, is 0.5 m. Also, protective shielding like barrier walls will further reduce the impact of blast.

Acknowledgements The authors are grateful to the grants received from MHRD through NIT Hamirpur and IIT Roorkee.

Note Additional simulation data for better understanding can be obtained from the author on request.

Appendix 1. Mesh Convergence Study

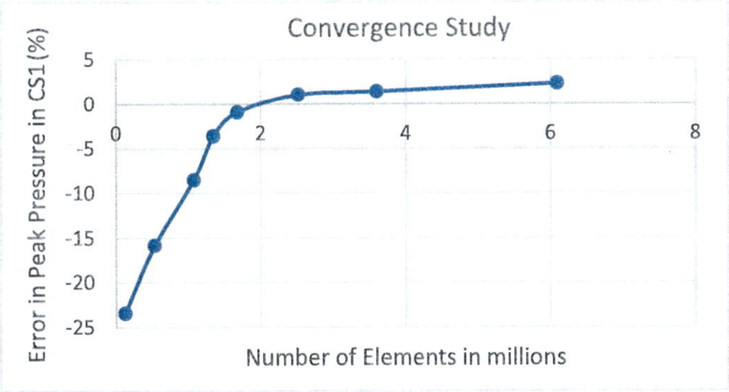

References

1. Xu K, Lu Y (2006) Numerical simulation study of spallation in reinforced concrete plates subjected to blast loading. Comput Struct 84(5–6):431–438. https://doi.org/10.1016/j.compstruc.2005.09.029
2. Lin X, Zhang YX, Hazell PJ (2014) Modelling the response of reinforced concrete panels under blast loading. Mater Des 56:620–628. https://doi.org/10.1016/j.matdes.2013.11.069
3. Zhou XQ, Kuznetsov VA, Hao H, Waschl J (2008) Numerical prediction of concrete slab response to blast loading. Int J Impact Eng 35(10):1186–1200. https://doi.org/10.1016/j.ijimpeng.2008.01.004
4. Wang W, Zhang D, Lu F, Wang SC, Tang F (2013) Experimental study and numerical simulation of the damage mode of a square reinforced concrete slab under close-in explosion. Eng Fail Anal 27:41–51. https://doi.org/10.1016/j.engfailanal.2012.07.010
5. Schenker A et al (2008) Full-scale field tests of concrete slabs subjected to blast loads. Int J Impact Eng 35(3):184–198. https://doi.org/10.1016/j.ijimpeng.2006.12.008
6. Ngo T, Mendis P, Gupta A, Ramsay J (2007) Blast loading and blast effects on structures—an overview. Electron J Struct Eng 7:76–91
7. Kumar V, Kartik KV, Iqbal MA (2020) Experimental and numerical investigation of reinforced concrete slabs under blast loading. Eng Struct 206:110125. https://doi.org/10.1016/j.engstruct.2019.110125
8. Nagy N, Mohamed M, Boot JC (2010) Nonlinear numerical modelling for the effects of surface explosions on buried reinforced concrete structures. Geomech Eng 2(1):1–18. https://doi.org/10.12989/gae.2010.2.1.001
9. Thai DK, Kim SE (2015) Numerical simulation of reinforced concrete slabs under missile impact. Struct Eng Mech 53(3):455–479. https://doi.org/10.12989/sem.2015.53.3.455
10. Elsanadedy HM, Almusallam TH, Abbas H, Al-Salloum YA, Alsayed SH (2011) Effect of blast loading on CFRP-retrofitted RC columns—a numerical study. Lat Am J Solids Struct 8(1):55–81. https://doi.org/10.1590/S1679-78252011000100004

11. Kong X, Qi X, Gu Y, Lawan IA, Qu Y (2018) Numerical evaluation of blast resistance of RC slab strengthened with AFRP. Constr Build Mater 178:244–253. https://doi.org/10.1016/j.conbuildmat.2018.05.081
12. Abaqus User Manual (2020)
13. Johnson GR, Cook WH (1983) A computational constitutive model and data for metals subjected to large strain, high strain rates and high pressures. In: Seventh International Symposium on Ballist, pp 541–547
14. Karlos V, Solomos G (2013) Calculation of blast loads for application to structural components. Administrative arrangement No. JRC 32253-2011 with DG-HOMEActivity A5—blast simulation technology development
15. D. of Army, Navy (1990) Tm_5_1300_1990.Pdf
16. Sastry, YB Sudhir et al (2014) Studies on ballistic impact of the composite panels. Theoret Appl Fract Mech 72:2–12. https://doi.org/10.1016/j.tafmec.2014.07.010